FIELDS AND PARTICLES

LECTURE NOTES AND SUPPLEMENTS IN PHYSICS

John David Jackson and David Pines, *Editors*

Introduction to Dispersion Techniques in Field Theory	*Gabriel Barton (Sussex)*
Notes on Quantum Mechanics	*Gordon Baym (Illinois)*
Intermediate Quantum Mechanics Second Edition	*H. A. Bethe (Cornell)* *R. W. Jackiw (Harvard)*
The Special Theory of Relativity	*David Bohm (Birkbeck)*
Matrix Methods in Optical Instrument Design	*W. Brouwer (Diffraction Limited)*
Relativistic Kinematics of Scattering and Spin	*R. Hagedorn (CERN)*
Mathematics for Quantum Mechanics	*J. D. Jackson (Illinois)*
Symmetry in the Solid State	*Robert S. Knox, Albert Gold (Rochester)*
Fields and Particles	*K. Nishijima (Tokyo)*
Introduction to Strong Interactions	*David Park (Williams)*
Elementary Excitations in Solids	*David Pines (Illinois)*

FIELDS AND PARTICLES

FIELD THEORY AND DISPERSION RELATIONS

K. NISHIJIMA
University of Tokyo

1969
THE BENJAMIN/CUMMINGS PUBLISHING COMPANY, INC.
ADVANCED BOOK PROGRAM

Reading, Massachusetts
London • Amsterdam • Don Mills, Ontario • Sydney • Tokyo

FIELDS AND PARTICLES

First printing, 1969
Second printing, 1971
Third printing, with revisions, 1974
Fourth printing, 1980

Library of Congress Catalog Card Number: 69-12567
International Standard Book Number: 0-8053-7398-5
0-8053-7397-7 (pbk.)

Manufactured in the United States of America

EFGHIJK-AL-89876543

PREFACE

This book is based on lectures delivered at the University of Illinois, and is written as an introduction to field theory.

Field theory is a sort of language that describes the properties of elementary particles. There are, roughly, two alternative approaches. One is the customary field theory that reached its pinnacle about twenty years ago through its success in quantum electrodynamics. The other is the S-matrix theory, or dispersion theory, that has been developed in the last decade and is suited for describing the properties of strong interactions. In the former approach the field aspect of various phenomena is emphasized, while in the latter the particle aspect is abstracted from the former by eliminating the field concept in the theory. They now look as if they had been developed independently from the start.

One of the primary purposes of this book is to clarify their relationship by giving elementary examples. As we find in the text, they are complementary rather than contradictory. That is, given a Lagrangian, the customary field theory is complete in the sense that we can evaluate arbitrary transition amplitudes, at least in principle. In practice, however, its use is limited to perturbation theory since this is the only approximation for which the renormalization prescription is known to work. On the other hand, dispersion theory automatically leads to renormalized transition amplitudes that can be compared directly with experimental results. Thus, we can use it in approximations more powerful than perturbation theory. A drawback to this approach, however, is that we do not have a complete set of dispersion relations that determines the dynamics of a given system.

In this connection an important subject is the so-called asymptotic condition that relates fields with particles. This condition can, in a sense, replace the old-fashioned canonical quantization prescription in the presence of interactions. Because of the importance of this condition in clarifying the intimate relationship between the two alternative approaches mentioned above, an emphasis has been laid on this subject.

These are reasons why we have chosen the title *Fields and Particles* for this book. We hope that this book serves the readers in clarifying these subjects and that it is useful as a handbook in carrying out various practical calculations.

I would like to express my gratitude to Miss C. Carter and Mrs. G. Hikida for their patient typing of the original notes, and to Mr. W. H. Weihofen for his pertinent assistance in editing and proofreading. Without their assistance this book would not have been completed.

K. Nishijima

Tokyo, Japan
August, 1968

CONTENTS

CHAPTER 1

CANONICAL QUANTIZATION OF FIELDS

The purpose of quantized field theory consists in the application of the principles of quantum mechanics to systems with infinite degrees of freedom. First we shall briefly review particle quantum mechanics to see how one can generalize the formalism. The most important applications of quantized field theory are found in those phenomena in which the number of particles changes. We assume familiarity with such phenomena to some extent, that is, in the study of old-fashioned quantum electrodynamics where only the radiation field is quantized. That theory is, however, not complete since the electron field is not yet quantized. There are many important phenomena in which the particle number changes; for instance,

beta decay: $n \to p + e^- + \nu$

pion production: $N + N \to N + N + \pi$

$\gamma + p \to N + \pi$

meson decay: $\pi^\circ \to 2\gamma$

$\pi^+ \to \mu^+ + \nu$ etc.

In order to discuss such occurrences one cannot use particle quantum mechanics; one is forced to introduce quantized field theory.

Quantum theory of fields is based on relativity and quantum mechanics, so that it is convenient to use the so-called natural units in which we use $\hbar$ and c, the light velocity in vacuum, as units.[1] Since any quantity can be measured in units of a certain power of length multiplied by requisite powers of $\hbar$ and c, we can measure a physical quantity in units of a certain power of length once the natural units are employed. Since $e^2/\hbar c$ is a dimensionless quantity, e^2 reduces to a dimensionless quantity in the natural units.

In accordance with the above conventions we shall use the length $x_0 = ct$ of the light path as the time coordinate as well as the imaginary time coordinate $x_4 = ix_0 = ict$. Thus tensor indices designated by Greek letters run from 1 to 4 and involve the imaginary time coordinate. In this connection we shall use a special rule for complex (Hermitian) conjugation. For quantities with an index zero a star (dagger) means the complex (Hermitian) conjugate in the ordinary sense, but in general we shall mean by $\varphi^*_{\alpha\beta\ldots}$ the complex conjugate of $\varphi^*_{\alpha\beta\ldots}$ multiplied by $(-1)^n$ where n is the number of 4's among α, β, ... For instance, $\varphi_4 = i\varphi_0$ and $\varphi_4^* = i\varphi_0^*$. This convention is convenient, since the imaginary unit i originates from relativity but not from quantum theory.

Also, throughout this book we shall define the scalar product of two four-dimensional vectors a and b by

$$ab = a_\lambda b_\lambda = a_1 b_1 + a_2 b_2 + a_3 b_3 - a_0 b_0$$

1-1 VARIATIONAL PRINCIPLE

In particle quantum mechanics there are two different approaches which are mathematically equivalent: the Schrödinger approach and the Heisenberg approach.

Schrödinger Approach

Given the Hamiltonian of a system as a function of coordinates q_r and their canonical conjugate momenta p_r, $H(q_r, p_r)$, we set up an equation of motion called the Schrödinger equation:

$$i \frac{\partial}{\partial t} \Psi(t) = H\left(q_r, \frac{1}{i}\frac{\partial}{\partial q_r}\right)\Psi(t) \qquad (1\text{-}1)$$

The expectation value of a dynamical quantity $\mathcal{O}$, which is a function of q_r and p_r, is given at time t by

$$\langle \mathcal{O} \rangle = \int \Psi^*(t)\, \mathcal{O}\, \Psi(t)\, dq, \qquad dq = \rho\, dq_1 \ldots dq_N$$

where N denotes the degrees of freedom and ρ is an appropriate density.

Heisenberg Approach

Unlike the Schrödinger approach all operators are time dependent but the state vector Ψ is time independent. The dynamics of the system is not governed by the Schrödinger equation but by a set of equations called canonical equations:

$$\frac{dq_r(t)}{dt} = i\left[H, q_r(t)\right]; \qquad \frac{dp_r(t)}{dt} = i\left[H, p_r(t)\right] \qquad (1\text{-}2)$$

with the canonical commutation relations

$$\left[q_r(t), p_s(t)\right] = i\delta_{rs} \qquad (1\text{-}3)$$

The expectation value of $\mathcal{O}$ at time t is given by

$$\langle \mathcal{O} \rangle = \int \Psi^* \mathcal{O}(t) \Psi \, dq \tag{1-4}$$

with

$$\mathcal{O}(t) = \mathcal{O}\left[q_r(t), p_r(t)\right] \tag{1-5}$$

Both approaches give the same result for $\langle \mathcal{O} \rangle$, since these two representations are related to one another through

$$\Psi(t) = e^{-iHt} \Psi \tag{1-6}$$

$$\mathcal{O}(t) = e^{iHt} \mathcal{O} e^{-iHt} \tag{1-7}$$

As a logical basis the Heisenberg approach is more convenient than the Schrödinger approach, while the latter is more convenient for practical applications. We shall start from the Heisenberg approach.

One can write the canonical equations as

$$\frac{dq_r(t)}{dt} = \frac{\partial H}{\partial p_r(t)} \qquad \frac{dp_r(t)}{dt} = -\frac{\partial H}{\partial q_r(t)} \tag{1-8}$$

if appropriate care is taken with regard to the order of noncommuting operators. The equivalence between the commutator form and the classical form of the canonical equations follows from the relations

$$\frac{\partial P}{\partial p_r} = i[P, q_r] \qquad \frac{\partial P}{\partial q_r} = -i[P, p_r] \tag{1-9}$$

where P is a polynominal in p_r and q_r. Here again the same caution must be exercised with regard to the order of operators.

As far as the canonical equations are concerned there is no difference from classical theory in the form, and we can thoroughly utilize the principles of classical field theory in this representation. The only difference occurs in the presence or absence of the commutation relations, so we may formulate quantized field theory by *(classical field theory)* + *(commutation relations)*.

Action Principle

Assume that a function of q_r and $\dot{q}_r$, the Lagrangian $L(q_r, \dot{q}_r)$, is given. The classical equations of motion are derived from the action principle which may be expressed in the form of a variational principle:

$$\delta \int_{t_0}^{t_1} dt \, L(q_r, \dot{q}_r) = 0 \tag{1-10}$$

In order to quantize the system one has to introduce the canonical conjugate variables. We define p_r by

$$p_r = \frac{\partial L}{\partial \dot{q}_r} \tag{1-11}$$

Then Lagrange's equations, as well as the canonical commutation relations, follow:

$$\frac{d}{dt}\left(\frac{\partial L}{\partial \dot{q}_r}\right) - \frac{\partial L}{\partial q_r} = 0 \tag{1-12}$$

$$\left[q_r(t),\ p_s(t)\right] = i\,\delta_{rs} \tag{1-13}$$

This is the standard way of quantizing a dynamical system starting from a Lagrangian. We next define the Hamiltonian:

$$H = \sum_r p_r \dot{q}_r - L \tag{1-14}$$

Then Eqs. (1-12) — (1-14) can be combined to yield the canonical equations of the form

$$\frac{dq_r}{dt} = i\left[H,\ q_r\right] \qquad \frac{dp_r}{dt} = i\left[H,\ p_r\right] \tag{1-15}$$

We shall now proceed from particle theory to field theory. Let us assume that a dynamical system consists of a finite number of particles and that there are no interactions at all among the particles. In such a case the total Lagrangian is given by the sum of Lagrangians corresponding to these free particles, namely,

$$L = \sum_{n=1}^{N} L_n \tag{1-16}$$

and the action principle reads

$$\delta \int \sum_{n=1}^{N} L_n(q_n,\ \dot{q}_n)\, dt = 0 \tag{1-17}$$

In such a case it is easy to extend the result to a system with infinite degrees of freedom. When the degrees of freedom thus constitute a continuum, we replace the subscript n by a variable x and regard the dynamical variable as a function of x.

$$q_n \rightarrow \varphi(x) \tag{1-18}$$

We refer to $\varphi(x)$ as a field. The action principle then assumes the form

$$\delta \int L\,dt = \delta \int dt \left[\int dx\, \mathcal{L}\, [\varphi(x),\ \dot{\varphi}(x)] \right] = 0 \tag{1-19}$$

$\mathcal{L}$ is called the Lagrangian density. In three-dimensional problems dx denotes d^3x; hence, using $d^4x = d^3x\ dt$,

$$\delta \int \mathcal{L}\, [\varphi(x),\ \dot{\varphi}(x)]\, d^4x = 0 \tag{1-20}$$

When there are several fields we use an additional subscript α to distinguish among different fields.

$$\delta \int \mathcal{L} \left[\varphi_\alpha(x),\ \dot{\varphi}_\alpha(x)\right] d^4x = 0 \tag{1-21}$$

This is the general form of the Lagrangian density when there are no interactions between fields at different points. However, this constraint is overly restrictive in that it cannot cover propagation of waves; we must therefore introduce interactions of fields *at different space points.*

In relativistic field theory we assume that a field interacts only with its *infinitesimal neighbors* as contrasted with the idea of action at a distance. This assumption means that $\mathcal{L}$ is a function of $\varphi_\alpha(x)$, $\dot{\varphi}_\alpha(x)$, and $\varphi_\alpha(x + dx)$. Alternatively, instead of the last quantity, it is better to use $\partial\varphi_\alpha/\partial x_k$ (k = 1, 2, 3) since

$$\varphi_\alpha(x + dx) = \varphi_\alpha(x) + \frac{\partial \varphi_\alpha(x)}{\partial x_k}\, dx_k \tag{1-22}$$

This gradient together with $\dot{\varphi}$ may be denoted by

$$\frac{\partial \varphi_\alpha(x)}{\partial x_\mu} \qquad (\mu = 1,\ 2,\ 3,\ 4) \tag{1-23}$$

So the general form of the Lagrangian density in relativistic field theory is given by

$$\mathcal{L} \left[\varphi_\alpha(x),\ \frac{\partial \varphi_\alpha(x)}{\partial x_\mu} \right] \tag{1-24}$$

and the action principle is

$$\delta \int \mathcal{L} \left[\varphi_\alpha(x),\ \frac{\partial \varphi_\alpha(x)}{\partial x_\mu}\right] d^4x = 0 \tag{1-25}$$

Quantization

We now apply the quantization procedure. First consider the relation

$$\sum_s \left[q_r, p_s \right] A_s = i A_r \tag{1-26}$$

where A_r is an arbitrary quantity. This relation is more convenient to generalize than the original commutation relations. We go from the discrete subscript to the continuous variable:

$$q_s \to \varphi_\alpha(x) \tag{1-27a}$$

$$p_s = \frac{\partial L_s}{\partial \dot{q}_s} \to \pi_\alpha(x) = \frac{\partial \mathcal{L}\left[\varphi_\alpha(x), \partial \varphi_\alpha(x)/\partial x_\mu\right]}{\partial \dot{\varphi}_\alpha(x)} \tag{1-27b}$$

$$\sum_s \to \int d^3x \tag{1-27c}$$

with these rules for the transition to the continuum, Eq. (1-26) becomes

$$\int d^3x' \left[\varphi_\alpha(x), \pi_\beta(x')\right] f(x') = i f(x)\, \delta_{\alpha\beta} \qquad \text{for} \quad x_0 = x_0' \tag{1-28}$$

which implies

$$\left[\varphi_\alpha(x), \pi_\beta(x')\right] = i\, \delta_{\alpha\beta}\, \delta^3(x - x') \qquad \text{for} \quad x_0 = x_0' \tag{1-29}$$

Corresponding to $[q_r, q_s] = [p_r, p_s] = 0$ we also have

$$\left[\varphi_\alpha(x), \varphi_\beta(x')\right] = \left[\pi_\alpha(x), \pi_\beta(x')\right] = 0 \qquad \text{for} \quad x_0 = x_0' \tag{1-30}$$

Quantization of fields in the above form was first discussed by Heisenberg and Pauli.

Field Equations

Once the action principle is established it is not hard to derive the field equations from

$$\delta \int_\Omega \mathcal{L}\, d^4x = 0 \tag{1-31}$$

The integral is to be performed over a four-dimensional domain Ω and we assume that the variations $\delta\varphi_\alpha$ vanish on the boundary surface of Ω. Then we get

$$\delta \int_\Omega \mathcal{L}\left[\varphi_\alpha(x), \frac{\partial\varphi_\alpha(x)}{\partial x_\mu}\right] d^4x$$

$$= \int_\Omega \left[\delta\varphi_\alpha(x)\frac{\partial\mathcal{L}(x)}{\partial\varphi_\alpha(x)} + \delta\left(\frac{\partial\varphi_\alpha(x)}{\partial x_\mu}\right)\frac{\partial\mathcal{L}(x)}{\partial\left(\frac{\partial\varphi_\alpha(x)}{\partial x_\mu}\right)}\right] d^4x$$

$$= \int_\Omega \delta\varphi_\alpha(x)\left[\frac{\partial\mathcal{L}(x)}{\partial\varphi_\alpha(x)} - \frac{\partial}{\partial x_\mu}\left(\frac{\partial\mathcal{L}(x)}{\partial\left(\frac{\partial\varphi_\alpha(x)}{\partial x_\mu}\right)}\right)\right] d^4x$$

$$+ \int_{\text{surface}} \delta\varphi_\alpha(x) \frac{\partial\mathcal{L}(x)}{\partial\left(\frac{\partial\varphi_\alpha(x)}{\partial x_\mu}\right)} d\sigma_\mu \tag{1-32}$$

where use has been made of the divergence theorem

$$\int_\Omega \frac{\partial A_\mu}{\partial x_\mu} d^4x = \int_{\text{surface}} A_\mu \, d\sigma_\mu \qquad \text{with} \quad d\sigma_\mu = \frac{d^4x}{dx_\mu} \tag{1-33}$$

Since $\delta\varphi_\alpha(x)$ are assumed to vanish on the surface of Ω, the last surface integral vanishes. The variations $\delta\varphi_\alpha$ are otherwise completely arbitrary so that the variational principle leads to the so-called Euler equation

$$\frac{\partial\mathcal{L}(x)}{\partial\varphi_\alpha(x)} - \frac{\partial}{\partial x_\mu}\left[\frac{\partial\mathcal{L}(x)}{\partial\left(\frac{\partial\varphi_\alpha(x)}{\partial x_\mu}\right)}\right] = 0 \tag{1-34}$$

Example (1-1) Real spinless field and the Klein-Gordon equation

We want to derive the Klein-Gordon equation

$$(\Box - m^2)\,\varphi = 0 \tag{1-35}$$

from the action principle. If we choose

$$\mathcal{L} = -\frac{1}{2}\left[\left(\frac{\partial\varphi}{\partial x_\mu}\right)^2 + m^2\varphi^2\right] \tag{1-36}$$

the Euler equation gives at once

$$\frac{\partial\mathcal{L}}{\partial\varphi} - \frac{\partial}{\partial x_\mu}\left(\frac{\partial\mathcal{L}}{\partial\left(\frac{\partial\varphi}{\partial x_\mu}\right)}\right) = -m^2\varphi - \frac{\partial}{\partial x_\mu}\left(-\frac{\partial\varphi}{\partial x_\mu}\right)$$

$$= (\Box - m^2)\varphi = 0 \tag{1-37}$$

The minus sign in $\mathcal{L}$ is chosen to make the Hamiltonian or the total energy of the system positive definite.

Hamiltonian

The Hamiltonian of a dynamical system is given by

$$H = \sum_r p_r \dot{q}_r - L \tag{1-38}$$

In field theory, therefore, a corresponding expression is given by

$$H = \int d^3x \sum_\alpha \pi_\alpha(x)\,\dot{\varphi}_\alpha(x) - \int d^3x\,\mathcal{L}(x) \equiv \int d^3x\,\mathcal{H}(x) \tag{1-39}$$

where $\mathcal{H}(x)$ is called the Hamiltonian density and is given by

$$\mathcal{H}(x) = \sum_\alpha \pi_\alpha(x)\,\dot{\varphi}_\alpha(x) - \mathcal{L}(x) \tag{1-40}$$

Example (1-2) Real spinless field and the Hamiltonian

We have chosen, in Example (1-1), the Lagrangian density (1-36) to obtain the correct field equation, (1-35),

$$\mathcal{L} = \frac{1}{2}\left[\dot{\varphi}^2 - (\nabla\varphi)^2 - m^2\varphi^2\right] \tag{1-41}$$

Hence

$$\pi = \frac{\partial\mathcal{L}}{\partial\dot{\varphi}} = \dot{\varphi} \tag{1-42}$$

The Hamiltonian density is given by

$$\mathcal{H} = \pi\dot{\varphi} - \mathcal{L} = \frac{1}{2}\left[\pi^2 + (\nabla\varphi)^2 + m^2\varphi^2\right] \tag{1-43}$$

which is a positive definite expression as it should be.

1-2 CANONICAL EQUATIONS

We have derived the field equations from the variational principle. There is, however, another method to derive quantum mechanical field equations, that is, from the canonical equations:

$$\frac{\partial\varphi_\alpha(x)}{\partial t} = i\left[H, \varphi_\alpha(x)\right] \tag{1-44a}$$

$$\frac{\partial\pi_\alpha(x)}{\partial t} = i\left[H, \pi_\alpha(x)\right] \tag{1-44b}$$

Now we must prove that the canonical equations are equivalent to the Euler equations. This proof follows, provided that the classical canonical formalism is applicable to the Lagrangian in question. Unless it is possible to convert the classical theory into a canonical form, no consistent way of quantizing the theory is known; in noncanonical theories, both sets of equations, if they exist, are generally different. We may diagram our procedural question as follows:

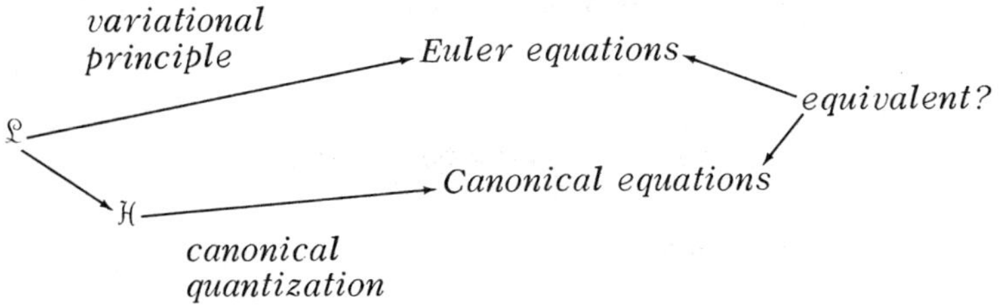

We begin the proof by calculating the commutator[2]

$$\left[H, \pi_\alpha(x)\right] = \int_{x_0'=x_0} d^3x' \left[\mathcal{H}(x'), \pi_\alpha(x)\right] \tag{1-45}$$

First we notice that x_0 or t in $H = \int d^3x\, \mathcal{H}(x)$ can be chosen arbitrarily, since we have in the canonical formalism the identity

$$\frac{d}{dt} H = i\left[H, H\right] = 0 \tag{1-46}$$

Next we use the equal-time canonical commutation relations to evaluate the above commutator. For the definition of the Hamiltonian density given in Section 1-1 we get

$$\left[H, \pi_\alpha(x)\right] = \int_{x_0' = x_0} d^3x' \left\{ \sum_\beta \pi_\beta(x') \left[\frac{\partial \varphi_\beta(x')}{\partial x_0'}, \pi_\alpha(x)\right] - \left[\mathcal{L}(x'), \pi_\alpha(x)\right] \right\} \tag{1-47}$$

The second term is given by

$$\begin{aligned} \left[\mathcal{L}(x'), \pi_\alpha(x)\right] &= \frac{\partial \mathcal{L}(x')}{\partial \varphi_\alpha(x')}\, i\delta^3(x - x') \\ &+ \sum_{k=1}^{3} \frac{\partial \mathcal{L}(x')}{\partial \left(\dfrac{\partial \varphi_\alpha(x')}{\partial x_k'}\right)}\, i \frac{\partial}{\partial x_k'}\, \delta^3(x - x') \\ &+ \sum_\beta \frac{\partial \mathcal{L}(x')}{\partial \dot{\varphi}_\beta(x')} \left[\frac{\partial \varphi_\beta(x')}{\partial x_0'}, \pi_\alpha(x)\right] \end{aligned} \tag{1-48}$$

since we have the following formula provided appropriate care is taken with regard to the order of noncommuting operators:

$$\left[F(A, B, \ldots), \pi_\alpha\right] = \frac{\partial F}{\partial A}\left[A, \pi_\alpha\right] + \frac{\partial F}{\partial B}\left[B, \pi_\alpha\right] + \cdots \tag{1-49}$$

Thus the canonical equation for π_α may be written

$$\begin{aligned} \frac{\partial \pi_\alpha(x)}{\partial x_0} &= i\left[H, \pi_\alpha(x)\right] \\ &= \frac{\partial \mathcal{L}}{\partial \varphi_\alpha(x)} - \sum_{k=1}^{3} \frac{\partial}{\partial x_k} \frac{\partial \mathcal{L}}{\partial \left(\dfrac{\partial \varphi_\alpha(x)}{\partial x_k}\right)} \\ &+ i \int_{x_0' = x_0} d^3x' \sum_\beta \left(\pi_\beta(x') - \frac{\partial \mathcal{L}(x')}{\partial \dot{\varphi}_\beta(x')}\right) \left[\frac{\partial \varphi_\beta(x')}{\partial x_0'}, \pi_\alpha(x)\right] \end{aligned} \tag{1-50}$$

Similarly we shall evaluate the commutator $[H, \varphi_\alpha(x)]$.

$$[H, \varphi_\alpha(x)] = \int_{x_0' = x_0} d^3x' \left\{ -i \frac{\partial \varphi_\alpha(x')}{\partial x_0'} \delta^3(x - x') + \sum_\beta \pi_\beta(x') \right.$$

$$\left. \times \left[\frac{\partial \varphi_\beta(x')}{\partial x_0'}, \varphi_\alpha(x) \right] - [\mathfrak{L}(x'), \varphi_\alpha(x)] \right\} \tag{1-51}$$

where

$$[\mathfrak{L}(x'), \varphi_\alpha(x)] = \sum_\beta \frac{\partial \mathfrak{L}(x')}{\partial \dot{\varphi}_\beta(x')} \left[\frac{\partial \varphi_\beta(x')}{\partial x_0'}, \varphi_\alpha(x) \right] \tag{1-52}$$

So the canonical equation for φ_α reads

$$\frac{\partial \varphi_\alpha(x)}{\partial x_0} = \frac{\partial \varphi_\alpha(x)}{\partial x_0} + i \sum_\beta \int_{x_0'=x_0} d^3x' \left[\pi_\beta(x') - \frac{\partial \mathfrak{L}(x')}{\partial \dot{\varphi}_\beta(x')} \right]$$

$$\times \left[\frac{\partial \varphi_\beta(x')}{\partial x_0'}, \varphi_\alpha(x) \right] \tag{1-53}$$

The second term on the right-hand side has to vanish, and this is consistent with the definition of π_β.

$$\pi_\beta = \frac{\partial \mathfrak{L}}{\partial \dot{\varphi}_\beta} \tag{1-54}$$

Substitution of this definition into Eq. (1-50) yields

$$\frac{\partial \mathfrak{L}}{\partial \varphi_\alpha} - \sum_{k=1}^{3} \frac{\partial}{\partial x_k} \left(\frac{\partial \mathfrak{L}}{\partial \left(\dfrac{\partial \varphi_\alpha}{\partial x_k} \right)} \right) - \frac{\partial}{\partial x_0} \left(\frac{\partial \mathfrak{L}}{\partial \dot{\varphi}_\alpha} \right) = 0 \tag{1-55}$$

or in a relativistically invariant form

$$\frac{\partial \mathfrak{L}}{\partial \varphi_\alpha} - \frac{\partial}{\partial x_\mu} \left(\frac{\partial \mathfrak{L}}{\partial \left(\dfrac{\partial \varphi_\alpha}{\partial x_\mu} \right)} \right) = 0 \tag{1-56}$$

This concludes the proof of the equivalence of the canonical equations to the Euler equations.

In quantum mechanics H represents the total energy of the system. Setting $P_4 = iP_0 = iH$, we already have

$$[P_4 , F(x)] = i \frac{\partial F(x)}{\partial x_4} \tag{1-57}$$

provided that F depends on x only through $\varphi_\alpha(x)$ and their derivatives. Now introduce P_k (k = 1, 2, 3) by

$$P_k = - \int d^3 x \sum_\alpha \pi_\alpha (x) \frac{\partial \varphi_\alpha(x)}{\partial x_k} \tag{1-58}$$

See Ryder, eq. (3.20), p. 88 (energy-momentum tensor) or this book, eq (1.84), p. 16

then we get the following commutation relations:

$$[P_k , \varphi_\alpha(x)] = i \int d^3x' \; \delta^3 (x - x') \frac{\partial \varphi_\alpha(x')}{\partial x_k'} = i \frac{\partial \varphi_\alpha(x)}{\partial x_k} \tag{1-59a}$$

$$[P_k , \pi_\alpha(x)] = -\int d^3x' \; \pi_\alpha(x) \frac{\partial}{\partial x_k'} \, i\delta^3 (x - x') = i \frac{\partial \pi_\alpha(x)}{\partial x_k} \tag{1-59b}$$

Therefore if F(x) is a function of $\varphi_\alpha(x)$ and their derivatives, we generally get

$$[P_k, F(x)] = i \frac{\partial F(x)}{\partial x_k} \tag{1-60}$$

This result may be combined with the one for P_4 to yield

$$[P_\mu, F(x)] = i \frac{\partial F(x)}{\partial x_\mu} \qquad (\mu = 1, 2, 3, 4) \tag{1-61}$$

This is a covariant equation; hence P_μ must represent a four-vector whose fourth component is the energy. Therefore the spatial components of P_μ must represent the total momentum of the system.

It is important that the four components of P_μ commute with each other so that they can be simultaneously diagonalized:

$$[P_\mu, P_\nu] = 0 \tag{1-62}$$

Let us employ such a representation and introduce two states a and b corresponding to eigenvalues p_μ and q_μ. Then from the commutation relations between p_μ and F(x) we get

$$i \frac{\partial}{\partial x_\mu} \langle a | F(x) | b \rangle = \langle a | [P_\mu, F(x)] | b \rangle$$

$$= (p_\mu - q_\mu) \langle a | F(x) | b \rangle \tag{1-63}$$

Integration of the differential equations above gives

$$\langle a | F(x) | b \rangle = \langle a | F(0) | b \rangle e^{-i(p-q)x} \tag{1-64}$$

This relation giving the x dependence of the matrix elements of the Heisenberg operator F(x)— the operator in the Heisenberg representation—proves to be very important as we shall see later.

Next we shall prove that, given the equal-time commutation relations for a certain fixed time, the same commutation relations hold for an arbitrary time; that is, if

$$[\varphi_\alpha(x), \pi_\beta(x')] = i\delta_{\alpha\beta}\delta^3(x - x') \tag{1-65}$$

is true for $x_0 = x_0' = t_0$, then it is true for $x_0 = x_0' = t$, where t is an arbitrary time. In order to prove this statement we use mathematical induction.

(1) It is true for $t = t_0$.

(2) If it is true for t, it is true for t + dt also.

We shall prove the second statement, using boldface letters to designate the spatial coordinates.

$$[\varphi_\alpha(\mathbf{x}, t + dt), \pi_\beta(\mathbf{x}', t + dt)]$$

$$= [\varphi_\alpha(\mathbf{x}, t), \pi_\beta(\mathbf{x}', t)] + ([\dot{\varphi}_\alpha(\mathbf{x}, t), \pi_\beta(\mathbf{x}', t)]$$

$$+ [\varphi_\alpha(\mathbf{x}, t), \dot{\pi}_\beta(\mathbf{x}', t)])dt + \mathcal{O}(dt^2).$$

Therefore, what we have to prove is

$$[\dot{\varphi}_\alpha(\mathbf{x}, t), \pi_\beta(\mathbf{x}', t)] + [\varphi_\alpha(\mathbf{x}, t), \dot{\pi}_\beta(\mathbf{x}', t)] = 0 \tag{1-66}$$

From the canonical equations of motion the left-hand side is equal to

$$i\Big[[H, \varphi_\alpha(\mathbf{x}, t)]\,\pi_\beta(\mathbf{x}', t)\Big] + i\Big[\varphi_\alpha(\mathbf{x}, t), [H, \pi_\beta(\mathbf{x}', t)]\Big]$$

$$= i\Big[H, [\varphi_\alpha(\mathbf{x}, t), \pi_\beta(\mathbf{x}', t)]\Big]$$

$$= i\Big[H, \text{c number}\Big]$$

$$= 0$$

so that if the canonical commutation relations are valid at a certain time they are always valid.

It is clear that the canonical commutation relations are invariant under translations in space-time and also under spatial rotations. Since they are written for equal time, however, a special time axis has to be chosen. If another time axis is chosen, corresponding to a new frame of reference moving relative to the original one, it is not obvious whether or not one can again get commutation relations of the same form. For this reason it is important to investigate the transformation properties of the theory.

1-3 VARIATIONAL PRINCIPLE AND CONSERVATION LAWS

It is important to recognize that most conservation laws can be derived from the invariance properties of the action integral under infinitesimal variations without directly studying the field equations. This problem is called the invariant variational problem.[3]

Assume that the action integral

$$I = \int_\Omega \mathcal{L}[\varphi_\alpha(x), \varphi_{\alpha:\mu}(x)]\, d^4x \qquad \left(\varphi_{\alpha:\mu} = \frac{\partial \varphi_\alpha}{\partial x_\mu}\right) \tag{1-67}$$

is invariant under the infinitesimal transformation

$$\varphi_\alpha(x) \to \varphi'_\alpha(x') = \varphi_\alpha(x) + \delta\varphi_\alpha(x) \tag{1-68}$$

$$x_\mu \to x'_\mu = x_\mu + \delta x_\mu \tag{1-69}$$

without asking the field variables φ_α to satisfy the Euler equations. The term $\varphi'_\alpha(x)$ means a change in the functional form of $\varphi_\alpha(x)$ in its dependence on x. The variation $\delta\varphi_\alpha(x)$ is defined by

$$\delta\varphi_\alpha(x) = \varphi'_\alpha(x') - \varphi_\alpha(x) \tag{1-70}$$

but it is convenient to define another variation without changing x,

$$\delta^* \varphi_\alpha(x) = \varphi'_\alpha(x) - \varphi_\alpha(x) \tag{1-71}$$

This is the variation used in deriving the Euler equation. The relation between these two variations is given by

$$\begin{aligned}
\delta^* \varphi_\alpha(x) &= \delta\varphi_\alpha(x) + \varphi'_\alpha(x) - \varphi'_\alpha(x') \\
&= \delta\varphi_\alpha(x) + \frac{\partial \varphi'_\alpha(x)}{\partial x_\mu}(x - x')_\mu \\
&= \delta\varphi_\alpha(x) - \frac{\partial \varphi'_\alpha(x)}{\partial x_\mu}\delta x_\mu \\
&= \delta\varphi_\alpha(x) - \varphi_{\alpha:\mu}(x)\,\delta x_\mu + \mathcal{O}(\delta^2)
\end{aligned} \tag{1-72}$$

For the δ^* variations we have already used the following relationship in deriving the Euler equation:

$$\delta^* \varphi_{\alpha:\mu} = \frac{\partial}{\partial x_\mu}\delta^* \varphi_\alpha \tag{1-73}$$

From this result we get

$$\delta\varphi_{\alpha:\mu} = \partial_\mu \delta\varphi_\alpha - \varphi_{\alpha:\nu}\frac{\partial \delta x_\nu}{\partial x_\mu} \tag{1-74}$$

Now we shall study the consequences of the assumption $\delta I = 0$.

$$\begin{aligned}
\delta I = &\int_{\Omega'} \mathcal{L}\left[\varphi'_\alpha(x'),\ \varphi'_{\alpha:\mu}(x')\right] d^4x' \\
&- \int_\Omega \mathcal{L}\left[\varphi_\alpha(x),\ \varphi_{\alpha:\mu}(x)\right] d^4x
\end{aligned} \tag{1-75}$$

The domain of integration is transformed from Ω to Ω' with the transformation $x_\mu \to x_\mu'$. In order to write the expression under one integral we transform the integration over x' back to one over x by using

$$d^4x' = \frac{\partial(x'_0, \ldots, x'_3)}{\partial(x_0, \ldots, x_3)}\, d^4x \tag{1-76}$$

Hence we may write

$$\delta I = \int_\Omega d^4x \left(\mathcal{L}[\varphi_\alpha(x) + \delta\varphi_\alpha(x),\ \varphi_{\alpha:\mu}(x) + \delta\varphi_{\alpha:\mu}(x)] \frac{\partial(x'_0 \ldots x'_3)}{\partial(x_0 \ldots x_3)} - \mathcal{L}[\varphi_\alpha(x),\ \varphi_{\alpha:\mu}(x)] \right) \tag{1-77}$$

For infinitesimal transformations the Jacobian reduces to

$$\frac{\partial(x'_0 \ldots x'_3)}{\partial(x_0 \ldots x_3)} = 1 + \frac{\partial(\delta x_\nu)}{\partial x_\nu} \tag{1-78}$$

and we may write δI, suppressing the summation over α, as

$$\delta I = \int_\Omega \left[\frac{\partial \mathcal{L}}{\partial \varphi_\alpha} \delta\varphi_\alpha + \frac{\partial \mathcal{L}}{\partial \varphi_{\alpha:\mu}} \delta\varphi_{\alpha:\mu} + \mathcal{L} \frac{\partial(\delta x_\nu)}{\partial x_\nu} \right] d^4x = 0 \tag{1-79}$$

This invariance should hold independent of the choice of Ω, so that the assumption $\delta I = 0$ leads to

$$\delta \mathcal{L} + \mathcal{L} \frac{\partial(\delta x_\nu)}{\partial x_\nu} = 0 \tag{1-80}$$

where

$$\delta \mathcal{L} = \frac{\partial \mathcal{L}}{\partial \varphi_\alpha} \delta\varphi_\alpha + \frac{\partial \mathcal{L}}{\partial \varphi_{\alpha:\mu}} \delta\varphi_{\alpha:\mu} \tag{1-81}$$

In terms of $\delta^* \varphi_\alpha$ our condition on $\mathcal{L}$ reads

$$[\mathcal{L}]_{\varphi_\alpha} \delta^* \varphi_\alpha + \frac{\partial}{\partial x_\mu} \left[\frac{\partial \mathcal{L}}{\partial \varphi_{\alpha:\mu}} \delta^* \varphi_\alpha + \mathcal{L}\, \delta x_\mu \right] = 0 \tag{1-82}$$

where $[\mathcal{L}]_{\varphi_\alpha}$ is the Euler derivative defined by

$$[\mathcal{L}]_{\varphi_\alpha} = \frac{\partial \mathcal{L}}{\partial \varphi_\alpha} - \frac{\partial}{\partial x_\mu} \left(\frac{\partial \mathcal{L}}{\partial \varphi_{\alpha:\mu}} \right) \tag{1-83}$$

Let us define the canonical energy-momentum tensor by

$$T_{\mu\nu} = - \frac{\partial \mathcal{L}}{\partial \varphi_{\alpha:\mu}} \varphi_{\alpha:\nu} + \delta_{\mu\nu} \mathcal{L} \tag{1-84}$$

Then the invariance condition (1-82) in terms of this quantity may be expressed by

$$[\mathfrak{L}]_{\varphi_\alpha} \delta^* \varphi_\alpha + \frac{\partial}{\partial x_\mu}\left[\frac{\partial \mathfrak{L}}{\partial \varphi_{\alpha:\mu}} \delta\varphi_\alpha + T_{\mu\nu}\delta x_\nu\right] = 0 \qquad (1\text{-}85)$$

If there is an infinitesimal transformation group with N parameters that leaves the action integral I invariant, that is,

$$\delta x_\mu = x'_\mu - x_\mu = \sum_{r=1}^{N} a_{\mu(r)}(x)\epsilon_r \qquad (1\text{-}86)$$

$$\delta\varphi_\alpha(x) = \varphi'_\alpha(x') - \varphi_\alpha(x) = \sum_{r=1}^{N} C_{\alpha(r)}(\varphi, x)\,\epsilon_r \qquad (1\text{-}87)$$

then we get, by using the Euler equations (1-34), N different conservation laws

$$\frac{\partial}{\partial x_\mu}\left[\frac{\partial \mathfrak{L}}{\partial \varphi_{\alpha:\mu}} C_{\alpha(r)} + T_{\mu\nu}\, a_{\nu(r)}\right] = 0$$

$$(r = 1, 2, \ldots, N) \qquad (1\text{-}88)$$

Example (1-3) — Translation

The action integral is invariant, provided that the Lagrangian density does not depend explicitly on the coordinates, under the four-parametric group of translations:

$$x_\mu \to x'_\mu = x_\mu + \epsilon_\mu \qquad \text{or} \qquad a_{\mu(r)} = \delta_{\mu r} \qquad (1\text{-}89)$$

$$\delta\varphi_\alpha = 0 \qquad \text{or} \qquad C_{\alpha(r)} = 0 \qquad (1\text{-}90)$$

The corresponding conservation laws are given by

$$\frac{\partial T_{\mu r}}{\partial x_\mu} = 0 \qquad \text{for } r = 1, 2, 3, 4 \qquad (1\text{-}91)$$

This represents exactly the energy-momentum conservation, and the following vector is a constant of motion:

$$P_\nu = \int d^3x\, T_{0\nu}(x) \qquad (1\text{-}92)$$

When the Lagrangian density depends, not only implicitly through the field variables, but also explicitly, on the coordinates x, we get

$$\frac{\partial T_{\mu\nu}}{\partial x_\mu} = \frac{\partial \mathcal{L}}{\partial (x_\nu)} \tag{1-93}$$

where $\partial/\partial(x_\nu)$ designates the derivative with respect to the explicit x dependence through an external field. In such a case, the energy-momentum conservation fails to hold.

Example (1-4) — Phase transformation

Let us consider general fields φ_α which are not real but complex; then in order to make $\mathcal{L}$ and hence $\mathcal{H}$ real we have to introduce their complex conjugate fields φ^*_α.

$$I = \int \mathcal{L}[\varphi_\alpha(x), \varphi^*_\alpha(x), \varphi_{\alpha:\mu}(x), \varphi^*_{\alpha:\mu}(x)]\, d^4x \tag{1-94}$$

Suppose that the Lagrangian density is invariant under the infinitesimal phase transformation

$$\varphi_\alpha \to \varphi'_\alpha = e^{i\lambda}\varphi_\alpha = \varphi_\alpha + i\lambda\varphi_\alpha \tag{1-95a}$$

$$\varphi^*_\alpha \to \left(\varphi^*_\alpha\right)' = e^{-i\lambda}\varphi^*_\alpha = \varphi^*_\alpha - i\lambda\varphi^*_\alpha \tag{1-95b}$$

which is true when $\mathcal{L}$ involves φ and φ^* in the combination $\varphi^*\varphi$ or their derivatives but not in the combination like $\varphi + \varphi^*$. The latter transformation may be expressed by

$$\delta\varphi_\alpha = i\varphi_\alpha\lambda \quad \text{or} \quad C_\alpha = i\varphi_\alpha \tag{1-96a}$$

$$\delta\varphi^*_\alpha = -i\varphi^*_\alpha\lambda \quad \text{or} \quad C^*_\alpha = -i\varphi^*_\alpha \tag{1-96b}$$

so that we get

$$\frac{\partial}{\partial x_\mu}\left[\frac{\partial \mathcal{L}}{\partial \varphi_{\alpha:\mu}}\varphi_\alpha - \frac{\partial \mathcal{L}}{\partial \varphi^*_{\alpha:\mu}}\varphi^*_\alpha\right] = 0 \tag{1-97}$$

which expresses the conservation of a "current."

Example (1-5) — Lorentz transformation

Let us assume that the action integral is invariant under the infinitesimal Lorentz transformation

$$x_\mu \to x'_\mu = x_\mu + \epsilon_{\mu\nu} x_\nu \quad \text{with} \quad \epsilon_{\nu\mu} = -\epsilon_{\mu\nu} \tag{1-98}$$

$$\varphi_\alpha(x) \to \varphi'_\alpha(x') = \varphi_\alpha(x) + \frac{1}{2}\epsilon_{\mu\nu}(S_{\mu\nu})_{\alpha\beta}\,\varphi_\beta(x) \tag{1-99}$$

In proving the Lorentz invariance of the Dirac equation we learn that the Dirac spinors have to transform according to

$$\psi_r \to \psi'_r = (1 + \frac{1}{2}\epsilon_{\mu\nu} S_{\mu\nu})_{rs}\,\psi_s \tag{1-100}$$

where

$$S_{\mu\nu} = \frac{1}{4}(\gamma_\mu\gamma_\nu - \gamma_\nu\gamma_\mu) \tag{1-101}$$

The form (1-99) is the generalization of the transformation of the Dirac spinors. In fact, $S_{\mu\nu}$ is given for spin $0, \frac{1}{2}$, and 1 fields by

$$S_{\mu\nu} = 0 \qquad \text{spin } 0 \tag{1-102a}$$

$$S_{\mu\nu} = \frac{1}{4}(\gamma_\mu\gamma_\nu - \gamma_\nu\gamma_\mu) \qquad \text{spin } \frac{1}{2} \tag{1-102b}$$

$$S_{\mu\nu} = a_{\mu\nu} \qquad \text{spin } 1 \tag{1-102c}$$

The infinitesimal Lorentz transformation for the coordinates can be expressed by

$$\begin{aligned} \delta x_\mu &= \epsilon_{\mu\nu} x_\nu \\ &= \frac{1}{2}\epsilon_{\rho\sigma}(a_{\rho\sigma})_{\mu\nu} x_\nu \end{aligned} \tag{1-103}$$

with

$$(a_{\rho\sigma})_{\mu\nu} = \delta_{\mu\rho}\,\delta_{\sigma\nu} - \delta_{\mu\sigma}\,\delta_{\rho\nu} \tag{1-104}$$

$a_{\rho\sigma}$ represents a 4 x 4 matrix. The set of matrices $a_{\rho\sigma}$ is characterized by the commutation relations

$$[a_{\rho\sigma}, a_{\lambda\tau}] = \delta_{\sigma\lambda} a_{\rho\tau} - \delta_{\sigma\tau} a_{\rho\lambda} - \delta_{\rho\lambda} a_{\sigma\tau} + \delta_{\rho\tau} a_{\sigma\lambda} \tag{1-105}$$

Since the transformation of a field is a "representation" of the Lorentz transformation, the matrices S should satisfy the same commutation relations as a, that is,

$$[S_{\rho\sigma}, S_{\lambda\tau}] = \delta_{\sigma\lambda} S_{\rho\tau} - \delta_{\sigma\tau} S_{\rho\lambda} - \delta_{\rho\lambda} S_{\sigma\tau} + \delta_{\rho\tau} S_{\sigma\lambda} \tag{1-106}$$

In any case, if the action integral is invariant under the Lorentz transformation with the appropriate choice of S, the corresponding conservation law for every pair of subscripts as ρ and σ is given by

$$\frac{\partial}{\partial x_\mu}\left[\frac{\partial \mathfrak{L}}{\partial \varphi_{\alpha:\mu}} (S_{\rho\sigma})_{\alpha\beta}\, \varphi_\beta + (T_{\mu\rho} x_\sigma - T_{\mu\sigma} x_\rho)\right] = 0 \tag{1-107}$$

This equation expresses the conservation of angular momentum. The first term in the bracket corresponds to the spin angular momentum density, and the second to the orbital angular momentum density.

Define

$$\mathfrak{M}_{\mu\sigma\rho} = x_\sigma T_{\mu\rho} - x_\rho T_{\mu\sigma} + \frac{\partial \mathfrak{L}}{\partial \varphi_{\alpha:\mu}} (S_{\rho\sigma})_{\alpha\beta}\, \varphi_\beta \tag{1-108}$$

then

$$M_{\sigma\rho} = \int d^3x\, \mathfrak{M}_{0\sigma\rho} \tag{1-109}$$

represents an angular momentum component corresponding to a rotation in the $\sigma\rho$ plane, for instance, $M_{12} = J_z$.

The energy-momentum tensor $T_{\mu\nu}$ is not symmetric in general, and for certain purposes it is necessary to introduce a symmetric expression called the symmetric energy-momentum tensor $\Theta_{\mu\nu}$ satisfying the following conditions:

$$P_\nu = \int d^3x\, T_{0\nu} = \int d^3x\, \Theta_{0\nu}$$

$$M_{\sigma\rho} = \int d^3x\, \mathfrak{M}_{0\sigma\rho} = \int d^3x\, (x_\sigma \Theta_{0\rho} - x_\rho \Theta_{0\sigma}) \tag{1-110}$$

We shall not enter into details, however, since we shall not need it in this book.

REFERENCES

1. Regarding the conventions introduced in this book, refer to W. Pauli, Rev. Mod. Phys. **13**, 203 (1941).
2. G. Källén, *Encyclopedia of Physics* (Julius Springer, Berlin, Göttingen, Heidelberg, 1958), Vol. 5, Part 1, p. 173.
3. E. Noether, Gött. Nachr. p. 235 (1918). See also ref. 1.

CHAPTER 2
QUANTIZATION OF FREE FIELDS

Now that we have developed a general theory of field quantization we shall apply it to the quantization of special fields.[1]

2-1 REAL SPINLESS FIELDS

The Lagrangian density for a real spinless field, referred to as a neutral scalar field in a less rigorous terminology, has been discussed already and is given by

$$\mathcal{L} = -\frac{1}{2}\left[\left(\frac{\partial\varphi}{\partial x_\mu}\right)^2 + m^2\varphi^2\right] \tag{2-1}$$

In order to introduce various fields, however, it is necessary to list the criteria for choosing an appropriate Lagrangian density.

(1) $\mathcal{L}$ is Lorentz invariant.

(2) $\mathcal{H}$, calculated from $\mathcal{L}$, is positive definite.

(3) In the absence of interactions, field operators have to satisfy the Klein-Gordon equations. This condition further requires $\mathcal{L}_{free}$ to be bilinear in the field variables.

Apart from a trivial numerical factor which expresses the normalization of field variables, the form of $\mathcal{L}$ given in (2-1) is unique. In the neutral scalar theory we know

$$\pi(x) = \dot{\varphi}(x) \tag{2-2}$$

$$\mathcal{H}(x) = \frac{1}{2}\left\{\dot{\varphi}^2(x) + [\nabla\varphi(x)]^2 + m^2\varphi(x)^2\right\} \tag{2-3}$$

The equal-time commutation relations are given by

$$[\varphi(x), \varphi(x')] = [\pi(x), \pi(x')] = 0 \tag{2-4}$$

$$[\varphi(x), \pi(x')] = i\,\delta^3(x - x') \qquad \text{for} \quad x'_0 = x_0 \tag{2-5}$$

In order to find the physical interpretation of these commutation relations it is convenient to Fourier-analyze the field variables.

Suppose that the field is enclosed in a cube of volume $V = L^3$; then the exponential functions $e^{i\mathbf{px}}$, with $p_i = n_i\,(2\pi/L)$ ($i = 1, 2, 3$), form a complete set (n_i = an integer). Thus we can expand $\varphi(x)$ as

$$\varphi(x) = V^{-1/2} \sum_{\mathbf{p}} e^{i\mathbf{px}}\, c(\mathbf{p}, t) \tag{2-6}$$

Substituting this expansion into the Klein-Gordon equation we get

$$\ddot{c}(\mathbf{p}, t) = -(\mathbf{p}^2 + m^2)\, c(\mathbf{p}, t) \tag{2-7}$$

With $p_0 = (\mathbf{p}^2 + m^2)^{1/2}$ we get

$$c(\mathbf{p}, t) = c_1(\mathbf{p}) \exp(-ip_0 t) + c_2(\mathbf{p}) \exp(ip_0 t) \tag{2-8}$$

Hence

$$\varphi(x) = V^{-1/2} \sum_{\mathbf{p}} e^{i\mathbf{px}}\,[c_1(\mathbf{p}) \exp(-ip_0 t) + c_2(\mathbf{p}) \exp(ip_0 t)] \tag{2-9}$$

Since $\varphi(x)$ is assumed to be a real field, it is Hermitian, so that

$$c_2(\mathbf{p}) = c_1^{\dagger}(-\mathbf{p}) \tag{2-10}$$

Hence, using only $c(\mathbf{p}) = c_1(\mathbf{p})$, we may expand $\varphi(x)$ and $\pi(x)$ as

$$\varphi(x) = V^{-1/2} \sum_{\mathbf{p}} [\exp(i\mathbf{px} - ip_0 t)\, c(\mathbf{p}) + \exp(-i\mathbf{px} + ip_0 t)\, c^{\dagger}(\mathbf{p})] \tag{2-11}$$

$$\pi(x) = V^{-1/2} \sum_{\mathbf{p}} [-ip_0 \exp(i\mathbf{px} - ip_0 t)\, c(\mathbf{p})$$

$$+\, ip_0 \exp(-i\mathbf{px} + ip_0 t)\, c^{\dagger}(\mathbf{p})] \tag{2-12}$$

From here on we shall denote $\mathbf{px} - p_0 t$ by px. Substituting these expansions into the equal-time commutation relations between two φ's and between two π's, we find from the linear independence of Fourier components

$$[c(\mathbf{p}), c(\mathbf{q})] = [c^{\dagger}(\mathbf{p}), c^{\dagger}(\mathbf{q})] = 0 \tag{2-13}$$

$$[c(\mathbf{p}), c^{\dagger}(\mathbf{q})] + [c^{\dagger}(\mathbf{p}), c(\mathbf{q})] = 0 \tag{2-14}$$

The commutation relation

equal times

$$[\varphi(x), \pi(x')] = i V^{-1} \sum_{p} \exp[ip(x-x')] \qquad \text{for} \quad x_0' = x_0 \tag{2-15}$$

implies

$$ip_0 [c(p), c^\dagger(q)] - ip_0 [c^\dagger(p), c(q)] = \begin{cases} i & \text{for} \quad p = q \\ 0 & \text{for} \quad p \neq q \end{cases} \tag{2-16}$$

or

$$[c(p), c^\dagger(q)] - [c^\dagger(p), c(q)] = p_0^{-1} \delta_{p,q} \tag{2-17}$$

Combining Eq. (2-17) with Eq. (2-14), we see that

$$[c(p), c^\dagger(q)] = (2p_0)^{-1} \delta_{p,q} \tag{2-18}$$

Setting

$$c(p) = (2p_0)^{-1/2} a(p) \qquad c^\dagger(p) = (2p_0)^{-1/2} a^\dagger(p) \tag{2-19}$$

The commutation relations can be expressed in a simpler form,

$$[a(p), a(q)] = [a^\dagger(p), a^\dagger(q)] = 0 \tag{2-20}$$

$$[a(p), a^\dagger(q)] = \delta_{p,q} \tag{2-21}$$

The field operator $\varphi(x)$ is expanded in the following form:

$$\varphi(x) = \sum_{p} (2p_0 V)^{-1/2} [e^{ipx} a(p) + e^{-ipx} a^\dagger(p)] \tag{2-22}$$

Next we shall express the energy and momentum of the system in terms of a and $a^\dagger$. (use 2-22)

(2-3)
$$H = \int \mathcal{H}(x)\, d^3x = \frac{1}{2} \sum_{p} p_0 [a(p) a^\dagger(p) + a^\dagger(p) a(p)] \tag{2-23}$$

$$P_k = -\int \pi(x) \frac{\partial \varphi(x)}{\partial x_k} d^3x = \frac{1}{2} \sum_{p} p_k [a(p) a^\dagger(p) + a^\dagger(p) a(p)] \tag{2-24}$$

See p. 12

In the calculation of P_k, terms like

$$\sum_{\mathbf{p}} p_k\, a(\mathbf{p})\, a(-\mathbf{p}) \exp[-ip_0(t + t')] \tag{2-25}$$

vanish because of the odd symmetry of the summand.

Defining $n(\mathbf{p})$ as

$$n(\mathbf{p}) = a^\dagger(\mathbf{p})\, a(\mathbf{p}) \tag{2-26}$$

and making use of $[a(\mathbf{p}), a^\dagger(\mathbf{p})] = 1$, we get

$$H = \sum_{\mathbf{p}} p_0 [n(\mathbf{p}) + \tfrac{1}{2}] \tag{2-27}$$

$$P_k = \sum_{\mathbf{p}} p_k [n(\mathbf{p}) + \tfrac{1}{2}] \tag{2-28}$$

From the commutation relation $[a, a^\dagger] = 1$ we can prove that the eigenvalues of $n = a^\dagger a$ are nonnegative integers: 0, 1, 2, ...

Proof: Let us take a matrix representation in which n is diagonal. Since $[a, a^\dagger] = 1$ with a given $\mathbf{p}$ assumed,

$$[n, a] = -a \tag{2-29a}$$

$$[n, a^\dagger] = a^\dagger \tag{2-29b}$$

Let n_α' and n_β' be the two eigenvalues of n for states α and β. Taking the matrix elements of the above relation between states α and β, we get

$$\langle \alpha | [n, a^\dagger] | \beta \rangle = (n'_\alpha - n'_\beta) \langle \alpha | a^\dagger | \beta \rangle = \langle \alpha | a^\dagger | \beta \rangle \tag{2-30}$$

Hence $\langle \alpha | a^\dagger | \beta \rangle \neq 0$ only when $n'_\alpha = n'_\beta + 1$. Since n is a positive definite operator, there must be a minimum eigenvalue n'_0. Then for the states having nonvanishing matrix elements of $a^\dagger$ and a, and hence n, the values of n' are given by

$$n'_0,\ n'_0 + 1,\ n'_0 + 2,\ \ldots$$

Defining an eigenstate $|n'\rangle$ as

$$n|n'\rangle = n'|n'\rangle \tag{2-31}$$

then

$$\langle n'|n|n'\rangle = \sum_m \langle n'|a^\dagger|m\rangle \langle m|a|n'\rangle$$
$$= \langle n'|a^\dagger|n' - 1\rangle\langle n' - 1|a|n'\rangle$$
$$= |\langle n' - 1|a|n'\rangle|^2 = n' \tag{2-32}$$

so that, apart from trivial phase factors, we find

$$\langle n' - 1|a|n'\rangle = \langle n'|a^\dagger|n' - 1\rangle = (n')^{1/2} \tag{2-33}$$

In order to find the smallest eigenvalue, n_0', we note from the above commutation relations of n and a that

$$na|n'\rangle = (n' - 1)\, a|n'\rangle \qquad \text{for all states } |n'\rangle \tag{2-34}$$

and in particular

$$na|n_0'\rangle = (n_0' - 1)\, a|n_0'\rangle \tag{2-35}$$

However, since n_0' is the smallest eigenvalue of n, then $a|n_0'\rangle$ must be equal to zero, that is,

$$a|n_0'\rangle = 0 \tag{2-36}$$

Hence

$$|a|n_0'\rangle|^2 = \langle n_0'|a^\dagger a|n_0'\rangle$$
$$= \langle n_0'|n|n_0'\rangle$$
$$= n_0'$$
$$= 0 \tag{2-37}$$

Thus we have shown that the possible values of n' are nonnegative integers: 0, 1, 2, ...

Now we express the energy and momentum of the system in terms of the number operators n(p):

$$H = \sum_p p_0\,[n(p) + \tfrac{1}{2}] = \sum_p p_0 n(p) + \frac{1}{2}\sum_p p_0 \tag{2-38}$$

$$P_k = \sum_p p_k\,[n(p) + \tfrac{1}{2}] = \sum_p p_k n(p) + \frac{1}{2}\sum_p p_k \tag{2-39}$$

The second term in H is divergent but a constant. We drop this term inasmuch as we cannot observe the absolute magnitude of the energy eigenvalues, but only their differences. In addition, the transition from classical theory to quantum theory is not unique, as the order of operators in a product can be chosen arbitrarily without altering classical theory, but such is not the case in quantum theory. In fact, if we were to modify H within the classical theory as follows, we would get only the first term:

$$H = \frac{1}{2}\sum_p p_0[a(p)a^\dagger(p) + a^\dagger(p)a(p)]$$
$$\longrightarrow \sum_p p_0 a^\dagger(p)a(p) \tag{2-40}$$

Thus we shall keep only the first terms in H and P_k, and we may write

$$P_\mu = \sum_p p_\mu n(p) \tag{2-41}$$

The physical interpretation of this formula is clear: n(p) expresses the number of particles with momentum p and energy $p_0 = (p^2 + m^2)^{1/2}$ and for this reason is called the *number operator*. Consequently a(p) and $a^\dagger(p)$ denote operators which respectively decrease and increase by unity the number of particles with energy momentum p_μ, and they are called *destruction* and *creation operators*, respectively. The field variables φ_α and π_α are thus quantum mechanical operators which change the number of particles. The particle so obtained is called the quantum of the field φ_α, and φ_α and π_α are called *field operators*.

We know the commutation relations between field operators only for equal times, but with the help of field equations we can derive the commutation relations for unequal times.

Since we know the commutation relations between a and $a^\dagger$ in the expansion (2-22) we can calculate $[\varphi(x), \varphi(y)]$ even for $x_0 \neq y_0$.

$$[\varphi(x), \varphi(y)] = \sum_p \sum_q (2p_0V)^{-1/2}(2q_0V)^{-1/2}$$
$$\left\{ e^{ipx - iqy}[a(p), a^\dagger(q)] + e^{-ipx + iqy}[a^\dagger(p), a(q)] \right\}$$
$$= \sum_p (2p_0V)^{-1}(e^{ip(x-y)} - e^{-ip(x-y)}) \tag{2-42}$$

Substituting

$$\sum_{p} \rightarrow \frac{V}{(2\pi)^3}\int d^3p \quad \text{(element of phase space)} \tag{2-43}$$

the commutator becomes

$$\frac{1}{(2\pi)^3}\int \frac{d^3p}{2p_0}\left(e^{ip(x-y)} - e^{-ip(x-y)}\right)$$

$$= \frac{i}{(2\pi)^3}\int \frac{d^3p}{p_0} \sin p(x-y)$$

$$\equiv i\Delta(x-y) \tag{2-44}$$

The function Δ was introduced originally by Jordan and Pauli[2], but the one defined here differs from theirs by a minus sign.

Let us study the properties of the Δ function,

$$\Delta(x) = \frac{-i}{(2\pi)^3}\int \frac{d^3p}{2p_0}\left(e^{ipx} - e^{-ipx}\right) \tag{2-45}$$

(1) $\Delta(x)$ is a Lorentz invariant function; namely, if $x'_\mu = a_{\mu\nu}x_\nu$ is a proper Lorentz transformation, excluding reflections, we get

$$\Delta(x') = \Delta(x) \tag{2-46}$$

Proof: $\delta(p^2 + m^2)$ is a manifestly invariant function, which gives p_0 implicitly as a function of $\mathbf{p}$. The graph of this function is a hyperbola with two branches, one for $p_0 < 0$ and one for $p_0 > 0$.

$$p_0{}^2 = |\mathbf{p}|^2 + m^2; \quad p_0^2 - |\vec{p}|^2 = m^2 \text{: mass-shell condition, or free particle}$$

A point on one branch is transformed into another point on the same branch by a proper Lorentz transformation. Since reflections are excluded these two branches are not mixed. We select the upper branch characterized by

$$p_0 > 0$$

In order to do this we define

$$\theta(p_0) = \begin{cases} 1 & \text{if } p_0 > 0 \\ 0 & \text{if } p_0 < 0 \end{cases} \tag{2-47}$$

and

$$\epsilon(p_0) = \theta(p_0) - \theta(-p_0) = \begin{cases} 1 & \text{if} \quad p_0 > 0 \\ -1 & \text{if} \quad p_0 < 0 \end{cases} \tag{2-48}$$

Both $\theta(p_0)\,\delta(p^2 + m^2)$ and $\theta(-p_0)\,\delta(p^2 + m^2)$ are invariant under proper Lorentz transformations, and they consist of the positive and negative energy branches of the hyperbola, respectively. Thus their difference

$$[\theta(p_0) - \theta(-p_0)]\,\delta(p^2 + m^2)$$

$$= \epsilon(p_0)\,\delta(p^2 + m^2) \tag{2-49}$$

is also Lorentz invariant although it incorporates both branches. Therefore it is clear that we can make an invariant function by using

$$\int d^4p\,\epsilon(p_0)\,\delta(p^2 + m^2)\,e^{ipx}$$

$$= \int d^3p\,dp_0\,\epsilon(p_0)\,\delta(p^2 + m^2)\exp(i\mathbf{px} - ip_0x_0)$$

$$= \frac{1}{2}\int d^3p\,dp_0\,(\mathbf{p}^2 + m^2)^{-1/2}\,[\delta(p_0 - (\mathbf{p}^2 + m^2)^{1/2})$$

$$- \delta(p_0 + (\mathbf{p}^2 + m^2)^{1/2})]\exp(i\mathbf{px} - ip_0x_0)$$

$$= \frac{1}{2}\int d^3p\,(\mathbf{p}^2 + m^2)^{-1/2}\,[\exp(i\mathbf{px} - i(\mathbf{p}^2 + m^2)^{1/2}x_0)$$

$$- \exp(i\mathbf{px} + i(\mathbf{p}^2 + m^2)^{1/2}x_0)]$$

From here on we shall use $p_0 = (\mathbf{p}^2 + m^2)^{1/2}$ Since we can use $-\mathbf{p}$ for $\mathbf{p}$ in the integration, the above integral can be written as

$$\int \frac{d^3p}{2p_0}\,[\exp(i\mathbf{px} - ip_0x_0) - \exp(-i\mathbf{px} + ip_0x_0)]$$

$$= \int \frac{d^3p}{2p_0}\left(e^{ipx} - e^{-ipx}\right) \tag{2-50}$$

Hence

$$\Delta(x) = \frac{-i}{(2\pi)^3} \int d^4p\ \epsilon(p_0)\delta(p^2 + m^2)e^{ipx} \qquad (2\text{-}51)$$

is Lorentz invariant.

(2) $\Delta(x)$ is an odd function: $\Delta(-x) = -\Delta(x)$. This is clear from the definition of $\Delta(x)$.

Combining these two properties one can prove the following:

(3) $\Delta(x) = 0$ if x is spacelike.

If x is spacelike one can choose a coordinate system in which $x_0' = 0$. Lorentz invariance of $\Delta(x)$, then, requires $\Delta(x) = \Delta(x')$. If x' is a vector of the form $(\mathbf{x}', x_0' = 0)$, one can always find a spatial rotation of 180° which transforms $\mathbf{x}'$ into $-\mathbf{x}'$, so that

$$\Delta(x) = \Delta(x') = \Delta(-x') = -\Delta(x') = 0 \qquad (2\text{-}52)$$

This means that $\varphi(x)$ and $\varphi(y)$ commute when the points x and y are separated by a spacelike distance. This property is called the *microscopic causality condition.* Physically this condition means that any disturbance cannot propagate faster than the velocity of light so that observations of two phenomena separated by a spacelike distance are not hindered.

The Lorentz invariance of the Δ function is also important. It means that the canonical quantization method which refers to a special time axis yields a completely covariant method of quantization. This means that if the canonical commutation relations hold in one frame of reference, they also hold in other frames.

(4) The following relation is valid:

$$(\Box - m^2)\Delta(x) = 0 \qquad (2\text{-}53)$$

This is clear from

$$(\Box_x - m^2)[\varphi(x), \varphi(0)] = [(\Box - m^2)\varphi(x), \varphi(0)] = 0$$

(5) The following relations hold:

$$\Delta(x) = 0 \qquad \frac{\partial}{\partial x_0}\Delta(x) = -\delta^3(x) \qquad \text{for } x_0 = 0 \qquad (2\text{-}54)$$

The first relation follows from property (3). The second one is verified by a direct calculation or by using the equal-time commutation relation $[\dot{\varphi}(x), \varphi(0)] = -i\delta^3(x)$.

2-2 THE MEANING OF FIELD QUANTIZATION—NONRELATIVISTIC EXAMPLES

So far we have discussed the formal generalization of particle quantum mechanics to field theory. However, we can show that one also can apply the idea of field quantization to particle mechanics.

Consider the simplest quantum mechanical problem, namely, the nonrelativistic Schrödinger equation,

$$i\frac{\partial\varphi}{\partial t} = H\varphi = \frac{p^2}{2m}\varphi + V(x)\varphi \tag{2-55}$$

This equation describes the motion of a particle moving under the influence of an external potential V(x). As far as this Hamiltonian is concerned, there always will be just one particle if there was only one initially.

Suppose that a particle with initial momentum p_i is scattered by the potential V(x) and then found with momentum p_f. There are two alternative languages to describe this process.

(1) A particle with momentum $\mathbf{p}_i$ is scattered into another momentum state $\mathbf{p}_f$.
(2) The number of particles with momentum **p** is $\delta_{\mathbf{p},\mathbf{p}_i}$ in the initial state and $\delta_{\mathbf{p},\mathbf{p}_f}$ in the final state.

In the first language the dynamical variable that specifies a state is the *momentum*, and in the second language it is the *number of particles* with a given momentum. For this simple problem there is no difference between these two languages and we have no clue to judge the superiority of one language over the other.

To obtain such a clue let us proceed to the two-body problem,

$$H = \frac{p_1^{\,2}}{2m} + V(x_1) + \frac{p_2^{\,2}}{2m} + V(x_2) \tag{2-56}$$

We assume that two particles are *identical* and describe the scattering process in two languages.

(1) A particle with momentum $\mathbf{p}_i^{(1)}$ is scattered into $\mathbf{p}_f^{(1)}$, and a particle with momentum $\mathbf{p}_i^{(2)}$ is scattered into $\mathbf{p}_f^{(2)}$; or the particle with momentum $\mathbf{p}_i^{(1)}$ is scattered into $\mathbf{p}_f^{(2)}$, and $\mathbf{p}_i^{(2)}$ into $\mathbf{p}_f^{(1)}$. Namely, we have either $\mathbf{p}_i^{(1)} \to \mathbf{p}_f^{(1)}$ and $\mathbf{p}_i^{(2)} \to \mathbf{p}_f^{(2)}$, or $\mathbf{p}_i^{(1)} \to \mathbf{p}_f^{(2)}$ and $\mathbf{p}_i^{(2)} \to \mathbf{p}_f^{(1)}$.

(2) The number of particles with momentum p is $\delta_{\mathbf{p},\mathbf{p}_i^{(1)}} + \delta_{\mathbf{p},\mathbf{p}_i^{(2)}}$ in the initial state and $\delta_{\mathbf{p},\mathbf{p}_f^{(1)}} + \delta_{\mathbf{p},\mathbf{p}_f^{(2)}}$ in the final state.

In the second language it is not necessary to distinguish between the two alternatives. Thus in two-body scattering we see that the second language is more convenient, but the evidence is not yet conclusive. Finally we shall give a decisive example which definitely favors the second language. Suppose that a particle decays into other particles. How can we express this process in the first language? The answer is that the vocabulary of the first language is too poor to describe this process. Into what state does the particle, initially present, go? Nowhere. In the second language we can simply say that the number of particles of the kind initially present changes from 1 to 0.

For this reason we use the second language to describe even the simplest system. In the second language, the state vector is specified by dynamical variables n(p), where n(p) designates the number of particles with momentum p. The simple scattering process $p_i \to p_f$ is described in the second language by the statement that a particle with momentum p_i is destroyed and a particle with momentum p_f is created.

To describe the scattering process mathematically in the second language we have to introduce a field operator ψ. In this section we use φ to denote the Schrödinger wave function and ψ to denote the corresponding operator. The field equation is given by

$$i\frac{\partial \psi}{\partial t} = -\frac{\Delta}{2m}\psi + V(x)\psi \tag{2-57}$$

This is of the same form as the single particle Schrödinger equation, but the interpretation is completely different. The Lagrangian density that reproduces the field equation by means of the action principle is given by

$$\mathcal{L} = i\psi^\dagger \frac{\partial \psi}{\partial t} - \frac{(\nabla \psi^\dagger)(\nabla \psi)}{2m} - V\psi^\dagger \psi \tag{2-58}$$

In applying the variational principle, $\delta\psi$ and $\delta\psi^\dagger$ should be treated as independent variations. The Hamiltonian density is given by

$$\mathcal{H} = \pi \frac{\partial \psi}{\partial t} - \mathcal{L}$$

$$= \frac{(\nabla \psi^\dagger)(\nabla \psi)}{2m} + V\psi^\dagger \psi \tag{2-59}$$

where

$$\pi = \frac{\partial \mathcal{L}}{\partial \dot{\psi}} = i\psi^\dagger \tag{2-60}$$

Let us assume that ψ describes a Bose field, then the canonical commutation relations are valid.

$$[\psi(x), \pi(x')] = i\delta^3(x - x') \tag{2-61}$$

or

$$[\psi(x), \psi^\dagger(x')] = \delta^3(x - x') \qquad \text{for} \quad t' = t \tag{2-62}$$

We quantize this system at t = 0 and solve the scattering problem in the Heisenberg representation.

$$\psi(x) = V^{-1/2} \sum_p e^{ipx} a(p, t) \tag{2-63a}$$

$$\psi^\dagger(x) = V^{-1/2} \sum_p e^{-ipx} a^\dagger(p, t) \tag{2-63b}$$

The equal-time commutation relation between ψ and $\psi^\dagger$ requires that

$$[a(p, t), a^\dagger(q, t)] = \delta_{p,q} \tag{2-64}$$

Introduce the momentum operator **P**,

$$\begin{aligned} P_k &= -\int d^3x\, \pi(x) \frac{\partial\psi(x)}{\partial x_k} \\ &= \frac{1}{i} \int d^3x\, \psi^\dagger(x) \frac{\partial\psi(x)}{\partial x_k} \\ &= \sum_p p_k\, a^\dagger(p, t)\, a(p, t) \end{aligned} \tag{2-65}$$

The number operator n(p, t) in the Heisenberg representation reads

$$n(p, t) = a^\dagger(p, t)\, a(p, t) \tag{2-66}$$

In the present problem **P** does not commute with H so that the number operators become time dependent.

The vacuum state Ψ_0 is defined as

$$\Psi_0 = \left| n(p) = 0 \qquad \text{for all } p \right\rangle \tag{2-67a}$$

or

$$a(p, t)\, \Psi_0 = 0 \qquad \text{for all } p \tag{2-67b}$$

since a is a destruction operator. Then $a^\dagger(p, t)\, \Psi_0$ represents a state in which a particle with momentum p is present at time t, since

$$\begin{aligned} &\mathbf{P}(t)(a^\dagger(p, t)\, \Psi_0) \\ &\quad = [\mathbf{P}(t), a^\dagger(p, t)]\, \Psi_0 + a^\dagger(p, t)\, \mathbf{P}(t)\, \Psi_0 \\ &\quad = p(a^\dagger(p, t)\, \Psi_0) \end{aligned} \tag{2-68}$$

where use has been made of $\mathbf{P}(t)\,\Psi_0 = 0$. The operator expressing the total number of particles

$$N = \sum_{p} n(p, t) \tag{2-69}$$

commutes with H, and we easily can show that

$$N(a^{\dagger}(p, t)\,\Psi_0) = a^{\dagger}(p, t)\,\Psi_0 \tag{2-70}$$

Therefore, $a^{\dagger}(p, t)\,\Psi_0$ represents a one-particle state with momentum p prepared at time t.

There is a peculiar situation for quantized nonrelativistic theory. In relativistic theory a field operator is decomposed into destruction and creation operators, whereas in nonrelativistic theory only one kind of operator appears. In relativistic theory, the Fourier component $c(p, t)$ satisfies a second-order differential equation in time,

$$\ddot{c}(p, t) = -(p^2 + m^2)\,c(p, t) \tag{2-71}$$

so that it is a superposition of two frequencies, one positive and the other negative, corresponding to destruction and creation parts, respectively. In nonrelativistic theory $a(p, t)$ satisfies a first-order differential equation,

$$-i\,\dot{a}(p, t) = \frac{p^2}{2m}\,a(p, t) \qquad \text{(We put } V = 0 \text{ here.)} \tag{2-72}$$

so that it has only a destruction part.

Assume that there is only one particle with momentum p_i at $t = 0$. The state vector is given by

$$\Psi_i = a^{\dagger}(p_i, 0)\Psi_0 \tag{2-73}$$

If Ψ_0 is properly normalized, namely, $(\Psi_0, \Psi_0) = 1$, then Ψ_i is also normalized to unity, since

$$\begin{aligned}(\Psi_i, \Psi_i) &= (\Psi_0, a(p_i, 0)\,a^{\dagger}(p_i, 0)\,\Psi_0)\\ &= (\Psi_0, [a(p_i, 0), a^{\dagger}(p_i, 0)]\Psi_0)\\ &= (\Psi_0, \Psi_0)\end{aligned} \tag{2-74}$$

In the scattering problem we want to know the expectation value of the number of particles with momentum p_f at time t.

$$\begin{aligned}n(p_f, t) &= (\Psi_i, n(p_f, t)\,\Psi_i)\\ &= (\Psi_i, a^{\dagger}(p_f, t)\,a(p_f, t)\,\Psi_i)\\ &= \sum_{n} (\Psi_i, a^{\dagger}(p_f, t)\,\Psi_n)\,(\Psi_n, a(p_f, t)\,\Psi_i)\end{aligned} \tag{2-75}$$

Since a is a destruction operator decreasing the number of particles by one and Ψ_i represents a one-particle state, the expression above survives only for $\Psi_n = \Psi_0$. Hence

$$\langle n(p_f, t)\rangle = |(\Psi_0, a(p_f, t)\Psi_i)|^2 = |\varphi_i(p_f, t)|^2 \tag{2-76}$$

where

$$\varphi_i(p_f, t) = (\Psi_0, a(p_f, t)\Psi_i) \tag{2-77}$$

We also define $\varphi_i(\mathbf{x}, t)$ as follows:

$$\varphi_i(\mathbf{x}, t) = (\Psi_0, \psi(\mathbf{x}, t)\Psi_i)$$

$$= V^{-1/2}\sum_{\mathbf{p}} e^{ipx}\varphi_i(p, t) \tag{2-78}$$

This function $\varphi_i(\mathbf{x}, t)$ can be identified as the Schrödinger wave function satisfying the proper initial condition.

(1) Initial condition

$$\varphi_i(p, 0) = (\Psi_0, a(p, 0)\, a^\dagger(p_i, 0)\Psi_0) = \delta_{p, p_i} \tag{2-79}$$

so that

$$\varphi_i(\mathbf{x}, 0) = V^{-1/2}\exp(ip_i\mathbf{x}) \tag{2-80}$$

(2) Schrödinger equation

From the definition of $\varphi_i(\mathbf{x}, t)$ and the field equation, it is clear that $\varphi_i(\mathbf{x}, t)$ satisfies the Schrödinger equation:

$$i\frac{\partial}{\partial t}\varphi_i(\mathbf{x}, t) = -\frac{\Delta}{2m}\varphi_i(\mathbf{x}, t) + V(\mathbf{x})\varphi_i(\mathbf{x}, t) \tag{2-81}$$

Thus $\varphi_i(\mathbf{x}, t)$ is interpreted as the ordinary wave function, and $\varphi_i(p, t)$ is its Fourier transform, Therefore, in a single-particle system the expectation value $\langle n(p_f, t)\rangle$ is equal to the probability of finding the particle in the momentum p_f state at time t.

From the above analysis we learn that the ordinary Schrödinger wave function $\varphi(\mathbf{x}, t)$ corresponds to the matrix element.

$$(\Psi_0, \psi(\mathbf{x}, t)\Psi) \tag{2-82}$$

The usefulness of the quantized theory consists in the point that the same Hamiltonian can be used to describe the many-body problem. In fact, $(\Psi_0, \psi(x_1, t)\, \psi(x_2, t)\, \Psi)$ satisfies the Schrödinger equation for two particles. The quantized field equations cover a more general system, the one-particle system being a special solution. Furthermore, the two-body wave function is automatically symmetric, corresponding to Bose statistics.

2-3 INVARIANT GREEN'S FUNCTIONS AND PEIERLS' METHOD OF QUANTIZATION

Invariant Functions

We have introduced an invariant function $\Delta(x)$ in Section 2-1, but it is convenient to introduce several related functions. As has been shown $\Delta(x)$ vanishes for spacelike x, namely, outside the light cone. $\Delta(x)$ is nonzero only inside the future or past light cone as shown in Fig. 2-1. Now let us define a new function which is equal to $-\Delta(x)$ inside the future light cone and vanishes otherwise: the retarded Δ function, Δ^R or Δ^{ret}.

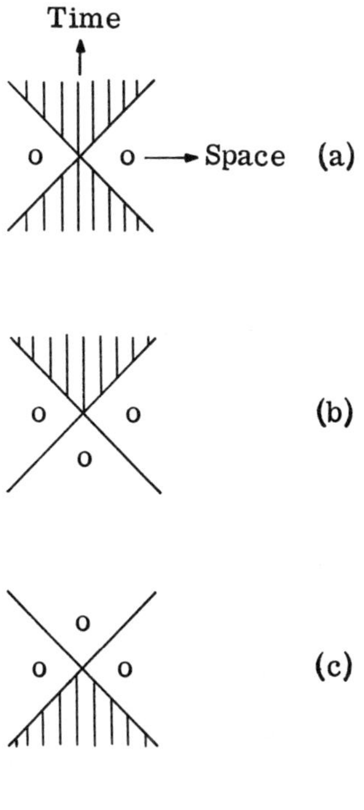

Figure 2-1. (a) Support of $\Delta(x)$, (b) Support of $\Delta^R(x)$, (c) Support of $\Delta^A(x)$.

$$\Delta^R(x) = -\theta(x_0)\Delta(x) = \begin{cases} -\Delta(x) & \text{for} \quad x_0 > 0 \\ 0 & \text{for} \quad x_0 < 0 \end{cases} \tag{2-83}$$

In a similar way we define $\Delta^A(x)$, the advanced Δ function, as

$$\Delta^A(x) = \theta(-x_0)\Delta(x) = \begin{cases} 0 & \text{for} \quad x_0 > 0 \\ \Delta(x), & \text{for} \quad x_0 < 0 \end{cases} \tag{2-84}$$

Clearly we have

$$\Delta(x) = \Delta^A(x) - \Delta^R(x), \qquad \Delta^A(x) = \Delta^R(-x) \tag{2-85}$$

We also can introduce other invariant functions with special properties.

$$\bar{\Delta}(x) = -\frac{1}{2}\epsilon(x_0)\Delta(x) = -\frac{1}{2}(\theta(x_0) - \theta(-x_0))\Delta(x)$$

$$= \frac{1}{2}\left[\Delta^R(x) + \Delta^A(x)\right] \tag{2-86}$$

This is an even function of x.

Let us recall the definition of $\Delta(x)$:

$$\Delta(x) = \frac{-i}{(2\pi)^3}\int \frac{d^3p}{2p_0}\left(e^{ipx} - e^{-ipx}\right) \tag{2-87}$$

The first term involves only positive frequencies, and this part of $\Delta(x)$ is called $\Delta^{(+)}(x)$, while the second part, involving only negative frequencies, is called $\Delta^{(-)}(x)$.

$$\Delta^{(+)}(x) = \frac{-i}{(2\pi)^3}\int \frac{d^3p}{2p_0} e^{ipx}$$

$$= \frac{-i}{(2\pi)^3}\int d^4p\, \theta(p_0)\delta(p^2 + m^2)e^{ipx} \tag{2-88}$$

$$\Delta^{(-)}(x) = \frac{i}{(2\pi)^3}\int \frac{d^3p}{2p_0} e^{-ipx}$$

$$= \frac{i}{(2\pi)^3}\int d^4p\, \theta(p_0)\delta(p^2 + m^2)e^{-ipx}$$

$$= \frac{i}{(2\pi)^3}\int d^4p\, \theta(-p_0)\delta(p^2 + m^2)e^{ipx} \tag{2-89}$$

Then we can immediately recognize the following relationships:

$$\Delta^{(+)}(-x) = -\Delta^{(-)}(x) \tag{2-90a}$$

$$\Delta^{(-)}(-x) = -\Delta^{(+)}(x) \tag{2-90b}$$

We define $\Delta^{(1)}(x)$, a real even function, by

$$\Delta^{(1)}(x) = i[\Delta^{(+)}(x) - \Delta^{(-)}(x)]$$

$$= \frac{1}{(2\pi)^3} \int d^4p\, \delta(p^2 + m^2)\, e^{ipx} \tag{2-91}$$

Obviously $\Delta^{(1)}(x)$ satisfies the Klein-Gordon equation:

$$K_x \Delta^{(1)}(x) = (\Box_x - m^2)\,\Delta^{(1)}(x) = 0 \tag{2-92}$$

and therefore so do $\Delta^{(+)}(x)$ and $\Delta^{(-)}(x)$, since we have

$$\Delta^{(+)}(x) = \frac{1}{2}[\Delta(x) - i\Delta^{(1)}(x)] \tag{2-93a}$$

$$\Delta^{(-)}(x) = \frac{1}{2}[\Delta(x) + i\Delta^{(1)}(x)] \tag{2-93b}$$

Next let us prove the equations,

$$K_x \Delta^R(x) = K_x \Delta^A(x) = K_x \overline{\Delta}(x) = -\delta^4(x) \tag{2-94}$$

In order to prove the first two, it suffices to prove the last,

$$K_x \overline{\Delta}(x) = -\delta^4(x) \tag{2-95}$$

since

$$\Delta^R(x) = \overline{\Delta}(x) - \frac{1}{2}\Delta(x) \tag{2-96a}$$

$$\Delta^A(x) = \overline{\Delta}(x) + \frac{1}{2}\Delta(x) \tag{2-96b}$$

The proof can be accomplished by calculating the Fourier transform of $\overline{\Delta}(x)$, using

$$\epsilon(x_0) = \frac{x_0}{|x_0|} = \frac{2}{\pi} \int_0^\infty \frac{d\tau}{\tau} \sin(\tau x_0)$$

$$= \frac{1}{i\pi} P \int_{-\infty}^{\infty} \frac{d\tau}{\tau} \exp\ i\tau x_0 \tag{2-97}$$

From this follows the integral representation

$$\overline{\Delta}(x) = \frac{-1}{2\pi i} P \int \frac{d\tau}{\tau} \exp(i\tau x_0) \frac{(-i)}{(2\pi)^3} \int d^4k\, \epsilon(k_0)\, \delta(k^2 + m^2)\, e^{ikx}$$

$$= \frac{1}{(2\pi)^4} \int d^4p\, e^{ipx}\, P \int \frac{d\tau}{\tau} \delta(p^2 - (p_0 + \tau)^2 + m^2) \frac{p_0 + \tau}{|p_0 + \tau|}$$

$$= \frac{1}{(2\pi)^4} P \int \frac{d^4p}{p^2 + m^2} e^{ipx} \tag{2-98}$$

in which we made a transformation of the variables of integration

$$\mathbf{p} = \mathbf{k} \qquad p_0 = k_0 - \tau$$

Hence we have

$$K_x \overline{\Delta}(x) = -\frac{1}{(2\pi)^4} \int d^4p\, e^{ipx} = -\delta^4(x) \tag{2-99}$$

Finally we introduce Feynman's Δ function, or Stueckelberg's causal function, denoted by $\Delta_F(x)$,

$$\Delta_F(x) = \frac{1}{2} \Delta^{(1)}(x) - i\overline{\Delta}(x)$$

$$= \frac{-i}{(2\pi)^4} \int d^4p\, e^{ipx} \left(\frac{P}{p^2 + m^2} + i\pi\delta(p^2 + m^2) \right) \tag{2-100}$$

This function satisfies

$$K_x \Delta_F(x) = i\delta^4(x) \tag{2-101}$$

Peierls' Method of Quantization

In Section 1-2 we proved that the canonical equations are equivalent to the field equations provided that the canonical commutation relations are valid, that is,

canonical commutation relations + canonical equations → field equations

We shall solve the inverse problem: assuming that the canonical equations are equivalent to the field equations, we derive the commutation relations. The following arguments are based essentially on Peierls' idea[3] but are not necessarily the same.

Suppose that the Lagrangian density is given by

$$\mathcal{L}[\varphi_\alpha, \varphi_\alpha^\dagger, \varphi_{\alpha:\mu}, \varphi_{\alpha:\mu}^\dagger] \tag{2-102}$$

where φ_α's are complex fields. In the absence of interactions, $\mathcal{L}$ is a bilinear form in φ_α, $\varphi_\alpha^\dagger$ and their derivatives, and the Euler equations are given in the following form:

$$[\mathcal{L}]_{\varphi_\alpha^\dagger} = D_{\alpha\beta}(\partial)\varphi_\beta = 0 \tag{2-103}$$

where D is a function of differential operators and is a matrix. For instance, in the case of a free real spinless field, we get

$$[\mathcal{L}]_\varphi = (\Box_x - m^2)\varphi = 0 \tag{2-104}$$

so that

$$D(\partial) = \Box_x - m^2 = K_x \tag{2-105}$$

In this case D is a one-dimensional matrix since there is only one component.

For a technical reason, we next introduce an additional term in the Lagrangian density, expressing the interaction of fields φ_α with an external c-number field Q(x), namely,

$$\mathcal{L} \to \mathcal{L} - Q(x)\,\varphi_\alpha^\dagger(x)\varphi_\alpha(x) \tag{2-106}$$

Then the field equations become

$$D_{\alpha\beta}(\partial)\varphi_\beta - Q\varphi_\alpha = 0 \tag{2-107}$$

If φ is real we get an additional factor of 2 with Q. Let us assume that Q is nonzero only in a finite space-time domain Ω. To be more specific, let us assume that φ's satisfy the free field equations without the Q terms before T_1 and after T_2. (See Fig. 2-2.)

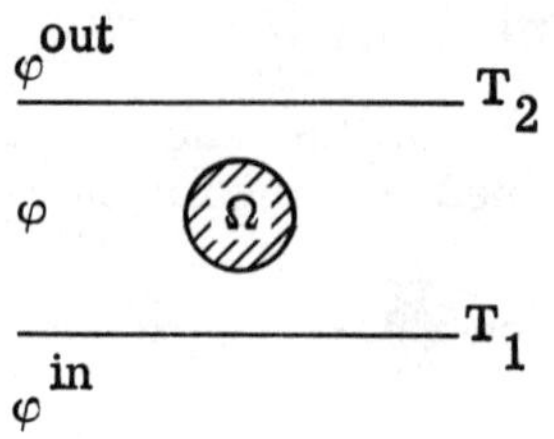

Figure 2-2. The space-time domain Ω which is the support of the external field $Q(x)$.

The φ_α before T_1 and after T_2 are designated as φ_α^{in} and φ_α^{out}, respectively.

$$D_{\alpha\beta}(\partial)\varphi_\beta^{in} = 0 \qquad \text{for} \quad t < T_1 \tag{2-108a}$$

$$D_{\alpha\beta}(\partial)\varphi_\beta^{out} = 0 \qquad \text{for} \quad t > T_2 \tag{2-108b}$$

If $Q = 0$ even in Ω, we may write

$$\varphi_\alpha = \varphi_\alpha^{in} = \varphi_\alpha^{out} \tag{2-109}$$

We next try to solve for φ in terms of φ^{in} and set

$$\varphi_\alpha = \varphi_\alpha^{in} + \varphi'_\alpha \tag{2-110}$$

It is clear, then, that we have $\varphi'_\alpha = 0$ for $t < T_1$. Applying D to φ_α, we get

$$D_{\alpha\beta}(\partial)\varphi_\beta = D_{\alpha\beta}(\partial)\varphi_\beta^{in} + D_{\alpha\beta}(\partial)\varphi'_\beta = Q\varphi_\alpha \tag{2-111}$$

or

$$D_{\alpha\beta}(\partial)\varphi'_\beta = Q\varphi_\alpha \tag{2-112}$$

since φ_β^{in} satisfies the free field equation. Since we know the boundary condition for φ', we can solve Eq. (2-112) with the aid of the retarded Green's functions. Let us consider an equation

$$D_{\alpha\beta}(\partial)\Delta^R_{\beta\gamma}(x) = -\delta_{\alpha\gamma}\cdot\delta^4(x) \tag{2-113}$$

with the boundary condition that Δ^R vanishes for $x_0 < 0$. The $\Delta^R_{\alpha\beta}(x)$ is the generalization of the retarded Δ function for a multicomponent field. If $\Delta^R_{\alpha\beta}(x)$ is obtained as the solution of the above equation, we get

$$\varphi'_\alpha(x) = -\int_\Omega d^4x'\, \Delta^R_{\alpha\beta}(x - x')\, Q(x')\, \varphi_\beta(x') \tag{2-114}$$

Thus $\varphi'_\alpha(x)$ vanishes for $x_0 < T_1$, from the properties of Δ^R and of Ω. Hence we get an integral equation for φ_α.

$$\varphi_\alpha(x) = \varphi_\alpha^{in}(x) - \int_\Omega d^4x'\, \Delta^R_{\alpha\beta}(x - x')\, Q(x')\, \varphi_\beta(x') \tag{2-115}$$

In principle this integral equation can be solved to give $\varphi_\alpha(x)$ in terms of $\varphi_\alpha^{in}(x)$ and $Q(x)$, but that solution is not needed for the present purpose.

By using φ^{out} we obtain a similar set of equations,

$$\varphi_\alpha(x) = \varphi_\alpha^{out}(x) - \int d^4x'\, \Delta^A_{\alpha\beta}(x - x')\, Q(x')\, \varphi_\beta(x') \tag{2-116}$$

where $\Delta^A_{\alpha\beta}$ satisfies the same equation as $\Delta^R_{\alpha\beta}$ does, although the boundary conditions are different.

Equations (2-115) and (2-116) are derived from the action principle alone; we have not yet used the quantization procedure.

We next solve the same problem (for φ) by using the canonical formalism

$$\mathcal{H} = \frac{\partial \mathcal{L}}{\partial \dot{\varphi}_\alpha}\, \dot{\varphi}_\alpha - \mathcal{L} \tag{2-117}$$

where we already used the summation convention over α. Introduction of Q modifies $\mathcal{H}(x)$ to

$$\mathcal{H}(x) \to \mathcal{H}(x) + Q(x)\varphi_\alpha^\dagger(x)\varphi_\alpha(x) = \mathcal{H}(x) + \mathcal{H}'(x) \tag{2-118}$$

Thus the Hamiltonian is given by

$$H(t) = \int_{x_0 = t} [\mathcal{H}(x) + \mathcal{H}'(x)]\, d^3x = H_0 + H' \tag{2-119}$$

The presence of the external field Q(x) makes H time dependent since we have, in general,

$$\frac{dF}{dt} = \frac{\partial F}{\partial(t)} + i[H(t), F] \tag{2-120}$$

The canonical equation for φ_α is given by

$$\frac{\partial\varphi_\alpha(x)}{\partial x_0} = i[H(x_0),\ \varphi_\alpha(x)] \tag{2-121}$$

Because of the absence of Q, we have for $x_0 < T_1$

$$\frac{\partial\varphi_\alpha^{in}(x)}{\partial x_0} = i\left[H_0\ [\varphi_\alpha^{in}(x),\ \ldots],\ \varphi_\alpha^{in}(x)\right] \tag{2-122}$$

To solve the canonical equations in the presence of Q, we set

$$\varphi_\alpha(x) = U(x_0)^{-1}\ \varphi_\alpha^{in}(x)\ U(x_0) \tag{2-123}$$

Because the φ_α's and φ_α^{in}'s should satisfy the same equal-time commutation relations, we can always find such a unitary matrix U.

Since $\varphi_\alpha = \varphi_\alpha^{in}$ for $x_0 < T_1$, we get $U = 1$ for $x_0 < T_1$. Now the canonical equation becomes

$$\begin{aligned}
&\frac{\partial}{\partial x_0}[\,U(x_0)^{-1}\ \varphi_\alpha^{in}(x)U(x_0)\,] \\
&= i\left[H[U(x_0)^{-1}\ \varphi_\alpha^{in}(x)U(x_0),\ \ldots],\ U(x_0)^{-1}\varphi_\alpha^{in}(x)U(x_0)\right] \\
&= iU(x_0)^{-1}\left[H[\varphi_\alpha^{in}(x),\ \ldots],\ \varphi_\alpha^{in}(x)\right]U(x_0) \\
&= iU(x_0)^{-1}\left[H_0[\varphi_\alpha^{in}(x),\ \ldots],\varphi_\alpha^{in}(x)\right]U(x_0) \\
&\quad + iU(x_0)^{-1}\left[H'[\varphi_\alpha^{in}(x),\ \ldots],\ \varphi_\alpha^{in}(x)\right]U(x_0) \\
&= U(x_0)^{-1}\frac{\partial\varphi_\alpha^{in}(x)}{\partial x_0}U(x_0) + iU(x_0)^{-1}\left[H'[\varphi_\alpha^{in}(x),\ \ldots],\right. \\
&\quad \left.\varphi_\alpha^{in}(x)\right]U(x_0)
\end{aligned} \tag{2-124}$$

Calculating the left-hand side, from $U(x_0)^{-1}\ U(x_0) = 1$, we have

$$\frac{\partial U(x_0)^{-1}}{\partial x_0} = -U(x_0)^{-1}\frac{\partial U(x_0)}{\partial x_0}U(x_0)^{-1} \tag{2-125}$$

so that

$$\frac{\partial}{\partial x_0}[U(x_0)^{-1}\varphi_\alpha^{in}(x)\,U(x_0)]$$

$$= U(x_0)^{-1}\,\frac{\partial\varphi_\alpha^{in}(x)}{\partial x_0}\,U(x_0)$$

$$+ U(x_0)^{-1}\,\varphi_\alpha^{in}(x)\frac{\partial U(x_0)}{\partial x_0} - U(x_0)^{-1}\,\frac{\partial U(x_0)}{\partial x_0}\,U(x_0)^{-1}\,\varphi_\alpha^{in}(x)U(x_0) \tag{2-126}$$

Hence Eq. (2-124) reduces to

$$\varphi_\alpha^{in}(x)\,\frac{\partial U(x_0)}{\partial x_0}\,U(x_0)^{-1} - \frac{\partial U(x_0)}{\partial x_0}\,U(x_0)^{-1}\,\varphi_\alpha^{in}(x)$$

$$= i\left[H'\,[\varphi_\alpha^{in}(x),\,\ldots],\,\varphi_\alpha^{in}(x)\right] \tag{2-127}$$

If we write this equation in the form

$$\left[iH'\,[\varphi_\alpha^{in}(x),\,\ldots] + \frac{\partial U(x_0)}{\partial x_0}\,U(x_0)^{-1},\,\varphi_\alpha^{in}(x)\right] = 0 \tag{2-128}$$

it is easy to see that the equation for $U(x_0)$ reduces to

$$iH'\,[\varphi_\alpha^{in}(x),\,\ldots] + \frac{\partial U(x_0)}{\partial x_0}\,U(x_0)^{-1} = 0 \tag{2-129}$$

or

$$i\,\frac{\partial U(x_0)}{\partial x_0} = H'\,[\varphi_\alpha^{in}(x),\,\ldots]U(x_0) \tag{2-130}$$

where

$$H'[\varphi_\alpha^{in}(x),\,\ldots] = \int d^3x\,Q(x)\varphi_\alpha^{\dagger\,in}(x)\varphi_\alpha^{in}(x) \tag{2-131}$$

The differential equation for $U(x_0)$ can be cast in an integral form by taking account of the boundary condition $U(x_0) = 1$ for $x_0 < T_1$, namely,

$$U(x_0) = 1 - i\int_{-\infty}^{x_0} d^4x'\; Q(x')\varphi_\beta^{\dagger in}(x')\varphi_\beta^{in}(x')U(x_0') \tag{2-132}$$

Let us next assume that Q is infinitesimally small; then

$$U(x_0) = 1 - i\int_{-\infty}^{x_0} d^4x'\; Q(x')\varphi_\beta^{\dagger\, in}(x')\varphi_\beta^{in}(x') \tag{2-133}$$

and

$$\varphi_\alpha(x) = U(x_0)^{-1}\,\varphi_\alpha^{in}(x)\;U(x_0)$$

$$= \varphi_\alpha^{in}(x) - i\int_{-\infty}^{-\infty} d^4x'\;\theta(x_0 - x_0')[\varphi_\alpha^{in}(x),\;\varphi_\beta^{\dagger\, in}(x')\varphi_\beta^{in}(x')]Q(x') \tag{2-134}$$

We replace $\varphi_\beta(x')$ by $\varphi_\beta^{in}(x')$ in the integral, since Q is assumed to be infinitesimal; then

$$\varphi_\alpha(x) = \varphi_\alpha^{in}(x) - \int d^4x'\;\Delta_{\alpha\beta}^{R}(x - x')\varphi_\beta^{in}(x')Q(x') \tag{2-135}$$

Comparing (2-134) and (2-135), we get

$$\theta(x_0 - x_0')\,[\varphi_\alpha^{in}(x),\;\varphi_\beta^{\dagger\, in}(x')\varphi_\beta^{in}(x')] = -i\Delta_{\alpha\beta}^{R}(x - x')\varphi_\beta^{in}(x') \tag{2-136}$$

since $Q(x')$ is arbitrary. In the absence of Q, therefore,

$$\theta(x_0 - x_0')\;[\varphi_\alpha(x),\;\varphi_\beta^{\dagger}(x')\varphi_\beta(x')] = -i\Delta_{\alpha\beta}^{R}(x - x')\varphi_\beta(x') \tag{2-137}$$

In a similar way we find, starting from the future, the relation

$$\theta(x_0' - x_0)[\varphi_\alpha(x),\;\varphi_\beta^{\dagger}(x')\varphi_\beta(x')] = i\Delta_{\alpha\beta}^{A}(x - x')\;\varphi_\beta(x') \tag{2-138}$$

Define $\Delta_{\alpha\beta}(x - x') = \Delta^{A}_{\alpha\beta}(x - x') - \Delta^{R}_{\alpha\beta}(x - x')$; then

$$[\varphi_\alpha(x), \varphi^\dagger_\beta(x')\varphi_\beta(x')] = i\Delta_{\alpha\beta}(x - x')\,\varphi_\beta(x') \tag{2-139}$$

This equation covers all times, both $x_0 > x'_0$ and $x_0 < x'_0$.

As has been remarked earlier, the right-hand side must be multiplied by a factor of 2 for real fields, so that

(1) Real fields:

$$[\varphi_\alpha(x), \varphi_\beta(x')\varphi_\beta(x')] = 2i\Delta_{\alpha\beta}(x - x')\,\varphi_\beta(x') \tag{2-140}$$

(2) Complex fields:

$$[\varphi_\alpha(x), \varphi^\dagger_\beta(x')\varphi_\beta(x')] = i\Delta_{\alpha\beta}(x - x')\varphi_\beta(x') \tag{2-141}$$

Example (2-1) Real spinless field

The invariant Green's functions reduce to

$$\Delta^{A}_{\alpha\beta} \to \Delta^{A} \qquad \Delta^{R}_{\alpha\beta} \to \Delta^{R} \qquad \text{and} \qquad \Delta_{\alpha\beta} \to \Delta$$

The resulting commutation relation is

$$[\varphi(x), \varphi(x')\varphi(x')] = 2i\Delta(x - x')\varphi(x') \tag{2-142}$$

From this relation we easily can obtain

$$[\varphi(x), \varphi(x')] = i\Delta(x - x') \tag{2-143}$$

In the case of complex fields, there are two alternative solutions.

$$(1)\quad [\varphi_\alpha(x), \varphi_\beta(x')] = 0 \qquad [\varphi^\dagger_\alpha(x), \varphi^\dagger_\beta(x')] = 0 \tag{2-144}$$

and

$$[\varphi_\alpha(x), \varphi^\dagger_\beta(x')] = i\Delta_{\alpha\beta}(x - x') \tag{2-145}$$

$$(2)\quad \left\{\varphi_\alpha(x), \varphi_\beta(x')\right\} = 0 \qquad \left\{\varphi^\dagger_\alpha(x), \varphi^\dagger_\beta(x')\right\} = 0 \tag{2-146}$$

and

$$\left\{\varphi_\alpha(x), \varphi^\dagger_\beta(x')\right\} = i\Delta_{\alpha\beta}(x - x') \tag{2-147}$$

The canonical commutation relations correspond to the first solution. We shall learn later that we have to apply the first method to quantize a Bose-Einstein field and the second one for a Fermi-Dirac field. In other words, if we apply the first method the quanta of the field should necessarily obey Bose-Einstein statistics while a method using the second kind of relation leads to Fermi-Dirac statistics.

The most important conclusion we can draw here is that there are two possible methods of quantization which guarantee the equivalence of the canonical equations to the Euler equations or the field equations.

2-4 COMPLEX SPINLESS FIELDS

Lagrangian Density

We again return to the quantization of a special field. In the case of a complex spinless field we have

$$\mathcal{L} = -\left(\frac{\partial\varphi^\dagger}{\partial x_\mu} \cdot \frac{\partial\varphi}{\partial x_\mu} + m^2\varphi^\dagger\varphi\right) \tag{2-148}$$

This is the only possible Lagrangian density that leads to (1) a positive definite Hamiltonian, (2) a Lorentz invariant theory, and (3) the Klein-Gordon equation.

As an example of the application of Peierls' method let us first discuss the quantization of this field. The field equation is

$$[\mathcal{L}]_{\varphi^\dagger} = (\Box_x - m^2)\,\varphi(x) = 0 \tag{2-149}$$

Therefore, we get

$$[\varphi(x), \varphi^\dagger(x')] = i(\Delta^A(x - x') - \Delta^R(x - x')) = i\Delta(x - x') \tag{2-150}$$

or

$$\{\varphi(x), \varphi^\dagger(x')\} = i\Delta(x - x') \tag{2-151}$$

We have to choose the first possibility, (2-150), for the reason to be stated later. From the Lagrangian density, (2-148), we have

$$\pi^\dagger = \frac{\partial\mathcal{L}}{\partial\dot{\varphi}} = \dot{\varphi}^\dagger \tag{2-152a}$$

$$\pi = \frac{\partial\mathcal{L}}{\partial\dot{\varphi}^\dagger} = \dot{\varphi} \tag{2-152b}$$

and

$$\begin{aligned} \mathcal{H} &= \dot{\varphi}^\dagger\dot{\varphi} + \nabla\varphi^\dagger \cdot \nabla\varphi + m^2\varphi^\dagger\varphi \\ &= \pi^\dagger\pi + \nabla\varphi^\dagger \cdot \nabla\varphi + m^2\varphi^\dagger\varphi \end{aligned} \tag{2-153}$$

which is positive definite, as required.

Commutation Relations

The complete sets of commutation relations in the two alternative cases are given, respectively, by

(1) $[\varphi(x), \varphi(x')] = [\varphi^\dagger(x), \varphi^\dagger(x')] = 0$ (2-154)

$[\varphi(x), \varphi^\dagger(x')] = [\varphi^\dagger(x), \varphi(x')] = i\Delta(x - x')$ (2-155)

(2) $\{\varphi(x), \varphi(x')\} = \{\varphi^\dagger(x), \varphi^\dagger(x')\} = 0$ (2-156)

$\{\varphi(x), \varphi^\dagger(x')\} = -\{\varphi^\dagger(x), \varphi(x')\} = i\Delta(x - x')$ (2-157)

The second set is inconsistent, however, as we shall see soon,

$$\{\varphi(x), \varphi^\dagger(x')\} = \mathcal{C}\{\varphi^\dagger(x), \varphi(x')\}\mathcal{C}^{-1}$$

requires that this anticommutator be a q number. Since Δ is a c number function, however, $i\Delta(x - x')$ cannot be a q number. Therefore, the second set is inconsistent and must be abandoned.

There is a general theorem by Pauli[4] on the connection between spin and statistics:

integral spin (scalar, vector, etc.) $\leftrightarrows$ Bose-Einstein statistics

half-integral spin (spinor, etc.) $\leftrightarrows$ Fermi-Dirac statistics

We do not give the proof of this theorem here, but we shall check which quantization method we should apply case by case. In the present case we have proved that spinless fields obey Bose-Einstein statistics.

Phase Transformation

In the case of a complex field we usually require invariance of the theory under phase transformations,

$$\varphi \to \varphi' = e^{i\lambda}\varphi \qquad \varphi^\dagger \to \varphi^{\dagger\prime} = e^{-i\lambda}\varphi^\dagger \tag{2-158}$$

Taking a conservation law as discussed in Section 1-3,

$$\frac{\partial}{\partial x_\mu}\left[\frac{\partial \mathcal{L}}{\partial \varphi_{:\mu}}\varphi - \frac{\partial \mathcal{L}}{\partial \varphi^\dagger_{:\mu}}\varphi^\dagger\right] = 0 \tag{2-159}$$

and setting

$$j_\mu = -i\left(\frac{\partial \mathcal{L}}{\partial \varphi_{:\mu}}\varphi - \frac{\partial \mathcal{L}}{\partial \varphi^\dagger_{:\mu}}\varphi^\dagger\right)$$

$$= -i\left(\frac{\partial \varphi}{\partial x_\mu}\varphi^\dagger - \frac{\partial \varphi^\dagger}{\partial x_\mu}\varphi\right) \tag{2-160}$$

then the conservation law is given by

$$\frac{\partial j_\mu}{\partial x_\mu} = 0 \tag{2-161}$$

and we have a conserved quantity or a quantum number N defined by

$$N = \int j_0 \, d^3x = -i \int (\pi^\dagger \varphi - \pi\varphi^\dagger)\, d^3x \tag{2-162}$$

Fourier Expansion

We can expand $\varphi(x)$ and $\varphi^\dagger(x)$ into plane waves. Using the procedure described in the quantization of real field in Section 2-1, we get the following results:

$$\varphi(x) = \sum_p (2p_0 V)^{-1/2} \left(e^{ipx}\, a(p) + e^{-ipx}\, b^\dagger(p)\right) \tag{2-163a}$$

$$\varphi^\dagger(x) = \sum_p (2p_0 V)^{-1/2} \left(e^{-ipx}\, a^\dagger(p) + e^{ipx}\, b(p)\right) \tag{2-163b}$$

The coefficients satisfy

$$[a(p), a^\dagger(q)] = \delta_{p,q} \tag{2-164a}$$

$$[b(p), b^\dagger(q)] = \delta_{p,q} \tag{2-164b}$$

and all other commutators vanish.

If we set

$$\varphi = 2^{-1/2} (\varphi_1 - i\varphi_2), \quad \varphi^\dagger = 2^{-1/2} (\varphi_1 + i\varphi_2) \tag{2-165}$$

then φ_1 and φ_2 are real fields and we can quantize them by the method introduced in Section 2-1 after writing the Lagrangian density as

$$-\left(\frac{\partial\varphi^\dagger}{\partial x_\mu}\frac{\partial\varphi}{\partial x_\mu} + m\,\varphi^\dagger\varphi\right)$$

$$= -\frac{1}{2}\left[\left(\frac{\partial\varphi_1}{\partial x_\mu}\right)^2 + m^2\varphi_1^2\right] - \frac{1}{2}\left[\left(\frac{\partial\varphi_2}{\partial x_\mu}\right)^2 + m^2\varphi_2^2\right] \tag{2-166}$$

In this way we naturally get the same result as in Eq. (2-155). This trick is used in the theory of pions.

Inserting the Fourier expansion of the field operators into

$$H = \int d^3x\, \mathcal{H}(x) = \int d^3x(\dot{\varphi}^\dagger\dot{\varphi} + \nabla\varphi^\dagger\nabla\varphi + m^2\varphi^\dagger\varphi) \tag{2-167}$$

and

$$P_k = -\int d^3x \left(\pi^\dagger \frac{\partial \varphi}{\partial x_k} + \pi \frac{\partial \varphi^\dagger}{\partial x_k} \right) \tag{2-168}$$

we find

$$H = \frac{1}{2} \sum_p p_0 \left[a^\dagger(p)\, a(p) + a(p)a^\dagger(p) + b(p)b^\dagger(p) + b^\dagger(p)\, b(p) \right] \tag{2-169}$$

$$P_k = \frac{1}{2} \sum_p p_k \left[a^\dagger(p)\, a(p) + a(p)a^\dagger(p) + b(p)b^\dagger(p) + b^\dagger(p)\, b(p) \right] \tag{2-170}$$

$$N = \frac{1}{2} \sum_p \left[a^\dagger(p)\, a(p) + a(p)a^\dagger(p) - b(p)b^\dagger(p) - b^\dagger(p)\, b(p) \right] \tag{2-171}$$

Defining $n_+(p) = a^\dagger(p)\, a(p)$, $n_-(p) = b^\dagger(p)\, b(p)$, we get

$$H = \sum_p p_0 \left[n_+(p) + n_-(p) \right] + \sum_p p_0 \tag{2-172}$$

$$P_k = \sum_p p_k \left[n_+(p) + n_-(p) \right] \tag{2-173}$$

$$N = \sum_p \left[n_+(p) - n_-(p) \right] \tag{2-174}$$

We can drop $\sum p_0$ again for the same reason as mentioned before. The a and b particles are distinguished by the quantum number N; they are called antiparticles of each other.

Particle-Antiparticle Conjugation (Charge Conjugation)

We have seen that the theory is symmetric in a and b. We shall see later that if a particles have charge q, then b particles have charge -q. It is also possible that neutral particles are described by complex field operators.

We consider the following transformation

$$a(p) \leftrightarrows b(p) \qquad a^{\dagger}(p) \leftrightarrows b^{\dagger}(p) \tag{2-175}$$

This is called *particle-antiparticle conjugation*, or more often *charge conjugation*. Since the commutation relations are not changed by this transformation, there exists a unitary transformation $\mathcal{C}$ satisfying

$$\mathcal{C}\, a(p)\, \mathcal{C}^{-1} = b(p) \qquad \mathcal{C}\, b(p)\, \mathcal{C}^{-1} = a(p) \tag{2-176a}$$

$$\mathcal{C}\, a^{\dagger}(p)\, \mathcal{C}^{-1} = b^{\dagger}(p) \qquad \mathcal{C}\, b^{\dagger}(p)\mathcal{C}^{-1} = a^{\dagger}(p) \tag{2-176b}$$

These relationships indicate that $\mathcal{C}^2$ should be a c number. The field operators should transform as

$$\mathcal{C}\, \varphi(x)\, \mathcal{C}^{-1} = \varphi^{\dagger}(x) \tag{2-177a}$$

$$\mathcal{C}\, \varphi^{\dagger}(x)\mathcal{C}^{-1} = \varphi(x) \tag{2-177b}$$

and we also get

$$\mathcal{C}\, P_{\mu}\, \mathcal{C}^{-1} = P_{\mu} \qquad \mathcal{C}\, N\, \mathcal{C}^{-1} = -N \tag{2-178}$$

In particular, if we take the fourth component of the first equation, of (2-178) we have

$$\left[\mathcal{C}, H\right] = 0 \tag{2-179}$$

and hence $\mathcal{C}$ is a quantum number; then $\mathcal{C}^2$ is a c number and we choose a phase convention so that

$$\mathcal{C}^2 = 1 \tag{2-180}$$

The eigenvalues of $\mathcal{C}$ are +1 and -1, both infinitely degenerate. Invariance under $\mathcal{C}$ is called *charge conjugation invariance*.

We have discussed here, for simplicity, the transformations (2-177), but there is an alternative choice,

$$\mathcal{C}\, \varphi(x)\, \mathcal{C}^{-1} = -\varphi^{\dagger}(x) \tag{2-181a}$$

$$\mathcal{C}\, \varphi^{\dagger}(x)\, \mathcal{C}^{-1} = -\varphi(x) \tag{2-181b}$$

The free field is invariant under either pair of transformations, but when interactions are present only one pair makes the theory invariant.

In the case of real fields, the antiparticle is identical with the particle itself, and we have

$$\mathcal{C}\,\varphi(x)\,\mathcal{C}^{-1} = \varphi(x) \qquad (2\text{-}182a)$$

or

$$\mathcal{C}\,\varphi(x)\,\mathcal{C}^{-1} = -\varphi(x) \qquad (2\text{-}182b)$$

This invariance principle is useful to derive some selection rules. We shall discuss this problem later.

A state vector, which is an eigenstate of all n(p), is denoted by $\Phi\,[n'(p_1), n'(p_2), \ldots]$. When all terms n' are zero, this state is the vacuum state, Φ_0. We fix $\mathcal{C}$ by requiring that the vacuum state be even under $\mathcal{C}$:

$$\mathcal{C}\,\Phi_0 = \Phi_0 \qquad (2\text{-}183)$$

2-5 COMPLEX SPIN-1 FIELD

We introduce a complex vector field $\varphi_\mu(x)$ (μ = 1, 2, 3, 4). Hence the Lagrangian density is a function of φ_μ, $\varphi_\mu^\dagger$ and their derivatives.

Lagrangian Density and Hamiltonian Density

Our first task is to find the correct Lagrangian density. One of the requirements is that each component satisfies the Klein-Gordon equation,

$$K_x\varphi_\mu = (\Box - m^2)\varphi_\mu = 0 \qquad (2\text{-}184)$$

In the discussion that follows we assume $m \neq 0$. Furthermore, φ_μ has to satisfy

$$\frac{\partial \varphi_\mu}{\partial x_\mu} = 0 \qquad (2\text{-}185)$$

which is called the *irreducibility condition.* If this quantity does not vanish it represents a scalar field, and φ_μ then represents a field describing a mixture of spin-0 and spin-1 particles.

Let us make our first trial with

$$\mathcal{L} = -\left[\frac{\partial \varphi_\mu^\dagger}{\partial x_\nu}\cdot\frac{\partial \varphi_\mu}{\partial x_\nu} + m^2\varphi_\mu^\dagger\,\varphi_\mu\right] \qquad (2\text{-}186)$$

which leads to the Klein-Gordon equations,

$$(\square - m^2)\,\varphi_\mu = 0 \qquad \text{for} \quad \mu = 1, 2, 3, 4 \tag{2-187}$$

The corresponding Hamiltonian density, however, is not given by a positive definite form.

$$\mathcal{H} = \pi_\mu^\dagger \pi_\mu + \nabla\varphi_\mu^\dagger \cdot \nabla\varphi_\mu + m^2 \varphi_\mu^\dagger \varphi_\mu \tag{2-188}$$

Notice here that φ_0 and $\varphi_0^\dagger$ are Hermitian conjugates, but φ_4 and $\varphi_4^\dagger$ are anti-Hermitian conjugates:

$$\varphi_4 = i\varphi_0 \qquad \varphi_4^\dagger = i\varphi_0^\dagger \tag{2-189}$$

Hence the Hamiltonian density is given, in terms of φ_0 and $\varphi_0^\dagger$, by

$$\mathcal{H} = \sum_{k=1}^{3} \left(\pi_k^\dagger \pi_k + \nabla\varphi_k^\dagger \nabla\varphi_k + m^2 \varphi_k^\dagger \varphi_k\right)$$

$$- \left(\pi_0^\dagger \pi_0 + \nabla\varphi_0^\dagger \cdot \nabla\varphi_0 + m^2 \varphi_0^\dagger \varphi_0\right) \tag{2-190}$$

which is clearly not a positive definite form.

Therefore, we have to find another form, and as a possible choice let us take

$$\mathcal{L} = -\frac{1}{2}\left(\frac{\partial\varphi_\nu^\dagger}{\partial x_\mu} - \frac{\partial\varphi_\mu^\dagger}{\partial x_\nu}\right)\left(\frac{\partial\varphi_\nu}{\partial x_\mu} - \frac{\partial\varphi_\mu}{\partial x_\nu}\right) - m^2 \varphi_\mu^\dagger \varphi_\mu \tag{2-191}$$

This form is suggested by the Lagrangian density for the electromagnetic field. Let us introduce the *field strength* by

$$F_{\mu\nu} = \frac{\partial\varphi_\nu}{\partial x_\mu} - \frac{\partial\varphi_\mu}{\partial x_\nu} \tag{2-192}$$

then the Euler equation becomes

$$\frac{\partial \mathcal{L}}{\partial\varphi_\nu^\dagger} - \frac{\partial}{\partial x_\mu}\left(\frac{\partial \mathcal{L}}{\partial\varphi_{\nu:\mu}^\dagger}\right) = -m^2\varphi_\nu + \frac{\partial}{\partial x_\mu}\left(\frac{\partial\varphi_\nu}{\partial x_\mu} - \frac{\partial\varphi_\mu}{\partial x_\nu}\right) = 0 \tag{2-193}$$

or

$$\frac{\partial F_{\mu\nu}}{\partial x_\mu} = m^2 \varphi_\nu \tag{2-194}$$

This is the same as Maxwell's equation in the limit $m^2 = 0$. From this equation we get

$$m^2 \frac{\partial \varphi_\nu}{\partial x_\nu} = \frac{\partial^2 F_{\mu\nu}}{\partial x_\mu \partial x_\nu} = 0 \tag{2-195}$$

because $F_{\mu\nu}$ is antisymmetric in μ and ν. Since $m^2 \neq 0$ we get the Lorentz condition

$$\frac{\partial \varphi_\nu}{\partial x_\nu} = 0 \tag{2-196}$$

The original Euler equation reads

$$\frac{\partial}{\partial x_\mu}\left(\frac{\partial \varphi_\nu}{\partial x_\mu} - \frac{\partial \varphi_\mu}{\partial x_\nu}\right) = m^2 \varphi_\nu \tag{2-197}$$

Using the Lorentz condition we find

$$(\Box - m^2)\varphi_\nu = 0 \tag{2-198}$$

We now must verify that the Hamiltonian density is positive definite. For this purpose we have to introduce π, the canonical conjugate operators to φ.

$$\pi_\nu^\dagger = \frac{\partial \mathcal{L}}{\partial \dot{\varphi}_\nu} = \frac{1}{i}\left(\frac{\partial \varphi_4^\dagger}{\partial x_\nu} - \frac{\partial \varphi_\nu^\dagger}{\partial x_4}\right) = \frac{1}{i} F_{\nu 4}^\dagger \tag{2-199a}$$

$$\pi_\nu = \frac{\partial \mathcal{L}}{\partial \dot{\varphi}_\nu^\dagger} = \frac{1}{i}\left(\frac{\partial \varphi_4}{\partial x_\nu} - \frac{\partial \varphi_\nu}{\partial x_4}\right) = \frac{1}{i} F_{\nu 4} \tag{2-199b}$$

Here we encounter a new kind of difficulty, although it can be avoided. The trouble is that π_4 and $\pi_4^\dagger$ vanish identically:

$$\pi_4 = \pi_4^\dagger = 0 \tag{2-200}$$

The difficulty occurs if we try to quantize the theory using the canonical commutation relation which now does not hold for π_4; indeed we get $[\varphi_\nu^\dagger(x),\ \pi_4(x')] = 0$ for all ν. The way to get out of this trouble is to eliminate the fourth component from the theory by expressing it in terms of the other components. This procedure appears to destroy the Lorentz invariance of the theory, but we shall find later that it does not do so. First, from the field equation, (2-193), we get

$$\varphi_4 = \frac{1}{m^2}\sum_{k=1}^{3}\frac{\partial}{\partial x_k}\left(\frac{\partial\varphi_4}{\partial x_k} - \frac{\partial\varphi_k}{\partial x_4}\right) = \frac{i}{m^2}\operatorname{div}\boldsymbol{\pi} \tag{2-201a}$$

$$\varphi_4^\dagger = \frac{i}{m^2}\operatorname{div}\boldsymbol{\pi}^\dagger \tag{2-201b}$$

Thus φ_4 is expressible in terms of the other components. From Eq. (2-199) which expresses π in terms of $\mathbf{F}$, we can express $\dot{\varphi}$ in terms of π and φ:

$$\dot{\varphi}_k = \pi_k + i\,\frac{\partial\varphi_4}{\partial x_k} \tag{2-202a}$$

$$\dot{\varphi}_k^\dagger = \pi_k^\dagger + i\,\frac{\partial\varphi_4^\dagger}{\partial x_k} \tag{2-202b}$$

and thus eliminate all fourth components from the Hamiltonian density:

$$\mathcal{H} = \pi_\mu^\dagger\dot{\varphi}_\mu + \pi_\mu\dot{\varphi}_\mu^\dagger - \mathcal{L} \tag{2-203}$$

We further add to $\mathcal{H}$ a three-dimensional divergence,

$$-i\,\frac{\partial}{\partial x_k}\left(\pi_k^\dagger\varphi_4 + \pi_k\varphi_4^\dagger\right) \tag{2-204}$$

without altering $H = \int \mathcal{H}(x)d^3x$. Then the new Hamiltonian density is given by

$$\mathcal{H} = \pi_k^\dagger\pi_k + \frac{1}{2}\mathbf{F}_{ij}^\dagger\mathbf{F}_{ij} + m^2\varphi_k^\dagger\varphi_k + m^2\varphi_4^\dagger\varphi_4 - i\left(\frac{\partial\pi_k^\dagger}{\partial x_k}\varphi_4 + \frac{\partial\pi_k}{\partial x_k}\varphi_4^\dagger\right)$$

$$= \pi_k^\dagger\pi_k + \frac{1}{m^2}\frac{\partial\pi_i^\dagger}{\partial x_i}\frac{\partial\pi_k}{\partial x_k} + m^2\varphi_k^\dagger\varphi_k + \frac{1}{2}\mathbf{F}_{ij}^\dagger\mathbf{F}_{ij}$$

$$= (\boldsymbol{\pi}^\dagger\boldsymbol{\pi}) + \frac{1}{m^2}(\operatorname{div}\boldsymbol{\pi}^\dagger)(\operatorname{div}\boldsymbol{\pi}) + m^2(\boldsymbol{\varphi}^\dagger\boldsymbol{\varphi}) + (\operatorname{rot}\boldsymbol{\varphi}^\dagger)\cdot(\operatorname{rot}\boldsymbol{\varphi})$$

(2-205)

In this way we have completely eliminated the fourth component and obtained a positive definite Hamiltonian density. It may seem tricky to introduce a three-dimensional divergence; there is, however, a simpler and more elegant method which we shall discuss later.

Canonical Quantization

Now let us impose the quantization conditions,

$$[\pi_j^\dagger(x), \varphi_k(x')] = [\pi_j(x), \varphi_k^\dagger(x')] = -i\delta_{jk}\delta^3(x - x'), \quad (x_0' = x_0) \tag{2-206}$$

$$\text{all other commutators} = 0 \tag{2-207}$$

Then we see that the canonical equations are equivalent to the original field equations in spite of the trick of eliminating φ_4,

$$\begin{aligned}\dot{\varphi}_j(x) &= i\int d^3x' \, [\mathcal{H}(x'), \varphi_j(x)] \\ &= \int d^3x' \, [\pi_j(x')\delta^3(x' - x) + \frac{1}{m^2}\operatorname{div}\boldsymbol{\pi}(x') \cdot \frac{\partial}{\partial x_j'}\delta^3(x' - x)]\end{aligned} \tag{2-208}$$

or

$$\dot{\boldsymbol{\varphi}}(x) = \boldsymbol{\pi}(x) - \frac{1}{m^2}\operatorname{grad}\operatorname{div}\boldsymbol{\pi}(x) \tag{2-209}$$

In a similar way

$$\dot{\boldsymbol{\pi}} = -m^2\boldsymbol{\varphi} - \operatorname{rot}\operatorname{rot}\boldsymbol{\varphi} \tag{2-210}$$

From these equations we further get the Klein-Gordon equations:

$$\ddot{\boldsymbol{\varphi}} = -m^2\boldsymbol{\varphi} - \operatorname{rot}\operatorname{rot}\boldsymbol{\varphi} + \operatorname{grad}\operatorname{div}\boldsymbol{\varphi} = -m^2\boldsymbol{\varphi} + \Delta\boldsymbol{\varphi} \tag{2-211}$$

$$\ddot{\boldsymbol{\pi}} = -m^2\boldsymbol{\pi} - \operatorname{rot}\operatorname{rot}\boldsymbol{\pi} + \operatorname{grad}\operatorname{div}\boldsymbol{\pi} = -m^2\boldsymbol{\pi} + \Delta\boldsymbol{\pi} \tag{2-212}$$

In the canonical formalism we can *define* φ_4 and $\varphi_4^\dagger$ as

$$\varphi_4 = \frac{i}{m^2}\operatorname{div}\boldsymbol{\pi}, \quad \varphi_4^\dagger = \frac{i}{m^2}\operatorname{div}\boldsymbol{\pi}^\dagger \tag{2-213}$$

then φ_4 and $\varphi_4^\dagger$ satisfy the Klein-Gordon equation, and the Lorentz condition:

$$\frac{1}{i}\dot{\varphi}_4 = \frac{1}{m^2}\operatorname{div}\dot{\boldsymbol{\pi}} = -\operatorname{div}\boldsymbol{\varphi} \tag{2-214}$$

Thus we have reproduced the original Euler equation. It should be emphasized that the assumption $m^2 \neq 0$ plays an essential role in the preceding derivation.

We have established the equal-time commutation relations so that with the help of the field equations we can derive the unequal-time commutation relations

$$[\varphi_\mu(x, t), \varphi_\nu(x', t)] = [\varphi_\mu^\dagger(x, t), \varphi_\nu^\dagger(x', t)] = 0 \tag{2-215}$$

$$[\varphi_j(x, t), \pi_k^\dagger(x', t)] = [\varphi_j^\dagger(x, t), \pi_k(x', t)] = i\delta_{jk}\,\delta^3(x-x') \tag{2-216}$$

In order to derive the commutation relations we recall that the commutator $[\varphi, \varphi^\dagger]$ satisfies the Klein-Gordon equation. If we know the commutator and its time derivative for equal times, we can completely determine the commutator

$$[\varphi_j(x, t), \varphi_k^\dagger(x', t)] = 0 \tag{2-217}$$

$$[\dot{\varphi}_j(x, t), \varphi_k^\dagger(x', t)] = [\pi_j(x, t) - \frac{1}{m^2}\frac{\partial}{\partial x_j}\operatorname{div}\boldsymbol{\pi}(x, t), \varphi_k^\dagger(x', t)]$$

$$= -i\left[\delta_{jk}\,\delta^3(x-x') - \frac{1}{m^2}\frac{\partial^2}{\partial x_j\,\partial x_k}\delta^3(x-x')\right] \tag{2-218}$$

$$K_x[\varphi_j(x, t), \varphi_k^\dagger(x', t')] = 0 \tag{2-219}$$

The solution of the last equation is given by

$$[\varphi_j(x), \varphi_k^\dagger(x')] = i\left(\delta_{jk} - \frac{1}{m^2}\frac{\partial^2}{\partial x_j\,\partial x_k}\right)\Delta(x-x') \tag{2-220}$$

where x and x' denote the four-dimensional coordinates.

Since φ_4 is expressible in terms of the other components we can calculate the commutator between φ_4 and other components. The generalized result is given by

$$[\varphi_\mu(x), \varphi_\nu^\dagger(x')] = i\left(\delta_{\mu\nu} - \frac{1}{m^2}\frac{\partial^2}{\partial x_\mu \partial x_\nu}\right)\Delta(x - x') \qquad (2\text{-}221)$$

with $\mu, \nu = 1, 2, 3, 4$. In spite of the apparently noncovariant treatment of φ_4, we obtain covariant commutation relations.

We have seen that the canonical quantization procedure is rather complicated; we shall now give a simpler method.

Application of Peierls' Method

We use the operator $D(\partial)$ as defined in Section 2-3:

$$[\mathfrak{L}]_{\varphi_\alpha^\dagger} = D_{\alpha\beta}(\partial)\,\varphi_\beta = 0 \qquad (2\text{-}222)$$

For a complex spin-1 field,

$$[\mathfrak{L}]_{\varphi_\alpha^\dagger} = \frac{\partial}{\partial x_\beta}\left(\frac{\partial\varphi_\alpha}{\partial x_\beta} - \frac{\partial\varphi_\beta}{\partial x_\alpha}\right) - m^2\varphi_\alpha$$

$$= (\Box - m^2)\,\varphi_\alpha - \frac{\partial^2}{\partial x_\alpha \partial x_\beta}\,\varphi_\beta$$

$$\equiv D_{\alpha\beta}\varphi_\beta \qquad (2\text{-}223)$$

so that

$$D_{\alpha\beta} = \delta_{\alpha\beta}(\Box - m^2) - \frac{\partial^2}{\partial x_\alpha \partial x_\beta} \qquad (2\text{-}224)$$

We know that φ satisfies the Klein-Gordon equation as a consequence of the field equation, (2-222), which indicates the existence of an operator C satisfying

$$D_{\alpha\beta}C_{\beta\gamma} = C_{\alpha\beta}D_{\beta\gamma} = \delta_{\alpha\gamma}(\Box - m^2) \qquad (2\text{-}225)$$

The solution of the equation

$$D_{\alpha\beta}\,\Delta_{\beta\gamma}^{(R,A)}(x) = -\delta_{\alpha\gamma}\,\delta^4(x) \qquad (2\text{-}226)$$

is then given by

$$\Delta^{(R,A)}_{\alpha\beta}(x) = C_{\alpha\beta}(\partial)\ \Delta^{(R,A)}(x) \tag{2-227}$$

and hence

$$\Delta_{\alpha\beta}(x) = C_{\alpha\beta}(\partial)\ \Delta(x) \tag{2-228}$$

This is the general procedure to find the commutation relations based on Peierls' method. Let us find $C_{\alpha\beta}$ in the present problem.

$$D_{\beta\gamma} = \delta_{\beta\gamma}(\Box - m^2) - \frac{\partial^2}{\partial x_\beta \partial x_\gamma} \tag{2-229}$$

then

$$\frac{\partial^2}{\partial x_\alpha \partial x_\beta} D_{\beta\gamma} = \frac{\partial^2}{\partial x_\alpha \partial x_\gamma}(\Box - m^2) - \Box \frac{\partial^2}{\partial x_\alpha \partial x_\gamma} = -m^2 \frac{\partial^2}{\partial x_\alpha \partial x_\gamma} \tag{2-230}$$

Hence

$$D_{\alpha\gamma} - \frac{1}{m^2}\frac{\partial^2}{\partial x_\alpha \partial x_\beta} D_{\beta\gamma} = \delta_{\alpha\gamma}(\Box - m^2) \tag{2-231}$$

This means that Eq. (2-225) is satisfied if we choose

$$C_{\alpha\beta} = \delta_{\alpha\beta} - \frac{1}{m^2}\frac{\partial^2}{\partial x_\alpha \partial x_\beta} \tag{2-232}$$

Terms D and C are essentially the inverse matrices of one another so that they must commute, namely,

$$D_{\alpha\beta} C_{\beta\gamma} = C_{\alpha\beta} D_{\beta\gamma} = \delta_{\alpha\gamma}(\Box - m^2)$$

The commutation relations are given now by

$$\begin{aligned}\left[\varphi_\alpha(x),\ \varphi^\dagger_\beta(x')\right] &= iC_{\alpha\beta}\,\Delta(x - x') \\ &= i\left(\delta_{\alpha\beta} - \frac{1}{m^2}\frac{\partial^2}{\partial x_\alpha\, \partial x_\beta}\right)\Delta(x - x')\end{aligned} \tag{2-233}$$

This agrees with the previous result (2-221) and the derivation is simpler. This method needs no trick; everything comes out completely automatically. The assumption of the existence of C is equivalent to another asumption that all free-field operators satisfy the Klein-Gordon equation as a result of of the original field equations.

Since our Lagrangian density is invariant under phase transformations, we get the current conservation law as discussed in Section 1-3.

$$j_\nu = -i\left(\frac{\partial \mathcal{L}}{\partial \varphi_{\mu:\nu}}\varphi_\mu - \frac{\partial \mathcal{L}}{\partial \varphi^\dagger_{\mu:\nu}}\varphi^\dagger_\mu\right)$$

$$= i(F^\dagger_{\nu\mu}\varphi_\mu - F_{\nu\mu}\varphi^\dagger_\mu) \tag{2-234}$$

and the conservation of this current j is expressed by

$$\frac{\partial j_\nu}{\partial x_\nu} = 0 \tag{2-235}$$

Fourier Expansion

Let us expand φ and $\varphi^\dagger$ into Fourier series:

$$\varphi_k(x) = \sum_p (2p_0 V)^{-1/2}(e^{ipx} a_k(p) + e^{-ipx} b^\dagger_k(p)) \tag{2-236a}$$

$$\varphi^\dagger_k(x) = \sum_p (2p_0 V)^{-1/2}(e^{ipx} b_k(p) + e^{-ipx} a^\dagger_k(p)) \tag{2-236b}$$

$(k = 1, 2, 3)$

If we use $\varphi_{\mu:\mu} = \varphi^\dagger_{\mu:\mu} = 0$, we can express φ_4 and $\varphi^\dagger_4$ in terms of the a and b given in Eqs. (2-236). To find the commutation relations for a and b, we use

$$\left[\varphi_j(x), \varphi^\dagger_k(x')\right] = i\left(\delta_{jk} - \frac{1}{m^2}\frac{\partial^2}{\partial x_j \partial x_k}\right)\Delta(x - x') \tag{2-237}$$

We express the left-hand side in terms of $[a, a^\dagger]$ and $[b, b^\dagger]$. Comparing the Fourier components of both sides, we get

$$\left[a_j(\mathbf{p}),\ a_k^\dagger(\mathbf{p})\right] = \delta_{jk} + \frac{\mathbf{p}_j\,\mathbf{p}_k}{m^2} \tag{2-238a}$$

$$\left[b_j(\mathbf{p}),\ b_k^\dagger(\mathbf{p})\right] = \delta_{jk} + \frac{\mathbf{p}_j\,\mathbf{p}_k}{m^2} \tag{2-238b}$$

$$\text{all other commutators} = 0 \tag{2-238c}$$

If we put, for $\mu = 1, 2, 3, 4$,

$$\varphi_\mu(x) = \sum_{\mathbf{p}} (2p_0V)^{-1/2}\,(e^{ipx}\,a_\mu(\mathbf{p}) + e^{-ipx}\,b_\mu^\dagger(\mathbf{p})) \tag{2-239a}$$

$$\varphi_\mu^\dagger(x) = \sum_{\mathbf{p}} (2p_0V)^{-1/2}\,(e^{ipx}\,b_\mu(\mathbf{p}) + e^{-ipx}\,a_\mu^\dagger(\mathbf{p})) \tag{2-239b}$$

we get for this general case

$$\left[a_\mu(\mathbf{p}),\ a_\nu^\dagger(\mathbf{p})\right] = \left[b_\mu(\mathbf{p}),\ b_\nu^\dagger(\mathbf{p})\right] = \delta_{\mu\nu} + \frac{p_\mu p_\nu}{m^2} \tag{2-240}$$

From $\varphi_{\mu:\mu} = \varphi^\dagger_{\mu:\mu} = 0$ we derive

$$p_\mu a_\mu(\mathbf{p}) = p_\mu b_\mu(\mathbf{p}) = p_\mu a_\mu^\dagger(\mathbf{p}) = p_\mu b_\mu^\dagger(\mathbf{p}) = 0 \tag{2-241}$$

From these equations we get, for instance, a_0 in terms of other components.

$$p_0 a_0(\mathbf{p}) = \mathbf{p}\,\mathbf{a}(\mathbf{p}) \tag{2-242}$$

or

$$a_0(\mathbf{p}) = \frac{\mathbf{p}}{p_0}\,\mathbf{a}(\mathbf{p}), \qquad \text{etc.} \tag{2-243}$$

The zeroth or fourth component is not independent of the other components. We therefore discuss only the space components. For a given $\mathbf{p}$, we introduce three unit vectors orthogonal to each other;

$$\mathbf{e}_1 = \frac{\mathbf{p}}{|\mathbf{p}|} \qquad \mathbf{e}_2 \qquad \mathbf{e}_3 = \mathbf{e}_1 \times \mathbf{e}_2 \tag{2-244}$$

Then we write

$$\mathbf{a}(p) = \mathbf{e}_1 a^{(1)}(p) + \mathbf{e}_2 a^{(2)}(p) + \mathbf{e}_3 a^{(3)}(p), \qquad \text{etc.} \tag{2-245}$$

and the commutation relations among $a^{(i)}$'s are

$$\left[a^{(1)}, a^{(1)\dagger}\right] = 1 + \frac{p_1^2}{m^2} = 1 + \frac{\mathbf{p}^2}{m^2} = \frac{p_0^2}{m^2} \tag{2-246}$$

$$\left[a^{(2)}, a^{(2)\dagger}\right] = \left[a^{(3)}, a^{(3)\dagger}\right] = 1 \tag{2-247}$$

The $a^{(1)}$ can be referred to as the longitudinal component, and $a^{(2)}$ and $a^{(3)}$ as transverse components. Therefore, if we redefine the $a^{(i)}$'s by

$$\mathbf{a}(p) = \mathbf{e}_1 \left(\frac{p_0}{m}\right) a^{(1)}(p) + \mathbf{e}_2\, a^{(2)}(p) + \mathbf{e}_3\, a^{(3)}(p) \tag{2-248}$$

we get

$$\left[a^{(j)}, a^{(k)\dagger}\right] = \delta_{jk} \tag{2-249}$$

Inserting the results into the expressions for H, P_k, and N, we obtain, after dropping terms like $\sum_{\mathbf{p}} p_0$,

$$H = \sum_{\mathbf{p}} \sum_{j} p_0 \left(n_+^{(j)}(p) + n_-^{(j)}(p)\right) \tag{2-250}$$

$$P_k = \sum_{\mathbf{p}} \sum_{j} p_k \left(n_+^{(j)}(p) + n_-^{(j)}(p)\right) \tag{2-251}$$

$$N = \int j_0(x) d^3x = \sum_{\mathbf{p}} \sum_{j} \left(n_+^{(j)}(p) - n_-^{(j)}(p)\right) \tag{2-252}$$

with n_+ and n_- defined, respectively, as

$$n_+^{(j)}(p) = a^{(j)}(p)^\dagger a^{(j)}(p), \qquad n_-^{(j)}(p) = b^{(j)}(p)^\dagger b^{(j)}(p) \tag{2-253}$$

where n_+ denotes the number of particles with $N = 1$ and n_- the number of particles of negative N.

Next we have to discuss the physical meaning of the superscript (j), which is related to the particle spin.

Spin

In Section 1-3 the expression for the angular momentum operator has been given. Here we shall discuss only the spin angular momentum.

$$(M_{\rho\sigma})_{\text{spin}} = -\int \pi_\alpha^\dagger(x)\,(S_{\rho\sigma})_{\alpha\beta}\,\varphi_\beta(x)d^3x - \int \pi_\alpha(x)(S_{\rho\sigma})_{\alpha\beta}\,\varphi_\beta^\dagger(x)d^3x \tag{2-254}$$

Notice that we replaced $\partial\mathfrak{L}/\partial\dot{\varphi}_\alpha$ by $\pi_\alpha^\dagger$ in the spin angular momentum density. In the case of vector fields we can prove that the Lorentz covariance of the field equation implies

$$(S_{\rho\sigma})_{\alpha\beta} = \delta_{\rho\alpha}\delta_{\sigma\beta} - \delta_{\rho\beta}\delta_{\sigma\alpha} \tag{2-255}$$

The proof is analogous to the one for the Dirac equation. Substituting the expression (2-255) for S into $M_{\rho\sigma}$, we find

$$(M_{\rho\sigma})_{\text{spin}} = \int \left[\pi_\sigma^\dagger(x)\,\varphi_\rho(x) - \pi_\rho^\dagger(x)\varphi_\sigma(x)\right] dx$$

$$+\int \left[\pi_\sigma(x)\,\varphi_\rho^\dagger(x) - \pi_\rho(x)\,\varphi_\sigma^\dagger(x)\right] d^3x \tag{2-256}$$

where π and $\pi^\dagger$ have already been defined as canonical conjugates of $\varphi^\dagger$ and φ. The subscripts ρ and σ run from 1 to 3.

Substitution of the Fourier representations of the field operators into the integrals of (2-256) leads to a sum of many terms, but we select only those terms corresponding to particles at rest.

$$\mathcal{S}_{\rho\sigma} = (M_{\rho\sigma})_{\text{spin}}(p=0) = i(-a_\rho^\dagger a_\sigma + a_\sigma^\dagger a_\rho - b_\rho^\dagger b_\sigma + b_\sigma^\dagger b_\rho) \tag{2-257}$$

where $a_\rho = a_\rho(p=0)$, and so on. Let us consider the x component of the spin angular momentum, denoted simply by $\mathcal{S}_1$; then we have

$$\mathcal{S}_1 = \mathcal{S}_{23} = -i(a_2^\dagger a_3 - a_3^\dagger a_2 + b_2^\dagger b_3 - b_3^\dagger b_2) \tag{2-258}$$

In order to find the eigenvalue of $\mathcal{S}_1$, we introduce the following combinations:

$$a_+ = 2^{-1/2}(a_2 - ia_3) \qquad a_+^\dagger = 2^{-1/2}(a_2^\dagger + ia_3^\dagger) \tag{2-259a}$$

$$a_- = 2^{-1/2}(a_2 + ia_3) \qquad a_-^\dagger = 2^{-1/2}(a_2^\dagger - ia_3^\dagger) \tag{2-259b}$$

Then we find that

$$\left[a_+, a_+^\dagger\right] = \left[a_-, a_-^\dagger\right] = 1 \tag{2-260}$$

Set $n_+ = a_+^\dagger a_+$, $n_- = a_-^\dagger a_-$; then

$$n_+ + n_- = n_1 + n_2 \tag{2-261}$$

$$\mathcal{S}_1 = n_+ - n_- \tag{2-262}$$

The eigenvalues of $\mathcal{S}_1$ for the three one-particle states,

$$\Phi_+ = a_+^\dagger \Phi_0 \qquad \Phi_1 = a_1^\dagger \Phi_0 \qquad \Phi_- = a_-^\dagger \Phi_0 \tag{2-263}$$

can be found at once:

$$\mathcal{S}_1 \Phi_+ = \Phi_+ \qquad \mathcal{S}_1 \Phi_1 = 0 \qquad \mathcal{S}_1 \Phi_- = -\Phi_- \tag{2-264}$$

The possible eigenvalues of $\mathcal{S}_1$ for a one-particle state at rest are, therefore, 1, 0, -1. This means that the quantum of the field has spin 1.

2-6 ELECTROMAGNETIC FIELD

The Method of Heisenberg and Pauli [5]

The electromagnetic field is described by a real vector field, the four potential A_μ. One might think that the quantization would be more or less similar to the one discussed in the previous section. There is, however, an essential difference, and the quantization procedure of the preceding section does not work at all when m = 0. In order to see this we start from the Lagrangian density given in the preceding section, and make the appropriate changes for a real massless field.

$$\mathcal{L} = -\frac{1}{4} F_{\mu\nu} F_{\mu\nu} \qquad F_{\mu\nu} = \left(\frac{\partial A_\nu}{\partial x_\mu} - \frac{\partial A_\mu}{\partial x_\nu}\right) \tag{2-265}$$

or in terms of **E** and **H**, where $F_{4k} = - F_{k4} = iE_k$, $F_{12} = - F_{21} = H_3$, and so on, we have

$$\mathcal{L} = \frac{1}{2}\left(\mathbf{E}^2 - \mathbf{H}^2\right) \tag{2-266}$$

In the classical theory, the Lagrangian density is given by

$$\mathcal{L} = \frac{1}{8\pi}\left(\mathbf{E}^2 - \mathbf{H}^2\right) \tag{2-267}$$

so that there is a change of scale. The units we are using are called Heaviside units, and the classical units are called Gaussian units. In quantum field theory it is convenient to use the Heaviside units, which we shall use throughout this volume. So much for the units; let us now discuss the field equations. The Euler equation is

$$\frac{\partial}{\partial x_\mu}\left(\frac{\partial A_\nu}{\partial x_\mu} - \frac{\partial A_\mu}{\partial x_\nu}\right) = \Box A_\nu - \frac{\partial^2 A_\mu}{\partial x_\mu \partial x_\nu} = 0 \tag{2-268}$$

or

$$\frac{\partial F_{\mu\nu}}{\partial x_\mu} = 0 \tag{2-269}$$

Namely, we have Maxwell's equations for the electromagnetic field. As is well known the classical theory is invariant under the general gauge transformations

$$A_\mu \to A_\mu + \frac{\partial \Lambda}{\partial x_\mu} \tag{2-270}$$

Both the field strengths, the $F_{\mu\nu}$, and field equations are invariant. In the classical theory we impose the Lorentz condition

$$\frac{\partial A_\mu}{\partial x_\mu} = 0 \tag{2-271}$$

Then the gauge function Λ under which the theory is invariant is restricted to those functions satisfying

$$\Box \Lambda = 0 \tag{2-272}$$

With the Lorentz condition, the field equation becomes

$$\Box A_\mu = 0 \tag{2-273}$$

In order to quantize this theory we first have to introduce operators canonically conjugate to the potential,

$$\pi_\mu = \frac{\partial \mathcal{L}}{\partial \dot{A}_\mu} = iF_{4\mu} = -E_\mu \tag{2-274}$$

This means that the canonical conjugate of A_4 vanishes identically, and to summarize the difficulties:

(1) We do not get the Klein-Gordon equation for the A's, $\Box A_\mu = 0$, unless we impose the Lorentz condition.

(2) The operator canonically conjugate to A_4 vanishes identically.

In the preceding section there was no serious difficulty, since φ_4 is not independent of the other components because $\partial \varphi_\mu / \partial x_\mu = 0$. In the present case, $\partial A_\mu / \partial x_\mu = 0$ does not follow from the field equations, owing to the fact that $m = 0$. Therefore, this Lagrangian density is not suitable for our purpose. In semiclassical treatments of radiation it can be used if we choose a special gauge called the Coulomb gauge, in which A_4 is eliminated by means of a gauge transformation. This method, although complicated and apparently noncovariant, gives the correct results. It is convenient in the sense that it gives intuitive pictures of the electromagnetic field, but the apparent Lorentz invariance of the theory is completely lost and hence it is not appropriate for the systematic discussion of higher-order corrections.

Fermi's Method [6]

A way out of this difficulty was devised by Fermi, who modified the Lagrangian density in the following way:

$$\mathcal{L} = -\frac{1}{4} F_{\mu\nu} F_{\mu\nu} - \frac{1}{2}\left(\frac{\partial A_\mu}{\partial x_\mu}\right)^2 \tag{2-275}$$

He added the last term to the conventional Lagrangian density. The additional term is equal to zero if the Lorentz condition is imposed as in the classical theory. Therefore, this modification does not change the classical theory in the Lorentz gauge. In quantum theory, however, the Lorentz condition cannot be treated as a field equation, so that essential modifications are necessary.

(1) The canonical conjugates of the field operators are given by

$$\pi_k = iF_{4k} \qquad \pi_4 = i\frac{\partial A_\mu}{\partial x_\mu} \tag{2-276}$$

and we get a nonvanishing π_4 in this case.

(2) The field equations are given by

$$[\mathcal{L}]_{A_\nu} = \frac{\partial \mathcal{L}}{\partial A_\nu} - \frac{\partial}{\partial x_\mu}\left(\frac{\partial \mathcal{L}}{\partial A_{\nu:\mu}}\right)$$

$$= \frac{1}{2}\frac{\partial}{\partial x_\mu}\left(F_{\mu\nu} - F_{\nu\mu}\right) + \frac{\partial}{\partial x_\nu}\left(\frac{\partial A_\rho}{\partial x_\rho}\right)$$

$$= \frac{\partial F_{\mu\nu}}{\partial x_\mu} + \frac{\partial}{\partial x_\nu}\left(\frac{\partial A_\mu}{\partial x_\mu}\right)$$

$$= \Box A_\nu = 0 \tag{2-277}$$

so we obtain the Klein-Gordon equation naturally.

We proceed to find the commutation relations using Peierls' method; setting

$$[\mathcal{L}]_{A_\mu} = D_{\mu\nu}(\partial)A_\nu = 0 \tag{2-278}$$

we find

$$D_{\mu\nu}(\partial) = \delta_{\mu\nu}\Box \tag{2-279}$$

and for the operator C

$$D_{\alpha\beta}C_{\beta\gamma} = \delta_{\alpha\gamma}\Box \rightarrow C_{\beta\gamma} = \delta_{\beta\gamma} \tag{2-280}$$

and we get the commutation relations for the A's,

$$[A_\alpha(x), A_\beta(y)] = i\Delta_{\alpha\beta}(x - y)$$

$$= iC_{\alpha\beta}\Delta(x - y)$$

$$= i\delta_{\alpha\beta}\Delta(x - y) \tag{2-281}$$

It is customary to use the symbol D(x) instead of $\Delta(x)$ when $m = 0$. Hence, for free fields we get

$$[A_\mu(x), A_\nu(y)] = i\delta_{\mu\nu}D(x - y) \tag{2-282}$$

The following properties of the D functions are associated with m = 0.

$$D(x) = 0 \qquad \frac{\partial}{\partial x_0} D(x) = -\delta^3(x)$$

$$\text{for} \quad x_0 = 0 \tag{2-283}$$

$$D(x) = -\frac{1}{2\pi}\epsilon(x_0)\delta(x^2) \tag{2-284}$$

$$D^{(1)}(x) = \frac{1}{2\pi^2} P \frac{1}{x^2} \tag{2-285}$$

where, in the last equation, P stands for Cauchy's principal value.

Using the unequal-time commutation relations, we easily can prove the canonical commutation relations for equal times:

$$[A_\mu(x), \pi_\nu(y)] = i\delta_{\mu\nu}\delta^3(x - y) \qquad \text{for} \quad x_0 = y_0 \tag{2-286}$$

Another important relation which follows directly from the unequal-time commutation relation is

$$\left[\frac{\partial A_\mu(x)}{\partial x_\mu}, \frac{\partial A_\nu(y)}{\partial y_\nu}\right] = 0 \tag{2-287}$$

This relation is important for the self-consistency of Fermi's method, as we shall see in the next section.

Notice that Fermi's Lagrangian density can be written in the following form:

$$-\frac{1}{4}F_{\nu\mu}F_{\mu\nu} - \frac{1}{2}\left(\frac{\partial A_\mu}{\partial x_\mu}\right)^2 = -\frac{1}{2}\frac{\partial A_\mu}{\partial x_\nu}\frac{\partial A_\mu}{\partial x_\nu} + \frac{1}{2}\frac{\partial}{\partial x_\nu}\left(A_\mu \frac{\partial A_\nu}{\partial x_\mu} - A_\nu \frac{\partial A_\mu}{\partial x_\mu}\right) \tag{2-288}$$

For the variational principle and the field equations, therefore, Fermi's Lagrangian density is equivalent to the following simple expression:

$$\mathfrak{L} = -\frac{1}{2}\frac{\partial A_\mu}{\partial x_\nu} \cdot \frac{\partial A_\mu}{\partial x_\nu} \tag{2-289}$$

As we have discussed in the beginning of Section 2-5, this type of Lagrangian leads to the Klein-Gordon equations as well as to the negative-energy difficulty.

Fourier Representation

Let us introduce the Fourier representation of the potential,

$$A_\mu(x) = \sum_k (2k_0V)^{-1/2} \left[e^{ikx} A_\mu(k) + e^{-ikx} A_\mu^\dagger(k)\right] \tag{2-290}$$

For every k, there are four linearly-independent polarization vectors; we shall express the four-vector $A_\mu(k)$ in terms of unit polarization vectors, $e^{(\lambda)}$;

$$A_\mu(k) = \sum_\lambda e_\mu^{(\lambda)} \; a(k, \lambda) \tag{2-291a}$$

$$A_\mu^\dagger(k) = \sum_\lambda e_\mu^{(\lambda)} \; a^\dagger(k, \lambda) \tag{2-291b}$$

The most convenient choice of the unit vectors $e^{(\lambda)}$ is given by

$$\begin{aligned} e_4^{(1)} &= e_4^{(2)} = e_4^{(3)} = 0 \\ e_4^{(4)} &= 1 \\ e^{(3)} &= k/k_0 \\ e^{(1)}k &= e^{(2)}k = 0 \end{aligned} \tag{2-292}$$

The photons corresponding to $\lambda = 1$ and 2 are called transversely polarized, and those corresponding to $\lambda = 3$ and 4 are longitudinal and scalar photons, respectively.

From the orthogonality condition, we have

$$e_\mu^{(\lambda)} e_\mu^{(\lambda')} = \delta_{\lambda\lambda'} \tag{2-293}$$

and completeness of the polarization states is expressed by

$$e_\mu^{(\lambda)} e_\nu^{(\lambda)} = \delta_{\mu\nu} \tag{2-294}$$

The Fourier transform of the electromagnetic field is given by

$$A_\mu(x) = \sum_{k,\lambda} (2k_0V)^{-1/2} \; (e_\mu^{(\lambda)} \; a(\mathbf{k}, \lambda) \; e^{ikx} + e_\mu^{(\lambda)} \; a^\dagger(\mathbf{k}, \lambda) \; e^{-ikx}) \tag{2-295}$$

Notice that $a(\mathbf{k}, \lambda)$ and $a^\dagger(\mathbf{k}, \lambda)$ are Hermitian conjugates of each other for $\lambda = 1, 2, 3$; and $ia(\mathbf{k}, 4)$ and $ia^\dagger(\mathbf{k}, 4)$ are Hermitian conjugates.

Substituting the above expression into the commutation relations for the A's, we find

$$[a(k, \lambda), \; a^\dagger(k', \lambda')] = \delta_{\lambda\lambda'} \cdot \delta_{kk'} \tag{2-296}$$

$$\text{all other commutators} = 0 \tag{2-297}$$

Hamiltonian for the Free Electromagnetic Field

The Hamiltonian of the system is given by

$$H = \frac{1}{2} \int d^3x \left[\frac{\partial A_\mu}{\partial x_0} \frac{\partial A_\mu}{\partial x_0} + \frac{\partial A_\mu}{\partial x_k} \frac{\partial A_\mu}{\partial x_k} \right] \tag{2-298}$$

where we have used the modified form of Fermi's Lagrangian density. Inserting the Fourier expansion of the A's into this integral, we find

$$H = \frac{1}{2} \sum_{\mathbf{k},\lambda} k_0 \; \{a(\mathbf{k}, \lambda), \; a^\dagger(\mathbf{k}, \lambda)\} \tag{2-299}$$

Let us introduce operators p and q for each $\mathbf{k}$ by

$$q^{(\lambda)} = (2k_0)^{-1/2} \; (a(\mathbf{k}, \lambda) + a^\dagger(\mathbf{k}, \lambda)) \tag{2-300}$$

$$\lambda = 1, \; 2, \; 3$$

$$p^{(\lambda)} = -i(k_0/2)^{1/2} \; (a(\mathbf{k}, \lambda) - a^\dagger(\mathbf{k}, \lambda)) \tag{2-301}$$

and

$$q^{(4)} = (2k_0)^{-1/2} \; (a(\mathbf{k}, 4) - a^\dagger(\mathbf{k}, 4)) \tag{2-302}$$

$$p^{(4)} = i(k_0/2)^{1/2} \; (a(\mathbf{k}, 4) + a^\dagger(\mathbf{k}, 4)) \tag{2-303}$$

All p's and q's are *Hermitian* and satisfy the canonical commutation relations

$$\left[p^{(\lambda)},\ q^{(\lambda')}\right] = -i\delta_{\lambda\lambda'} \tag{2-304}$$

The Hamiltonian is given in terms of these operators by

$$H = \frac{1}{2} \sum_{\mathbf{k}} \left\{ \sum_{\lambda=1}^{3} [p^{(\lambda)}(\mathbf{k})^2 + k_0{}^2 q^{(\lambda)}(\mathbf{k})^2] - [p^{(4)}(\mathbf{k})^2 + k_0{}^2 q^{(4)}(\mathbf{k})^2] \right\} \tag{2-305}$$

The free electromagnetic field is mathematically equivalent to the superposition of an infinite set of harmonic oscillators. Therefore, we get, by omitting the zero-point energies of the oscillators, the energy eigenvalues:

$$E = \sum_{\mathbf{k}} \left\{ \sum_{\lambda=1}^{3} n(\mathbf{k}, \lambda) - n(\mathbf{k}, 4) \right\} k_0 \tag{2-306}$$

where $n(\mathbf{k}, \lambda)$ stands for the number of photons with momentum $\mathbf{k}$ and polarization λ. An important point is that E is not positive definite, as was mentioned already. There is, however, a method to avoid this difficulty which we shall discuss in the next section.

Now let us consider the following Hamiltonian:

$$H = \frac{1}{2} k_0 \{a(\mathbf{k}, \lambda),\ a^\dagger(\mathbf{k}, \lambda)\} + \text{constant} \tag{2-307}$$

then

$$\begin{aligned} Ha^\dagger|n\rangle &= [H, a^\dagger]|n\rangle + a^\dagger H|n\rangle \\ &= k_0 a^\dagger|n\rangle + a^\dagger H|n\rangle \end{aligned} \tag{2-308}$$

For $\lambda = 1, 2, 3$, we have $H|n\rangle = nk_0|n\rangle$, whereas for $\lambda = 4$, we have $H|n\rangle = -nk_0|n\rangle$, so that we find

$$Ha^\dagger|n\rangle = \begin{cases} (1+n)k_0 a^\dagger|n\rangle & \lambda = 1, 2, 3 \\ (1-n)k_0 a^\dagger|n\rangle & \lambda = 4 \end{cases} \tag{2-309}$$

The $a^\dagger|n\rangle$ is also an eigenstate of H if $|n\rangle$ is. The $a^\dagger$ is a creation operator for positive energy photons ($\lambda = 1, 2, 3$), and a destruction operator for negative energy photons ($\lambda = 4$).

Let $|0\rangle$ be the vacuum state with no photons; then any state can be constructed by applying creation operators to it, for instance,

$$|n^{(\lambda)}\rangle = c_n^{(\lambda)} [a^\dagger(\mathbf{k}, \lambda)]^n|0\rangle \qquad \lambda = 1, 2, 3 \tag{2-310a}$$

$$|n^{(4)}\rangle = c_n^{(4)} [a(\mathbf{k}, 4)]^n|0\rangle \tag{2-310b}$$

The c_n is the normalization constant and is subject to the condition

$$\langle n^{(\lambda)}|n^{(\lambda)}\rangle = 1 \tag{2-311}$$

By expressing c_n in terms of c_{n-1} we can find c_n inductively:

$$\begin{aligned}\langle n^{(\lambda)}|n^{(\lambda)}\rangle &= |c_n^{(\lambda)}|^2\langle 0|a^n(a^\dagger)^n|0\rangle \\ &= |c_n^{(\lambda)}|^2\langle 0|a^n a^\dagger (a^\dagger)^{n-1}|0\rangle \\ &= |c_n^{(\lambda)}|^2\langle 0|[a^n, a^\dagger](a^\dagger)^{n-1}|0\rangle \\ &= n|c_n^{(\lambda)}|^2\langle 0|a^{n-1}(a^\dagger)^{n-1}|0\rangle = 1\end{aligned} \tag{2-312}$$

while we also have

$$|c_{n-1}^{(\lambda)}|^2\langle 0|a^{n-1}(a^\dagger)^{n-1}|0\rangle = 1 \tag{2-313}$$

hence we find

$$n|c_n^{(\lambda)}|^2 = |c_{n-1}^{(\lambda)}|^2 \tag{2-314}$$

By using $|c_1^{(\lambda)}|^2 = 1$ and the recursion relation, (2-314), we find

$$|c_n^{(\lambda)}|^2 = (n!)^{-1} \qquad \text{or} \qquad |c_n^{(\lambda)}| = (n!)^{-1/2} \tag{2-315}$$

For $\lambda = 4$, we find the same result although we have to start from

$$\langle n^{(4)}|n^{(4)}\rangle = |c_n^{(4)}|^2\langle 0|(a^\dagger)^n a^n|0\rangle (-1)^n \tag{2-316}$$

Therefore, the most general normalized eigenvector is given by

$$\prod_{\mathbf{k}} \left(\prod_{\lambda=1}^{3} \frac{[a^{\dagger}(\mathbf{k}, \lambda)]^{n(\mathbf{k}, \lambda)}}{[n(\mathbf{k}, \lambda)!]^{1/2}} \right) \left(\frac{[a(\mathbf{k}, 4)]^{n(\mathbf{k}, 4)}}{[n(\mathbf{k}, 4)!]^{1/2}} \right) | 0 \rangle \qquad (2\text{-}317)$$

We have quantized the electromagnetic field. Only with the quantized radiation field can we understand the hypothesis of light quanta.

The matrix elements of the a's are given by

$$\langle n | a(\mathbf{k}, \lambda) | n + 1 \rangle = \langle n + 1 | a^{\dagger} (\mathbf{k}, \lambda) | n \rangle = (n + 1)^{1/2}$$

$$\lambda = 1, 2, 3 \qquad (2\text{-}318a)$$

$$\langle n + 1 | a(\mathbf{k}, 4) | n \rangle = - \langle n | a^{\dagger} (\mathbf{k}, 4) | n + 1 \rangle = (n+ 1)^{1/2} \qquad (2\text{-}318b)$$

It is clear that $a(\mathbf{k}, \lambda)$ is a destruction and $a^{\dagger}(\mathbf{k}, \lambda)$ a creation operator for $\lambda = 1, 2, 3$, but $a(\mathbf{k}, 4)$ is a creation and $a^{\dagger}(\mathbf{k}, 4)$ a destruction operator for negative energy photons. Hence we may summarize the results by saying that $a(\mathbf{k}, \lambda)$ is a destruction operator for energy and $a^{\dagger}(\mathbf{k}, \lambda)$ a creation operator for energy for $\lambda = 1, 2, 3, 4$, while $N(\mathbf{k}, \lambda) = a^{\dagger}(\mathbf{k}, \lambda)\, a(\mathbf{k}, \lambda)$ for $\lambda = 1, 2, 3$ and $N(\mathbf{k}, 4) = -a(\mathbf{k}, 4)\, a^{\dagger}(\mathbf{k}, 4)$ are the number operators. Using these operators and dropping the zero-point energies, we may write the Hamiltonian as

$$H = \sum_{\mathbf{k}} k_0 \left(\sum_{\lambda=1}^{3} N(\mathbf{k}, \lambda) - N(\mathbf{k}, 4) \right) \qquad (2\text{-}319)$$

The corresponding expression for P_k is

$$P_k = - \int d^3x \left[i \left(\frac{\partial A_1}{\partial x_4} - \frac{\partial A_4}{\partial x_1} \right) \frac{\partial A_1}{\partial x_k} + i \frac{\partial A_\nu}{\partial x_\nu} \cdot \frac{\partial A_4}{\partial x_k} \right]$$

$$= - \int d^3x \frac{\partial A_\mu}{\partial x_0} \cdot \frac{\partial A_\mu}{\partial x_k}$$

so that

$$\mathbf{P} = \frac{1}{2} \sum_{\mathbf{k}, \lambda} \left\{ a(\mathbf{k}, \lambda), a^{\dagger}(\mathbf{k}, \lambda) \right\} \mathbf{k}$$

$$= \sum_{\mathbf{k}} \mathbf{k} \left[\sum_{\lambda=1}^{3} N(\mathbf{k}, \lambda) - N(\mathbf{k}, 4) \right] \qquad (2\text{-}320)$$

From this result we conclude that N(**k**, 4) gives the number of photons with energy $-k_0$ and momentum $-\mathbf{k}$.

Spin of the Photon

The angular momentum density is given by

$$\mathfrak{M}_{0j\ell} = x_j T_{0\ell} - x_\ell T_{0j} + i \frac{\partial \mathfrak{L}}{\partial A_{\alpha:4}} (S_{j\ell})_{\alpha\beta} A_\beta \qquad (2\text{-}321)$$

By making use of the explicit expression S for the vector field

$$(S_{\rho\sigma})_{\alpha\beta} = \delta_{\rho\alpha}\delta_{\sigma\beta} - \delta_{\rho\beta}\delta_{\sigma\alpha} \qquad (2\text{-}322)$$

we get the following expression for the total angular momentum:

$$M_{j\ell} = \int \mathfrak{M}_{0j\ell}\, d^3x$$

$$= -\int d^3x \left[x_j \frac{\partial A_\nu}{\partial x_0} \frac{\partial A_\nu}{\partial x_\ell} - x_\ell \frac{\partial A_\nu}{\partial x_0} \frac{\partial A_\nu}{\partial x_j} + \frac{\partial A_j}{\partial x_0} A_\ell - \frac{\partial A_\ell}{\partial x_0} A_j \right] \qquad (2\text{-}323)$$

We then insert the Fourier expansion of the A's into Eq. (2-323). The first two terms, because $e_\nu^{(\lambda)} e_\nu^{(\lambda')} = \delta_{\lambda\lambda'}$, are independent of the polarization directions. This corresponds to the orbital angular momentum. The last two terms yield

$$-\int d^3x \left[\frac{\partial A_j}{\partial x_0} A_\ell - \frac{\partial A_\ell}{\partial x_0} A_j \right]$$

$$= -\frac{i}{2} \sum_{\mathbf{k},\lambda,\lambda'} e_j^{(\lambda')} e_\ell^{(\lambda)} \left[\left\{ a^\dagger(\mathbf{k},\lambda'),\ a(\mathbf{k},\lambda) \right\} - \left\{ a(\mathbf{k},\lambda'),\ a^\dagger(\mathbf{k},\lambda) \right\} \right]$$

$$= -i \sum_{\mathbf{k},\lambda,\lambda'} e_j^{(\lambda')} e_\ell^{(\lambda)} \left[a^\dagger(\mathbf{k},\lambda') a(\mathbf{k},\lambda) - a^\dagger(\mathbf{k},\lambda) a(\mathbf{k},\lambda') \right]$$

Let us take one particular k and choose the z axis in that direction and, using $a(\mathbf{k}, \lambda) = a^{(\lambda)}$, define

$$a_+ = 2^{-1/2} (a^{(1)} - ia^{(2)}) \tag{2-324a}$$

$$a_- = 2^{-1/2} (a^{(1)} + ia^{(2)}) \tag{2-324b}$$

then the $a_\pm$ satisfy

$$[a_+, a_+^\dagger] = [a_-, a_-^\dagger] = 1 \tag{2-325}$$

with all other commutators vanishing; we may write

$$E(\mathbf{k}) = k_0 (a_+^\dagger a_+ + a_-^\dagger a_- + a^{(3)\dagger} a^{(3)}$$

$$+ a^{(4)} a^{(4)\dagger}) \tag{2-326}$$

and the spin angular momentum in the z direction is given by

$$\mathcal{S}_3 = J_3(\mathbf{k}) = J_{12}(\mathbf{k}) = a_+^\dagger a_+ - a_-^\dagger a_- \tag{2-327}$$

This means that $a_+^\dagger$ creates a photon with momentum **k** and spin angular momentum $\mathcal{S}_3 = 1$, and $a_-^\dagger$ creates a photon with momentum **k** and spin angular momentum $\mathcal{S}_3 = -1$. The former is called the right circularly-polarized photon and the latter the left circularly-polarized photon. For the longitudinal photon, $\mathcal{S}_3 = 0$.

Summary

We have discussed both the method of Heisenberg and Pauli and that of Fermi. We tabulate their properties for comparison.

	Heisenberg–Pauli	*Fermi*
Lagrangian density	$-\frac{1}{4} F_{\mu\nu} F_{\mu\nu} = \frac{1}{2}(E^2 - H^2)$	$-\frac{1}{4} F_{\mu\nu} F_{\mu\nu} - \frac{1}{2}\left(\frac{\partial A_\mu}{\partial x_\mu}\right)^2$
Euler equation	Maxwell	Klein-Gordon
Quantization	$\pi_4 = 0$ so that A_4 is eliminated from the beginning by means of a gauge transformation. This destroys the formal Lorentz covariance	Canonical and formally Lorentz covariant
Energy density	$\frac{1}{2}(E^2 + H^2) \geq 0$	Not positive definite

The H-P method suffers from no serious difficulty except a lack of formal Lorentz covariance, but this turns out to be serious in higher-order calculations. Therefore, we shall discuss how to improve the formally Lorentz-invariant Fermi theory.

2-7 LORENTZ CONDITION

The Fermi theory first was intended to properly quantize the electromagnetic field without destroying the Lorentz invariance and it has succeeded in this regard; but there are two difficulties:

(1) The Hamiltonian is not positive definite.

(2) The field equations are different from Maxwell's equations.

In classical theory the equivalence between the field equations $\Box A_\mu = 0$ and Maxwell's equations $\partial F_{\mu\nu}/\partial x_\mu = 0$ follows from the Lorentz condition

$$\frac{\partial A_\mu}{\partial x_\mu} = 0 \tag{2-328}$$

since

$$\frac{\partial F_{\mu\nu}}{\partial x_\mu} = \Box A_\nu - \frac{\partial}{\partial x_\nu}\left(\frac{\partial A_\mu}{\partial x_\mu}\right) \tag{2-329}$$

Therefore, we examine the possibility of introducing the Lorentz condition into quantum theory. It is clear that one cannot regard this condition as an operator equation, since

$$[A_4(x), \pi_4(x')] = i\delta^3(x - x') \qquad \text{for} \quad x_0 = x'_0 \tag{2-330}$$

where

$$\pi_4(x) = i \frac{\partial A_\mu}{\partial x_\mu} \tag{2-331}$$

It is not necessary, however, to regard it as an operator equation, because that which corresponds to the classical quantity is not the operator but rather the expectation value of the operator.

We postulate that the state vectors that are realized in nature are those satisfying the subsidiary condition

$$\left(\frac{\partial A_\mu}{\partial x_\mu}\right) \Psi = 0 \tag{2-332}$$

Then Maxwell's equations follow from the field equations for the expectation values:

$$\frac{\partial}{\partial x_\mu}\left(\Psi,\ F_{\mu\nu}(x)\,\Psi\right) = 0 \tag{2-333}$$

Thus with this method, originally introduced by Fermi, we solve the second difficulty. It so happens, however, that the Lorentz condition, (2-332), also removes the first difficulty concerned with the negative energy states.

The subsidiary condition, (2-332), to be referred to simply as the Lorentz condition hereafter, is expressed in terms of the Fourier coefficients as

$$[a(\mathbf{k}, 3) + ia(\mathbf{k}, 4)] \quad \Psi = 0 \tag{2-334a}$$

$$[a^\dagger(\mathbf{k}, 3) + ia^\dagger(\mathbf{k}, 4)]\,\Psi = 0 \tag{2-334b}$$

Therefore, the Lorentz condition refers only to the longitudinal and scalar photons, and says nothing about the transverse photons. Let us decompose Ψ into a direct product of state vectors for transverse, and longitudinal and scalar photons, Ψ_T and Φ, respectively.

$$\Psi = \Psi_T \times \prod_{\mathbf{k}} \Phi_{\mathbf{k}} \tag{2-335}$$

where $\Phi_{\mathbf{k}}$ is considered as a superposition of states with $n^{(3)}$ longitudinal photons and $n^{(4)}$ scalar photons and with a weighting factor $\alpha_{n^{(3)},\,n^{(4)}}$, for a given momentum $\mathbf{k}$, that is ,

$$\Phi_{\mathbf{k}} = \sum_{n^{(3)},\,n^{(4)}} \alpha_{n^{(3)},\,n^{(4)}} \left| n^{(3)}\ n^{(4)} \right\rangle$$

$$= \sum_{n^{(3)},\,n^{(4)}} \alpha_{n^{(3)},\,n^{(4)}} \frac{[a^\dagger(\mathbf{k}, 3)]^{n^{(3)}}\,[a(\mathbf{k}, 4)]^{n^{(4)}}}{\left(n^{(3)}!\ n^{(4)}!\right)^{1/2}} \left|0\right\rangle \tag{2-336}$$

The Lorentz condition for k reads

$$[a(\mathbf{k}, 3) + ia(\mathbf{k}, 4)]\,\Phi_{\mathbf{k}} = 0 \tag{2-337a}$$

$$[a^{\dagger}(\mathbf{k}, 3) + ia^{\dagger}(\mathbf{k}, 4)]\,\Phi_{\mathbf{k}} = 0 \tag{2-337b}$$

Substitution of the expansion for $\Phi_{\mathbf{k}}$ into the above set leads to

$$\left(n^{(3)} + 1\right)^{1/2} \alpha_{n^{(3)}+1,\, n^{(4)}} + i\left(n^{(4)}\right)^{1/2} \alpha_{n^{(3)},\, n^{(4)}-1} = 0 \tag{2-338}$$

$$\left(n^{(3)}\right)^{1/2} \alpha_{n^{(3)}-1,\, n^{(4)}} - i\left(n^{(4)} + 1\right)^{1/2} \alpha_{n^{(3)},\, n^{(4)}+1} = 0 \tag{2-339}$$

The solution of the above set of recursion relations is uniquely given by

$$\alpha_{n^{(3)},\, n^{(4)}} = c\,\delta_{n^{(3)},\, n^{(4)}}\,(-i)^{n^{(3)}} \tag{2-340}$$

where c is a normalization constant to be determined from

$$(\Phi_{\mathbf{k}}, \Phi_{\mathbf{k}}) = 1 \tag{2-341}$$

Therefore, the subsidiary condition requires that there be an equal number of longitudinal and scalar photons. There is now no remaining freedom for these photons, because $\Phi = \Pi_{\mathbf{k}}\Phi_{\mathbf{k}}$ is uniquely determined and common to all the physically realizable states. In other words, the state of longitudinal and scalar photons is frozen.

$$\Psi = \Psi_T \times \Phi \tag{2-342}$$

Furthermore, since there is always the same number of longitudinal and scalar photons with the same magnitude of energy and momentum but with opposite signs, the total observed energy of the system is positive definite:

$$\begin{aligned}(\Psi, H\Psi) &= \sum_{\mathbf{k}} k_0 \Big[\big(\Psi, [N(\mathbf{k}, 1) + N(\mathbf{k}, 2)]\,\Psi\big) + \big(\Psi, [N(\mathbf{k}, 3) - N(\mathbf{k}, 4)]\,\Psi\big)\Big] \\ &= \sum_{\mathbf{k}} k_0 \Big[\big(\Psi_T, [N(\mathbf{k}, 1) + N(\mathbf{k}, 2)]\,\Psi_T\big) \\ &\quad + \big(\Phi, [N(\mathbf{k}, 3) - N(\mathbf{k}, 4)]\,\Phi\big)\Big] \\ &= \sum_{\mathbf{k}} k_0 \big(\Psi_T, [N(\mathbf{k}, 1) + N(\mathbf{k}, 2)]\,\Psi_T\big)\end{aligned} \tag{2-343}$$

The total energy is completely determined by the state of the transverse photons alone, and with the introduction of the Lorentz condition we can remove the two difficulties present in the original theory.

2-8 DIFFICULTIES WITH THE LORENTZ CONDITION

The Lorentz condition gives rise to a new kind of difficulty. From the free-field commutation relations we see at once that

$$\left[\frac{\partial A_\mu(x)}{\partial x_\mu}, A_\nu(y)\right] = i \frac{\partial}{\partial x_\nu} D(x - y) \tag{2-344}$$

so that

$$\left(\Psi, \left[\frac{\partial A_\mu(x)}{\partial x_\mu}, A_\nu(y)\right] \Psi\right) = i \frac{\partial}{\partial x_\nu} D(x - y) \neq 0 \tag{2-345}$$

even when Ψ satisfies the Lorentz condition (2-332). This contradiction arises from the fact that the state vector Φ is not normalizable, namely, $(\Phi_{\mathbf{k}}, \Phi_{\mathbf{k}}) = \infty$ for all $\mathbf{k}$.

Similar examples are familiar from quantum mechanics; for example,

$$(\Psi, [p, q] \Psi) = -i \tag{2-346}$$

even if Ψ satisfies $p\Psi = 0$. This is because Ψ is not normalizable if it satisfies $p\Psi = 0$.

Although the state vector Φ is not normalizable, it is common to all the physically realizable states so that this difficulty is not very serious. In fact, if we can describe the theory by using only the transverse part of the state vectors there is no difficulty. This is possible, however, only for vector fields with $\mu \neq 0$; therefore, we shall formulate electrodynamics as a limiting case of the neutral vector field in the next section.

The Stueckelberg Field[7]

We have learned that the Lagrangian density for the electromagnetic field (e. m.) leads to a nonpositive-definite Hamiltonian and that this difficulty can be removed by the introduction of the Lorentz condition. Therefore, we may try to work with a Lagrangian density for a real vector field leading to a nonpositive-definite Hamiltonian.

We recall that the Lagrangian density

$$\mathcal{L} = -\frac{1}{2}\left(\frac{\partial A_\mu}{\partial x_\nu}\right)\left(\frac{\partial A_\mu}{\partial x_\nu}\right) \tag{2-347}$$

is equivalent to that of Fermi. Therefore, we take, for a real vector field with a nonvanishing mass μ, the Lagrangian density

$$\mathcal{L} = -\frac{1}{2}\left(\frac{\partial A_\mu}{\partial x_\nu} \cdot \frac{\partial A_\mu}{\partial x_\nu} + \mu^2 A_\mu A_\mu\right) \tag{2-348}$$

This had been rejected in Section 2-5 because of a resulting nonpositive-definite Hamiltonian. We now attempt to remove this difficulty by introducing a supplementary condition. This problem was solved by Stueckelberg; he started with a Lagrangian density for a mixture of vector and scalar fields,

$$\mathcal{L} = -\frac{1}{2}\left(\frac{\partial A_\mu}{\partial x_\nu} \cdot \frac{\partial A_\mu}{\partial x_\nu} + \mu^2 A_\mu A_\mu\right) - \frac{1}{2}\left(\frac{\partial B}{\partial x_\mu} \cdot \frac{\partial B}{\partial x_\mu} + \mu^2 BB\right) \tag{2-349}$$

and introduced a supplementary condition:

$$\left(\frac{\partial A_\mu}{\partial x_\mu} + \mu B\right)\Psi = 0 \tag{2-350}$$

In the limit $\mu \to 0$, the supplementary condition becomes

$$\left(\frac{\partial A_\mu}{\partial x_\mu}\right)\Psi = 0 \tag{2-351}$$

and the Lagrangian density becomes

$$\mathcal{L}_{\text{e.m.}} + \mathcal{L}_B \tag{2-352}$$

When there is no interaction between fields A and B we can discuss them separately, and the theory is equivalent to electrodynamics in this limit. The important point is that the state vector satisfying the supplementary condition (2-350) has a positive energy in the present case, too. Let us express the supplementary condition in terms of Fourier coefficients as we did in Section 2-7.

$$[ika(k, 3) + ik_4 a(k, 4) + \mu b(k)]\Psi = 0 \tag{2-353a}$$

$$[-ika^\dagger(k, 3) - ik_4 a^\dagger(k, 4) + \mu b^\dagger(k)]\Psi = 0 \tag{2-353b}$$

where $k = k_3 = |k|$. If we set

$$k_4 c(k) = ika(k, 3) + \mu b(k) \tag{2-354}$$

then the supplementary condition becomes

$$[c(k) + ia(k, 4)]\, \Psi = 0 \tag{2-355a}$$

$$[c^\dagger(k) + ia^\dagger(k, 4)]\, \Psi = 0 \tag{2-355b}$$

We find that this condition is of exactly the same form as the Lorentz condition in electrodynamics. In this connection, notice

$$[c(k), c^\dagger(k)] = 1 \tag{2-356}$$

The Ψ is still unnormalizable, but one can prove that the state obeying the supplementary condition (2-350) has a positive energy, and that Stueckelberg's theory reduces to electrodynamics in the limit $\mu \to 0$. Introducing

$$k_4 d(k) = \mu a(k, 3) + ikb(k) \tag{2-357}$$

we find the commutation relations

$$[c(k), d(k)] = [c^\dagger(k), d^\dagger(k)] = [c(k), d^\dagger(k)]$$

$$= [c^\dagger(k), d(k)] = 0 \tag{2-358}$$

We then use a (k, 1), a(k, 2), d(k), c(k), and a(k, 4) instead of a(k, 1), a(k, 2), a(k, 3), b(k), and a(k, 4). In the supplementary condition (2-355) only c(k) and a(k, 4) occur, namely,

$$\Psi = \Psi_T \times \Phi \tag{2-359}$$

where Ψ_T contains only "1," "2," and d mesons, and Φ contains "4" and c mesons. The structure of Stueckelberg's theory is the same as electrodynamics except that there is an additional degree of freedom represented by d. In the limit $\mu \to 0$, however, the d mesons no longer interact with the 1 and 2 mesons.

Reduction of the Stueckelberg Field to the Ordinary Vector Field

We can prove that Stueckelberg's theory with the subsidiary condition is equivalent to the conventional formulation for vector fields.

$$[A_\mu(x), A_\nu(y)] = i\,\delta_{\mu\nu}\,\Delta(x - y) \tag{2-360}$$

$$[B(x), B(y)] = i\,\Delta(x - y) \tag{2-361}$$

$$\left[\frac{\partial A_\mu(x)}{\partial x_\mu} + \mu B(x)\right] \Psi = 0 \tag{2-362}$$

In the free Stueckelberg theory, we can get rid of the negative-energy difficulty. In order to avoid this difficulty in the presence of interactions, we have to impose the "gauge" invariance on the theory. We impose a restriction by requiring that the theory be invariant under the transformation.

$$A_\mu \to A_\mu + \frac{\partial \Lambda}{\partial x_\mu} \qquad B \to B - \mu\Lambda \tag{2-363}$$

for any Λ satisfying the Klein-Gordon equation $K_x \Lambda(x) = 0$.

Consequently, appearing in the interaction with other fields are invariant combinations of the A_μ's and B, such as

$$U_\mu(x) = A_\mu(x) + \frac{1}{\mu}\frac{\partial B(x)}{\partial x_\mu} \qquad \text{or} \qquad F_{\mu\nu}(x) \tag{2-364}$$

We can prove that

$$\left[U_\mu(x), \frac{\partial A_\nu(y)}{\partial y_\nu} + \mu B(y)\right] = 0 \tag{2-365}$$

Therefore, if Ψ satisfies the supplementary condition (2-350), so does $F(U_\mu)\Psi$, too, namely,

$$\left(\frac{\partial A_\mu}{\partial x_\mu} + \mu B\right) F(U_\mu)\Psi = 0 \tag{2-366}$$

for any functional F of U_μ. Thus with $\Psi = \Psi_T \times \Phi$, the $F(U_\mu)$ effectively operates only on Ψ_T since Φ is fixed by the supplementary condition.

$$F(U_\mu)\Psi = \Psi' = \Psi'_T \times \Phi \tag{2-367}$$

Therefore, in order to project the operator U_μ into the subspace of the Hilbert space containing only the 1, 2, and d mesons, let us define φ_μ.

$$\varphi_\mu(x) = \left(\Phi, U_\mu(x)\, \Phi\right)$$

Notice that $\varphi_\mu(x)$ still is an operator in the space represented by Ψ_T. Since U_μ commutes with $(\partial A_\mu/\partial x_\mu) + \mu B$, we may write

$$F(\varphi_\mu) = \left(\Phi, F(U_\mu)\, \Phi\right) \tag{2-368}$$

This is an important relation which guarantees that φ forms an algebra isomorphic to that of U. We then can obtain the commutation relations and field equations for φ_μ in the subspace mentioned above.

commutation relations:

$$[\varphi_\mu(x), \ \varphi_\nu(y)] = (\Phi, \ [U_\mu(x), \ U_\nu(y)] \ \Phi)$$

$$= i\left(\delta_{\mu\nu} - \frac{1}{\mu^2}\frac{\partial^2}{\partial x_\mu \ \partial x_\nu}\right) \ \Delta(x-y) \qquad (2\text{-}369)$$

field equations:

$$(\Box - \mu^2) \ \varphi_\mu(x) = (\Phi, \ (\Box - \mu^2) \ U_\mu(x) \ \Phi) = 0 \qquad (2\text{-}370)$$

$$\frac{\partial \varphi_\mu}{\partial x_\mu} = \left(\Phi, \frac{\partial U_\mu}{\partial x_\mu} \Phi\right) = \left(\Phi, \left(\frac{\partial A_\mu}{\partial x_\mu} + \frac{1}{\mu} \Box B\right) \ \Phi\right)$$

$$= \left(\Phi, \left(\frac{\partial A_\mu}{\partial x_\mu} + \mu B\right) \ \Phi\right) = 0 \qquad (2\text{-}371)$$

Thus we have established that φ_μ is equivalent to the ordinary operator for the vector field. The use of a real vector field with a nonvanishing mass has many benefits for other purposes, as in dealing with infrared divergences.

Another method of avoiding this difficulty is provided by the introduction of an indefinite metric, the Gupta-Bleuler method.

The Gupta-Bleuler Method [8]

As the Lorentz condition we have taken $\left(\frac{\partial A_\mu}{\partial x_\mu}\right) \Psi = 0$, but we can use a weaker condition such as

$$\left(\Psi, \ \frac{\partial A_\mu}{\partial x_\mu} \ \Psi\right) = 0 \qquad (2\text{-}372)$$

since this is what is actually needed in deriving Maxwell's equations. This is satisfied if, for instance, Ψ satisfies

$$[a(k, 3) + \ ia(k, 4)] \ \Psi = 0 \qquad (2\text{-}373)$$

but not

$$[a^\dagger(k, 3) + \ ia^\dagger(k, 4)] \ \Psi = 0 \qquad (2\text{-}374)$$

The difficulty associated with these equations is that they involve creation operators as well as destruction operators.

In order to avoid this difficulty we adopt the following representations:

$$\langle n+1 | a^{\dagger(4)} | n \rangle = \langle n | a^{(4)} | n+1 \rangle = (n+1)^{1/2} \qquad (2\text{-}375)$$

so that all four $a^{\dagger(\lambda)}$'s are Hermitian conjugates of $a^{(\lambda)}$, and hence all the four components of the potential, $A_\mu(x)$, themselves are Hermitian. This contradicts the classical reality condition of the electromagnetic potential since $A_4 = i\varphi$ is regarded as a real quantity, but we can make everything consistent by introducing a metric operator η. We define

$$\text{norm of } \Psi = (\Psi, \eta\Psi) \tag{2-376}$$

In order to make the norm real, η must be Hermitian.

$$\eta^\dagger = \eta \tag{2-377}$$

The expectation value of an operator F is defined by

$$\langle F \rangle = (\Psi, \eta F \Psi) \tag{2-378}$$

This value is not necessarily real unless F commutes with η. The norm of a state vector is thus not always positive; there are three possibilities: (1) positive norm, (2) negative norm, (3) zero norm. We require that all observable state vectors belong to the first class so that we can apply the probability interpretation of the state vectors.

In order to make the expectation value of $A_k(x)$ real and that of $A_4(x)$ pure imaginary, in conformity with the classical reality condition, we assume

$$(\Psi, \eta A_k(x)\Psi) = (\Psi, A_k(x)\eta\Psi)$$

$$\text{or} \qquad [A_k(x), \eta] = 0 \tag{2-379}$$

and

$$(\Psi, \eta A_4(x)\Psi) = -(\Psi, A_4(x)\eta\Psi)$$

$$\text{or} \qquad \{A_4(x), \eta\} = 0 \tag{2-380}$$

In momentum space we find

$$[a(k, \lambda), \eta] = 0 \qquad \text{for} \quad \lambda \neq 4 \tag{2-381}$$

and

$$\{a(k, 4), \eta\} = 0 \tag{2-382}$$

Hence η is diagonal in the factor Hilbert space corresponding to transverse and longitudinal photons and is equal to the unit matrix. In the factor Hilbert space corresponding to scalar photons, however, we have

$$\langle n^{(4)} | \eta | n^{(4)\prime} \rangle = \delta_{n^{(4)},\, n^{(4)\prime}} (-1)^{n^{(4)}} \tag{2-383}$$

or generally we can write η as

$$\eta = (-1)^{N(4)} \tag{2-384}$$

where N(4) is the number of scalar photons.

The number operators are defined by

$$N(k, \lambda) = a^{\dagger}(k, \lambda)\, a(k, \lambda) \qquad \text{for} \quad \lambda = 1, 2, 3, 4 \tag{2-385}$$

Then we have

$$H = \sum_{k} k_0 \sum_{\lambda=1}^{4} N(k, \lambda) \tag{2-386}$$

$$P = \sum_{k} k \sum_{\lambda=1}^{4} N(k, \lambda) \tag{2-387}$$

In this formulation the scalar photons carry *positive energies*, but the expectation value of the energy of a scalar photon becomes negative again because of the indefinite metric.

The modified Lorentz condition is given by

$$[a(k, 3) + ia(k, 4)]\, \Psi = 0 \tag{2-388}$$

or

$$\frac{\partial A_{\nu}^{(+)}(x)}{\partial x_{\nu}} \Psi = 0 \tag{2-389}$$

where $A^{(+)}(x)$ refers to the positive frequency components or the destructive part of A(x), and similarly $A^{(-)}(x)$ to the negative frequency components. With the aid of η, we have from the complex conjugate of the above equation

$$0 = \left(\Psi, \left(\frac{\partial A_{k}^{(-)}(x)}{\partial x_{k}} - \frac{\partial A_{4}^{(-)}(x)}{\partial x_{4}}\right)\eta\right.$$

$$= \left(\Psi, \eta \frac{\partial A_{\nu}^{(-)}(x)}{\partial x_{\nu}}\right. \tag{2-390}$$

Therefore,

$$\left(\Psi,\ \eta \frac{\partial A_\nu}{\partial x_\nu} \Psi\right) = \left(\Psi,\ \eta \frac{\partial A_\nu^{(-)}}{\partial x_\nu} \Psi\right) + \left(\Psi,\ \eta \frac{\partial A_\nu^{(+)}}{\partial x_\nu} \Psi\right) = 0 \tag{2-391}$$

Previously the Lorentz condition had a unique solution, but such is not the case here since the modified Lorentz condition involves only destruction operators. We use, as in Section 2-7, the form

$$\Psi = \Psi_T \times \Phi \qquad \Phi = \prod_k \Phi_k . \tag{2-392}$$

The solutions of the equation for a fixed k,

$$(a^{(3)} + ia^{(4)})\ \Phi_k = 0 \tag{2-393}$$

are given by

$$\Phi^{(0)} = |0,\ 0\rangle$$

$$\Phi^{(1)} = |1,\ 0\rangle + i|0,\ 1\rangle$$

$$\vdots$$

$$\Phi^{(n)} = \sum_{r=0}^{n} i^r \left(\frac{n!}{r!\ (n-r)!}\right)^{1/2} |n-r,\ r\rangle \tag{2-394}$$

where n - r and r in $|n-r,\ r\rangle$ denote the numbers of the longitudinal and scalar photons, respectively. Clearly all vectors are orthogonal to each other.

$$(\Phi^{(n)},\ \eta\Phi^{(n')}) = 0 \qquad \text{for} \quad n \neq n' \tag{2-395}$$

$$(\Phi^{(n)},\ \eta\Phi^{(n)}) = \sum_{r=0}^{n} (-1)^r \frac{n!}{r!\ (n-r)!} = \delta_{n,0} \tag{2-396}$$

Let us require that Φ be normalized to unity; then the general solution of the modified Lorentz condition is given by

$$\Phi_k = \Phi^{(0)} + \sum_{n=1}^{\infty} c^{(n)}(k)\, \Phi^{(n)} \tag{2-397}$$

for each k. The $c^{(n)}$'s are arbitrary constants, since the norm of $\Phi^{(n)}$ is zero for $n \neq 0$. Let us calculate the expectation values of the components of the electromagnetic potential in the above state. We set

$$\Psi_T = |\text{no transverse photon}\rangle \tag{2-398}$$

Then we have

$$\begin{aligned}(\Psi, \eta A_\mu(x) \Psi) = \sum_k (2k_0 V)^{-1/2} \Big[& e^{ikx} e_\mu^{(3)} (\Phi_k, \eta\, a(k, 3)\, \Phi_k) \\ & + e^{-ikx} e_\mu^{(3)} (\Phi_k, \eta\, a^\dagger(k, 3)\, \Phi_k) \\ & + e^{ikx} e_\mu^{(4)} (\Phi_k\, \eta\, a(k, 4)\, \Phi_k) \\ & + e^{-ikx} e_\mu^{(4)} (\Phi_k, \eta\, a^\dagger(k, 4) \Phi_k)\end{aligned} \tag{2-399}$$

By using $a^{(3)} \Phi^{(n)} = n^{1/2} \Phi^{(n-1)}$ and $a^{(4)} \Phi^{(n)} = i\, n^{1/2} \Phi^{(n-1)}$, we find

$$(\Phi_k, \eta\, a(k, 3)\, \Phi_k) = (\Phi^{(0)}, \eta \Phi^{(0)})\, c^{(1)}(k) = c^{(1)}(k) \tag{2-400}$$

$$(\Phi_k, \eta\, a(k, 4)\, \Phi_k) = i c^{(1)}(k) \tag{2-401}$$

and finally

$$(\Psi, \eta A_\mu(x) \Psi) = \frac{\partial \Lambda(x)}{\partial x_\mu} \tag{2-402}$$

where

$$\Lambda(x) = \sum_k (2k_0 V)^{-1/2} \left(\frac{c^{(1)}(k)\, e^{ikx} - c^{(1)\dagger}(k)\, e^{-ikx}}{ik_0} \right)$$

$$\Box \Lambda(x) = 0 \tag{2-403}$$

Thus we find that the arbitrariness of the solution of the modified Lorentz condition corresponds to the arbitrariness of the gauge. Therefore, we expect that observable and, hence, gauge invariant quantities have expectation values independent of the c's.

When c's do not all vanish, then Φ_k is *not* an eigenstate of H, but the expectation value of H is given by

$$(\Psi, \eta H \Psi) = \sum_k k_0 \,(n(k, 1) + n(k, 2)) \tag{2-404}$$

where $n(k, \lambda)$ is the expectation value of $N(k, \lambda)$. This result is the same as (2-343) in the preceding section.

For practical calculations it is convenient to fix Φ_k, although all the observable quantities are independent of the choice of the c's. We take a gauge in which all the c's are zero so that $\Phi_k = \Phi_k^{(0)}$. In this gauge we have

$$\langle 0 | \eta A_\mu (x) | 0 \rangle = \langle 0 | A_\mu (x) | 0 \rangle = 0 \tag{2-405}$$

and

$$\begin{aligned} \langle 0 | \eta \{ A_\mu(x), A_\nu (x') \} | 0 \rangle &= \langle 0 | \{ A_\mu (x), A_\nu(x') \} | 0 \rangle \\ &= \delta_{\mu\nu} D^{(1)} (x - x') \end{aligned} \tag{2-406}$$

2-9 DIRAC FIELDS

We shall now quantize the Dirac field, which corresponds to particles with spin $\frac{1}{2}$.

The Dirac Equation and Spinors

The Dirac equation

$$(\gamma_\mu \frac{\partial}{\partial x_\mu} + m) \psi = 0 \qquad \text{or simply}$$

$$(\gamma \partial + m) \psi = 0 \tag{2-407}$$

follows from the Lagrangian density

$$\mathcal{L} = -\bar{\psi} (\gamma_\mu \frac{\partial}{\partial x_\mu} + m)\psi \tag{2-408}$$

by use of the action principle. The γ matrices are four-by-four matrices and are Hermitian, and they satisfy the commutation relations

$$\gamma_\mu \gamma_\nu + \gamma_\nu \gamma_\mu = 2\delta_{\mu\nu} \tag{2-409}$$

The plane-wave solutions of the Dirac equation may be written as

$$\psi_\alpha(x) = u_\alpha(q) \exp(i\mathbf{qx} - iq_0 x_0) \tag{2-410}$$

where $q_0 = \pm E = \pm(\mathbf{q}^2 + m^2)^{1/2}$. For a given $\mathbf{q}$, there are four solutions $u^{(r)}$, ($r = 1, 2, 3, 4$), called the Dirac spinors, which satisfy the following relations:

$$\sum_r u_\alpha^{(r)*} u_\beta^{(r)} = \delta_{\alpha\beta} \qquad \text{(completeness)} \tag{2-411}$$

$$\sum_\alpha u_\alpha^{(r)*} u_\alpha^{(s)} = \delta_{rs} \qquad \text{(orthogonality)} \tag{2-412}$$

$$\sum_{r=1}^{2} \bar{u}_\alpha^{(r)} u_\beta^{(r)} = -\frac{1}{2E}(i\gamma q^{(+)} - m)_{\beta\alpha} \qquad \text{(positive-energy states)} \tag{2-413}$$

$$\sum_{r=3}^{4} \bar{u}_\alpha^{(r)} u_\beta^{(r)} = \frac{1}{2E}(i\gamma q^{(-)} - m)_{\beta\alpha} \qquad \text{(negative-energy states)} \tag{2-414}$$

where $q^{(\pm)}$ means that $q_0 = \pm E$ and $\bar{u}(q)$ is defined by

$$\bar{u}(q) = u^\dagger(q)\gamma_4 \tag{2-415}$$

Quantization

In quantizing the Dirac field we apply Peierls' method. The operator $D(\partial)$ is defined by

$$[\mathfrak{L}]_{\psi_\alpha^\dagger} = D_{\alpha\beta}(\partial)\,\psi_\beta = -[\gamma_4(\gamma\partial + m)]_{\alpha\beta}\,\psi_\beta \tag{2-416}$$

Set

$$C_{\alpha\beta}(\partial) = -[(\gamma\partial - m)\,\gamma_4]_{\alpha\beta} \tag{2-417}$$

then

$$D_{\alpha\beta} C_{\beta\gamma} = [\gamma_4(\gamma\partial + m)(\gamma\partial - m)\gamma_4]_{\alpha\gamma}$$

$$= \delta_{\alpha\gamma}(\Box - m^2) \tag{2-418}$$

Hence the commutation relations are given by

$$[\psi_\alpha(x), \psi_\beta(y)]_\pm = [\psi_\alpha^\dagger(x), \psi_\beta^\dagger(y)]_\pm = 0 \tag{2-419}$$

$$[\psi_\alpha(x), \psi_\beta^\dagger(y)]_\pm = iC_{\alpha\beta}(\partial)\,\Delta(x - y)$$

$$= -i\,[(\gamma\partial - m)\gamma_4]_{\alpha\beta}\,\Delta(x - y) \tag{2-420}$$

or

$$[\psi_\alpha(x), \overline{\psi}_\beta(y)]_\pm = -i\,(\gamma\partial - m)_{\alpha\beta}\Delta(x - y)$$

$$\equiv -iS_{\alpha\beta}(x - y) \tag{2-421}$$

where $[A, B]_\pm$ denotes $AB \pm BA$.

We have at present no way to determine which type of quantization we should apply. This problem is discussed below.

The operator π_α, canonically conjugate to ψ_α, is defined by

$$\pi_\alpha = \frac{\partial \mathcal{L}}{\partial \dot{\psi}_\alpha} = i\,(\overline{\psi}\gamma_4)_\alpha = i\psi_\alpha^\dagger \tag{2-422}$$

The Hamiltonian density is therefore given by

$$\mathcal{H} = i\overline{\psi}\gamma_4\dot{\psi} + \overline{\psi}(\gamma\partial + m)\psi$$

$$= \psi^\dagger\gamma_4\left(\gamma_k \frac{\partial}{\partial x_k} + m\right)\psi$$

$$= \psi^\dagger\left(\frac{1}{i}\alpha_k \frac{\partial}{\partial x_k} + m\beta\right)\psi \tag{2-423}$$

where $\alpha_k = i\gamma_4\gamma_k$ $\beta = \gamma_4$. We recognize the Dirac Hamiltonian in the unquantized theory in the above expression.

Invariance of the action integral under phase transformations leads to a current conservation law.

$$\frac{\partial j_\mu}{\partial x_\mu} = 0 \qquad \text{with} \quad j_\mu = i\,\bar{\psi}\gamma_\mu\psi \tag{2-424}$$

Let us expand ψ into a Fourier series.

$$\psi_\alpha(x) = V^{-1/2} \sum_q \left[\exp\left[i\,(qx - Ex_0)\right] \sum_{r=1}^{2} u_\alpha^{(r)}(q)\; a^{(r)}(q) \right.$$

$$\left. + \exp\left[i\,(qx + Ex_0)\right] \sum_{r=3}^{4} u_\alpha^{(r)}(q)\; a^{(r)}(q) \right] \tag{2-425}$$

where $u^{(1)}$ and $u^{(2)}$ are positive-energy solutions of the free Dirac equation, and $u^{(3)}$ and $u^{(4)}$ are negative-energy solutions. Then the total energy of the system is given by

$$H = \int d^3x\, \mathcal{H}(x) = \sum_q E \left[\sum_{r=1}^{2} a^{(r)\dagger}(q)\, a^{(r)}(q) \right.$$

$$\left. - \sum_{r=3}^{4} a^{(r)\dagger}(q)\; a^{(r)}(q) \right] \tag{2-426}$$

where use has been made of the relation

$$(\alpha_k q_k + m\beta)\, u^{(r)}(q) = \pm E u^{(r)}(q) \qquad \begin{cases} + & \text{for} \quad r = 1, 2 \\ - & \text{for} \quad r = 3, 4 \end{cases} \tag{2-427}$$

The charge operator is given by

$$N = -i \int j_4(x)\, d^3x = \sum_q \sum_{r=1}^{4} a^{(r)\dagger}(q)\; a^{(r)}(q) \tag{2-428}$$

The equal-time commutation relations follow from

$$[\psi_\alpha(x), \bar{\psi}_\beta(y)]_\pm = -i\,(\gamma\partial - m)_{\alpha\beta}\, \Delta(x - y) \tag{2-429}$$

For $x_0 = y_0$, only the time derivative of Δ survives on the right-hand side. Since the right-hand side is given by

$$-i\gamma_4 \frac{\partial}{\partial x_4} \Delta (x - y) = \gamma_4 \delta^3 (x - y) \qquad \text{for} \quad x_0 = y_0 \tag{2-430}$$

the equal-time commutation relation may be written as

$$[\psi_\alpha(x), \psi_\beta^\dagger (y)]_\pm = \delta_{\alpha\beta} \; \delta^3 (x - y) \qquad \text{for} \quad x_0 = y_0 \tag{2-431}$$

where we replaced $\bar{\psi}(y) = \psi^\dagger(y)\gamma 4$ by $\psi^\dagger(y)$ to simplify the commutator. Substitution of the Fourier series of $\psi_\alpha(x)$ and $\psi^\dagger_\beta$ (y) into the above equal-time commutation relation yields

$$[\psi_\alpha(x), \psi_\beta^\dagger (y)]_\pm$$

$$= \frac{1}{V}\Bigg[\sum_{q}\Bigg(\exp[i(qx - Ex_0)]\sum_{r=1}^{2} u_\alpha^{(r)}(q)\, a^{(r)}(q)$$

$$+ \exp[i(qx + Ex_0)]\sum_{r=3}^{4} u_\alpha^{(r)}(q)\, a^{(r)}(q)\Bigg),$$

$$\sum_{q'}\Bigg(\exp[-i(q'y - E'x_0)]\sum_{s=1}^{2} u_\beta^{(s)*}(q')\, a^{(s)\dagger}(q')$$

$$+ \exp[-i(q'y + E'x_0)]\sum_{s=3}^{4} u_\beta^{(s)*}(q')\, a^{(s)\dagger}(q')\Bigg)\Bigg]_\pm$$

$$= V^{-1}\, \delta_{\alpha\beta} \sum_{q} e^{iq(x-y)} \tag{2-432}$$

The commutation relations for the Fourier coefficients are given by

$$[a^{(r)}(q), a^{(s)\dagger}(q')]_\pm = \delta_{rs}\, \delta_{q,q'} \tag{2-433}$$

and from the other commutation relations we can obtain

$$[a^{(r)}(q), a^{(s)}(q')]_{\pm} = [a^{(r)\dagger}(q), a^{(s)\dagger}(q')]_{\pm} = 0 \tag{2-434}$$

If we apply the (-) type commutation relations, we find that the Hamiltonian is not positive definite since

$$H = \sum_q E\left(\sum_{r=1}^{2} n(q, r) - \sum_{r=3}^{4} n(q, r)\right) \tag{2-435}$$

In order to obtain a positive definite energy we must apply the (+) type commutation relations. Then we get

$$-\sum_{r=3}^{4} a^{(r)\dagger}(q)\, a^{(r)}(q) = \sum_{r=3}^{4} a^{(r)}(q)\, a^{(r)\dagger}(q) - 2 \tag{2-436}$$

and substitution of this expression into the Hamiltonian yields

$$H = \sum_q E\left(\sum_{r=1}^{2} a^{(r)\dagger}(q)\, a^{(r)}(q) + \sum_{r=3}^{4} a^{(r)}(q)\, a^{(r)\dagger}(q)\right) - 2\sum_q E \tag{2-437}$$

The additional constant term is negative and divergent, but we can drop it since it is a constant c number. The operator part of H is then positive definite. The operator a's are interpreted as

$a^{(r)\dagger}(q)$	$(r = 1, 2)$	creation of a positive-energy electron
$a^{(r)}(q)$	$(r = 1, 2)$	destruction of a positive-energy electron
$a^{(r)\dagger}(q)$	$(r = 3, 4)$	creation of a negative-energy electron
$a^{(r)}(q)$	$(r = 3, 4)$	destruction of a negative-energy electron.

Let us recall Dirac's hole theory which was introduced to avoid the negative energy difficulty. He assumed that all the negative energy states are occupied in the physical vacuum state. If a negative-energy electron state is not occupied, it will be observed as a positively-charged particle with a positive energy. In this sense, the destruction of a negative-energy electron means the creation of a positively-charged particle with a positive energy. Therefore, we set

$$b^{(1)}(q) = a^{(4)\dagger}(-q) \tag{2-438a}$$

$$b^{(2)}(q) = a^{(3)\dagger}(-q) \tag{2-438b}$$

Then we have

$$H = \sum_{q} E \sum_{r=1}^{2} (N_+(q, r) + N_-(q, r)) \tag{2-439}$$

where

$$N_+(q, r) = a^{(r)\dagger}(q)\, a^{(r)}(q) \tag{2-440a}$$

$$N_-(q, r) = b^{(r)\dagger}(q)\, b^{(r)}(q) \tag{2-440b}$$

For the charge operator we have

$$N = \sum_{q} \sum_{r=1}^{2} (N_+(q, r) - N_-(q, r)) \tag{2-441}$$

In the above formulas we discarded the c-number contributions from the electrons in the negative-energy sea. Let us investigate the eigenvalues of the number operator $N_\pm(q, r)$, which is denoted simply by N.

$$N^2 = a^\dagger a a^\dagger a = a^\dagger (1 - a^\dagger a)\, a = a^\dagger a - (a^\dagger)^2 a^2 = a^\dagger a = N \tag{2-442}$$

Note that $(a^\dagger)^2 = a^2 = 0$ because $\{a, a\} = \{a^\dagger, a^\dagger\} = 0$. Thus we have a simple equation for N, namely,

$$N^2 = N \tag{2-443}$$

and the only possible eigenvalues of N are 0 and 1. This means that a state can be occupied by, at most, one particle; this statement is called the Pauli exclusion principle. Therefore, the (+) type quantization leads to Fermi-Dirac statistics.[9]

Spin

Let us now consider the total angular momentum carried by the Dirac field:

$$M_{j\ell} = \int d^3x (x_j \; T_{0\ell} - x_\ell \; T_{0j} \; + i \frac{\partial \mathcal{L}}{\partial \psi_{\alpha:4}} (S_{j\ell})_{\alpha\beta} \psi_\beta)$$

$$= \int d^3x (x_j \; T_{0\ell} - x_\ell \; T_{0j} \; - i\psi^\dagger (S_{j\ell}) \psi) \tag{2-444}$$

where

$$S_{j\ell} = \frac{1}{4}(\gamma_j \gamma_\ell - \gamma_\ell \gamma_j) = \frac{i}{2}\sigma_{j\ell} \tag{2-445}$$

Therefore, the spin angular momentum is given by

$$\mathcal{S}_{j\ell} = \frac{1}{2} \int d^3x \, \psi^\dagger \, \sigma_{j\ell} \, \psi \tag{2-446}$$

By using the Fourier expansion of the Dirac fields we can obtain an expression for $\mathcal{S}_{j\ell}$ in terms of the operator a's. If we select a particular component corresponding to **q** = 0, we find

$$\mathcal{S}_3 = \mathcal{S}_{12} = \frac{1}{2}(a^{(1)\dagger} a^{(1)} - a^{(2)\dagger} a^{(2)} + b^{(1)\dagger} b^{(1)} - b^{(2)\dagger} b^{(2)}) \tag{2-447}$$

where we use the convention $\sigma_3 u^{(r)} = u^{(r)}$ for r = 1, 3 and $-u^{(r)}$ for r = 2, 4. This expression shows that the eigenvalue of $\mathcal{S}_3$ for a one-particle state specified by $a^{(r)\dagger}|0\rangle$ or $b^{(r)\dagger}|0\rangle$ is

$$\begin{aligned} \mathcal{S}_3 |r\rangle &= +\frac{1}{2}|r\rangle \qquad \text{for} \quad r = 1 \\ &= -\frac{1}{2}|r\rangle \qquad \text{for} \quad r = 2 \end{aligned} \tag{2-448}$$

From this result we conclude that the Dirac particle has spin $\frac{1}{2}$.

Charge Conjugation[10]

The Dirac equation is invariant under charge conjugation. We have introduced the Dirac equation for the electron and interpreted ψ as representing the destruction operator of an electron and $\psi^\dagger$ the destruction operator of a positron. However, since we could state that ψ represents the destruction operator of a positron instead of an electron, there is an arbitrariness in the interpretation of operator ψ. Physically, this means that the theory should be invariant under the transformation

$$\text{electron} \leftrightarrows \text{positron}$$

This transformation has been discussed in connection with the complex spinless field in Section 2-4 and is denoted by $\mathcal{C}$

$$\mathcal{C}\,\varphi(x)\,\mathcal{C}^{-1} = \varphi^\dagger(x) \quad \text{or generally} \quad e^{i\alpha}\,\varphi^\dagger(x) \tag{2-449a}$$

$$\mathcal{C}\,\varphi^\dagger(x)\,\mathcal{C}^{-1} = \varphi(x) \quad \text{or generally} \quad e^{-i\alpha}\,\varphi(x) \tag{2-449b}$$

We fix the phase of the unitary transformation $\mathcal{C}$ by requiring that

$$\mathcal{C}\,\Phi_{\text{vac}} = \Phi_{\text{vac}} \tag{2-450}$$

As far as free fields are concerned α is arbitrary, but if interactions are introduced α can be determined to some extent. We usually write the charge conjugate of ψ as

$$\psi_\alpha(x) \to \psi'_\alpha(x) = C_{\alpha\beta}\,\bar{\psi}_\beta(x) \tag{2-451}$$

The C is a Dirac matrix and should be distinguished from the quantized unitary operator $\mathcal{C}$. Unlike the spinless field, ψ has four components; so the above transformation is of the most general form.

$$\mathcal{C}\,\psi_\alpha(x)\,\mathcal{C}^{-1} = C_{\alpha\beta}\,\bar{\psi}_\beta(x) \tag{2-452}$$

We shall study the properties of this matrix C.

From the definition of charge conjugation we require

$$\mathcal{C}\,N\,\mathcal{C}^{-1} = -N \tag{2-453}$$

since N expresses the number of electrons minus that of positrons.

$$N = -i\int j_4(x)\, d^3x$$

The current operator $j_\mu(x)$ has been defined by $i\bar{\psi}(x)\gamma_\mu\psi(x)$, but in the quantized theory we find it more convenient to define $j_\mu(x)$ by

$$j_\mu(x) = \frac{i}{2}[\bar{\psi}(x),\ \gamma_\mu\psi(x)] = \frac{i}{2}[\bar{\psi}_\alpha(x), (\gamma_\mu)_{\alpha\beta}\ \psi_\beta(x)] \tag{2-454}$$

This differs from the old form by a divergent constant as seen from

$$\frac{i}{2}[\bar{\psi}(x), \gamma_\mu \psi(x)] = \frac{i}{2}[\bar{\psi}(x)\gamma_\mu\psi(x) - (\gamma_\mu)_{\alpha\beta}\, \psi_\beta(x)\, \bar{\psi}_\alpha(x)]$$

$$= \frac{i}{2}\Big[\bar{\psi}(x)\gamma_\mu\psi(x) + \bar{\psi}(x)\gamma_\mu\psi(x)$$

$$- (\gamma_\mu)_{\alpha\beta}\left\{\psi_\beta(x), \bar{\psi}_\alpha(x)\right\}\Big]$$

$$= i\bar{\psi}(x)\gamma_\mu\psi(x) - \frac{1}{2}\mathrm{Tr}\,[\gamma_\mu S(0)] \tag{2-455}$$

Now the anticommutativity between N and $\mathcal{C}$ implies that

$$\mathcal{C} j_\mu(x)\, \mathcal{C}^{-1} = -j_\mu(x) \tag{2-456}$$

or

$$\mathcal{C}[\bar{\psi}(x), \gamma_\mu\psi(x)]\,\mathcal{C}^{-1} = -[\bar{\psi}(x), \gamma_\mu\psi(x)] \tag{2-457}$$

This is a necessary condition that $\mathcal{C}$ has to satisfy.

We return to the Dirac equations,

$$(\gamma\partial + m)\psi = 0\ , \quad (\gamma^T\partial - m)\bar{\psi} = 0 \tag{2-458}$$

where $\gamma_\mu{}^T$ is the transposition of γ_μ. We require that the Dirac equations be invariant under $\mathcal{C}$:

$$(\gamma\partial + m)\psi' = (\gamma\partial + m)\,\mathcal{C}\psi\mathcal{C}^{-1} = 0 \tag{2-459}$$

or

$$(\gamma\partial + m)\, C\bar{\psi} = 0 \tag{2-460}$$

Multiplication of the inverse matrix C^{-1} by this equation yields

$$(C^{-1}\gamma_\mu C\ \partial_\mu + m)\bar{\psi} = 0 \tag{2-461}$$

Comparing this equation with the original equation for $\bar{\psi}$, (2-458), we have

$$C^{-1}\gamma_\mu C = -\gamma_\mu^T \tag{2-462}$$

or

$$C\gamma_\mu^T C^{-1} = -\gamma_\mu \tag{2-463}$$

Since the $(-\gamma_\mu^T)$'s satisfy the same commutation relations as the γ_μ's do, the matrix C can be chosen to be unitary.

Let us take the transpose of the Eq. (2-462),

$$C^T\gamma_\mu^T(C^T)^{-1} = -\gamma_\mu \tag{2-464}$$

and substitute for γ_μ^T the expression in Eq. (2-462)

$$C^T(C^{-1}\gamma_\mu C)(C^T)^{-1} = \gamma_\mu$$

or

$$(C^TC^{-1})\gamma_\mu(C^TC^{-1})^{-1} = \gamma_\mu \tag{2-465}$$

This means that C^TC^{-1} should commute with all the Dirac matrices, so that it must be a constant multiple of the unit matrix. Furthermore, this constant is independent of the representation of the Dirac matrices. Thus if S is a matrix which changes the representation of the γ_μ's to

$$\gamma'_\mu = S\gamma_\mu S^{-1} \tag{2-466}$$

then the matrix C in this new representation is given by

$$C' = SCS^T \tag{2-467}$$

and hence

$$C'^T(C')^{-1} = SC^TC^{-1}S^{-1} = C^TC^{-1} \tag{2-468}$$

Therefore, once C^TC^{-1} is determined in a special representation we can use the same constant in any representation. For this purpose we choose Pauli's representation in which γ_2 and γ_4 are symmetric, while γ_1 and γ_3 are antisymmetric; that is,

$$\begin{aligned} \gamma_\mu^T &= \gamma_\mu \qquad \text{for } \mu = 2, 4 \\ &= -\gamma_\mu \qquad \text{for } \mu = 1, 3 \end{aligned} \tag{2-469}$$

In this case $C = \gamma_2 \gamma_4$ satisfies the equations for C, and in this representation we easily can calculate $C^T C^{-1}$.

$$C^T = (\gamma_2 \gamma_4)^T = \gamma_4^T \gamma_2^T = \gamma_4 \gamma_2 = -\gamma_2 \gamma_4 = -C \tag{2-470}$$

Hence $C^T C^{-1} = -1$ in any representation, so that C is an *antisymmetric* matrix.

Taking the Hermitian conjugate of $\mathcal{C}\, \psi_\alpha(x)\, \mathcal{C}^{-1} = C_{\alpha\beta}\, \bar{\psi}_\beta(x)$, and utilizing the properties of the matrix C, we easily can find that

$$\mathcal{C}\, \bar{\psi}_\alpha(x)\, \mathcal{C}^{-1} = (C^{-1})_{\alpha\beta}\, \psi_\beta(x) \tag{2-471}$$

With these results we can investigate the transformation properties of bilinear forms

$$\mathcal{C}\, [\bar{\psi}(x),\ \mathcal{O}\, \psi(x)]\, \mathcal{C}^{-1} \tag{2-472}$$

where $\mathcal{O}$ is a Dirac matrix.

$$\begin{aligned} \mathcal{C}\, [\bar{\psi}(x),\ \mathcal{O}\, \psi(x)]\, \mathcal{C}^{-1} &= \mathcal{O}_{\alpha\beta} [\mathcal{C} \bar{\psi}_\alpha(x)\, \mathcal{C}^{-1},\ \mathcal{C} \psi_\beta(x)\, \mathcal{C}^{-1}] \\ &= \mathcal{O}_{\alpha\beta} [C^{-1}_{\alpha\lambda}\, \psi_\lambda(x),\ C_{\beta\mu}\, \bar{\psi}_\mu(x)] \\ &= C^{-1}_{\alpha\lambda}\, \mathcal{O}_{\alpha\beta}\, C_{\beta\mu} [\psi_\lambda(x),\ \bar{\psi}_\mu(x)] \\ &= -C^{-1}_{\lambda\alpha}\, \mathcal{O}_{\alpha\beta}\, C_{\beta\mu} [\psi_\lambda(x),\ \bar{\psi}_\mu(x)] \\ &= (C^{-1} \mathcal{O} C)_{\lambda\mu} [\bar{\psi}_\mu(x),\ \psi_\lambda(x)] \\ &= [\bar{\psi}(x),\ \mathcal{O}'\, \psi(x)] \end{aligned} \tag{2-473}$$

where

$$\mathcal{O}' = (C^{-1} \mathcal{O}\, C)^T = C\, \mathcal{O}^T C^{-1} = \eta\, \mathcal{O} \tag{2-474}$$

The values of η can be determined by use of the fundamental relationship $C\gamma_\mu{}^T C^{-1} = -\gamma_\mu$ for any Dirac matrix $\mathcal{O}$. The results are given in Table 2-1.

Table 2-1

Transformation Properties of Dirac Matrices Under Charge Conjugation

Type of the Dirac matrix	Scalar	Vector	Tensor	Axial vector	Pseudo scalar
$\mathcal{O}$	1	γ_μ	$\sigma_{\mu\nu}$	$\gamma_\mu\gamma_5$	γ_5
η	1	-1	-1	1	1

Since the charge conjugation operator $\mathcal{C}$ commutes with H, it can be used to derive selection rules. Finally, it should be mentioned that the commutation relations are invariant under charge conjugation:

$$\begin{aligned} \mathcal{C}\left\{\psi_\alpha(x),\ \bar\psi_\beta(y)\right\}\mathcal{C}^{-1} &= \left\{\mathcal{C}\psi_\alpha(x)\,\mathcal{C}^{-1},\ \mathcal{C}\,\bar\psi_\beta(y)\,\mathcal{C}^{-1}\right\} \\ &= C_{\alpha\lambda}\,C^{-1}_{\beta\mu}\left\{\bar\psi_\lambda(x),\ \psi_\mu(y)\right\} \\ &= -iC_{\alpha\lambda}\,S_{\mu\lambda}(y-x)\,C^{-1}_{\beta\mu} \\ &= i(CS^T(y-x)\,C^{-1})_{\alpha\beta} \\ &= -iS_{\alpha\beta}(x-y) \\ &= \left\{\psi_\alpha(x),\ \bar\psi_\beta(y)\right\} \end{aligned} \quad (2\text{-}475)$$

Actually, given the transformation

$$\mathcal{C}\,\psi_\alpha(x)\ \mathcal{C}^{-1} = C_{\alpha\beta}\bar\psi_\beta(x) \quad (2\text{-}476)$$

the invariance of the commutation relations implies

$$\mathcal{C}\,\bar\psi_\alpha(x)\ \mathcal{C}^{-1} = C^{-1}_{\alpha\beta}\,\psi_\beta(x) \quad (2\text{-}477)$$

PROBLEMS

2-1 The ψ denotes a nonrelativistic Bose field, and the Lagrangian density is given by

$$\mathcal{L}(x) = i\psi^\dagger \frac{\partial \psi}{\partial x_0} - \frac{(\nabla \psi^\dagger)(\nabla \psi)}{2m}$$

$$-\frac{1}{2} \int_{y_0 = x_0} d^3y\, \psi^\dagger(x)\, \psi^\dagger(y)\, V(x-y)\, \psi(x)\, \psi(y)$$

(1) Prove that the field equation for ψ is given by

$$i\frac{\partial \psi(x)}{\partial x_0} = -\frac{\Delta}{2m}\psi(x) + \int_{y_0 = x_0} d^3y\, \psi^\dagger(y)\, \psi(y)\, V(x-y)\, \psi(x)$$

where

$$V(x-y) = V(|\mathbf{x} - \mathbf{y}|)$$

(2) Show that the Hamiltonian of this system is given by

$$H = \frac{1}{2m} \int d^3x (\nabla \psi^\dagger)(\nabla \psi)$$

$$+\frac{1}{2} \int_{y_0 = x_0} d^3x\, d^3y\, \psi^\dagger(x)\, \psi^\dagger(y)\, V(x-y)\, \psi(x)\, \psi(y)$$

(3) Define the symmetric function,

$$\varphi(\mathbf{x}_1, \ldots, \mathbf{x}_n, t) = (\Psi_0, \psi(\mathbf{x}_1, t) \ldots \psi(\mathbf{x}_n, t)\, \Psi_n)$$

where Ψ_n is an n-particle state. Prove that this φ satisfies

$$i\frac{\partial}{\partial t}\varphi = H_n \varphi$$

where

$$H_n = \sum_{r=1}^{n} \left(-\frac{\Delta_r}{2m}\right) + \sum_{i>j} V(\mathbf{x}_i - \mathbf{x}_j)$$

and that φ is normalized:

$$\frac{1}{n!} \int \varphi^* \varphi\, d^3x_1 \ldots d^3x_n = 1$$

The factorial n! is introduced to avoid counting the same state n! times.

2-2 The equation for the spin 3/2 field is called the Rarita-Schwinger equation.[11] The field operator carries a spinor index r and a vector μ, namely, $\psi_{r\mu}$, but the spinor index is usually suppressed, as is the case for the Dirac spinor. The Rarita-Schwinger equation is given by

$$D_{\mu\nu}\,\psi_\nu = 0$$

where

$$D_{\mu\nu} = (\gamma\partial + m)\,\delta_{\mu\nu} - \frac{1}{3}(\gamma_\mu\partial_\nu + \gamma_\nu\partial_\mu) + \frac{1}{3}\gamma_\mu(\gamma\partial - m)\gamma_\nu$$

(1) Prove the following relations:

$$\gamma_\mu\psi_\mu = 0 \qquad \partial_\mu\psi_\mu = 0 \qquad (\gamma\partial + m)\,\psi = 0$$

(2) Find the matrix C from the equation

$$C_{\lambda\mu}D_{\mu\nu} = D_{\lambda\mu}C_{\mu\nu} = \delta_{\lambda\nu}\,(\Box - m^2)$$

REFERENCES

1. See, for instance, G. Wentzel, *Quantum Theory of Fields*, (Interscience Publishers, Inc., New York, 1949).
2. P. Jordan and W. Pauli, Z. Physik **47**, 151 (1928).
3. R. Peierls, Proc. Roy. Soc. London **A214**, 143 (1952). For a unified treatment of field quantization, see Y. Takahashi, *Introduction to Field Quantization*, (Pergamon Press, London, (1968).
4. W. Pauli, Phys. Rev. 58, 716 (1940). For recent developments, see also R. F. Streater and A. S. Wightman, *PCT, Spin and Statistics, and All That*, (W. A. Benjamin, New York, 1964).
5. W. Heisenberg and W. Pauli, Z. Physik **56**, 1 (1926); **59**, 168 (1930).
6. E. Fermi, Rev. Mod. Phys. **4**, 125 (1932).
7. E. C. G. Stueckelberg, Helv. Phys. Acta, **11**, 225 (1938).
8. S. Gupta, Proc. Phys. Soc. London, **A63**, 681 (1950). K. Bleuler, Helv. Phys. Acta, **23**, 567 (1950). See also W. Heitler, *Quantum Theory of Radiation*, (Oxford University Press, 3rd ed., 1954).
9. P. Jordan and E. P. Wigner, Z. Physik **47**, 631 (1928).
10. W. Pauli, Ann. Inst. Henri Poincaré, **6**, 109 (1936).
11. W. Rarita and J. Schwinger, Phys. Rev. **60**, 61 (1941).

CHAPTER 3

QUANTIZATION OF INTERACTING FIELDS

If no interaction is present, we cannot make any observation of the field quanta. To be observed, a field must have interactions with other fields.

In the presence of interactions there is no general way of quantizing the system except by canonical quantization, that is,

$$\pi_\alpha = \frac{\partial \mathcal{L}}{\partial \dot{\varphi}_\alpha} \qquad \text{and} \qquad [\varphi_\alpha(x), \pi_\beta(y)]_\pm = i\delta_{\alpha\beta}\,\delta^3(x-y)$$

$$\text{for} \quad x_0 = y_0$$

In order to solve the field equations we regard these canonical commutation relations at a given time as the boundary condition to fix the field operators. There are, again, two methods of viewing the problem, namely, the Schrödinger picture and the Heisenberg picture.

3-1 SCHRÖDINGER EQUATION

The appropriate form of the Schrödinger equation for an interacting system of fields is

$$i\frac{\partial \Psi(t)}{\partial t} = H\,\Psi(t) \tag{3-1}$$

with $\Psi(t)$ representing the total state vector of the interacting system and H a functional of the field operators of the form,

$$H = \int d^3x\,\mathcal{H}[\varphi_\alpha(x), \pi_\alpha(x)] \tag{3-2}$$

In this representation, the field operators are time independent and they satisfy the canonical commutation relations,

$$[\varphi_\alpha(x), \pi_\beta(y)] = i\delta_{\alpha\beta}\,\delta^3(x-y) \tag{3-3}$$

Although Eqs. (3-1)-(3-3) give a well-defined description of the problem, the drawback of this representation is the apparent lack of Lorentz covariance. We have to keep a particular time axis throughout the treatment. The Heisenberg operators are defined as

$$\varphi_\alpha(\mathbf{x},\ t) = e^{itH}\ \varphi_\alpha(\mathbf{x})e^{-itH} \tag{3-4}$$

They satisfy covariant equations, but again the commutation relations, given by canonical quantization, apparently lack covariance.

In the case of free fields we could show that the canonical quantization method gives a covariant set of commutation relations by explicitly solving the field equations. In the presence of interactions, however, the field equations are nonlinear and we cannot solve them to check the covariance of the canonical commutation relations. A method of circumventing this difficulty was first proposed by S. Tomonaga in 1943.

3-2 MULTIPLE-TIME FORMULATION

Dirac's Multiple-Time Formulation[1]

We first introduce a formulation of electrodynamics in terms of multiple time. Dirac discussed the problem of a many-electron system interacting through the electromagnetic field. We start from the unquantized electron field.

Suppose we make an observation of N electrons to determine the state. We make an observation of the first electron at time t_1, the second at t_2, and so on. For the description of this system in the Schrödinger picture we can use the Schrödinger functional $\Psi(t)$, provided $t_1 = t_2 = \ldots = t_N$.

If this is not the case, we do not know how to express the results of our observation in terms of the single-time Schrödinger functional. For this purpose the ordinary Schrödinger equation may be extended so as to accommodate such observations by incorporating the electromagnetic Hamiltonian, H_{em}, and the electromagnetic field operator, $A(\mathbf{x}_n)$, in the following way.

The single-time Schrödinger equation for this system is

$$i\frac{\partial}{\partial t}\Psi(t) = \left(H_{em} + \sum_{n=1}^{N} H_n(\mathbf{x}_n,\ p_n,\ A(\mathbf{x}_n))\right)\Psi(t) \tag{3-5}$$

We introduce the unitary transformation

$$U(t) = \exp(it\,H_{em}) \tag{3-6}$$

and

$$A(\mathbf{x},\ t) = U(t)\,A(\mathbf{x})\,U(t)^{-1} \tag{3-7}$$

Then the potential $A(\mathbf{x}, t)$ is the Heisenberg operator in the absence of interactions, and U obviously commutes with the electron variables. Now set

$$\Phi(t) = U(t)\Psi(t) \tag{3-8}$$

then $\Phi(t)$ satisfies

$$i\frac{\partial}{\partial t}\Phi(t) = \left(\sum_{n=1}^{N} H_n(\mathbf{x}_n, \mathbf{p}_n, A(\mathbf{x}_n, t))\right)\Phi(t) \tag{3-9}$$

At this stage of the game we can introduce multiple times by defining $\Phi(\mathbf{x}_1, t_1; \mathbf{x}_2, t_2; \ldots; \mathbf{x}_N, t_N)$ as follows:

$$i\frac{\partial}{\partial t_n} \times \Phi(x_1 t_1; \ldots; \mathbf{x}_N t_N) = H_n(\mathbf{x}_n, \mathbf{p}_n, A(\mathbf{x}_n, t_n)) \times \Phi(\mathbf{x}_1 t_1; \ldots; \mathbf{x}_N t_N) \tag{3-10}$$

There are N such equations, and if we set $t_1 = t_2 \ldots = t_N = t$, we again obtain the original single-time Schrödinger equation. These generalized multiple-time equations enable us to describe the observation discussed at the beginning of this section.

Because there are N equations for one unknown Φ, we have to study the compatibility of these equations. We require that

$$\frac{\partial^2}{\partial t_1 \partial t_2}\Phi = \frac{\partial^2}{\partial t_2 \partial t_1}\Phi \qquad \text{etc.} \tag{3-11}$$

and obtain

$$(H_1 H_2 - H_2 H_1)\,\Phi(\mathbf{x}_1 t_1; \ldots; \mathbf{x}_N t_N) = 0 \qquad \text{etc.} \tag{3-12}$$

This is called the *integrability condition*. In order to discuss the physical meaning of this condition, let us consider the commutator $[H_1, H_2]$, where H_1 is a function of $\mathbf{x}_1$, $\mathbf{p}_1$, and $A(\mathbf{x}_1, t_1)$, and similarly H_2 is a function of variables with the subscript 2. In this commutator the only quantities that might not commute are the potentials. Therefore, we conclude that the integrability condition is satisfied if and only if

$$[A(\mathbf{x}_1, t_1), A(\mathbf{x}_2, t_2)] = 0 \qquad \text{etc.} \tag{3-13}$$

As we have seen in the preceding chapter, the latter condition is satisfied when all the N points are separated from each other by space-like distances, namely,

$$(\mathbf{x}_n - \mathbf{x}_{n'})^2 - (t_n - t_{n'})^2 > 0 \tag{3-14}$$

Bloch[2] has shown that if this condition is satisfied, then the relative probability that one will find the first electron at $\mathbf{x}_1$ at time t_1, the second at $\mathbf{x}_2$ at time t_2, and so on, is

$$W(\mathbf{x}_1, t_1; \ldots; \mathbf{x}_N, t_N) = |\Phi(\mathbf{x}_1, t_1; \ldots; \mathbf{x}_N, t_N)|^2 \tag{3-15}$$

We have seen that the integrability condition is concomitant with the microscopic casualty condition. The physical meaning of the integrability condition is clear: if $t_2 > t_1$ and $(t_2 - t_1)^2 - (\mathbf{x}_2 - \mathbf{x}_1)^2 > 0$, then the measurement at $(\mathbf{x}_1, t_1)$ influences the measurement at $(\mathbf{x}_2, t_2)$. Therefore, we cannot expect a unique solution for Φ, as W depends not only on the initial condition of the system but also on the measurement at $(\mathbf{x}_1, t_1)$.

Now we should recall the substitution that we have used to extend particle mechanics to field theory. This procedure enables us to translate the multiple-time formulation, introduced in the preceding, to the supermultiple-time formulation for fields.

Tomonaga's Supermultiple-Time Formulation[3]

The starting equation is the same as (3-1) in Dirac's theory,

$$i \frac{\partial \Psi(t)}{\partial t} = H\Psi(t) \tag{3-16}$$

where

$$H = \int d^3x\, \mathcal{H}(\mathbf{x}) = \int d^3x\, \mathcal{H}_f(\mathbf{x}) + \int d^3x\, \mathcal{H}_{int}(\mathbf{x}) = H_f + H_{int} \tag{3-17}$$

We treat electrons as quanta of the electron field, and H_f represents the free Hamiltonian for the electromagnetic *and* the electron fields.

Let us introduce a unitary transformation U(t) by

$$U(t) = \exp(itH_f) \tag{3-18}$$

and

$$\varphi_\alpha(\mathbf{x}, t) = U(t)\, \varphi_\alpha(\mathbf{x})\, U(t)^{-1} \tag{3-19}$$

where φ_α represents either the electron or the electromagnetic field; then $\varphi_\alpha(\mathbf{x}, t)$ satisfies the field equation for a free field. Therefore, we know the commutation relations for $\varphi_\alpha(x)$ completely. These commutation relations are, as we have shown in the preceding chapter, completely covariant relations.

A new state vector $\Phi(t)$, defined as

$$\Phi(t) = U(t)\, \Psi(t) \tag{3-20}$$

satisfies

$$i \frac{\partial}{\partial t} \Phi(t) = \left(U(t) \int \mathcal{H}_{int}(x)\, d^3x\, U(t)^{-1} \right) \Phi(t) \tag{3-21}$$

Inasmuch as $\mathcal{H}_{int}(x)$ is a polynomial of $\varphi_\alpha(x)$, we get

$$\begin{aligned} U(t)\, \mathcal{H}_{int}[\varphi_\alpha(x)] U(t)^{-1} &= \mathcal{H}_{int}[U(t)\, \varphi_\alpha(x)\, U(t)^{-1}] \\ &= \mathcal{H}_{int}[\varphi_\alpha(\mathbf{x}, t)] \end{aligned} \tag{3-22}$$

This is a hybrid between the Heisenberg and Schrödinger representations, called the *interaction representation* by Schwinger.[4]

We are left with the equation

$$i \frac{\partial}{\partial t} \Phi(t) = \int d^3x\, \mathcal{H}_{int}[\varphi_\alpha(x)]\, \Phi(t) \tag{3-23}$$

where x represents $\mathbf{x}$ and t. In the Dirac theory the Hamiltonian on the right-hand side is a sum, but it is replaced by an integral in the present case. Dirac considered each electron as having an individual time coordinate, while Tomonaga assigned an individual time coordinate to each point in space.

The correspondence between them is as follows:

Dirac	Tomonaga
$\sum_n$	$\int d^3x$
t_n	t_{xyz}

The set of points $t = t_{xyz}$ defines a three-dimensional hypersurface, σ. The time derivative is replaced by the so-called functional derivative:

$$\frac{\delta \Phi}{\delta t_{xyz}} = \lim_{\delta t \to 0} \frac{\Phi[t_{xyz} + \delta t_{xyz}] - \Phi[t_{xyz}]}{\int d^3x\ \delta t_{xyz}} \tag{3-24}$$

or with the state vector denoted by

$$\Phi = \Phi[t_{\mathbf{x}}, t_{\mathbf{x}'}, \ldots] = \Phi[\sigma] \tag{3-25}$$

we may write

$$\lim_{dV\to 0} \frac{\Phi[\sigma'] - \Phi[\sigma]}{dV} = \frac{\delta\Phi[\sigma]}{\delta\sigma(x)} \tag{3-26}$$

where dV is an infinitesimal four-dimensional volume between two hypersurfaces σ and σ', which overlap except at an infinitesimal vicinity of the point x. This derivative corresponds to $\partial\Phi/\partial t_n$ in the Dirac theory. Then Tomonaga decomposed the single-time equation into an infinite set of equations:

$$i\frac{\delta\Phi[\sigma]}{\delta\sigma(x)} = \mathcal{H}_{int}(x)\,\Phi[\sigma] \tag{3-27}$$

where $\mathcal{H}_{int}(x) = \mathcal{H}_{int}[\varphi_\alpha(x)]$. In the case of a flat hypersurface σ, namely, when σ is a hyperplane defined by t_{xyz} = constant, we can show

$$\frac{\partial\Phi}{\partial t} = \int d^3x\,\frac{\delta\Phi}{\delta\sigma(x)} \tag{3-28}$$

and we can rederive the original single-time Schrödinger equation. The relation (3-28) corresponds to the following relation in the Dirac theory:

$$\left(\frac{\partial\Phi}{\partial t}\right)_{t_1=t_2=\cdots=t} = \left(\sum_{n=1}^{N}\frac{\partial\Phi}{\partial t_n}\right)_{t_1=t_2=\cdots=t} \tag{3-29}$$

The integrability condition is now given by

$$\frac{\delta}{\delta\sigma(x)}\cdot\frac{\delta}{\delta\sigma(y)}\,\Phi[\sigma] = \frac{\delta}{\delta\sigma(y)}\cdot\frac{\delta}{\delta\sigma(x)}\,\Phi[\sigma] \tag{3-30}$$

or

$$[\mathcal{H}_{int}(x), \mathcal{H}_{int}(y)] = 0 \tag{3-31}$$

This condition is satisfied when $(x - y)^2 > 0$, or when all the points on the hypersurface σ are separated by spacelike distances. When this is the case, σ is called a spacelike hypersurface. The integrability condition then requires that the state vector $\Phi[\sigma]$ be determined only when σ is spacelike. The concept of a spacelike hypersurface is a generalization of the idea of the plane t = constant.

If the initial conditions are given in a certain Lorentz frame of reference, and if measurements are made by an observer in a different frame moving relative to the original frame, the single-time Schrödinger equation cannot provide enough information about that which is predicted by these measurements. This new supermultiple-time formulation, however, is expected to meet such a need.

We now can see that all the relationships are covariant in the interaction representation.

commutation relations: $[\varphi_\alpha(x), \varphi_\beta^\dagger(y)] = i\ C_{\alpha\beta}(\partial)\Delta(x - y)$ (3-32)

Schrödinger equation: $i\dfrac{\delta\Phi[\sigma]}{\delta\sigma(x)} = \mathcal{H}_{int}(x)\ \Phi[\sigma]$ (3-33)

The use of the multiple-time equation is new, but the interaction representation is not; it has long been used in perturbation theory.

3-3 TRANSFORMATION FUNCTIONAL—S MATRIX

The generalized Schrödinger equation, (3-27), introduced in the preceding section is called the Tomonaga-Schwinger equation — to be referred to as the T-S equation hereafter.

Let us consider the T-S equation (3-27) with the boundary condition $\Phi[\sigma_0] = \Phi$. The formal solution of this problem is given by

$$\Phi[\sigma] = U[\sigma, \sigma_0]\Phi \tag{3-34}$$

provided U satisfies the T-S equation

$$i\frac{\delta U[\sigma, \sigma_0]}{\delta\sigma(x)} = \mathcal{H}_{int}(x)\ U[\sigma, \sigma_0] \tag{3-35}$$

and the boundary condition

$$U[\sigma_0, \sigma_0] = 1 \tag{3-36}$$

This U is called the *generalized transformation functional* and is a unitary operator. The U has the decomposition property,

$$U[\sigma_2, \sigma_0] = U[\sigma_2, \sigma_1]\, U[\sigma_1, \sigma_0] \tag{3-37}$$

This is sometimes called the *casualty condition.*

In particular, when σ is a flat hyperplane corresponding to t = constant, we write U as $U(t_2, t_1)$.

The expectation value of a field operator in the interaction representation

$$\left(\Phi(t), \varphi_\alpha(x)\Phi(t)\right) = \left(\Phi,\ U(t, t_0)^{-1}\varphi_\alpha(\mathbf{x}, t)U(t, t_0)\Phi\right) \tag{3-38}$$

suggests that

$$U(t, t_0)^{-1}\varphi_\alpha(\mathbf{x}, t)U(t, t_0) = \varphi_\alpha^H(\mathbf{x}, t) \tag{3-39}$$

is a Heisenberg operator which is equal to the interaction operator φ_α at $t = t_0$. It is easily verified that the operator (φ_α) satisfies the field equations in the interaction representation. When the Heisenberg operators are fixed, we can define different interaction representations for different choices of t_0. Among them we choose two special interaction representations, labeled *in* and *out* as follows:

$$(1)\quad t_0 = -\infty \qquad \varphi_\alpha^{(H)}(\mathbf{x}, t) = U(t, -\infty)^{-1}\varphi_\alpha^{in}(\mathbf{x}, t)U(t, -\infty) \tag{3-40a}$$

$$(2)\quad t_0 = \infty \qquad \varphi_\alpha^{(H)}(\mathbf{x}, t) = U(t, \infty)^{-1}\varphi_\alpha^{out}(\mathbf{x}, t)U(t, \infty)$$

$$= U(\infty, t)\varphi_\alpha^{out}(\mathbf{x}, t)U(\infty, t)^{-1} \tag{3-40b}$$

These are incoming and outgoing fields, respectively. From these definitions we find

$$\varphi_\alpha^{out}(x) = S^{-1}\varphi_\alpha^{in}(x)\, S \tag{3-41}$$

with

$$S = U(\infty, -\infty) \tag{3-42}$$

The operator S is called the S matrix[5].

3-4 YANG-FELDMAN FORMALISM[6]

We have derived the covariant T-S equation (3-27) starting from the Schrödinger equation (3-16), but we can take an alternative approach by starting with the field equations for interacting fields in the Heisenberg representation.

In the presence of interactions we split the Lagrangian into free and interaction parts,

$$\mathfrak{L} = \mathfrak{L}_f + \mathfrak{L}_{int} \tag{3-43}$$

and write the Euler equation as

$$[\mathfrak{L}]_{\varphi_\alpha^\dagger} = [\mathfrak{L}_f]_{\varphi_\alpha^\dagger} + [\mathfrak{L}_{int}]_{\varphi_\alpha^\dagger}$$

$$= D_{\alpha\beta}(\partial)\varphi_\beta + \frac{\partial \mathfrak{L}_{int}}{\partial \varphi_\alpha^\dagger} = 0 \tag{3-44}$$

using the differential operator $D_{\alpha\beta}(\partial)$ introduced in Section 2-3. In Eq. (3-43) it is assumed that $_{\text{int}}$ does not depend on the derivatives of field operators, but when it does we should use the Euler derivative of $\mathcal{L}_{\text{int}}$, Eq. (3-44).

In order to solve this differential equation we recall Peierls' method of deriving the commutation relations for free fields. (See Section 2-3.) Our procedure here constitutes a generalization of that method. Furthermore, we assume that the interaction is present only in a finite space-time domain Ω as discussed in Section 2-4. In order to satisfy this assumption we have to treat the coupling constant e as a function of space-time; in particular, e is assumed to vanish for $t > T_2$ and $t < T_1$. It is implicitly postulated that the physical situation can be recovered by letting $T_1 \to -\infty$, $T_2 \to +\infty$.

The operator φ_α satisfies the free-field equations for $t > T_2$. and $t < T$. In accordance with the notation introduced in the preceding section, we set

$$\varphi_\alpha = \varphi_\alpha^{\text{in}} \qquad \text{for} \quad t < T_1 \tag{3-45a}$$

$$\varphi_\alpha = \varphi_\alpha^{\text{out}} \qquad \text{for} \quad t > T_2 \tag{3-45b}$$

then

$$D_{\alpha\beta}(\partial)\varphi_\beta^{\text{in}}(x) = D_{\alpha\beta}(\partial)\varphi_\beta^{\text{out}}(x) = 0 \tag{3-46}$$

Now we set

$$\varphi_\alpha(x) = \varphi_\alpha^{\text{in}}(x) + \varphi'_\alpha(x) \tag{3-47}$$

so that $\varphi'_\alpha = 0$ for $t < T_1$. Substituting this form into (3-46), the differential equation for φ_α, we get

$$D_{\alpha\beta}(\partial)\varphi_\beta(x) = D_{\alpha\beta}(\partial)\varphi_\beta^{\text{in}}(x) + D_{\alpha\beta}(\partial)\varphi'_\beta(x)$$

$$= D_{\alpha\beta}(\partial)\varphi'_\beta(x) = -\frac{\partial \mathcal{L}_{\text{int}}(x)}{\partial \varphi_\alpha^\dagger(x)} \tag{3-48}$$

In order to solve this equation we recall the invariant Green's function for $D_{\alpha\beta}(\partial)$:

$$D_{\alpha\beta}(\partial)\Delta^R_{\beta\gamma}(x) = -\delta_{\alpha\gamma}\delta^4(x) \tag{3-49}$$

Then the formal solution of the differential equation for φ' is given by

$$\varphi'_\alpha(x) = \int_\Omega d^4x' \, \Delta^R_{\alpha\beta}(x - x') \frac{\partial \mathcal{L}_{int}(x')}{\partial \varphi^\dagger_\beta(x')} \tag{3-50}$$

and hence

$$\varphi_\alpha(x) = \varphi_\alpha^{in}(x) + \int_\Omega d^4x' \, \Delta^R_{\alpha\beta}(x - x') \frac{\partial \mathcal{L}_{int}(x')}{\partial \varphi^\dagger_\beta(x')} \tag{3-51}$$

At this stage we let Ω expand to cover the entire space-time; to do this we change the form of the coupling constant as a function of the space-time coordinates adiabatically until e becomes a constant. The validity of this procedure is based on the hypothesis of adiabatic switching on and off of the interactions.

In this manner we get a set of integral equations for the φ_α's, which clearly satisfy the field equations. Inasmuch as we know the commutation relations for the $\varphi_\alpha{}^{in}$'s and have an implicit expression for φ_α in terms of the incoming field operators, we can, in principle, derive the commutation relations for the φ_α's. Starting from the future, we can also derive a set of integral equations which relates the Heisenberg operators to the outgoing field operators. To summarize our results, we have the following two sets of equations:

$$\varphi_\alpha(x) = \varphi_\alpha^{in}(x) + \int d^4x' \, \Delta^R_{\alpha\beta}(x - x') \frac{\partial \mathcal{L}_{int}(x')}{\partial \varphi^\dagger_\beta(x')} \tag{3-52}$$

and

$$\varphi_\alpha(x) = \varphi_\alpha^{out}(x) + \int d^4x' \, \Delta_{\alpha\beta}^A(x - x') \frac{\partial \mathcal{L}_{int}(x')}{\partial \varphi^\dagger_\beta(x')} \tag{3-53}$$

Taking the difference of these two equations we have

$$\varphi^{out}(x) = \varphi_\alpha^{in}(x) - \int d^4x' \, \Delta_{\alpha\beta}(x - x') \frac{\partial \mathcal{L}_{int}(x')}{\partial \varphi^\dagger_\beta(x')} \tag{3-54}$$

Suppose we can express φ_α in terms of the φ^{in}'s; then the right-hand side is expressed in terms of the φ^{in}'s and we are left with an equation of the form

$$\varphi_\alpha^{out}(x) = \varphi_\alpha^{in}(x) + F_\alpha[\varphi^{in}] \tag{3-55}$$

Comparing this equation with the relation between the incoming and outgoing fields, Eq. (3-41), we get an equation to determine S:

$$S^{-1}\varphi_\alpha^{in}(x)S = \varphi_\alpha^{in}(x) + F_\alpha[\varphi^{in}] \tag{3-56}$$

Example (3-1) — Quantum electrodynamics

In the absence of the interaction, the Lagrangian density for the system consisting of the electromagnetic field and the electron field is given by

$$\mathcal{L}_{em}[A] + \mathcal{L}_{el}[\psi, \bar{\psi}, \partial_\mu\psi, \partial_\mu\bar{\psi}] \tag{3-57}$$

The electromagnetic interaction is introduced by the substitution:

$$\partial_\mu\psi \rightarrow (\partial_\mu - ieA_\mu)\psi \tag{3-58}$$

$$\partial_\mu\bar{\psi} \rightarrow (\partial_\mu + ieA_\mu)\bar{\psi} \tag{3-59}$$

where e is the charge of an electron including the sign. This prescription leads to the following interaction Lagrangian density:

$$\mathcal{L}_{int} = j_\mu A_\mu \tag{3-60}$$

where $j_\mu = ie\bar{\psi}\gamma_\mu\psi$, or in a more rigorous treatment we have to use

$$j_\mu = \frac{1}{2}ie[\bar{\psi}, \gamma_\mu\psi] \tag{3-61}$$

Then the field equations are given by

$$\left(\gamma_\mu \frac{\partial}{\partial x_\mu} + m\right)\psi = ie\gamma_\mu A_\mu\psi \tag{3-62}$$

$$\frac{\partial\bar{\psi}}{\partial x_\mu}\gamma_\mu - m\bar{\psi} = -ie\bar{\psi}\gamma_\mu A_\mu \tag{3-63}$$

$$\Box A_\mu = -j_\mu \tag{3-64}$$

The corresponding integral equations—to be referred to as the Yang-Feldman equations or Y-F equations—are given by

$$A_\mu(x) = A_\mu^{in}(x) + \int d^4x' \; D^R(x - x')\, j_\mu(x') \tag{3-65}$$

$$\psi(x) = \psi^{in}(x) - ie \int d^4x' \, S^R(x - x')\, \gamma_\mu A_\mu (x')\psi(x') \tag{3-66}$$

$$\overline{\psi}(x) = \overline{\psi}^{in}(x) - ie \int d^4x' \, \overline{\psi}(x')\gamma_\mu A_\mu (x') S^A (x' - x) \tag{3-67}$$

The S functions are defined in terms of the Δ functions by

$$S^{(\;\;)}(x) = (\gamma\partial - m)\, \Delta^{(\;\;)}(x) \tag{3-68}$$

The relationship between the Yang-Feldman formalism and that of transformation functionals is rather clear. In the latter case, we have

$$\varphi_\alpha(x) = U\,[\sigma, -\infty]^{-1} \varphi_\alpha^{in}(x) U\,[\sigma, -\infty] \tag{3-69}$$

with

$$i \frac{\delta U\,[\sigma, -\infty]}{\delta\sigma(x)} = \mathcal{H}_{int}\,[\varphi_\alpha^{in}(x)]\; U\,[\sigma, -\infty] \tag{3-70}$$

If we express $U\,[\sigma, -\infty]$ in terms of φ^{in}'s and insert the result into Eq. (3-69), it should agree with the Yang-Feldman solution. We shall discuss this point in more detail in the next chapter.

REFERENCES

1. P.A.M. Dirac, Proc. Roy. Soc. London, **136**, 453 (1932).
2. F. Bloch, Phys. Z. USSR, 5, 301 (1943).
3. S. Tomonaga, Progr. Theor. Phys. (Kyoto) 1, 27 (1946).
4. J. Schwinger, Phys. Rev. **74**, 1439 (1948).
5. W. Heisenberg, Z. f. Physik, **120**, 513 (1943).
6. C. N. Yang and D. Feldman, Phys. Rev. **79**, 972 (1950).
   G. Källén, Arkiv för Fysik, Bd. 2, Nr. 37 (1950).

CHAPTER 4
COVARIANT PERTURBATION THEORY

In nonrelativistic quantum mechanics, perturbation theory is a useful approximation. In field theory it is desirable to carry out perturbation theory in a manifestly covariant way in order to avoid certain ambiguities arising from divergence difficulties.

4-1 EXAMPLES OF INTERACTING FIELDS

As the most familiar interactions in field theory we shall consider the interaction between the electron and electromagnetic fields, and the interaction between the nucleon and pion fields.

Electrodynamics

The free Lagrangian densities for the electron and electromagnetic fields are given, respectively, by

$$\mathcal{L}_{el} = -\bar{\psi}(\gamma\partial + m)\psi \tag{4-1}$$

$$\mathcal{L}_{em} = -\frac{1}{4}F_{\mu\nu}F_{\mu\nu} - \frac{1}{2}\left(\frac{\partial A_\mu}{\partial x_\mu}\right)^2 \tag{4-2}$$

The total Lagrangian density, including the interaction, is given by Dirac's substitution discussed in the preceding section;

$$\mathcal{L} = -\bar{\psi}[\gamma(\partial_\mu - ieA_\mu) + m]\psi - \frac{1}{4}F_{\mu\nu}F_{\mu\nu} - \frac{1}{2}\left(\frac{\partial A_\mu}{\partial x_\mu}\right)^2 \tag{4-3}$$

or by a more symmetrical form

$$\mathcal{L} = -\frac{1}{4}\left[\bar{\psi}, \left(\gamma \frac{\partial}{\partial x} + m\right)\psi\right] - \frac{1}{4}\left[-\frac{\partial \bar{\psi}}{\partial x} \cdot \gamma + m\bar{\psi}, \psi\right]$$

$$-\frac{1}{4} F_{\mu\nu}F_{\mu\nu} - \frac{1}{2}\left(\frac{\partial A_\mu}{\partial x_\mu}\right)^2 + \frac{ie}{2} A_\mu\left[\bar{\psi}, \gamma_\mu \psi\right] \tag{4-4}$$

If we set $j_\mu = \frac{ie}{2}\left[\bar{\psi}, \gamma_\mu \psi\right]$, the interaction Lagrangian density is given by

$$\mathcal{L}_{int} = j_\mu A_\mu \tag{4-5}$$

The field equations have been given in Section 3-4. The Lagrangian density is invariant under the following gauge transformation:

gauge transformations of the first kind

$$A_\mu \rightarrow A_\mu + \frac{\partial \Lambda}{\partial x_\mu} \tag{4-6}$$

gauge transformations of the second kind

$$\psi \rightarrow e^{ie\Lambda}\psi, \qquad \bar{\psi} \rightarrow e^{-ie\Lambda}\bar{\psi} \tag{4-7}$$

where the gauge function Λ is a c number satisfying $\Box\Lambda = 0$.

Pion-Nucleon Systems

Since the pion-nucleon interactions are charge independent, they should be invariant under rotations in charge space. The nucleon has two components in charge space, denoted by p and n, and hence is a spinor in charge space. The pion has three charge states, π^+, π^0, and π^-, and hence is a vector in charge space. Therefore, the pion-nucleon interaction must be expressed by a scalar in charge space.

Let φ be the field operator for the charged π, or the operator destroying a π^+ or creating a π^-, and introduce φ_1 and φ_2 by

$$\varphi = 2^{-1/2}(\varphi_1 - i\varphi_2) \qquad \varphi^\dagger = 2^{-1/2}(\varphi_1 + i\varphi_2) \tag{4-8}$$

as discussed in Section 2-4. We call the π^0 operator φ_3. Then φ_1, φ_2, and φ_3 are real fields, and the free Lagrangian density for pions is given by

$$\mathcal{L}_\pi = -\sum_{\alpha=1}^{3} \frac{1}{2}\left[\left(\frac{\partial \varphi_\alpha}{\partial x_\lambda}\right)^2 + \mu^2 \varphi_\alpha^2\right] \tag{4-9}$$

In a similar way the Lagrangian density for nucleons is given by

$$\mathcal{L}_N = -\sum_{\alpha=1}^{2} \bar{\psi}_\alpha (\gamma\partial + M)\psi_\alpha \tag{4-10}$$

with the proton field designated by ψ_1 and the neutron by ψ_2.

Now we introduce the τ matrices:

$$\tau_1 = \begin{pmatrix} 0 & 1 \\ 1 & 0 \end{pmatrix} \qquad \tau_2 = \begin{pmatrix} 0 & -i \\ i & 0 \end{pmatrix} \qquad \tau_3 = \begin{pmatrix} 1 & 0 \\ 0 & -1 \end{pmatrix} \qquad \mathbf{1} = \begin{pmatrix} 1 & 0 \\ 0 & 1 \end{pmatrix} \tag{4-11}$$

which are equivalent to the Pauli spin matrices for rotations in coordinate space. The τ matrices are operators in charge space. We adopt the Einstein convention also for the charge indices, and consider two kinds of bilinear forms:

(1) vector in charge space, $\bar{\psi}\,\mathcal{O}\tau_\alpha\,\psi$

(2) scalar in charge space, $\bar{\psi}\,\mathcal{O}\mathbf{1}\,\psi$

for any Dirac matrix $\mathcal{O}$. Under an infinitesimal rotation in charge space represented by an infinitesimal rotation vector **a**, the fields are transformed in a manner analogous to a rotation in coordinate space.

$$\psi \to \psi + \frac{i}{2}(\mathbf{a}\cdot\boldsymbol{\tau})\psi \tag{4-12}$$

$$\boldsymbol{\varphi} \to \boldsymbol{\varphi} - \mathbf{a} \times \boldsymbol{\varphi} \tag{4-13}$$

Therefore, it is clear that the free Lagrangian is invariant under the infinitesimal rotations in charge space. The expressions that we called vector and scalar in charge space are transformed as vector components and a scalar, respectively, as their names indicate.

Since $\boldsymbol{\varphi}$ is transformed as a vector, the expression

$$\bar{\psi}\,\mathcal{O}\boldsymbol{\tau}\psi\cdot\boldsymbol{\varphi}$$

is invariant under rotations in charge space. We introduce τ and $\tau^{\dagger}$ by

$$\tau_1 = \tau + \tau^{\dagger} \qquad \tau_2 = i(\tau - \tau^{\dagger}) \tag{4-14}$$

or

$$\tau = \begin{pmatrix} 0 & 0 \\ 1 & 0 \end{pmatrix} \qquad \tau^{\dagger} = \begin{pmatrix} 0 & 1 \\ 0 & 0 \end{pmatrix} \tag{4-15}$$

then the scalar product above can be written as

$$\bar{\psi}\,\mathcal{O}\boldsymbol{\tau}\psi\cdot\boldsymbol{\varphi} = 2^{1/2}\,(\bar{\psi}\,\mathcal{O}\tau\psi\cdot\varphi^{\dagger} + \bar{\psi}\,\mathcal{O}\,\tau^{\dagger}\psi\cdot\varphi) + \bar{\psi}\,\mathcal{O}\,\tau_3\psi\cdot\varphi_3 \tag{4-16}$$

Note that φ decreases and $\varphi^{\dagger}$ increases the charge of the pion field, while τ decreases and $\tau^{\dagger}$ increases the charge of the nucleon.

There are various possible interactions, depending on the form of the Dirac matrix $\mathcal{O}$.

(1) For a scalar interaction, $\mathcal{O} = 1$ and the interaction Lagrangian density is

$$-G\,\bar{\psi}\boldsymbol{\tau}\psi\cdot\boldsymbol{\varphi}$$

where G is a constant, referred to as a coupling constant.

(2) For a pseudoscalar interaction, $\mathcal{O} = \gamma_5$ and the Lagrangian density is

$$-iG\,\bar{\psi}\gamma_5\,\boldsymbol{\tau}\,\psi\cdot\boldsymbol{\varphi}$$

The factor of i is to make the Lagrangian density Hermitian.

In addition to these two there may be derivative couplings:

(3) For a vector interaction, $\mathcal{O} = \gamma_\mu$ and the Lagrangian density is

$$-iF\,\bar{\psi}\gamma_\mu\,\boldsymbol{\tau}\psi\cdot\frac{\partial\boldsymbol{\varphi}}{\partial x_\mu}$$

(4) For an axial vector interaction, $\mathcal{O} = \gamma_\mu\gamma_5$ and the Lagrangian density is

$$-iF\,\bar{\psi}\gamma_\mu\gamma_5\,\boldsymbol{\tau}\,\psi\cdot\frac{\partial\boldsymbol{\varphi}}{\partial x_\mu}$$

Notice that the G's are dimensionless in the natural units, but the coupling constants, the F's, have the dimension of length.

The pion is known to be a pseudoscalar particle so we take

$$\mathcal{L}_{\text{int}} = -iG\,\bar{\psi}\gamma_5\,\boldsymbol{\tau}\,\psi\cdot\boldsymbol{\varphi} \tag{4-17}$$

From experimental observations, it is known that

$$\frac{G^2}{4\pi} \approx 15 \tag{4-18}$$

This is to be compared with $e^2/4\pi = 1/137$ for electromagnetic interactions It is clear that one cannot legitimately apply perturbation theory to pion-nucleon interactions because the perturbation expansion would involve increasing powers of 15.

4-2 DECOMPOSITION OF FIELD OPERATORS

In order to understand the physical meaning of covariant perturbation theory it is important to introduce the decomposition of *free* field operators into positive and negative frequency parts.

$$\varphi_\alpha(x) = \varphi_\alpha^{(+)}(x) + \varphi_\alpha^{(-)}(x) \tag{4-19}$$

where $\varphi_\alpha^{(+)}$ is the part of φ_α that corresponds to destruction operators, and $\varphi_\alpha^{(-)}$ corresponds to creation.

$$\varphi_\alpha^{(+)} \sim e^{ipx} \qquad \varphi_\alpha^{(-)} \sim e^{-ipx} \tag{4-20}$$

In many calculations we have to evaluate the vacuum expectation values of products of field operators. Then we use this decomposition:

$$\begin{aligned} &\langle 0 | \varphi(x_1)\varphi(x_2) \cdots \varphi(x_n) | 0 \rangle \\ &\quad = \langle 0 | [\varphi^{(+)}(x_1) + \varphi^{(-)}(x_1)] [\varphi^{(+)}(x_2) + \varphi^{(-)}(x_2)] \cdots \\ &\quad [\varphi^{(+)}(x_n) + \varphi^{(-)}(x_n)] | 0 \rangle \end{aligned}$$

In order to evaluate this expression we make use of the relations

$$\varphi^{(+)}(x) | 0 \rangle = 0 \qquad \langle 0 | \varphi^{(-)}(x) = 0 \tag{4-21}$$

by reordering the operators in the following way: We bring all (-) operators to the left and all (+) operators to the right, so that the final form looks like

$$\varphi^{(-)}(x_1)\varphi^{(-)}(x_2) \cdots \varphi^{(+)}(x_1') \varphi^{(+)}(x_2') \cdots \tag{4-22}$$

Of course the φ's are not commutative, so we have to add corrections arising from the nonvanishing commutators.

Let us illustrate this procedure by a simple example. Let $\varphi(x)$ be a neutral scalar field, and consider the following product:

$$\begin{aligned}\varphi(x)\varphi(y) &= [\varphi^{(+)}(x) + \varphi^{(-)}(x)]\,[\varphi^{(+)}(y) + \varphi^{(-)}(y)] \\ &= \varphi^{(+)}(x)\varphi^{(+)}(y) + \varphi^{(-)}(x)\varphi^{(-)}(y) + \varphi^{(-)}(x)\varphi^{(+)}(y) \\ &\quad + \varphi^{(+)}(x)\varphi^{(-)}(y)\end{aligned}$$

The first three terms are already in the right order, and the last term can be written as

$$\varphi^{(+)}(x)\varphi^{(-)}(y) = \varphi^{(-)}(y)\varphi^{(+)}(x) + [\varphi^{(+)}(x),\ \varphi^{(-)}(y)]$$

We have therefore transformed $\varphi(x)\varphi(y)$ into the form:

$$\begin{aligned}\varphi(x)\varphi(y) &= \varphi^{(+)}(x)\varphi^{(+)}(y) + \varphi^{(-)}(x)\varphi^{(-)}(y) + \varphi^{(-)}(x)\varphi^{(+)}(y) \\ &\quad + \varphi^{(-)}(y)\varphi^{(+)}(x) + [\varphi^{(+)}(x),\ \varphi^{(-)}(y)]\end{aligned} \tag{4-23}$$

We denote the sum of the first four terms by $:\varphi(x)\,\varphi(y):$ and call it the normal product or the well-ordered product. The last term is a c-number commutator and is given explicitly by

$$\begin{aligned}&[\varphi^{(+)}(x),\ \varphi^{(-)}(y)] \\ &= \left[\sum_p (2p_0V)^{-1/2}\ e^{ipx}\, a(p),\ \sum_q (2q_0V)^{-1/2}\ e^{-iqy}\ a^\dagger(q)\right] \\ &= \sum_p\sum_q (2p_0V)^{-1/2} \cdot (2q_0V)^{-1/2}\ e^{ipx-iqy}\ \delta_{p,q} \\ &= \sum_p (2p_0V)^{-1}\ e^{ip(x-y)} \\ &= \frac{V}{(2\pi)^3}\int d^3p\ (2p_0V)^{-1}\ e^{ip(x-y)} \\ &= i\,\frac{(-i)}{(2\pi)^3}\int\frac{d^3p}{2p_0}\ e^{ip(x-y)} = i\Delta^{(+)}(x-y)\end{aligned} \tag{4-24}$$

The function $\Delta^{(+)}$ has been introduced in Section 2-3. Then

$$\varphi(x)\varphi(y) = :\varphi(x)\varphi(y): + i\Delta^{(+)}(x - y) \tag{4-25}$$

By definition it is clear that the normal product is symmetric in the operators:

$$:\varphi(x)\varphi(y): = :\varphi(y)\varphi(x): \tag{4-26}$$

and that the vacuum expectation value of the normal product vanishes.

$$\langle 0 |:\cdots:| 0\rangle = 0 \tag{4-27}$$

Therefore,

$$\langle 0 |\varphi(x)\varphi(y)| 0\rangle = i\Delta^{(+)}(x - y) \tag{4-28}$$

This means that the amplitude corresponding to creation of a quantum at y and subsequent destruction of it at x is given by $i\Delta^{(+)}(x - y)$. Then we can prove

$$\langle 0|\varphi(x_1)\varphi(x_2)\cdots\varphi(x_n)|0\rangle$$

$$= \sum_{\substack{k_1<k_2\\ k_3<k_4\\ \vdots}} i\Delta^{(+)}(x_{k_1} - x_{k_2})\, i\Delta^{(+)}(x_{k_3} - x_{k_4})\cdots i\Delta^{(+)}(x_{k_{n-1}} - x_{k_n}) \tag{4-29}$$

where $(k_1, k_2, \ldots, k_n)$ is a permutation of $(1, 2, \ldots, n)$ and n is assumed to be even. The vacuum expectation value of the product of an odd number of free field operators vanishes since the state $\varphi(x_1)\varphi(x_2)\ldots\varphi(x_n)|0>$ with n being odd represents a superposition of states of odd numbers of quanta.

The concept of the normal product has an application to quantum electrodynamics. The charge current density for a fermion field has been given by

$$j_\mu = \frac{1}{2}\, ie\, [\bar{\psi}, \gamma_\mu\psi] \tag{4-30}$$

The j_μ was modified to this form from $j_\mu = ie\bar{\psi}\gamma_\mu\psi$ because $\langle 0 |ie\bar{\psi}\gamma_\mu\psi| 0\rangle \neq 0$, thus giving rise to a nonvanishing charge in the vacuum. If we use the normal product we can write the current density as

$$j_\mu = ie:\bar{\psi}\gamma_\mu\psi: \tag{4-31}$$

so that its vacuum expectation value vanishes. In the discussion that follows we sometimes shall write $j_\mu = ie\bar{\psi}\gamma_\mu\psi$, but then it should always be understood in the sense of the normal product.

4-3 COVARIANT INTEGRATION OF THE TOMONAGA-SCHWINGER EQUATION

Let us discuss the integration of the T-S equation in perturbation theory.

$$i\,\frac{\delta U[\sigma, \sigma_0]}{\delta\sigma(x)} = \mathcal{H}_{int}(x)\; U[\sigma, \sigma_0] \tag{4-32}$$

In order to determine U let us set up a family of spacelike surfaces $\{\sigma\}$ in such a way as to fill the space-time domain between σ and σ_0. For every space-time point x in this domain there is one and only one spacelike surface $\sigma(x)$ belonging to this family. Then the T-S equation can be converted into an integral equation by setting

$$U[\sigma, \sigma_0] = 1 - i\int_{\sigma_0}^{\sigma} d^4x'\; \mathcal{H}_{int}(x')U[\sigma(x'), \sigma_0] \tag{4-33}$$

That this equation is equivalent to the original T-S equation can be seen, since (1) clearly, $U[\sigma_0, \sigma_0] = 1$, and (2)

$$\begin{aligned} i\,\frac{\delta}{\delta\sigma(x)}\, U[\sigma, \sigma_0] &= \frac{\delta}{\delta\sigma(x)}\int_{\sigma_0}^{\sigma} d^4x'\; \mathcal{H}_{int}(x') \\ &\qquad \times U[\sigma(x'), \sigma_0] \\ &= \frac{1}{dV}\left(\int_{\sigma_0}^{\sigma(x+dx)} - \int_{\sigma_0}^{\sigma(x)}\right) d^4x'\mathcal{H}_{int}(x') \\ &\qquad \times U[\sigma(x'), \sigma_0] \\ &= \frac{1}{dV}\int_{\sigma(x)}^{\sigma(x+dx)} d^4x'\; \mathcal{H}_{int}(x')\,U[\sigma(x'), \sigma_0] \\ &= \mathcal{H}_{int}(x)\,U[\sigma(x), \sigma_\sigma] \end{aligned} \tag{4-34}$$

where

$$dV = \int_{\sigma(x)}^{\sigma(x+dx)} d^4x \tag{4-35}$$

By iterating the integral equation we can express U as

$$U[\sigma, \sigma_0] = 1 - i \int_{\sigma_0}^{\sigma} d^4x' \mathcal{H}_{int}(x')$$

$$+ (-i)^2 \int_{\sigma_0}^{\sigma} d^4x' \mathcal{H}_{int}(x') \int_{\sigma_0}^{\sigma(x')} d^4x'' \mathcal{H}_{int}(x'') + \ldots$$

$$= 1 + \sum_{n=1}^{\infty} (-i)^n \int_{\sigma_0}^{\sigma} d^4x_1 \int_{\sigma_0}^{\sigma(x_1)} d^4x_2 \ldots$$

$$\int_{\sigma_0}^{\sigma(x_{n-1})} d^4x_n \mathcal{H}_{int}(x_1) \mathcal{H}_{int}(x_2) \ldots \mathcal{H}_{int}(x_n) \tag{4-36}$$

If we assume, in particular, that σ_0 and σ and all the members of the family $\{\sigma\}$ are flat surfaces parametrized by t = constant, we get

$$U(t, t_0) = 1 - i \int_{t_0}^{t} d^4x_1 \mathcal{H}_{int}(x_1)$$

$$+ (-i)^2 \int_{t_0}^{t} d^4x_1 \mathcal{H}_{int}(x_1) \int_{t_0}^{t_1} d^4x_2 \mathcal{H}_{int}(x_2) + \ldots \tag{4-37}$$

This can be written, using $H(t) = \int d^3x \, \mathcal{H}_{int}(x)$, as

$$U(t, t_0) = 1 - i \int_{t_0}^{t} dt_1 \, H(t_1) + (-i)^2 \int_{t_0}^{t} dt_1 \, H(t_1) \int_{t_0}^{t_1} dt_2 \, H(t_2) + \ldots \qquad (4\text{-}38)$$

This is precisely the form we get in noncovariant time-dependent perturbation theory. In the following discussions, however, we shall find that this form of the solution is not very convenient; let us transform it into another form, introduced by Dyson.

First we introduce the so-called P symbol, or *time-ordering operator P*.

$$\begin{aligned} P[A(t_1)B(t_2)] &= A(t_1)B(t_2) \qquad \text{for} \quad t_1 > t_2 \\ &= B(t_2)A(t_1) \qquad \text{for} \quad t_2 > t_1 \end{aligned} \qquad (4\text{-}39)$$

and more generally

$$P[A(t_1)B(t_2) \ldots Z(t_n)] = A(t_1)B(t_2) \ldots Z(t_n) \qquad \text{for} \quad t_1 > t_2 > \ldots > t_n \quad \text{etc.} \qquad (4\text{-}40)$$

The resulting expression is called a time-ordered product.

The expression $P[H(t_1) \ldots H(t_n)]$ is now a symmetric function of $t_1, \ldots, t_n$. If $f(x_1, \ldots, x_n)$ is a symmetric function, we may write

$$\int_a^b dx_1 \int_a^{x_1} dx_2 \ldots \int_a^{x_{n-1}} dx_n \, f(x_1, \ldots, x_n) = \frac{1}{n!} \int_a^b dx_1 \int_a^b dx_2 \ldots \int_a^b dx_n \, f(x_1, \ldots, x_n) \qquad (4\text{-}41)$$

By using Eq. (4-41) we can derive Dyson's formula; we can write

$$\int_{t_0}^{t} dt_1 \int_{t_0}^{t_1} dt_2 \ldots \int_{t_0}^{t_{n-1}} dt_n\, H(t_1) \ldots H(t_n)$$

$$= \int_{t_0}^{t} dt_1 \int_{t_0}^{t_1} dt_2 \ldots \int_{t_0}^{t_{n-1}} dt_n\, P[H(t_1) \ldots H(t_n)] \qquad (4\text{-}42)$$

Since the integrand is a symmetric function of $t_1, t_2, \ldots, t_n$, we can apply the integral formula (4-41), and (4-42) becomes

$$\frac{!}{n!} \int_{t_0}^{t} dt_1 \int_{t_0}^{t} dt_2 \ldots \int_{t_0}^{t} dt_n\, P\left[H(t_1) \ldots H(t_n)\right] \qquad (4\text{-}43)$$

Therefore, we may write

$$U(t, t_0) = 1 + \sum_{n=1}^{\infty} \frac{(-i)^n}{n!} \int_{t_0}^{t} \ldots \int_{t_0}^{t} dt_1 \ldots dt_n$$
$$\times\, P[H(t_1) \ldots H(t_n)] \qquad (4\text{-}44)$$

We may generalize this formula to the solution of the T-S equation, obtaining the so-called Dyson formula[1]

$$U[\sigma, \sigma_0] = 1 + \sum_{n=1}^{\infty} \frac{(-i)^n}{n!} \int_{\sigma_0}^{\sigma} \ldots \int_{\sigma_0}^{\sigma} d^4x_1 \ldots d^4x_n$$
$$\times\, P[\mathcal{H}_{int}(x_1) \ldots \mathcal{H}_{int}(x_n)] \qquad (4\text{-}45)$$

In this expression we have to be careful about the definition of the P symbol. It is defined with reference to a family of spacelike surfaces $\{\sigma\}$ filling up the space-time domain between σ and σ_0. When the surface $\sigma(x_1)$ lies in the future of $\sigma(x_2)$, let us denote the fact by writing $\sigma(x_1) > \sigma(x_2)$. Then the time-ordered product is defined by (see Fig. 4-1)

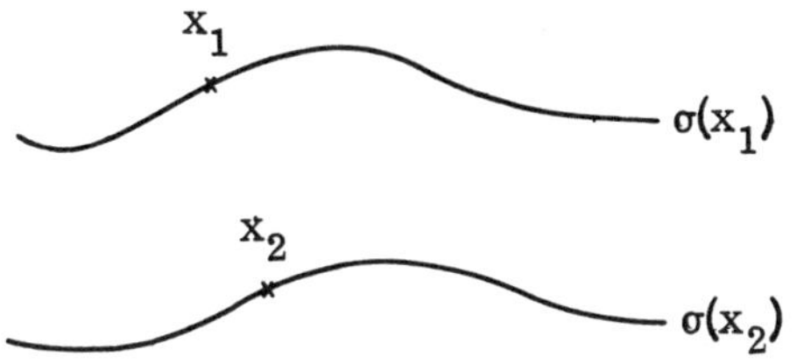

Figure 4-1. Illustration of a family of spacelike surfaces $\{\sigma\}$. For a given point x, there is always one and only one surface $\sigma(x)$, being a member of this family and passing through this point x.

$$P[\mathcal{H}_{int}(x_1)\,\mathcal{H}_{int}(x_2)] = \mathcal{H}_{int}(x_1)\,\mathcal{H}_{int}(x_2) \qquad \text{for} \quad \sigma(x_1) > \sigma(x_2)$$

$$= \mathcal{H}_{int}(x_2)\,\mathcal{H}_{int}(x_1) \qquad \text{for} \quad \sigma(x_2) > \sigma(x_1) \tag{4-46}$$

In order to have a unique meaning, however, U must be independent of the choice of a family. This can be proved by showing that the value of the P product itself does not depend on the choice of a family.

Take two families of spacelike surfaces and assume that $\sigma(x_1) > \sigma(x_2)$ in the family $\{\sigma\}_1$ and that $\sigma(x_2) > \sigma(x_1)$ in the second family $\{\sigma\}_2$x. Then $P[\mathcal{H}_{int}(x_1)\,\mathcal{H}_{int}(x_2)]$ is equal to $\mathcal{H}_{int}(x_1)\,\mathcal{H}_{int}(x_2)$ with reference to the first family, and is equal to $\mathcal{H}_{int}(x_2)\,\mathcal{H}_{int}(x_1)$ with reference to the second family.

However, such a situation can happen only when x_1 and x_2 are separated by a spacelike distance, as shown in Fig. 4-2. Then the integrability condition implies

$$[\mathcal{H}_{int}(x\ ),\ \mathcal{H}_{int}(x_2)] = 0, \qquad \text{for} \quad (x_1 - x_2)^2 > 0 \tag{4-47}$$

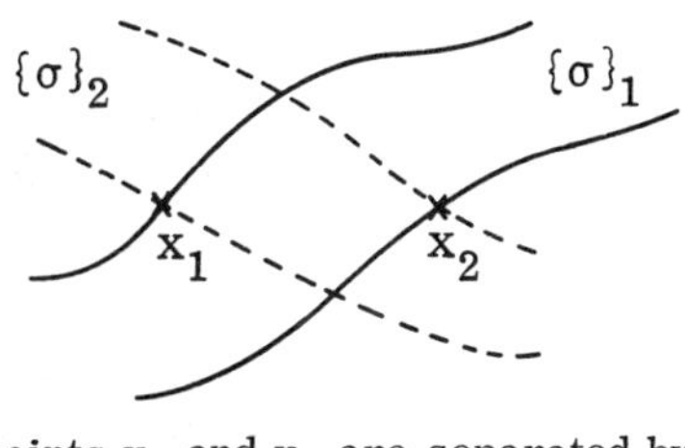

Figure 4-2. When two points x_1 and x_2 are separated by a spacelike distance, the future-past assignment depends on the choice of a family of spacelike surfaces. For the choice $\{\sigma\}_1$ the point x_1 lies in the future of x_2, but for the choice $\{\sigma\}_2$ the situation is reversed.

and therefore

$$\mathcal{H}_{int}(x_1)\,\mathcal{H}_{int}(x_2) = \mathcal{H}_{int}(x_2)\,\mathcal{H}_{int}(x_1) \tag{4-48}$$

Thus we learn that the integrability condition guarantees the family independence of the transformation functional U.

Finally let us show that the series expansion or Dyson's expression for U really satisfies the T-S equation. We use the formula valid for a symmetric function f:

$$\frac{d}{db}\int_a^b \cdots \int_a^b dx_1 \ldots dx_n\, f(x_1, \ldots, x_n)$$

$$= n\int_a^b \cdots \int_a^b dx_1 \ldots dx_{n-1}\, f(x_1, \ldots, x_{n-1}, b) \tag{4-49}$$

With the help of this formula we differentiate the series expansion of U.

$$i\,\frac{\delta}{\delta\sigma(x)}\,U[\sigma, \sigma_0] = \sum_{n=1}^{\infty}\frac{(-i)^{n-1}}{n!}\,\frac{\delta}{\delta\sigma(x)}\int_{\sigma_0}^{\sigma}\cdots\int_{\sigma_0}^{\sigma} d^4x_1 \ldots$$

$$d^4x_n\, P\,[\mathcal{H}_{int}(x_1) \ldots \mathcal{H}_{int}(x_n)]$$

$$= \sum_{n=1}^{\infty}\frac{(-i)^{n-1}}{(n-1)!}\int_{\sigma_0}^{\sigma}\cdots\int_{\sigma_0}^{\sigma} d^4x_1 \ldots d^4x_{n-1}$$

$$P[\mathcal{H}_{int}(x_1) \ldots \mathcal{H}_{int}(x_{n-1})\,\mathcal{H}_{int}(x)]$$

$$= \mathcal{H}_{int}(x)\sum_{n'=0}^{\infty}\frac{(-i)^{n'}}{n'!}\int_{\sigma}^{\sigma}\cdots\int_{\sigma}^{\sigma} d^4x_1 \ldots$$

$$d^4x_{n'}\, P[\mathcal{H}_{int}(x_1) \ldots \mathcal{H}_{int}(x_{n'})]$$

$$= \mathcal{H}_{int}(x)\,U[\sigma, \sigma_0] \tag{4-50}$$

where $n' = n - 1$ and use has been made of the fact that the points $x_1, \ldots, x_n$ are in the domain enclosed by σ_0 and σ, while x is on σ, so that

$$P[\mathcal{H}_{int}(x_1) \ldots \mathcal{H}_{int}(x_{n-1}) \mathcal{H}_{int}(x)]$$

$$= \mathcal{H}_{int}(x)\, P[\mathcal{H}_{int}(x_1) \ldots (x_{n-1})] \tag{4-51}$$

Thus we have established the series expansion of the solution of the T-S equation. In particular, the S matrix is given by

$$S = U(\infty, -\infty) = 1 + \sum_{n=1}^{\infty} \frac{(-i)^n}{n!} \int_{-\infty}^{\infty} \ldots \int_{-\infty}^{\infty} d^4x_1 \ldots d^4x_n\, P[\mathcal{H}_{int}(x_1)$$

$$\ldots \mathcal{H}_{int}(x_n)] \tag{4-52}$$

This is called Dyson's formula for the S matrix.[1]

4-4 PERTURBATION EXPANSION OF HEISENBERG OPERATORS

In Section 3-4 we have shown that the Heisenberg operator φ_α^{in} as

$$\varphi_\alpha(x) = U[\sigma, -\infty]^{-1} \varphi_\alpha^{in}(x) U[\sigma, -\infty] \tag{4-53}$$

where x is on the spacelike hypersurface σ. In this section we use a family of flat hypersurfaces or hyperplanes parametrized by t = constant.

It is not convenient, however, to insert here the series expansion of U. Instead we start from the differential equation for U, the single-time Schrödinger equation.

$$i \frac{\partial}{\partial t} U(t, -\infty) = \int d^3x\, \mathcal{H}^{in}(x, t)\, U(t, -\infty) \tag{4-54}$$

The $\mathcal{H}^{in}$ means that the interaction Hamiltonian is expressed in terms of incoming fields, for example,

$$\mathcal{H}^{in}(x) = -j_\mu^{in}(x) A_\mu^{in}(x) \tag{4-55}$$

What we have to evaluate is the expression

$$\varphi_\alpha(x) = U(t, -\infty)^{-1} \varphi_\alpha^{in}(x, t) U(t, -\infty) \tag{4-56}$$

and for this purpose we begin with a mathematical manipulation. We define $\mathcal{O}(t_0)$ as

$$\mathcal{O}(t_0) = U(t, t_0)^{-1} \mathcal{O}\, U(t, t_0) \tag{4-57}$$

where $U(t, t_0)$ is the transformation functional satisfying

$$\begin{aligned} i \frac{d}{dt_0} U(t, t_0)^{-1} &= i \frac{d}{dt_0} U(t_0, t) \\ &= H(t_0)\, U(t_0, t) \\ &= H(t_0)\, U(t, t_0)^{-1} \end{aligned} \tag{4-58}$$

Therefore, $\mathcal{O}(t_0)$ satisfies

$$i \frac{d}{dt} \mathcal{O}(t_0) = [H(t_0), \mathcal{O}(t_0)] \tag{4-59}$$

We convert this equation into an integral form,

$$\mathcal{O}(t_0) = \mathcal{O} - i \int_{t_0}^{t} dt_0' \, [\mathcal{O}(t_0'), H(t_0')] \tag{4-60}$$

and by iteration we find

$$\begin{aligned} \mathcal{O}(t_0) = {} & \mathcal{O} - i \int_{t_0}^{t} dt_0' \, [\mathcal{O}, H(t_0')] \\ & + (-i)^2 \int_{t_0}^{t} dt_0' \int_{t_0'}^{t} dt_0'' \, [[\mathcal{O}, H(t_0'')], H(t_0')] + \cdots \end{aligned} \tag{4-61}$$

By changing the variables of integration we may write

$$\mathcal{O}(t_0) = \mathcal{O} - i \int_{t_0}^{t} dt' \, [\mathcal{O}, H(t')]$$

$$+ (-i)^2 \int_{t_0}^{t} dt' \int_{t_0}^{t'} dt'' \, [[\mathcal{O}, H(t')], H(t'')] + \ldots \tag{4-62}$$

Identifying $\mathcal{O}(t_0)$, $\mathcal{O}$, and t_0 with $\varphi_\alpha(x)$, $\varphi_\alpha^{in}(x)$, and $-\infty$, respectively, we find

$$\varphi_\alpha(x) = U(t, -\infty)^{-1} \, \varphi_\alpha^{in}(\mathbf{x}, t) \, U(t, -\infty)$$

$$= \varphi_\alpha^{in}(\mathbf{x}, t) - i \int_{-\infty}^{t} dt' \, [\varphi_\alpha^{in}(\mathbf{x}, t), H(t')]$$

$$+ (-i)^2 \int_{-\infty}^{t} dt' \int_{-\infty}^{t'} dt'' \, [[\varphi_\alpha^{in}(\mathbf{x}, t), H(t')], H(t'')] + \ldots$$

$$= \varphi_\alpha^{in}(\mathbf{x}, t) + \sum_{n=1}^{\infty} (-i)^n \int_{-\infty}^{t} dt' \int_{-\infty}^{t'} dt'' \ldots \int_{-\infty}^{t^{(n-1)}} dt$$

$$[\ldots[\varphi_\alpha^{in}(\mathbf{x}, t), H(t')], H(t'')], \ldots, H(t^{(n)})] \tag{4-63}$$

where

$$H(t) = \int d^3x \, \mathcal{H}^{in}(\mathbf{x}, t).$$

We define the retarded or R product of operators as

$$R[A(x):A(x_1), \ldots, A(x_n)] = (-i)^n [\ldots [A(x), A(x_1')] \ldots$$

$$A(x_n')] \qquad \text{for} \quad t > t_1' > \ldots t_n' \tag{4-64}$$

where $(t'_1, \ldots, t'_n)$ is a permutation of $(t_1, \ldots, t_n)$.
If t is smaller than any t_k of $t_1, \ldots, t_n$, then $R[A(x): \ldots] = 0$.
Consequently,

$$(-i)^n \int_{-\infty}^{t} dt' \int_{-\infty}^{t'} dt'' \ldots \int_{-\infty}^{t^{(n-1)}} dt^{(n)} \quad [\ldots[\varphi_\alpha^{in}(\mathbf{x}, t),$$

$$H(t')], H(t'')] \ldots, H(t^{(n)})]$$

$$= \int_{-\infty}^{t} dt' \int_{-\infty}^{t'} dt'' \ldots \int_{-\infty}^{t^{(n-1)}} dt^{(n)} \; R[\varphi_\alpha^{in}(\mathbf{x}, t):H(t'), \ldots, H(t^{(n)})]$$

(4-65)

Since R makes the integrand symmetric in $t_1, \ldots, t_n$, the expression (4-65) equals

$$\frac{1}{n!} \int_{-\infty}^{t} dt' \int_{-\infty}^{t} dt'' \ldots \int_{-\infty}^{t} dt^{(n)} \; R[\varphi_\alpha^{in}(\mathbf{x}, t):H(t'), \ldots, H(t^{(n)})]$$

(4-66)

Furthermore, the integrand vanishes if any t_k of $t_1, \ldots, t_n$ becomes larger than t, so that we may write the expression (4-66) as

$$\frac{1}{n!} \int_{-\infty}^{\infty} dt' \int_{-\infty}^{\infty} dt'' \ldots \int_{-\infty}^{\infty} dt^{(n)} \; R[\varphi_\alpha^{in}(\mathbf{x}, t):H(t'), \ldots, H(t^{(n)})]$$

$$= \frac{1}{n!} \int_{-\infty}^{\infty} \ldots \int_{-\infty}^{\infty} d^4x_1 \ldots d^4x_n \, R[\varphi_\alpha^{in}(x): \mathcal{H}^{in}(x_1), \ldots, \mathcal{H}^{in}(x_n)]$$

(4-67)

Therefore, corresponding to Dyson's formula for U, we get

$$\varphi_\alpha(x) = \varphi_\alpha^{in}(x) + \sum_{n=1}^{\infty} \frac{1}{n!} \int_{-\infty}^{\infty} \cdots \int_{-\infty}^{\infty} d^4x_1 \ldots d^4x_n \, R[\varphi_\alpha^{in}(x) : \mathcal{H}^{in}(x_1), \ldots, \mathcal{H}^{in}(x_n)] \tag{4-68}$$

It should be mentioned here again that the *definition* of the R product depends on the choice of a family of spacelike hypersurfaces, but the *value* of the R product itself does not depend on that choice provided the integrability condition is satisfied. This is a situation very similar to the case of the P product.

4-5 EVALUATION OF THE S MATRIX – WICK'S THEOREM

In this and succeeding sections we shall use field operators in the sense of incoming field operators, unless stated otherwise. In order to evaluate cross sections and other observable quantities, we must evaluate S-matrix elements.

$$\begin{aligned} S_{ba} &= (\Phi_b, S\Phi_a) \\ &= (\Phi_b, \Phi_a) + \sum_{n=1}^{\infty} \frac{(-i)^n}{n!} \int_{-\infty}^{\infty} \cdots \int_{-\infty}^{\infty} d^4x_1 \ldots d^4x_n \\ &\quad \times (\Phi_b, P[(\mathcal{H}_{int}(x_1) \ldots \mathcal{H}_{int}(x_n)] \Phi_a) \end{aligned} \tag{4-69}$$

Therefore, we have to establish rules to evaluate matrix elements of P products of operators.

In quantum electrodynamics we use.

$$\mathcal{H}_{int}(x) = -j_\mu(x) A_\mu(x). \tag{4-70}$$

The P product can be factored:

$$\begin{aligned} P[\mathcal{H}_{int}(x_1) \ldots \mathcal{H}_{int}(x_n)] &= (-1)^n P[j_\mu(x_1) \ldots j_\sigma(x_n)] \\ &\quad \times P[A_\mu(x_1) \ldots A_\sigma(x_n)] \end{aligned} \tag{4-71}$$

since j's and A's always commute in the interaction representation. Decomposing the state vectors into direct products of electron and photon state vectors,

$$\Phi_a = \Phi_a^{el} \times \Phi_a^{ph} \tag{4-72}$$

$$\Phi_b = \Phi_b^{el} \times \Phi_b^{ph} \tag{4-73}$$

we may write each term in the power series expansion of S as

$$(\Phi_b, P[\mathcal{H}_{int}(x_1) \ldots \mathcal{H}_{int}(x_n)]\Phi_a)$$

$$= (-1)^n \left(\Phi_b^{el}, P[j_\mu(x_1) \ldots j_\sigma(x_n)]\, \Phi_a^{el}\right)$$

$$\times (\Phi_b^{ph}, P[A_\mu(x_1) \ldots A_\sigma(x_n)]\, \Phi_a^{ph}) \tag{4-74}$$

Let us first discuss the second factor. The photon state vector can be written as

$$\Phi_a^{ph} = a^\dagger(k_1, \lambda_1) \ldots a^\dagger(k_i, \lambda_i)\, \Phi_0 \tag{4-75a}$$

$$\Phi_b^{ph} = a^\dagger(k_1', \lambda_1') \ldots a^\dagger(k_f', \lambda_f')\Phi_0 \tag{4-75b}$$

We expand $P[A_\mu(x_1) \ldots A_\sigma(x_n)]$ into a sum of normal products and pick up those terms which involve i destruction operators and f creation operators. For simplicity we omit the component subscripts on the A's and the polarization indices, λ's in the a's and $a^\dagger$'s.

The expression we have to evaluate is

$$(\Phi_o, a(k_1') \ldots a(k_f')\, P[A(x_1) \ldots A(x_n)]$$

$$a^\dagger(k_1) \ldots a^\dagger(k_i)\Phi_o) \tag{4-76}$$

We try to eliminate all the a's and $a^\dagger$'s from this expression. To do this we first write (4-76) as

$$\left(\Phi_o, \left[a(k_1') \ldots a(k_f') \, P \, [A(x_1) \ldots A(x_n)] \right. \right.$$

$$\left. \left. a^\dagger (k_1) \;\ldots,\; a^\dagger (k_i) \right] \Phi_o \right) \tag{4-77}$$

which is valid since

$$(\Phi_o ,\; a^\dagger (k_i) \ldots \Phi_o) = (a (k_i) \, \Phi_o, \;\ldots\; \Phi_o) = 0 \tag{4-78}$$

To evaluate the commutator we use the formulas

$$[a (k', \lambda'), \; a^\dagger (k, \lambda)] = \delta_{k', \, k} \; \delta_{\lambda\lambda'} \tag{4-79}$$

and

$$[A_\mu(x), \; a^\dagger (k, \lambda)] = (2k_0 V)^{-1/2} \; e_\mu^{(\lambda)} \, e^{ikx}$$

$$= \langle 0 | \, A_\mu(x) \, | \; k, \; \lambda \rangle \tag{4-80}$$

with the conjugate formula for a. Use of (4-79) gives rise to terms in which the numbers of a's and $a^\dagger$'s are each reduced by one, but which retain the same form as in the original expression (4-76). These terms occur because the initial and final photon state vectors contain certain photons with the same **k** and λ; such photons do not participate in the physical processes of interest. Therefore, we assume that no pair of initial and final photons has the same k and λ, and hence the a's and $a^\dagger$'s all commute. Nonvanishing terms then arise from the application only of the second formula, (4-80), which reduces the same numbers of A's and $a^\dagger$'s or a's.

Thus we find

$$\left(\Phi_o, a(k'_1)\ldots a(k'_f)\,P[A(x_1)\ldots A(x_n)]\,a^\dagger(k_1)\ldots a^\dagger(k_i)\Phi_o\right)$$

$$=\sum_{j=1}^{n}\left(\Phi_o,\ a(k'_1)\ldots a(k'_f)\,P[A(x_1)\ldots A(x_{j-1})A(x_{j+1})\right.$$

$$\left.\ldots A(x_n)]\,a^\dagger(k_1)\ldots a^\dagger(k_{i-1})\Phi_o\right)$$

$$\times \langle 0|A(x_j)|k_i\rangle$$

$$\ldots$$

$$=\sum_{\text{comb}}\langle k'_1|A(x'_1)|0\rangle \quad \langle k'_f|A(x'_f)|0\rangle$$

$$\times \langle 0|P[A(x'_{f+i+1})\ldots A(x'_n)]|0\rangle$$

$$\times \langle 0|A(x'_{f+1})|k_1\rangle\ldots\langle 0|A(x'_{f+i})|k_i\rangle \qquad (4\text{-}81)$$

where $(x'_1, \ldots, x'_n)$ is a permutation of $(x_1, \ldots, x_n)$, and the summation has to be taken over all possible combinations of n variables divided into groups of f, n-f-i, and i, corresponding to the factors in (4-81). This expression can be written as

$$\sum_{\text{comb}}\langle b|\,:A(x'_1)\ldots A(x'_{i+f}):|a\rangle\,\langle 0|P[A(x'_{i+f+1})\ldots A(x'_n)]|0\rangle \qquad (4\text{-}82)$$

However, since a and b are arbitrary states and the vacuum expectation value is a c number, a general expression for the P product is

$$P[A(x_1)\ldots A(x_n)] = \sum_{\text{comb}} :A(x'_1)\ldots A(x'_j):\langle 0|P[A(x'_{j+1})$$

$$\ldots A(x'_n)]|0\rangle \qquad (4\text{-}83)$$

The sum should be taken over all possible ways to divide $x_1, \ldots, x_n$ into two groups, one in the normal product and the other in the vacuum expectation value. Although we have assumed that there is no common single photon state involved in both a and b, this restriction can be lifted in (4-83) where, namely, the operator identity is valid when its matrix element is taken between any pair of states.

Our next problem is the evaluation of the vacuum expectation values of P products. We begin with the simplest case.

$$
\begin{aligned}
&\langle 0 | P[A_\lambda(x)\, A_\mu(y)] | 0 \rangle \\
&= \theta(x_0 - y_0)\, \langle 0 | A_\lambda(x) A_\mu(y) | 0 \rangle + \theta(y_0 - x_0)\, \langle 0 | A_\mu(y)\, A_\lambda(x) | 0 \rangle \\
&= \frac{1}{2} \langle 0 | \left\{ A_\lambda(x),\, A_\mu(y) \right\} | 0 \rangle + \frac{1}{2}\, \epsilon(x_0 - y_0)\, \langle 0 | [A_\lambda(x)\, ,\, A_\mu(y)] | 0 \rangle \\
&= \frac{1}{2} \delta_{\lambda\mu}\, D^{(1)}(x - y) - i\delta_{\lambda\mu}\, \bar{D}(x - y) \\
&= \delta_{\lambda\mu}\, D_F(x - y)
\end{aligned}
\tag{4-84}
$$

where D_F denotes the Δ_F function for zero mass defined in Section 2-3. We also can write $D_F(x - y)$ in the form

$$
\begin{aligned}
D_F(x) &= i\theta(x_0)\, D^{(+)}(x) + i\theta(-x_0) D^{(+)}(-x) \\
&= \frac{-i}{(2\pi)^4} \int d^4k \, \frac{e^{ikx}}{k^2 - i\epsilon}
\end{aligned}
\tag{4-85}
$$

We can extend this result to prove the following theorem [2]:

$$
\langle 0 | P\, [A(x_1) \ldots A(x_n)] | 0 \rangle = \sum_{\text{comb}} \Delta_F(x'_1 - x'_2) \ldots \Delta_F(x'_{n-1} - x'_n), \quad \text{if} \quad n \text{ is even} \tag{4-86a}
$$

$$
= 0 \qquad \text{if} \quad n \text{ is odd} \tag{4-86b}
$$

The summation should be taken over all possible ways of dividing n variables into n/2 pairs. The series $(x'_1, \ldots, x'_n)$ is a permutation of $(x_1, \ldots, x_n)$.

In order to prove this theorem, assume a time ordering, $x'_1 > x'_2 > \ldots > x'_n$, then the left-hand side of (4-86) is given by

$$\langle 0|P\,[A(x_1)\ldots A(x_n)]\,|0\rangle = \langle 0|A(x'_1)\ldots A(x'_n)|0\rangle \tag{4-87}$$

Given $\Delta_F(x) = i\Delta^{(+)}(x)\theta(x_0) + i\Delta^{(+)}(-x)\theta(-x_0)$, and $x'_{k_1} > x'_{k_2}$ if $k_1 < k_2$, we find

$$\Delta_F(x'_{k_1} - x'_{k_2}) = i\Delta^{(+)}(x'_{k_1} - x'_{k_2}) \qquad \text{for} \quad k_1 < k_2 \tag{4-88}$$

The right-hand side of the theorem is given by

$$\sum_{\substack{k_1<k_2\\ \vdots}} i\Delta^{(+)}(x'_{k_1} - x'_{k_2})\ldots i\Delta^{(+)}(x'_{k_{n-1}} - x'_{k_n}) \tag{4-89}$$

As stated in Section 4-2 this is equal to

$$\langle 0|A(x'_1)\ldots A(x'_n)|0\rangle \tag{4-90}$$

which is equal to the left-hand side of the theorem.

Combining this theorem, (4-86), with the previous one for the P product, (4-83), we find

$$P\,[A(x_1)\ldots A(x_n)] = \sum :A(x'_1)\ldots A(x'_j):\,\Delta_F(x'_{j+1} - x'_{j+2})\ldots$$
$$\Delta_F(x'_{n-1} - x'_n) \tag{4-91}$$

Wick [3] used dots on the operators as the so-called contraction symbol to express Δ_F:

$$A^{\cdot}(x)\;A^{\cdot}(y) = \Delta_F(x - y)$$

Now we have demonstrated essentially the following formula, called Wick's theorem:

$$
\begin{aligned}
P[A(x_1)\ldots A(X_n)] = {} & :A(x_1)\ldots A(x_n): && \text{(no contraction)} \\
& + \sum_{j\neq k} :A(x_1)\ldots A^{\cdot}(x_j)\ldots A^{\cdot}(x_k)\ldots A(x_n): && \text{(one contraction)} \\
& + \sum_{j\neq k}\sum_{\ell\neq m} :A(x_1)\ldots A^{\cdot}(x_j)\ldots A^{\cdot}(x_k)\ldots \\
& \quad A^{\cdot\cdot}(x_\ell)\ldots A^{\cdot\cdot}(x_m)\ldots A(x_n): && \text{(two contractions)} \\
& + \ldots && \text{(more contractions)}
\end{aligned}
\tag{4-92}
$$

We can find similar formulas for the electron field, but for that case it is more convenient to introduce the T symbol, which incorporates the anticommutativity of the fermion fields.

Suppose A and B represent fermion operators, either ψ or $\overline{\psi}$. In contrast to the P product

$$
\begin{aligned}
P[A(t_1)B(t_2)] &= A(t_1)B(t_2) \qquad t_1 > t_2 \\
&= B(t_2)A(t_1) \qquad t_2 > t_1
\end{aligned}
\tag{4-93}
$$

we define the T product as

$$
\begin{aligned}
T[A(t_1)B(t_2)] &= A(t_1)B(t_2) \qquad t_1 > t_2 \\
&= -B(t_2)A(t_1) \qquad t_2 > t_1
\end{aligned}
\tag{4-94}
$$

If we take a product of many ψ's, we have the formula

$$\psi(x_1)\psi(x_2)\ldots\psi(x_n) = \epsilon_p\, \psi(x_1')\psi(x_2')\ldots\psi(x_n') \tag{4-95}$$

where $(x_1', \ldots, x_n')$ is a permutation of $(x_1, \ldots, x_n)$, and ϵ_P is equal to +1 or -1 if the permutation is even or odd, respectively. This formula results from the anticommutativity of the fermion fields. The T product keeps this property and we have

$$T[\psi(x_1)\psi(x_2)\ldots\psi(x_n)] = \epsilon_P\, T[\psi(x_1')\psi(x_2')\ldots\psi(x_n')] \tag{4-96}$$

This anticommutativity or the antisymmetry of the T product is valid when $\overline{\psi}$'s are present in addition to ψ's.

For boson fields there is no difference between P and T. In particular we see that

$$P(j(x_1)\ldots j(x_n)) = T(j(x_1)\ldots j(x_n)) \tag{4-97}$$

since j is bilinear in the electron fields. By carrying through the procedure in the same way as for electromagnetic fields, we can obtain the result that the contraction function for the electron field is given by

$$\psi^{\cdot}_{\alpha}(x)\bar{\psi}^{\cdot}_{\beta}(y) = \langle 0|T[\psi_{\alpha}(x)\bar{\psi}_{\beta}(y)]|0\rangle$$

$$= S_{F\alpha\beta}(x - y) = -\left(\gamma_{\mu}\frac{\partial}{\partial x_{\mu}} - m\right)_{\alpha\beta} \Delta_F(x - y) \tag{4-98}$$

In general the contraction function, or the Feynman function, satisfies

$$D_{\alpha\beta}(\partial)\, \Delta_{F\beta\gamma}(x) = i\delta_{\alpha\gamma}\, \delta^4(x) \tag{4-99}$$

$$\Delta_{F\alpha\beta}(x) = C_{\alpha\beta}(\partial)\, \Delta_F(x) \tag{4-100}$$

where $C_{\alpha\beta}(\partial)$ is the differential operator defined in Section 2-5, and $\Delta_F(x)$ has the Fourier representation

$$\Delta_F(x) = \frac{-i}{(2\pi)^4} \int d^4p\, \frac{e^{ipx}}{p^2 + m^2 - i\epsilon} \tag{4-101}$$

Note that $\psi^{\cdot}\psi^{\cdot} = \overline{\psi^{\cdot}}\,\overline{\psi^{\cdot}} = 0$. Now it is a relatively easy matter to formulate the generalized Wick expansion:

$$\begin{aligned} T(AB\ldots Z) &= :AB\ldots Z: \\ &+ \sum_{\text{one contr}} :A^{\cdot}B^{\cdot}\ldots Z: \\ &+ \sum_{\text{two contr}} :A^{\cdot}B^{\cdot}C^{\cdot\cdot}D^{\cdot\cdot}\ldots Z: \\ &+ \;\ldots \end{aligned} \tag{4-102}$$

4-6 FEYNMAN DIAGRAMS [4]

In order to get an intuitive insight into the structure of the S matrix, it is very useful to introduce Feynman diagrams. Let us consider quantum electrodynamics, then the n-th-order term in the expansion of the S matrix is given by

$$\frac{(-i)^n}{n!}\int d^4x_1 \ldots \int d^4x_n T[\mathcal{H}_{int}(x_1)\ldots\mathcal{H}_{int}(x_n)] \tag{4-103}$$

where

$$\mathcal{H}_{int}(x) = -j_\mu(x)\, A_\mu(x) = -ie:\bar{\psi}(x)\,\gamma_\mu\psi(x):A_\mu(x) \tag{4-104}$$

Therefore, from the discussion in the last section we can use

$$S^{(n)} = \frac{(-e)^n}{n!}\int d^4x_1 \ldots \int d^4x_n\; T\Big[:\bar{\psi}(x_1)\;\gamma_\mu\psi(x_1):$$

$$\ldots :\bar{\psi}(x_n)\;\gamma_\sigma\psi(x_n):\Big] \times T\Big[A_\mu(x_1)\;\ldots A_\sigma(x_n)\Big] \tag{4-105}$$

We shall replace $:\bar{\psi}\gamma_\mu\psi:$ simply by $\bar{\psi}\gamma_\mu\psi$ with the understanding that $\bar{\psi}$ and ψ at the same space-time point should not be contracted.

We expand the T products into normal products; then in each order of this expansion there is a certain number of contractions,

$$:AB^{\cdot}C^{\cdot}\ \ldots Z:$$

When B(x) and C(y) are contracted we draw a line connecting x and y. More graphically, we diagram contractions as in Figure 4-3.

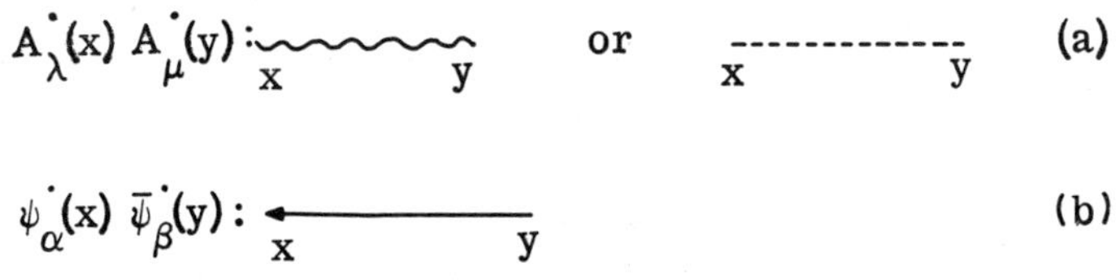

Figure 4-3. (a) The line representing propagation of a virtual photon.
(b) The line representing propagation of a virtual electron from y to x or of a virtual positron from x to y.

In the case of the electron field we draw a directed line from the argument of $\bar{\psi}$ to that of ψ. For uncontracted operators we also draw lines as in Figure 4-4.

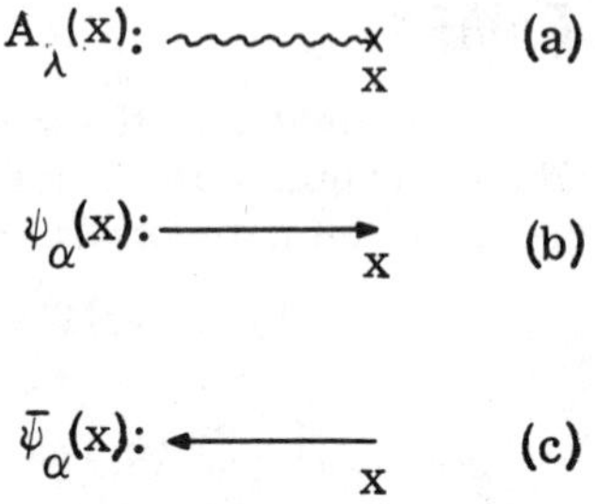

Figure 4-4. (a) The line representing creation or destruction of a real photon at a point x.
(b) The line representing destruction of a real electron or creation of a real positron at a point x.
(c) The line representing creation of a real electron or destruction of a real positron at a point x.

There is a one-to-one correspondence between each term of the S-matrix expansion and a Feynman diagram. For example, let us look at the terms which are of the second order in the S matrix. These terms are contained in the expression

$$S^{(2)} = \frac{(-e)^2}{2!} \int d^4x_1 \int d^4x_2\, T\,[:\bar{\psi}(x_1)\,\gamma_\mu\,\psi(x_1):\ :\bar{\psi}(x_2)\,\gamma_\nu\,\psi(x_2):]$$

$$\times\ T\,[A_\mu(x_1)\,A_\nu(x_2)] \tag{4-106}$$

The integrand may be separated into terms with different degrees of contraction [Eqs. (4-107) to (4-112)], in accordance with Wick's theorem; and the corresponding diagrams (Figures 4-5 to 4-10), illustrating the physical processes involved, may be drawn.

No contraction:

$$:\bar{\psi}(x_1)\,\gamma_\mu\psi(x_1)\ \bar{\psi}(x_2)\ \gamma_\nu\,\psi(x_2):\ :A_\mu(x_1)\,A_\nu(x_2): \tag{4-107}$$

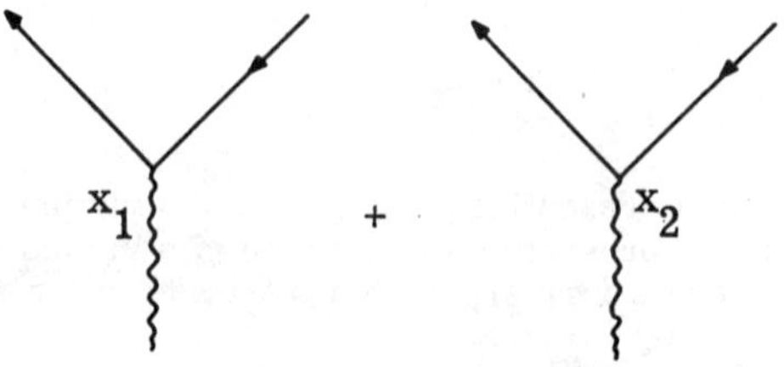

Figure 4-5. A diagram corresponding to (4-107).

This term corresponds by itself to no physical process.

The A's contracted:

$$:\bar{\psi}(x_1)\,\gamma_\mu\,\psi(x_1)\,\bar{\psi}(x_2)\,\gamma_\nu\,\psi(x_2):\;\delta_{\mu\nu}D_F(x_1 - x_2) \tag{4-108}$$

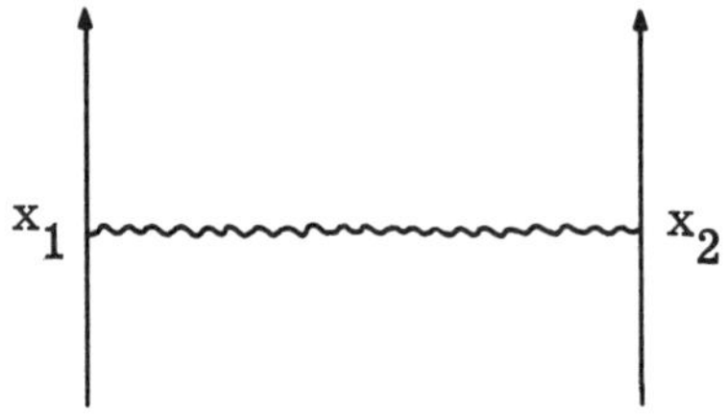

Figure 4-6. A diagram corresponding to (4-108).

This term corresponds to electron-electron, positron-positron, or electron-positron scattering.

One pair of ψ and $\bar{\psi}$ contracted:

$$(:\bar{\psi}(x_1)\;\gamma_\mu S_F(x_1 - x_2)\,\gamma_\nu\,\psi(x_2): + :\bar{\psi}(x_2)\,\gamma_\nu S_F(x_2 - x_1)\,\gamma_\mu\,\psi(x_1):)$$

$$:A_\mu(x_1)\,A_\nu(x_2): \tag{4-109}$$

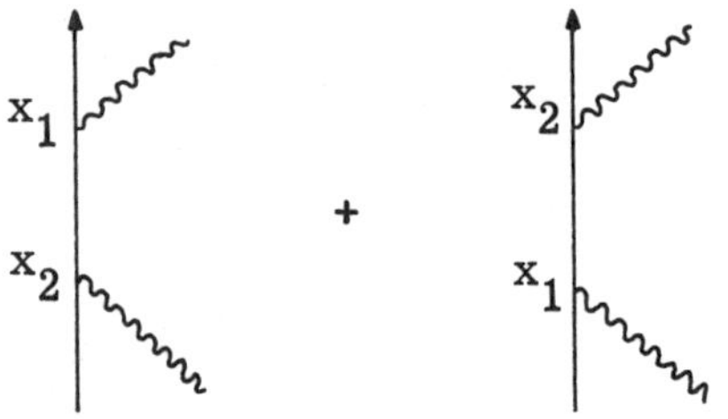

Figure 4-7. A diagram corresponding to (4-109).

These terms correspond to Compton scattering or electron-positron annihilation or creation.

One pair of ψ and $\bar{\psi}$ and the A's contracted:

$$[:\bar{\psi}(x_1)\,\gamma_\mu S_F(x_1 - x_2)\,\gamma_\nu\psi(x_2):+:\bar{\psi}(x_2)\,\gamma_\nu S_F(x_2 - x_1)\,\gamma_\mu\,\psi(x_1):]$$

$$\times\;\delta_{\mu\nu}D_F(x_1 - x_2) \tag{4-110}$$

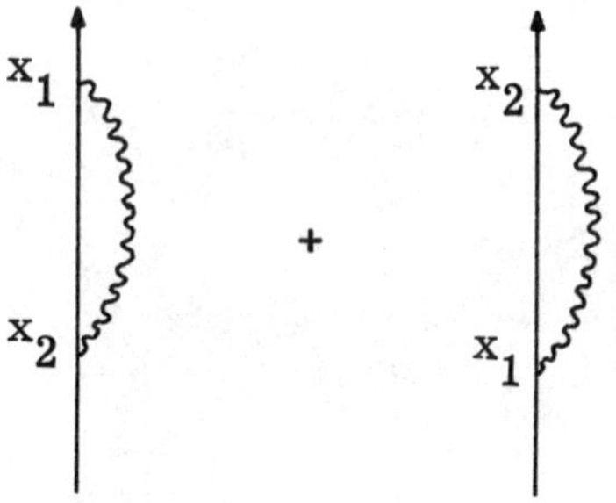

Figure 4-8. Diagrams corresponding to (4-110).

These terms correspond to the self-energy of the electron.

Two pairs of ψ and $\bar{\psi}$ contracted:

$$-\mathrm{Tr}[\gamma_\mu S_F(x_1 - x_2)\,\gamma_\nu S_F(x_2 - x_1)]{:}A_\mu(x_1)\,A_\nu(x_2){:} \quad (4\text{-}111)$$

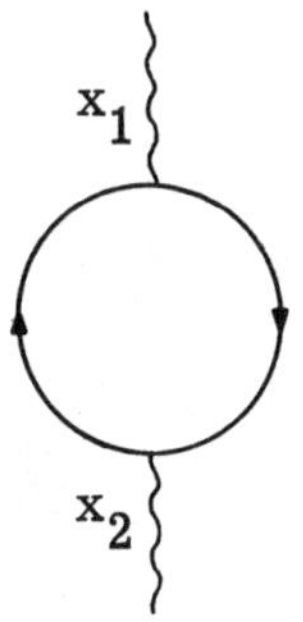

Figure 4-9. A diagram corresponding to (4-111).

This term corresponds to the vacuum polarization or the self-energy of the photon. Note the minus sign, which arises from commuting a ψ through $\bar{\psi}$ for purposes of contraction. All closed fermion loops acquire a minus sign in this manner.

All operators contracted:

$$-\mathrm{Tr}[\gamma_\mu S_F(x_1 - x_2)\,\gamma_\nu S_F(x_2 - x_1)]\,\delta_{\mu\nu} D_F(x_1 - x_2) \quad (4\text{-}112)$$

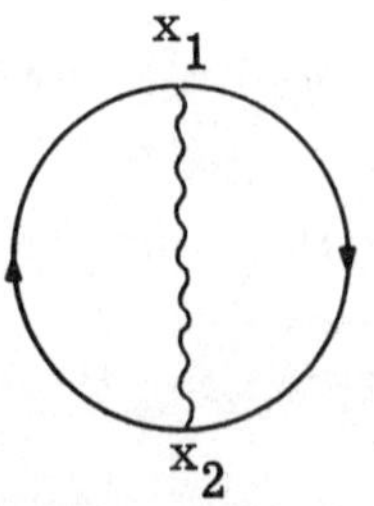

Figure 4-10. A diagram corresponding to (4-112).

This kind of diagram is called a bubble diagram and gives no contribution to observable quantities. It represents the so-called vacuum fluctuation and contributes only to the unobservable energy level of the vacuum state.

Example (4-1) S-Matrix Element for Compton Scattering

Let us examine in greater detail the part of the second-order S matrix describing Compton scattering:

$$S^{(2)}_{\text{Compton}} = \frac{(-e)^2}{2!}\int d^4x_1 \int d^4x_2 (:\bar{\psi}(x_1)\,\gamma_\mu S_F(x_1 - x_2)\,\gamma_\nu \psi(x_2):$$

$$+ :\bar{\psi}(x_2)\,\gamma_\nu S_F(x_2 - x_1)\,\gamma_\mu \psi(x_1):)$$

$$\times : A_\mu(x_1)\, A_\nu(x_2): \qquad (4\text{-}113)$$

The x_1, x_2, μ, and ν are dummy indices or variables of integration; by considering the substitutions $x_1 \leftrightarrows x_2$ and $\mu \leftrightarrows \nu$, we see that the two terms give exactly the same contribution to S. The factorial factor 2! in the denominator of $S^{(2)}$ is thus cancelled. The cancellation of n! in the denominator of $S^{(n)}$ is a general characteristic of the theory, but it sometimes occurs at a later stage when the matrix element is taken, for instance, for the vacuum polarization.

The S-matrix element for Compton scattering from a state $|p, k\rangle$ into another state $|p', k'\rangle$ is then

$$\langle p', k'|S^{(2)}|p, k\rangle = (-e)^2 \int d^4x_1 \int d^4x_2$$

$$\times \langle p'| :\bar{\psi}(x_1)\,\gamma_\mu S_F(x_1 - x_2)\,\gamma_\nu \psi(x_2):|p\rangle$$

$$\times \langle k'|:A_\mu(x_1)\, A_\nu(x_2):|k\rangle \qquad (4\text{-}114)$$

This expression is nonvanishing only if ψ destroys an electron with momentum p, $\bar{\psi}$ creates an electron with momentum p', and A_μ and A_ν destroy a photon with momentum k and create one with momentum k'. So we have

$$e^2 \int d^4x_1 \int d^4x_2 \langle p'|\bar{\psi}(x_1)|0\rangle\, \gamma_\mu S_F(x_1 - x_2)\,\gamma_\nu \langle 0|\psi(x_2)|p\rangle$$

$$\times \Big[\langle k'|A_\mu(x_1)|0\rangle\langle 0|A_\nu(x_2)|k\rangle$$

$$+ \langle k'|A_\nu(x_2)|0\rangle\langle 0|A_\mu(x_1)|k\rangle\Big] \qquad (4\text{-}115)$$

corresponding to the processes shown in Figure 4-11.

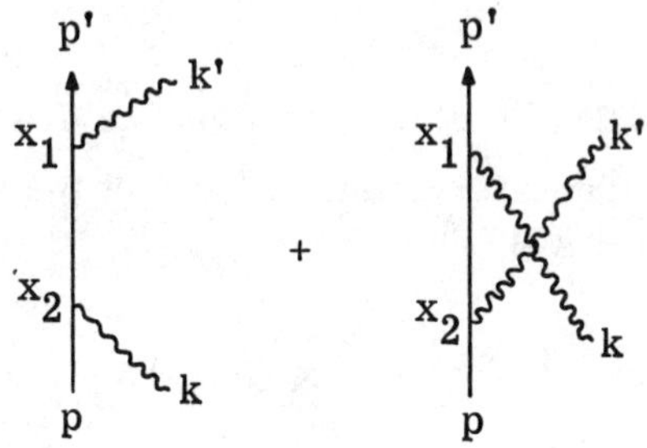

Figure 4-11. Diagrams for Compton scattering corresponding to (4-115).

Example (4-2) Closed Loops

We show here why we obtain a trace of Dirac matrices in the expression corresponding to a closed loop, such as is shown in the vacuum polarization diagram, Fig. 4-9. This trace arises from contractions of the electron fields.

$$\bar{\psi}^{\cdot}(x_1)\,\gamma_\mu\,\psi^{\cdot\cdot}(x_1)\,\bar{\psi}^{\cdot\cdot}(x_2)\,\gamma_\nu\,\psi^{\cdot}(x_2)$$

$$= \bar{\psi}^{\cdot}(x_1)\gamma_\mu S_F(x_1 - x_2)\,\gamma_\nu\,\psi^{\cdot}(x_2)$$

$$= \bar{\psi}^{\cdot}_\alpha(x_1)\left[\gamma_\mu S_F(x_1 - x_2)\,\gamma_\nu\right]_{\alpha\beta}\psi_\beta^{\cdot}(x_2)$$

$$= -\left[\gamma_\mu S_F(x_1 - x_2)\,\gamma_\nu\right]_{\alpha\beta}\psi_\beta^{\cdot}(x_2)\,\bar{\psi}_\alpha^{\cdot}(x_1)$$

$$= -\left[\gamma_\mu S_F(x_1 - x_2)\,\gamma_\nu\right]_{\alpha\beta}\left[S_F(x_2 - x_1)\right]_{\beta\alpha}$$

$$= -\mathrm{Tr}\left[\gamma_\mu S_F(x_1 - x_2)\,\gamma_\nu S_F(x_2 - x_1)\right]$$

This illustrates why we get a trace as well as a negative sign for a closed loop.

4-7 REMOVAL OF BUBBLE DIAGRAMS

Let us consider the amplitude corresponding to the transition from vacuum to vacuum. From the conservation of energy and momentum it is clear that

$$\left|\left\langle \text{vac}\middle| S \middle| \text{vac} \right\rangle\right|^2 = 1 \tag{4-116}$$

From now on we shall denote the vacuum state by $|0\rangle$; then the matrix element $\langle 0|S|0\rangle$ is

$$\langle 0|S|0\rangle = 1 + \sum_{n=1}^{\infty} \frac{(-i)^n}{n!} \int d^4x_1 \ldots \int d^4x_n$$
$$\times \langle 0|T[\mathcal{H}_{int}(x_1) \ldots \mathcal{H}_{int}(x_n)]|0\rangle \tag{4-117}$$

Examples of the corresponding diagrams are given in Fig. 4-12.

If we try to calculate the matrix element (4-117), we find that the result is divergent. But this kind of divergence is harmless, because $|\langle 0|S|0\rangle|^2 = 1$ and the occurrence of the divergence indicates that

$$\langle 0|S|0\rangle = e^{i\infty} \tag{4-118}$$

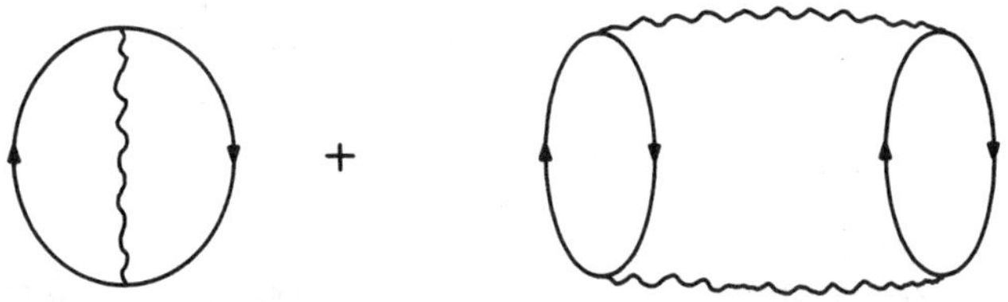

Figure 4-12. Examples of bubble diagrams.

The divergence in question can be absorbed into an irrelevant phase factor. It is convenient to normalize the phase factor by stipulating that

$$\langle 0|S'|0\rangle = 1 \tag{4-119}$$

Therefore we define the modified S matrix as

$$S' = \frac{S}{\langle 0|S|0\rangle} \tag{4-120}$$

If S is unitary, S' is also unitary; but in addition, all processes corresponding to bubble diagrams have been factored out of S', as the following will show.

Let us consider a process other than (vacuum → vacuum). The corresponding diagrams involve free-ended lines corresponding to incoming and outgoing particles, which are called *external lines*. In each Feynman diagram there are two kinds of partial diagrams: the part that is connected to at least one of the external lines is called the *connected part;* the part that is connected to none of the external lines is called a *bubble.* The bubble part gives a c-number factor multiplying the term from the connected part; thus

$$T[\mathcal{H}(x_1)\ldots\mathcal{H}(x_n)] = \sum_{\text{comb}} T_c[\mathcal{H}(x_1')\ldots\mathcal{H}(x_k')] \times T_b[\mathcal{H}(x_{k+1}')\ldots\mathcal{H}(x_n')] \tag{4-121}$$

with T_b representing the vacuum fluctuations from the bubble diagram,

$$T_b[\mathcal{H}(x_{k+1}')\ldots\mathcal{H}(x_n')] = \langle 0 | T[\mathcal{H}(x_{k+1}')\ldots\mathcal{H}(x_n')] | 0 \rangle \tag{4-122}$$

If there are no bubble diagrams, $T_b(\ ) = 1$. We define T_c by subtracting all the bubble contributions from T.

The S matrix now can be written in the form

$$S = 1 + \sum_{n=1}^{\infty} \frac{(-i)^n}{n!} \int d^4x_1 \ldots \int d^4x_n \sum_{\text{comb}} T_c(\mathcal{H}(x_1')\ldots\mathcal{H}(x_k')) \times T_b[\mathcal{H}(x_{k+1}')\ldots\mathcal{H}(x_n')] \tag{4-123}$$

Since all x's are dummy variables, the sum over all possible ways of dividing n variables into two groups, one entering into T_c and the other entering into T_b, gives the same expression repeatedly and thus provides a factor $n!/[k!(n-k)!]$. Then

$$S = \sum_{n=0}^{\infty} \frac{(-i)^n}{n!} \int d^4x_1 \ldots \int d^4x_k \int d^4x_{k+1} \ldots \int d^4x_n \times \sum_{k=0}^{n} \frac{n!}{k!(n-k)!} T_c[\mathcal{H}(x_1)\ldots\mathcal{H}(x_k)]\, T_b[\mathcal{H}(x_{k+1})\ldots (x_n)]$$

$$= \sum_{n,k} \frac{(-i)^k}{k!} \int d^4x_1 \ldots \int d^4x_k\, T_c[\mathcal{H}(x_1)\ldots\mathcal{H}(x_k)] \times \frac{(-i)^{n-k}}{(n-k)!} \int d^4x_{k+1} \ldots \int d^4x_n \langle 0 | T[\mathcal{H}(x_{k+1})\ldots\mathcal{H}(x_n)] | 0 \rangle$$

$$= \sum_{k=0}^{\infty} \frac{(-i)^k}{k!} \int d^4x_1 \ldots \int d^4x_k \, T_c \left[\mathcal{H}(x_1) \ldots \mathcal{H}(x_k) \right]$$

$$\times \sum_{m=0}^{\infty} \frac{(-i)^m}{m!} \int d^4x_1' \ldots \int d^4x_m' \langle 0 | T\left[\mathcal{H}(x_1') \ldots \mathcal{H}(x_m') \right] | 0 \rangle \tag{4-124}$$

where m = n - k, and the summation over n and k is replaced by a summation over k and m. Hence

$$S = \sum_{n=0}^{\infty} \frac{(-i)^n}{n!} \int d^4x_1 \cdots \int d^4x_n \, T_c \left[\mathcal{H}(x_1) \cdots \mathcal{H}(x_n) \right]$$

$$\times \langle 0 | S | 0 \rangle \tag{4-125}$$

and consequently

$$S' = \frac{S}{\langle 0 | S | 0 \rangle} = \sum_{n=0}^{\infty} \frac{(-i)^n}{n!} \int d^4x_1 \cdots \int d^4x_n \, T_c \left[\mathcal{H}(x_1) \cdots \mathcal{H}(x_n) \right] \tag{4-126}$$

This means that we can omit all bubbles in evaluating the matrix elements of S'. We shall use the symbol S in the sense of S' unless otherwise stated.

4-8 TRANSITION PROBABILITY

If the S-matrix element is known we immediately can evaluate the transition probability. Since the connection between them should be familiar we give here only the result.

For nonrelativistic calculations we may work with the noncovariant formalism. In terms of the transition element W_{fi}, which is equal to $(H_{int})_{fi}$ in the Born approximation, the S-matrix element is

$$\langle f | S | i \rangle = \delta_{fi} - 2\pi i \, \delta(E_f - E_i) \, W_{fi} \tag{4-127}$$

Then the probability for the transition i → f per unit time is given by

$$w_{fi} = 2\pi \left(\frac{V}{(2\pi)^3} \int d^3p_1 \right) \cdots \left(\frac{V}{(2\pi)^3} \int d^3p_n \right) \delta(E_f - E_i) \left| W_{fi} \right|^2 \tag{4-128}$$

where $p_1, \ldots, p_n$ are the momenta of the particles present in the final state. The cross section for a given reaction is related to the transition probability by

$$\sigma_{fi} = \frac{V}{v_{rel}} w_{fi} \tag{4-129}$$

where v_{rel} is the relative velocity between the colliding particles, that is, it is equal to $v_1 + v_2$.

In relativistic calculations we introduce a transition amplitude t_{fi} by

$$\langle f|S|i\rangle = \delta_{fi} - i\,\delta^4(P_f - P_i)t_{fi} \tag{4-130}$$

then the transition probability is given, in terms of t_{fi}, by

$$w_{fi} = \frac{V}{(2\pi)^4}\left(\frac{V}{(2\pi)^3}\int d^3p_1\right)\cdots\left(\frac{V}{(2\pi)^3}\int d^3p_n\right) \times \delta^4(P_f - P_i)\left|t_{fi}\right|^2 \tag{4-131}$$

The cross section is then given by

$$\sigma_{fi} = \frac{1}{v_{rel}}\cdot\frac{V^2}{(2\pi)^4}\left(\frac{V}{(2\pi)^3}\int d^3p_1\right)\cdots\left(\frac{V}{(2\pi)^3}\int d^3p_n\right) \times \delta^4(P_f - P_i)\left|t_{fi}\right|^2 \tag{4-132}$$

We shall see later that the cross section is independent of the choice of the volume of quantization V.

REFERENCES

1. F. J. Dyson, Phys. Rev. **75**, 1736 (1949).
2. We use Δ_F here instead of D_F to emphasize that this argument is valid for any field.
3. G. C. Wick, Phys. Rev. **80**, 268 (1950).
4. R. P. Feynman, Phys. Rev. **76**, 769 (1949).

CHAPTER 5
SIMPLE APPLICATIONS

In this chapter we shall illustrate the practical utility of the formalism developed in the foregoing chapters. Of primary interest in this respect are calculations of cross sections, both differential and total, for various reactions, and of decay rates. These calculations require a number of mathematical techniques and tricks, which we shall discuss in the course of doing the series of fundamental examples[1] that constitutes the sections of this chapter. Although several of the calculations may seem rather lengthy, the procedures are straightforward and will come easily with practice.

5-1 SCATTERING OF AN ELECTRON BY AN EXTERNAL FIELD

For our first example of the application of perturbation theory, we shall discuss the scattering of an electron by a fixed external potential.

The T-S equation for electrodynamics in the presence of an external field A^{ext} is given by

$$i \frac{\delta}{\delta\sigma(x)} U[\sigma, -\infty] = -j_\mu(x) \, [A_\mu(x) + A_\mu^{ext}(x)] \, U[\sigma, -\infty] \tag{5-1}$$

since here

$$\mathcal{H}_{int}(x) = -j_\mu(x) \, [A_\mu(x) + A_\mu^{ext}(x)] \tag{5-2}$$

In the Feynman diagram A^{ext} appears as a point. The diagram for this scattering, which is essentially Rutherford scattering, is shown in Fig. 5-1.

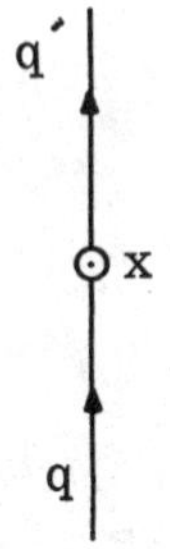

Figure 5-1. A diagram for Coulomb scattering. The small circle at the point x indicates an external field.

In the lowest-order approximation only the second term in the interaction Hamiltonian gives a nonvanishing contribution, and we obtain

$$S^{(1)} = -i \int d^4x \, \mathcal{H}^{ext}(x) \tag{5-3}$$

with

$$\mathcal{H}^{ext}(x) = -j_\mu(x) A_\mu^{ext}(x) \tag{5-4}$$

The S-matrix element for the scattering of an electron from a state with momentum q to another state with momentum q' is then

$$\begin{aligned} \langle q'|S|q\rangle &= -i \int d^4x \, (-ie)\langle q'| \, :\bar{\psi}(x)\gamma_\mu\psi(x): \, |q\rangle A_\mu^{ext}(x) \\ &= -e \int d^4x \, \langle q'|\bar{\psi}(x)|0\rangle \gamma_\mu \langle 0|\psi(x)|q\rangle A_\mu^{ext}(x) \end{aligned} \tag{5-5}$$

The electron wave functions are given by

$$\langle 0|\psi(x)|q, r\rangle = V^{-1/2} u^{(r)}(q) e^{iqx} \tag{5-6a}$$

$$\langle q', r'|\bar{\psi}(x)|0\rangle = V^{-1/2} \bar{u}^{(r')}(q') e^{-iq'x} \tag{5-6b}$$

Let us assume that A^{ext} is time independent, and define $A^{ext}(Q)$ as

$$A_\mu^{ext}(Q) = \int d^3x \, e^{iQx} A_\mu^{ext}(x) \tag{5-7}$$

The S-matrix element then becomes

$$\langle q'|S|q\rangle = -\frac{e}{V} \bar{u}^{(r')}(q')\gamma_\mu u^{(r)}(q) A_\mu^{ext}(\mathbf{q} - \mathbf{q}') \cdot 2\pi\delta(q_0 - q'_0) \tag{5-8}$$

Therefore, the transition probability is given by

$$w = 2\pi \frac{V}{(2\pi)^3} \int d^3q' \, \delta(q_0 - q'_0) \left| \frac{e}{V} \bar{u}^{(r')}(q')\gamma_\mu u^{(r)}(q) \right.$$

$$\left. \times \; A_\mu^{ext}(q - q') \right|^2 \qquad (5\text{-}9)$$

or, as a differential transition probability [let $h = \bar{u}^{(r')}(q')\gamma_\mu u^{(r)}(q) \times A_\mu^{ext}(q - q')$],

$$dw = 2\pi \frac{V}{(2\pi)^3} d^3q' \, \delta(q_0 - q'_0) \left| \frac{e}{V} h \right|^2 \qquad (5\text{-}10)$$

and the differential cross section is obtained from dw by dividing dw by the factor v_{rel}/V, where $v_{rel} = |q|/q_0$ is the velocity of the incident electron.

$$d\sigma = 2\pi \frac{V^2}{(2\pi)^3} \cdot \frac{d^3q'}{v_{rel}} \delta(q_0 - q'_0) \left| \frac{e}{V} h \right|^2 \qquad (5\text{-}11)$$

The cross section is, then, independent of the volume of quantization V.

$$d\sigma = 2\pi \frac{e^2}{(2\pi)^3} \frac{q_0 d^3q'}{|q|} \delta(q_0 - q'_0) \left| h \right|^2 \qquad (5\text{-}12)$$

The conservation of energy $q_0 = q'_0$ implies $|q| = |q'|$, so that

$$\int \frac{q_0 d^3q'}{|q|} \delta(q_0 - q'_0) \cdots = \int \frac{q_0 |q|^2 d|q'| \, d\Omega}{|q|} \delta(q_0 - q'_0) \cdots$$

$$= \int \frac{q_0 |q| \, q'_0 dq'_0 \, d\Omega}{|q|} \delta(q_0 - q'_0) \cdots$$

$$= q_0^2 \int d\Omega \cdots$$

Thus we obtain the differential cross section

$$\frac{d\sigma}{d\Omega} = \frac{e^2 q_0^2}{(2\pi)^2} |h|^2$$

$$= \frac{e^2 q_0^2}{(2\pi)^2} \left| \bar{u}^{(r')}(q') \gamma_\mu u^{(r)}(q) \cdot A_\mu^{ext}(q - q') \right|^2 \qquad (5\text{-}13)$$

Assuming that A_μ^{ext} is real for $\mu = 1, 2, 3$ and imaginary for $\mu = 4$ we find, since γ_k ($k = 1, 2, 3$) anticommutes with γ_4,

$$|h|^2 = -\bar{u}^{(r')}(q')\gamma_\mu u^{(r)}(q)\bar{u}^{(r)}(q)\gamma_\nu u^{(r')}(q')A_\mu^{ext}(q - q')$$

$$\times A_\nu^{ext}(q' - q)$$

We take the sum over final spin states and the average over the initial spin states. That is, we replace $|h|^2$ by a trace with the help of the Casimir trick.

$$\sum_{r=1}^{2} u_\alpha^{(r)}(q)\,\bar{u}_\beta^{(r)}(q) = \left(\frac{-iq\gamma + m}{2q_0}\right)_{\alpha\beta} \qquad (5\text{-}14)$$

where the summation is taken over the two spin states belonging to the positive energy. Then we find

$$|h|^2 \longrightarrow -\frac{1}{2}\sum_{r=1}^{2}\sum_{r'=1}^{2} \bar{u}^{(r')}(q')\gamma_\mu u^{(r)}(q)\bar{u}^{(r)}(q)\gamma_\nu u^{(r')}(q')$$

$$\times A_\mu^{ext}(q - q') \,.\, A_\nu^{ext}(q' - q)$$

$$= -\frac{1}{2}\,\mathrm{Tr}\left[\gamma_\mu \frac{i\gamma q - m}{2q_0}\;\gamma_\nu \frac{i\gamma q' - m}{2q_0}\right]$$

$$\times A_\mu^{ext}(q - q')\,A_\nu^{ext}(q' - q) \qquad (5\text{-}15)$$

The differential cross section is

$$\frac{d\sigma}{d\Omega} = -\frac{e^2}{8(2\pi)^2}\,\mathrm{Tr}\,[\gamma_\mu(i\gamma q - m)\,\gamma_\nu\,(i\gamma q' - m)]A_\mu^{ext}(q - q')$$

$$\times A_\nu^{ext}(q' - q)$$

$$= -\frac{e^2}{8\pi^2}\,[\delta_{\mu\nu}(qq' + m^2) - q_\mu q'_\nu - q'_\mu q_\nu]\,A_\mu^{ext}(q - q')$$

$$\times A_\nu^{ext}(q' - q) \qquad (5\text{-}16)$$

The evaluation of the trace of a product of γ matrices is discussed at the end of the section. As a practical case, we take for A^{ext} a Coulomb field of, say, a nucleus. Then

$$A_{\mu}^{ext}(x) = -\frac{Ze}{4\pi r} i\delta_{\mu 4} \tag{5-17}$$

where e is the electronic charge so that the nuclear charge is -Ze. The factor of 4π comes from the fact that we use Heaviside units, in which we have the fine structure constant given by

$$\frac{e^2}{4\pi} = \alpha = \frac{1}{137} \tag{5-18}$$

In this case we easily find

$$A_{\mu}^{ext}(Q) = -i\delta_{\mu 4} ZeQ^{-2} \tag{5-19}$$

where $\mathbf{Q}$ is the momentum transfer, which we can express in terms of the scattering angle θ and the magnitude of the momentum $|\mathbf{q}| = |\mathbf{q}'| = q$.

$$\mathbf{Q}^2 = (\mathbf{q} - \mathbf{q}')^2 = 2q^2(1 - \cos\theta) = 4q^2 \sin^2\theta/2 \tag{5-20}$$

We also find that

$$qq' + m^2 - 2q_4q_4 = 2q^2 \cos^2\theta/2 + 2m^2 \tag{5-21}$$

and we finally get

$$\frac{d\sigma}{d\Omega} = \left(\frac{Ze^2m}{8\pi q^2 \sin^2\theta/2}\right)^2 \left(1 + \frac{q^2}{m^2} \cos^2\theta/2\right) \tag{5-22}$$

In the nonrelativistic limit $q^2 \ll m^2$,

$$\frac{d\sigma}{d\Omega} = \left(\frac{Ze^2}{16\pi E \sin^2\theta/2}\right)^2 \tag{5-23}$$

where E is the kinetic energy $q^2/2m$. This agrees with the nonrelativistic Rutherford formula.

We give a general method of evaluation of traces of products of the Dirac γ matrices. From the commutation relations for γ matrices,

$$\gamma_\mu\gamma_\nu + \gamma_\nu\gamma_\mu = 2\delta_{\mu\nu} \tag{5-24}$$

the method of reducing the number of γ matrices in a trace follows:

$$\mathrm{Tr}(\gamma_{\nu_1}\gamma_{\nu_2}\cdots\gamma_{\nu_n}) = 2\delta_{\nu_1\nu_2}\,\mathrm{Tr}(\gamma_{\nu_3}\cdots\gamma_{\nu_n})$$

$$- \mathrm{Tr}(\gamma_{\nu_2}\gamma_{\nu_1}\gamma_{\nu_3}\cdots\gamma_{\nu_n})$$

$$= 2\sum_{i=2}^{n}\delta_{\nu_1\nu_i}(-1)^i\,\mathrm{Tr}[\gamma_{\nu_2}\cdots\gamma_{\nu_{i-1}}\gamma_{\nu_{i+1}}\cdots\gamma_{\nu_n}]$$

$$+ (-1)^{n-1}\,\mathrm{Tr}[\gamma_{\nu_2}\cdots\gamma_{\nu_n}\gamma_{\nu_1}] \tag{5-25}$$

Using the fundamental formula for the trace of a product

$$\mathrm{Tr}(AB) = \mathrm{Tr}(BA) \tag{5-26}$$

we find that for an even n

$$\mathrm{Tr}[\gamma_{\nu_1}\cdots\gamma_{\nu_n}] = \sum_{i=2}^{n}\delta_{\nu_1\nu_i}(-1)^i\,\mathrm{Tr}[\gamma_{\nu_2}\cdots\gamma_{\nu_{i-1}}\gamma_{\nu_{i+1}}\cdots\gamma_{\nu_n}] \tag{5-27}$$

If (5-27) is reiterated one can recognize a formal similarity to Wick's theorem. In this manner we can reduce the number of γ's in a matrix product. We also can show that the trace of the product of an odd number of γ matrices vanishes, by using

$$\gamma_5\gamma_{\nu_1}\cdots\gamma_{\nu_n}\gamma_5 = (-1)^n\,\gamma_{\nu_1}\cdots\gamma_{\nu_n} \tag{5-28}$$

which is a consequence of $\gamma_5\gamma_\mu + \gamma_\mu\gamma_5 = 0$ for $\mu = 1, 2, 3, 4$, and $\gamma_5^2 = 1$.

$$\mathrm{Tr}(\gamma_{\nu_1}\cdots\gamma_{\nu_n}) = \mathrm{Tr}(\gamma_5^2\gamma_{\nu_1}\cdots\gamma_{\nu_n})$$

$$= \mathrm{Tr}(\gamma_5\gamma_{\nu_1}\cdots\gamma_{\nu_n}\gamma_5)$$

$$= (-1)^n\,\mathrm{Tr}(\gamma_{\nu_1}\cdots\gamma_{\nu_n})$$

Hence

$$\mathrm{Tr}(\gamma_{\nu_1}\cdots\gamma_{\nu_n}) = 0 \quad \text{for an odd n} \tag{5-29}$$

Typical examples of the traces are

$$\mathrm{Tr}(\gamma_\mu \gamma_\nu) = 4\delta_{\mu\nu} \tag{5-30}$$

$$\mathrm{Tr}(\gamma_\lambda \gamma_\mu \gamma_\nu \gamma_\sigma) = 4(\delta_{\lambda\mu}\delta_{\nu\sigma} + \delta_{\mu\nu}\delta_{\lambda\sigma} - \delta_{\lambda\nu}\delta_{\mu\sigma}) \tag{5-31}$$

5-2 COMPTON SCATTERING

We discuss the process of the elastic scattering of a photon by an electron by using the techniques of covariant perturbation theory.

The lowest-order matrix element already has been given in Section 4-6.

$$\begin{aligned}\langle q', k'|S|q, k\rangle = e^2 \int d^4x' \int d^4x'' \langle q'|\bar{\psi}(x')|0\rangle \gamma_\mu S_F(x' - x'')\gamma_\nu \\ \times \langle 0|\psi(x'')|q\rangle \\ \times [\langle k'|A_\mu(x')|0\rangle \langle 0|A_\nu(x'')|k\rangle \\ + \langle k'|A_\nu(x'')|0\rangle \langle 0|A_\mu(x')|k\rangle]\end{aligned} \tag{5-32}$$

Let us set $k_0 = \omega$, $k'_0 = \omega'$; then the single particle wave functions in the integrand are given by

$$\langle 0|\psi(x)|q\rangle = V^{-1/2} u^{(r)}(q)\, e^{iqx} \tag{5-33a}$$

$$\langle q'|\bar{\psi}(x)|0\rangle = V^{-1/2} \bar{u}^{(r')}(q')\, e^{-iq'x} \tag{5-33b}$$

for the electrons, and for the photons we have

$$\langle 0|A_\mu(x)|k\rangle = (2\omega V)^{-1/2} e_\mu^{(\lambda)} e^{ikx} \tag{5-34a}$$

$$\langle k'|A_\mu(x)|0\rangle = (2\omega' V)^{-1/2} e_\mu^{(\lambda')} e^{-ik'x} \tag{5-34b}$$

The electron propagation function is given by

$$S_F(x) = -(\gamma\partial - m)\,\Delta_F(x)$$

$$= \frac{i}{(2\pi)^4} \int d^4p\, e^{ipx} \frac{ip\gamma - m}{p^2 + m^2 - i\epsilon} \tag{5-35}$$

Substituting these expressions into the S matrix element we have

$$\langle q', k'|S|q, k\rangle = (-e)^2 \int d^4x' \int d^4x'' \int d^4p$$

$$\times V^{-1/2}\, \bar{u}^{(r')}(q')\, \gamma_\mu \frac{i}{(2\pi)^4} \frac{ip\gamma - m}{p^2 + m^2 - i\epsilon}$$

$$\times \gamma_\nu u^{(r)}(q) V^{-1/2}$$

$$\times (2\omega V)^{-1/2} \cdot (2\omega' V)^{-1/2}\, e_\mu^{(\lambda')} e_\nu^{(\lambda)}$$

$$\times \exp\left[iqx'' + ip(x' - x'') - iq'x' + ikx'' - ik'x'\right]$$

$$+ (k \rightleftarrows -k', \lambda \rightleftarrows \lambda') \tag{5-36}$$

We first carry out x' and x'' integrations by using

$$\int d^4x\, e^{iQx} = (2\pi)^4\, \delta^4(Q) \tag{5-37}$$

Then

$$\langle q', k'|S|q, k\rangle = (-e)^2 (2\pi)^{8-4}\, i\, V^{-2} \cdot (4\omega\omega')^{-1/2}$$

$$\times \int d^4p\, \bar{u}^{(r')}(q')\, (\gamma_\mu e_\mu^{(\lambda')}) \frac{ip\gamma - m}{p^2 + m^2 - i\epsilon}$$

$$\times (\gamma_\nu e_\nu^{(\lambda)})\, u^{(r)}(q)$$

$$\times \delta^4(p - q' - k')\, \delta^4(q + k - p) + (k \rightleftarrows -k', \lambda \rightleftarrows \lambda') \tag{5-38}$$

The p integration is done easily by using

$$\int d^4p\, \delta(p - q' - k')\, \delta(q + k - p)\, f(p) = \delta^4(q + k - q' - k')\, f(q + k)$$

For simplicity $\gamma_\mu e_\mu^{(\lambda')}$ is written as $\gamma e'$, and $\gamma_\nu e_\nu^{(\lambda)}$ as γe; then

$$\langle q', k'|S|q, k\rangle = (ie^2/2V^2)(\omega\omega')^{-1/2}\, \bar{u}^{(r')}(q')$$

$$\times \left(\gamma e' \frac{i\gamma(q + k) - m}{(q + k)^2 + m^2} \gamma e + \gamma e \frac{i\gamma(q - k') - m}{(q - k')^2 + m^2} \gamma e'\right) u^{(r)}(q)$$

$$\times (2\pi)^4\, \delta^4(q + k - q' - k') \tag{5-39}$$

In order to simplify the matrix element we use

$$q^2 + m^2 = q'^2 + m^2 = k^2 = k'^2 = 0 \quad \text{(mass shell conditions)} \tag{5-40}$$

$$(q + k)^2 + m^2 = q^2 + m^2 + k^2 + 2qk = 2qk \tag{5-41a}$$

$$(q - k')^2 + m^2 = q^2 + m^2 + k'^2 - 2qk' = -2qk' \tag{5-41b}$$

Furthermore we can use the Dirac equation to reduce the number of γ matrices:

$$(i\gamma q + m)\, u^{(r)}(q) = 0 \tag{5-42a}$$

$$\bar{u}^{(r')}(q')\,(i\gamma q' + m) = 0 \tag{5-42b}$$

Now we choose the *laboratory system* to discuss this problem, namely we choose $\mathbf{q} = 0$, $q_0 = m$. For the transverse photons, the polarization vector e has no fourth component so that we can assume

$$qe = qe' = 0 \tag{5-43}$$

and the Lorentz condition is given by

$$ke = k'e' = 0 \tag{5-44}$$

With these points in mind, we proceed to reduce the Dirac matrix element. Using

$$(\gamma k)(\gamma e) = 2ke - (\gamma e)(\gamma k) = -(\gamma e)(\gamma k)$$

$$(\gamma q)(\gamma e) = 2qe - (\gamma e)(\gamma q) = -(\gamma e)(\gamma q)$$

we find that

$$\begin{aligned}\bar{u}(q')(\gamma e')\,(i\gamma(q + k) - m)\,(\gamma e)u(q) &= \bar{u}(q')(\gamma e')(\gamma e)\,(-i\gamma(q + k) - m)u(q)\\ &= \bar{u}(q')(\gamma e')(\gamma e)(-i\gamma k)u(q)\end{aligned}$$

In a similar way we find

$$\bar{u}(q')(\gamma e)(i\gamma(q - k') - m)(\gamma e')u(q) = \bar{u}(q')(\gamma e)(\gamma e')(i\gamma k')u(q)$$

Thus the S-matrix element reduces to

$$\langle q', k' | S | q, k \rangle = -(ie^2/2V^2)(\omega\omega')^{-1/2}\,\bar{u}^{(r')}(q')[(\gamma e')(\gamma e)(i\gamma k/2qk) + (\gamma e)(\gamma e')(i\gamma k'/2qk')]\,u^{(r)}(q) \times (2\pi)^4\,\delta^4(q + k - q' - k') = -i\delta^4(q + k - q' - k')\,t_{fi} \tag{5-45}$$

The cross section is then given by

$$\sigma = \frac{V^2}{(2\pi)^4 v_{rel}} \cdot \frac{V}{(2\pi)^3}\int d^3q' \frac{V}{(2\pi)^3}\int d^3k' \;\delta^4(q + k - q' - k')\,|t_{fi}|^2 \tag{5-46}$$

with $v_{rel} = 1$, the velocity of light. Obviously the result is V independent. To get the total cross section, we average over the initial spin states and sum over the final spin states as we did in the previous section:

$$|t_{fi}|^2 \to \frac{1}{2}\sum_{r=1}^{2}\sum_{r'=1}^{2} |t_{fi}|^2 \tag{5-47}$$

Thus we find

$$\frac{d\sigma}{d\Omega} = \frac{e^4}{128\pi^2}\frac{1}{m\omega}\int \frac{d^3q'}{q'_0}\int \omega'\,d\omega'\,\mathrm{Tr}[\cdots]\,\delta(\mathbf{k} - \mathbf{q}' - \mathbf{k}') \times \delta(m + \omega - q'_0 - \omega') \tag{5-48}$$

where

$$\mathrm{Tr}[\cdots] = \mathrm{Tr}\Big[\big((\gamma e')(\gamma e)(i\gamma k/2qk) + (\gamma e)(\gamma e')(i\gamma k'/2qk')\big)(i\gamma q - m) \times \big((i\gamma k'/2qk')(\gamma e')(\gamma e) + (i\gamma k/2qk)(\gamma e)(\gamma e')\big)(i\gamma q' - m)\Big] \tag{5-49}$$

On integrating the delta functions we find

$$\frac{d\sigma}{d\Omega} = \frac{e^4}{128\pi^2}\frac{\omega'^2}{m^2\omega^2}\,\mathrm{Tr}[\cdots] \tag{5-50}$$

In evaluating the trace we make use of

$$(m + \omega - \omega')^2 = m^2 + (\mathbf{k} - \mathbf{k}')^2 \qquad kk' = \mathbf{k}\,\mathbf{k}' - \omega\omega' = m(\omega' - \omega)$$

which result, respectively, from

$$q'^2 + m^2 = 0 \qquad (k - k')^2 = (q - q')^2$$

The relation between ω and ω' is given by the Compton formula,

$$\omega' = \frac{\omega m}{m + \omega\,(1 - \cos\theta)} \tag{5-51}$$

where θ is the scattering angle in the laboratory system.

Our next problem is the evaluation of the traces. To begin with, we introduce a vector a by

$$a = (k'/2qk') - (k/2qk) \tag{5-52}$$

so that $qa = 0$. With the help of this vector a we have

$$\begin{aligned}&(\gamma e')(\gamma e)(i\gamma k/2qk) + (\gamma e)(\gamma e')(i\gamma k'/2qk')\\ &= (\gamma e)(\gamma e')(i\gamma a) + 2(ee')\,(i\gamma k/2qk)\end{aligned} \tag{5-53}$$

then split up the trace into four parts,

$$\mathrm{Tr}[\cdots] = \mathrm{Tr}[\mathrm{I}] + \mathrm{Tr}[\mathrm{II}] + \mathrm{Tr}[\mathrm{III}] + \mathrm{Tr}[\mathrm{IV}] \tag{5-54}$$

where

$$\begin{aligned}\mathrm{Tr}[\mathrm{I}] = \mathrm{Tr}[&(\gamma e)(\gamma e')(i\gamma a)(i\gamma q - m)(i\gamma a)(\gamma e')(\gamma e)\\ &\times (i\gamma(q + k - k') - m)]\end{aligned} \tag{5-55a}$$

$$\begin{aligned}\mathrm{Tr}[\mathrm{II}] = 4(ee')^2\, \mathrm{Tr}[&(i\gamma k/2qk)(i\gamma q - m)(i\gamma k/2qk)\\ &\times (i\gamma(q + k - k') - m)]\end{aligned} \tag{5-55b}$$

$$\begin{aligned}\mathrm{Tr}[\mathrm{III}] = \mathrm{Tr}[\mathrm{IV}] = 2(ee')\,\mathrm{Tr}[&(\gamma e)(\gamma e')(i\gamma a)(i\gamma q - m)\\ &\times (i\gamma k/2qk)(i\gamma(q + k - k') - m)]\end{aligned} \tag{5-55c}$$

By use of the properties $qe = qe' = ke = k'e' = qa = 0$, the parts may be readily evaluated.

$$\begin{aligned}\mathrm{Tr}\,[\mathrm{I}] &= -\mathrm{Tr}\,[(i\gamma q + m)(\gamma e)(\gamma e')(i\gamma a)(i\gamma a)(\gamma e')(\gamma e) \\ &\qquad \times\ (i\gamma(q + k - k') - m)] \\ &= a^2\,\mathrm{Tr}\,[(i\gamma q + m)(i\gamma(q + k - k') - m)] \\ &= a^2\,\mathrm{Tr}\,[i\gamma q \cdot i\gamma(k - k')] \\ &= -4a^2\, q(k - k') \\ &= 2\,(\omega - \omega')^2/\omega\omega' \end{aligned} \tag{5-56a}$$

$$\begin{aligned}\mathrm{Tr}\,[\mathrm{II}] &= -4(ee')^2\,\mathrm{Tr}\,[(i\gamma k/2qk)\, i\gamma(q - k')] \\ &= 8\,(ee')^2\big(1 - (kk'/qk)\big)\end{aligned} \tag{5-56b}$$

$$\begin{aligned}\mathrm{Tr}\,[\mathrm{III} + \mathrm{IV}] &= 4(ee')\,\mathrm{Tr}\,[(i\gamma q + m)(\gamma e)(\gamma e')(i\gamma a) \\ &\qquad \times\ (i\gamma k/2qk)(i\gamma k')] \\ &= 4(ee')^2\,\mathrm{Tr}\,[i\gamma q\,(i\gamma k'/2qk')\,(i\gamma k/2qk)\, i\gamma k'] \\ &= 8(ee')^2(kk'/qk)\end{aligned} \tag{5-56c}$$

Finally

$$\mathrm{Tr}\,[\cdots] = 8(ee')^2 + 2\,(\omega - \omega')^2/\omega\omega' \tag{5-57}$$

and therefore we obtain the well-known Klein-Nishina formula[2]:

$$\frac{d\sigma}{d\Omega} = (e^2/4\pi)^2\,(1/4m^2)\,(\omega'/\omega)^2\,[(\omega - \omega')^2/\omega\omega' + 4(ee')^2] \tag{5-58}$$

Averaging over initial, and summing over final photon polarization states gives, for the last term,

$$(ee')^2 \to \frac{1}{2}\sum_{\lambda}\sum_{\lambda'}(e\,e')^2 = \frac{1}{2}(1 + \cos^2\theta) \tag{5-59}$$

and adds a factor of 2 to the other term. Therefore, we have

$$\frac{d\sigma}{d\Omega} = \frac{1}{2}\left(\frac{e^2}{4\pi}\right)^2 [m + \omega(1 - \cos\theta)]^{-2}\left[\frac{\omega^2\,(1 - \cos\theta)^2}{m\,[m + \omega(1 - \cos\theta)]} + 1 + \cos^2\theta\right] \tag{5-60}$$

We may prove the replacement of $(e\,e')^2$ by $\frac{1}{2}(1 + \cos^2\theta)$ as follows:

Choose the z axis in the direction of $\mathbf{k}'$, then

$$\sum_{\lambda'=1}^{2} (e\,e')^2 = e_x^{\;2} + e_y^{\;2} = e^2 - e_z^{\;2} = 1 - (e\,k')^2/k'^2$$

$$\frac{1}{2}\sum_{\lambda=1}^{2}\sum_{\lambda'=1}^{2}(e\,e')^2 = \frac{1}{2}\sum_{\lambda=1}^{2}\left[1 - (e\,k')^2/k'^2\right]$$

$$= 1 - (1/2k'^2)\sum_{\lambda=1}^{2}(e\,k')^2$$

$$= 1 - (1/2k'^2)\left(k'^2 - \frac{(k\,k')^2}{k^2}\right)$$

$$= \frac{1}{2}\left(1 + (k\,k')^2/k^2 k'^2\right)$$

$$= \frac{1}{2}(1 + \cos^2\theta) \tag{5-61}$$

At low energies, $\omega << m$ and the formula reduces to

$$\frac{d\sigma}{d\Omega} = \left(\frac{e^2}{4\pi m}\right)^2 \frac{1 + \cos^2\theta}{2}$$

$$= r_0^2\,\frac{1 + \cos^2\theta}{2} \tag{5-62}$$

where r_0 is the classical electronic radius $e^2/4\pi m$, and

$$r_0 = 2.818 \times 10^{-13}\ \text{cm} \tag{5-63}$$

Integration over the angular variable easily gives the total cross section

$$\sigma_{\text{total}} = (8\pi/3)r_0^2 = \sigma_{\text{Thomson}} = 6.65 \times 10^{-25}\ \text{cm}^2 \tag{5-64}$$

A more exact formula may be obtained by integrating the Klein-Nishina formula for the differential cross section:

$$\sigma_{\text{total}} = \int d\Omega\, r_0^2\, \frac{1 + \cos^2\theta}{2}\, \frac{1}{[1 + \gamma(1 - \cos\theta)]^2}$$

$$\times \left\{1 + \frac{\gamma^2(1 - \cos\theta)^2}{(1 + \cos^2\theta)\,[1 + \gamma(1 - \cos\theta)]}\right\}$$

$$= \sigma_{\text{Thomson}} \times \frac{3}{4}\left\{\frac{1 + \gamma}{\gamma^3}\left[\frac{2\gamma(1 + \gamma)}{1 + 2\gamma} - \log(1 + 2\gamma)\right]\right.$$

$$\left. + \frac{1}{2\gamma}\log(1 + 2\gamma) - \frac{1 + 3\gamma}{(1 + 2\gamma)^2}\right\} \tag{5-65}$$

where $\gamma = \omega/m$, the incident photon energy in units of the electron rest mass.

In the low-energy limit, $\gamma \ll 1$, we have

$$\sigma = \sigma_{\text{Thomson}}\,(1 - 2\gamma + (26/5)\gamma^2 + \ldots) \tag{5-66}$$

and in the high-energy limit, $\gamma \gg 1$, we have

$$\sigma = \sigma_{\text{Thomson}} \times (3/8\gamma)(\log(2\gamma) + 1/2) \tag{5-67}$$

The agreement of the Klein-Nishina formula with experiments is excellent.

5-3 PION-NUCLEON SCATTERING

As a problem analogous to Compton scattering we shall discuss pion-nucleon scattering. In quantum electrodynamics, perturbation theory gives an expansion in powers of the fine structure constant, $\alpha = e^2/4\pi = 1/137$ and in general the perturbation theory results show good agreement with experiments. In pion theory, however, the perturbation expansion is supposed to diverge since the expansion parameter $G^2/4\pi$ is nearly 15. Nevertheless it is worth while to evaluate scattering cross sections based on perturbation theory.

The interaction Hamiltonian density for the pion-nucleon system is given by

$$\mathcal{H}_{int} = iG\, \bar{\psi}\, \boldsymbol{\tau} \gamma_5\, \psi \cdot \boldsymbol{\varphi} + 4\pi\lambda\, (\boldsymbol{\varphi} \cdot \boldsymbol{\varphi})^2 \tag{5-68}$$

The second term is necessary for renormalization, and the parameter G is given by

$$\frac{G^2}{4\pi} \approx 15 \tag{5-69}$$

For discussing pion-nucleon scattering in the lowest-order approximation, only the first term gives a nonvanishing contribution, so we shall neglect the second term in this section. Then

$$\mathcal{H}_{int} = iG \left[2^{1/2} (\bar{\psi}_n \gamma_5 \psi_p \cdot \varphi^{\dagger} + \bar{\psi}_p \gamma_5 \psi_n \cdot \varphi) + \bar{\psi} \gamma_5 \tau_3 \psi \cdot \varphi_3\right] \tag{5-70}$$

The second-order S-matrix element for pion-nucleon scattering is given by

$$\langle p',\, q' | S | p,\, q \rangle = \frac{(-i)^2}{2!} \int d^4x' \int d^4x'' \langle p',\, q' | T[\mathcal{H}_{int}(x'), \mathcal{H}_{int}(x'')] | p, q \rangle \tag{5-71}$$

Let us discuss the following process for definiteness:

$$\pi^+ + p \rightarrow p + \pi^+ \tag{5-72}$$

The Feynman diagram corresponding to this process is given in Figure 5-2.

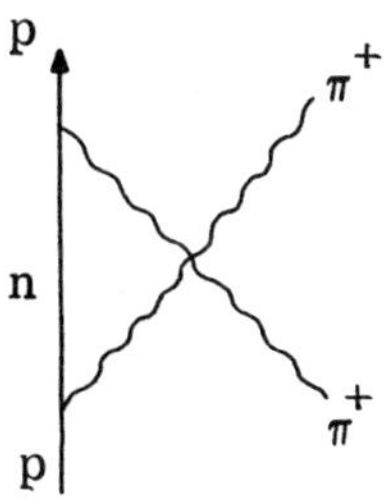

Figure 5-2. A diagram for the process $\pi^+ + p \rightarrow p + \pi^+$.

In the S-matrix element, we have two terms corresponding to two alternative possibilities, (1) either $\mathcal{H}(x')$ creates a π^+ and $\mathcal{H}(x'')$ destroys the incident π^+, or $\mathcal{H}(x'')$ creates a π^+ and $\mathcal{H}(x')$ destroys the incident π^+. These two terms give identical results since x' and x'' can be interchanged as variables of integration. Therefore, we keep only one of them, thereby omitting the factor 2! in the denominator.

$$\langle p', q'|S|p, q\rangle = (-i)^2 \int d^4x' \int d^4x''$$

$$\times (iG)^2 \langle p', q'|T[2^{1/2}\,\bar{\psi}_p(x')\gamma_5\,\psi_n(x')\varphi(x'),$$

$$2^{1/2}\,\bar{\psi}_n(x'')\gamma_5\psi_p(x'')\varphi^\dagger(x'')]\; p, q\rangle$$

$$= 2G^2 \int d^4x' \int d^4x'' \langle p'|\bar{\psi}_p(x')|0\rangle\, \gamma_5\, S_F(x'-x'')\gamma_5 \langle 0|\psi_p(x'')|p\rangle$$

$$\times \langle 0|\varphi(x')|q\rangle \quad \langle q'|\varphi^\dagger(x'')|0\rangle \tag{5-73}$$

where $S_F(x' - x'')$ results from the contraction of the virtual neutron operators, that is, $\psi^{\cdot}_n(x')\bar{\psi}^{\cdot}_n(x'')$. The single-particle wave functions in the integrand are given as before by

$$\langle 0|\psi_p(x)|p\rangle = V^{-1/2}\; u^{(r)}(p)e^{ipx} \qquad \text{etc.} \tag{5-74}$$

$$\langle 0|\varphi(x)|q\rangle = (2q_0V)^{-1/2}\; e^{iqx} \qquad \text{etc.} \tag{5-75}$$

Inserting these expressions and using procedures discussed in the preceding section we find

$$\langle p', q'|S|p, q\rangle = 2G^2(2\pi)^{8-4}(i)\delta^4(p+q-p'-q')\,V^{-2}\,(4q_0q_0')^{-1/2}$$

$$\times \bar{u}^{(r')}(p')\gamma_5 \frac{i\gamma(p-q')-m}{(p-q')^2+m^2}\,\gamma_5 u^{(r)}(p)$$

$$= -2iG^2(2\pi)^4\,\delta^4(p+q-p'-q')\,V^{-2}\,(4q_0q_0')^{-1/2}$$

$$\times \bar{u}^{(r')}(p')\frac{i\gamma(p-q')+m}{(p-q')^2+m^2}\,u^{(r)}(p) \tag{5-76}$$

where we have used $\gamma_5\;\gamma(p-q')\;\gamma_5 = -\gamma(p-q')$. Then we can also use

$$(i\gamma p + m)u^{(r)}(p) = 0,$$

$$(p-q')^2 + m^2 = p^2 + m^2 - 2pq' + q'^2 = -(2pq' + \mu^2)$$

The relevant momenta are shown in Figure 5-3.

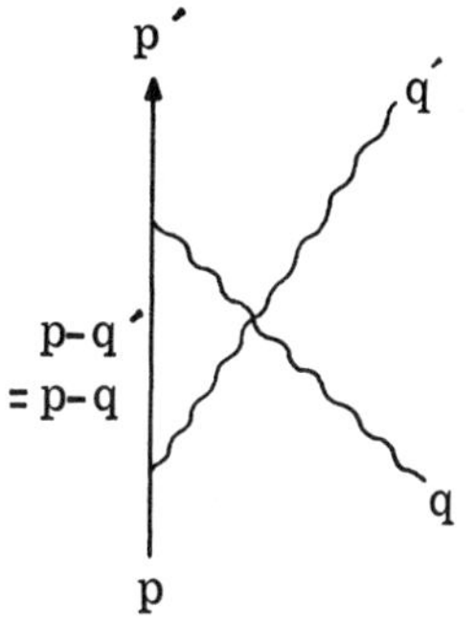

Figure 5-3. A diagram for $\pi^+ + p \to p + \pi^+$ labeled with momenta.

The S-matrix element reduces to

$$\langle p', q'|S|p, q\rangle = -2iG^2(2\pi)^4\delta^4(p + q - p' - q')V^{-2} \cdot (4q\, q')^{-1/2}$$

$$\times (2pq' + \mu^2)^{-1}\, \bar{u}^{(r')}(p')i\gamma q' u^{(r)}(p)$$

$$\equiv -i\delta^4(p + q - p' - q')t_{fi} \tag{5-77}$$

The cross section is given by

$$\sigma = \frac{V^2}{(2\pi)^4} \cdot \frac{1}{v_{rel}} \cdot \frac{V}{(2\pi)^3}\int d^3p' \frac{V}{(2\pi)^3}\int d^3q'$$

$$\times\, \delta^4(p + q - p' - q')\left|t_{fi}\right|^2 \tag{5-78}$$

In the laboratory system v_{rel} can be set equal to the velocity of the incident π^+ meson, namely, $v_{rel} = |\mathbf{q}|/q_0$. But here we shall calculate the cross section in the center of mass system such that $\mathbf{q} + \mathbf{p} = 0$. Then the relative velocity is given by

$$v_{rel} = (|\mathbf{q}|/q_0) + (|\mathbf{p}|/p_0) \tag{5-79}$$

and the cross section by

$$\sigma = \frac{1}{(2\pi)^2}\, \frac{1}{v_{rel}}\, 4G^4\int d^3p'\int d^3q'\, \delta^4(p + q - p' - q')$$

$$\times (4q_0q_0')^{-1}(2pq' + \mu^2)^{-2}\left|\bar{u}^{(r')}(p')i\gamma q' u^{(r)}(p)\right|^2 \tag{5-80}$$

In evaluating this expression in the center of mass system we introduce the following set of notations:

$$|\mathbf{q}| = |\mathbf{p}| \equiv q \qquad q_0 = \omega \qquad p_0 = E$$

$$|\mathbf{q}'| = |\mathbf{p}'| \equiv q' \qquad q'_0 = \omega' \qquad p'_0 = E' \tag{5-81}$$

Then

$$\begin{aligned}
&\int d^3p'\, d^3q'\, \delta^4(p' + q' - p - q) \ldots \\
&= \int d^3p'\, d^3q'\, \delta(\mathbf{p}' + \mathbf{q}')\, \delta(E' + \omega' - E - \omega) \ldots \\
&= \int d^3q'\, \delta(E' + \omega' - E - \omega) \ldots \\
&= \int q'^2 dq' d\Omega\, \delta(E' + \omega' - E - \omega) \ldots
\end{aligned} \tag{5-82}$$

The argument of the δ function vanishes for $q' = q$, since

$$\begin{aligned}
E' + \omega' - E - \omega = &\left(m^2 + q'^2\right)^{1/2} + \left(\mu^2 + q'^2\right)^{1/2} \\
&- \left(m^2 + q^2\right)^{1/2} - \left(\mu^2 + q^2\right)^{1/2}
\end{aligned} \tag{5-83}$$

so that the q' integration can be immediately carried out, leading to

$$\begin{aligned}
&\int \frac{q^2\, d\Omega}{\dfrac{\partial}{\partial q'}\left[(m^2 + q'^2)^{1/2} + (\mu^2 + q^2)^{1/2}\right]} \\
&= \int \frac{q^2\, d\Omega}{q(m^2 + q^2)^{-1/2} + q(\mu^2 + q^2)^{-1/2}} \\
&= \frac{E\omega q}{E + \omega} \int d\Omega \ldots
\end{aligned} \tag{5-84}$$

On the other hand, $1/v_{rel}$ is given by

$$\frac{1}{v_{rel}} = \left(\frac{q}{\omega} + \frac{q}{E}\right)^{-1} = \frac{E\,\omega}{q\,(E + \omega)} \tag{5-85}$$

so that

$$\frac{d\sigma}{d\Omega} = \frac{1}{(2\pi)^2} \frac{E\omega}{q(E+\omega)} \frac{E\omega q}{E+\omega} \frac{G^4}{\omega^2} \left(\frac{1}{2pq' + \mu^2}\right)^2$$

$$\times \left| \bar{u}^{(r')}(p')i\gamma q' u^{(r)}(p) \right|^2$$

$$= \frac{G^4}{(2\pi)^2} \frac{E^2}{(E+\omega)^2} \left(\frac{1}{2pq' + \mu^2}\right)^2 \left|\bar{u}^{(r')}(p')i\gamma q' u^{(r)}(p)\right|^2 \quad (5\text{-}86)$$

We next average over initial and sum over final proton spin states, so that

$$\left|\bar{u}^{(r')}(p')i\gamma q' u^{(r)}(p)\right|^2 = \bar{u}^{(r')}(p')i\gamma q' u^{(r)}(p)\bar{u}^{(r)}(p)i\gamma q' u^{(r')}(p')$$

is replaced by

$$\frac{1}{2} \operatorname{Tr} \left(\frac{i\gamma p' - m}{2E} \; i\gamma q' \; \frac{i\gamma p - m}{2E} \; i\gamma q'\right)$$

$$= \frac{1}{2E^2} [(p'q')(2pq') + \mu^2(pp' + m^2)] \quad (5\text{-}87)$$

The four-dimensional scalar products may be expressed in terms of the center of mass scattering angle θ and other quantities like q, ω, and E:

$$2pq' + \mu^2 = \mu^2 - 2E\omega - 2q^2 \cos\theta$$

$$p'q' = -q^2 - E\omega$$

$$pq' = -q^2 \cos\theta - E\omega$$

$$pp' = q^2 \cos\theta - E^2 \quad (5\text{-}88)$$

Hence

$$\frac{d\sigma}{d\Omega} = \frac{1}{(E+\omega)^2} \cdot \frac{G^4}{8\pi^2}$$

$$\cdot \frac{2(E\omega + q^2)^2 - q^2(2q^2 + 2E\omega + \mu^2)(1 - \cos\theta)}{(2E\omega - \mu^2 + 2q^2 \cos\theta)^2} \quad (5\text{-}89)$$

At low energies $q^2 << \mu^2$, and we find

$$\frac{d\sigma}{d\Omega} \approx \frac{G^4}{4\pi^2} \left(\frac{m}{m+\mu}\right)^2 \frac{1}{(2m-\mu)^2} \approx \left(\frac{G^2}{4\pi m}\right)^2 \tag{5-90}$$

neglecting the pion mass μ as compared with the nucleon mass m. This cross section may be compared with the differential cross section for Compton scattering

$$\frac{1}{2}\,(e^2/4\pi m)^2\,(1+\cos^2\theta)$$

We find

$$(G^2/4\pi m) \sim 3\times 10^{-13}\ \text{cm} \tag{5-91}$$

and

$$\sigma = 4\pi\frac{d\sigma}{d\Omega} \sim 1\ \text{barn} \quad (\text{barn} = 10^{-24}\ \text{cm}^2) \tag{5-92}$$

Experimentally, this cross section has been found to be of the order of a few mb (mb $= 10^{-27}$ cm^2), apparently confirming our initial doubts about the validity of perturbation theory for this process.

5-4 APPLICATION TO DECAY PROCESSES

Instead of discussing complicated processes in electrodynamics such as Bremsstrahlung, we study applications of perturbation theory to typical decay processes. The decay processes we study in this section are

$$\text{(a)}\ \Lambda^0 \rightarrow p + \pi^- \ \text{and (b)}\ \pi^0 \rightarrow 2\gamma$$

Λ^0 Decay

The interaction that gives rise to Λ decay may be written as

$$\mathcal{H}_{int} = if\,\bar{\psi}_p\,\mathcal{O}\,\psi_\Lambda \cdot \varphi + \text{Hermitian conjugate} \tag{5-93}$$

As we shall see later, the coupling constant f^2 is very small. In weak interactions such as this, parity is not conserved, and the general form of the interaction is therefore given by

$$\mathcal{H}_{int} = \bar{\psi}_p\,(f_S + f_P\,\gamma_5)\,\psi_\Lambda \cdot \varphi + \text{H. c.} \tag{5-94}$$

One also may add derivative couplings, as discussed in Section 4-1. For simplicity let us assume that $f_S = 0$, $f_P = f \neq 0$.

$$\mathcal{H}_{int} = f\, \bar{\psi}_p \gamma_5 \psi_\Lambda \cdot \varphi + \text{H. c.} \tag{5-95}$$

Then the S-matrix element for this decay in the lowest (first) order can be obtained as was done in the preceding sections.

$$\begin{aligned}\langle p\pi | S | \Lambda\rangle &= -i \int d^4x \;\langle p\pi^- | \mathcal{H}_{int}(x) | \Lambda\rangle \\ &= -if \int d^4x \;\langle p | \bar{\psi}_p(x) | 0\rangle \gamma_5 \langle 0 | \psi_\Lambda(x) | \Lambda\rangle \langle \pi^- | \varphi(x) | 0\rangle\end{aligned} \tag{5-96}$$

Designating the momenta of the Λ, the proton, and the π^- by P, p, and q, respectively, we can write the S-matrix element as

$$\langle pq | S | P\rangle = -i(2\pi)^4 \;\delta^4(P - p - q)\,(f/V)\,(2q_0 V)^{-1/2}\, \bar{u}_p(p)\, \gamma_5\, u_\Lambda(P) \tag{5-97}$$

Therefore, the transition probability is given by

$$\begin{aligned}w &= \frac{V}{(2\pi)^4}\, V \int \frac{d^3p}{(2\pi)^3}\, V \int \frac{d^3q}{(2\pi)^3}\; \delta^4(P - p - q) \\ &\quad \times (2\pi)^8\, \frac{f^2}{V^3}\, \frac{1}{2q_0}\; |\bar{u}_p(p)\gamma_5 u_\Lambda(P)|^2 \\ &= \frac{f^2}{(2\pi)^2} \int d^3p \int d^3q\;\; \delta^4(P - p - q)\; \frac{1}{2q_0} |\bar{u}_p(p)\, \gamma_5\, u_\Lambda(P)|^2\end{aligned} \tag{5-98}$$

The square of the spinor matrix element is written out as

$$|\bar{u}_p(p)\;\gamma_5 u_\Lambda(P)|^2 = -\bar{u}_p(p)\,\gamma_5 u_\Lambda(P)\,\bar{u}_\Lambda(P)\,\gamma_5 u_p(p)$$

and the average over the Λ spin states and sum over the possible proton spin states is carried out using the Casimir operators for each particle:

$$\begin{aligned}\frac{1}{2} \sum_{\Lambda\,\text{spin}} \sum_{p\,\text{spin}} |\bar{u}_p(p)\,\gamma_5 u_\Lambda(P)|^2 &= -\frac{1}{2}\,\text{Tr}\left[\frac{-ip\gamma + m}{2p_0}\,\gamma_5\, \frac{-iP\gamma + M}{2P_0}\,\gamma_5\right] \\ &= -\frac{1}{2p_0 P_0}\,(pP + mM)\end{aligned} \tag{5-99}$$

where m is the proton mass and M the Λ mass. We take a frame of reference in which the Λ^0 is at rest, then $\mathbf{P} = 0$, $P_0 = M$, and the decay probability is given by

$$w = \frac{f^2}{(2\pi)^2} \int \frac{d^3q}{2q_0} \, \delta(M - p_0 - q_0) \, \frac{-pP - mM}{2p_0 M} \tag{5-100}$$

Since $q = P - p$ and $q^2 = P^2 + p^2 - 2Pp$, we find

$$-Pp = \frac{1}{2}(q^2 - P^2 - p^2) = \frac{1}{2}(M^2 + m^2 - \mu^2) \tag{5-101}$$

so that

$$w = \frac{f^2}{(2\pi)^2} \left(\frac{(M - m)^2 - \mu^2}{8M} \right) \int \frac{d^3q}{q_0 p_0} \, \delta(M - p_0 - q_0) \tag{5-102}$$

In the rest frame we have $\mathbf{p} + \mathbf{q} = 0$ so that we introduce the magnitude of the center of mass momentum q by

$$|\mathbf{p}| = |\mathbf{q}| \equiv q$$

then the integration can be carried out as was done before:

$$\int d^3q \, \delta[M - (m^2 + q^2)^{1/2} - (\mu^2 + q^2)^{1/2}]$$

$$= \int q^2 dq d\Omega \, \delta\left[M - (m^2 + q^2)^{1/2} - (\mu^2 + q^2)^{1/2}\right]$$

$$= \frac{4\pi q^2}{\frac{\partial}{\partial q}\left[(m^2 + q^2)^{1/2} + (\mu^2 + q^2)^{1/2}\right]}$$

$$= 4\pi \, q_0 p_0 (q/M) \tag{5-103}$$

Thus the decay rate is

$$w = (f^2/4\pi)\,(q/M^2)\,\frac{1}{2}\,((M - m)^2 - \mu^2) \tag{5-104}$$

The magnitude q is fixed by the conservation of energy,

$$M = (m^2 + q^2)^{1/2} + (\mu^2 + q^2)^{1/2} \tag{5-105}$$

from which we find

$$q = (1/2M) \, [(M - m)^2 - \mu^2]^{1/2} \, [(M + m)^2 - \mu^2]^{1/2} \tag{5-106}$$

The decay rate or the decay width now can be written in terms of the masses of the three particles:

$$w = \Gamma(\Lambda^0 \to p + \pi^-) = (f^2/4\pi)\,(1/4M^3)\,[(M - m)^2 - \mu^2]^{3/2} \times [(M + m)^2 - \mu^2]^{1/2} \tag{5-107}$$

The lifetime τ is the inverse of w or Γ, and experimentally

$$\tau \sim 3 \times 10^{-10} \text{ sec} \tag{5-108}$$

This leads to the following value of the coupling constant:

$$f^2/4\pi \sim 10^{-11} \tag{5-109}$$

This is much smaller than $G^2/4\pi$ and $e^2/4\pi$.

Actually Λ^0 can also decay into $n + \pi^0$, so that the lifetime of the Λ^0 is given by

$$1/\tau = \Gamma(\Lambda^0 \to p + \pi^-) + \Gamma(\Lambda^0 \to n + \pi^0) \tag{5-110}$$

In converting 1/M in the natural units into seconds, we use

$$1/M \to \hbar/Mc^2 \sim 0.6 \times 10^{-24} \text{ sec} \tag{5-111}$$

π^0 Decay

We now shall attempt to evaluate the π^0 lifetime, but first let us study the related selection rules. We require that the pion-nucleon interaction, in particular the π^0-nucleon interaction,

$$iG\bar{\psi}\gamma_5\tau_3\psi \cdot \varphi_3 = iG(\bar{\psi}_p\gamma_5\psi_p - \bar{\psi}_n\gamma_5\psi_n)\,\varphi_3 \tag{5-112}$$

be invariant under charge conjugation. The nucleon current in this interaction transforms under charge conjugation as

$$\mathcal{C}(\bar{\psi}_p \gamma_5 \psi_p - \bar{\psi}_n \gamma_5 \psi_n)\,\mathcal{C}^{-1} = \bar{\psi}_p \gamma_5 \psi_p - \bar{\psi}_n \gamma_5 \psi_n \tag{5-113}$$

and hence invariance of the pion-nucleon interaction under $\mathcal{C}$ implies

$$\mathcal{C}\,\varphi_3\,\mathcal{C}^{-1} = \varphi_3 \tag{5-114}$$

Hence we learn that the π^0 state is even under $\mathcal{C}$. Now the one-photon state is odd under $\mathcal{C}$; thus we conclude that the π^0 can decay only into an even number of photons. This sort of statement is called the generalized Furry's theorem.

Let us now proceed to the process $\pi^0 \to 2\gamma$. The appropriate lowest-order Feynman diagram is given in Fig. 5-4.

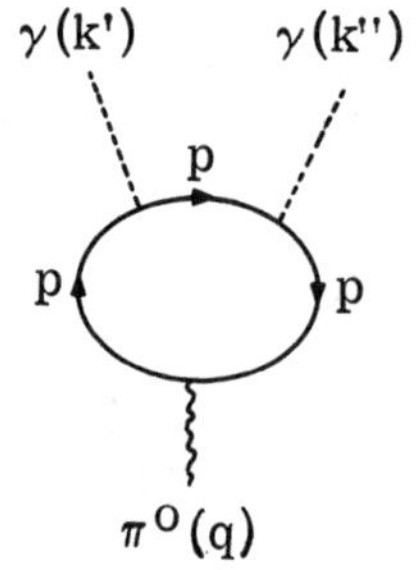

Figure 5-4. The lowest-order diagram for $\pi^0 \to 2\gamma$.

The interactions responsible for this diagram are

$$\mathcal{H}_{\pi^0} = iG\bar{\psi}_p \gamma_5 \psi_p \cdot \varphi_3 \tag{5-115}$$

$$\mathcal{H}_{em} = -ie\,\bar{\psi}_p \gamma_\mu \psi_p \cdot A_\mu \tag{5-116}$$

where e denotes the protonic charge. Hence the corresponding S-matrix element is

$$\langle k', k''|S|q\rangle = \frac{(-i)^3}{3!}\int d^4x' \int d^4x'' \int d^4x'''$$

$$\times \langle k', k'' \,|\, T\,[\mathcal{H}(x')\,\mathcal{H}(x'')\,\mathcal{H}(x''')\,]\,|\, q\rangle \tag{5-117}$$

where q, k', and k'' denote, respectively, the π^0 momentum and the photon momenta. Of the three Hamiltonians involved, two must be equal to $\mathcal{H}_{em}$ and the other one must be $\mathcal{H}_{\pi^0}$. There are three distinct ways to assign the variables of integration and all three cases contribute equally to the S-matrix element. Therefore we have

$$
\begin{aligned}
\langle k', k'' | S | q \rangle = \frac{(-i)^3}{2!} & \int d^4x' \int d^4x'' \int d^4x''' \\
& \times \langle k', k'' | T[\mathcal{H}_{em}(x'), \mathcal{H}_{em}(x''), \mathcal{H}_{\pi^0}(x''')] | q \rangle \\
= \frac{e^2 G}{2} & \int d^4x' \int d^4x'' \int d^4x''' \\
& \times \langle k', k'' | :A_\mu(x')A_\nu(x''): | 0 \rangle \langle 0 | \varphi_3(x''') | q \rangle \\
& \times \langle 0 | T[\bar{\psi}(x')\gamma_\mu\psi(x'), \bar{\psi}(x'')\gamma_\nu\psi(x''), \\
& \bar{\psi}(x''')\gamma_5\psi(x''')] | 0 \rangle
\end{aligned}
\tag{5-118}
$$

Now

$$
\begin{aligned}
\langle k', k'' | :A_\mu(x')A_\nu(x''): | 0 \rangle = {} & \langle k' | A_\mu(x') | 0 \rangle \langle k'' | A_\nu(x'') | 0 \rangle \\
& + \langle k'' | A_\mu(x') | 0 \rangle \langle k' | A_\nu(x'') | 0 \rangle
\end{aligned}
\tag{5-119}
$$

Since x', x'' and μ, ν are integration variables and dummy indices, it is easy to see that these two terms contribute equally to the S-matrix element, so that

$$
\begin{aligned}
\langle k', k'' | S | q \rangle = e^2 G^2 & \int d^4x' \int d^4x'' \int d^4x''' \langle k' | A_\mu(x') | 0 \rangle \\
& \times \langle k'' | A_\nu(x'') | 0 \rangle \langle 0 | \varphi_3(x''') | q \rangle \\
& \times \langle 0 | T[\psi(x')\gamma_\mu\psi(x'), \bar{\psi}(x'')\gamma_\nu\psi(x''), \\
& \bar{\psi}(x''')\gamma_5\psi(x''')] | 0 \rangle
\end{aligned}
\tag{5-120}
$$

There are two ways to contract the proton field operators in the vacuum expectation value; they are illustrated pictorially in Fig. 5-5. The only difference between these two diagrams consists in the sense of direction of the proton lines.

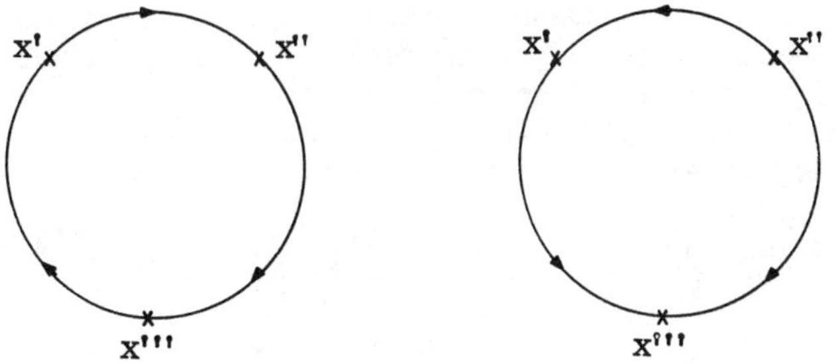

Figure 5-5. Two closed loops of the opposite directions.

$$\langle 0 | T[\cdots] | 0\rangle = -\mathrm{Tr}[\gamma_5 S_F(x''' - x'')\gamma_\nu S_F(x'' - x')\gamma_\mu S_F(x' - x''')]$$
$$-\mathrm{Tr}[\gamma_5 S_F(x''' - x')\gamma_\mu S_F(x' - x'')\gamma_\nu S_F(x'' - x''')] \tag{5-121}$$

These two terms are equal, as can be seen from the following argument. First we show that

$$C^{-1} S_F(x - y) C = S_F(y - x)^T \tag{5-122}$$

We recall that $S_F(x) = (-\gamma\partial + m)\Delta_F(x)$, that $\Delta_F(x)$ is even in the sense $\Delta_F(-x) = \Delta_F(x)$, and that $C^{-1}\gamma_\mu C = -\gamma_\mu^T$, $C^{-1}\gamma_5 C = \gamma_5^T$. Then

$$C^{-1} S_F(x - y) C = C^{-1}(-\gamma_\mu \frac{\partial}{\partial x_\mu} + m) C \Delta_F(x - y)$$
$$= (\gamma_\mu^T \frac{\partial}{\partial x_\mu} + m)\Delta_F(y - x)$$
$$= (-\gamma_\mu^T \frac{\partial}{\partial y_\mu} + m)\Delta_F(y - x)$$
$$= S_F(y - x)^T \tag{5-123}$$

Next we use the fact that $\mathrm{Tr}\, A^T = \mathrm{Tr}\, A$ with the above result to obtain the desired proof:

$$\mathrm{Tr}[\gamma_5 S_F(x''' - x'')\gamma_\nu S_F(x'' - x')\gamma_\mu S_F(x' - x''')]^T$$
$$= \mathrm{Tr}[S_F(x' - x''')^T \gamma_\mu^T S_F(x'' - x')^T \gamma_\nu^T S_F(x''' - x'')^T \gamma_5^T]$$
$$= \mathrm{Tr}[C^{-1} S_F(x''' - x') C \cdot C^{-1}(-\gamma_\mu) C \cdot C^{-1} S_F(x' - x'') C$$
$$\times C^{-1}(-\gamma_\mu) C \cdot C^{-1} S_F(x'' - x''') C \cdot C^{-1}\gamma_5 C]$$
$$= \mathrm{Tr}[\gamma_5 S_F(x''' - x')\gamma_\mu S_F(x' - x'')\gamma_\nu S_F(x'' - x''')] \tag{5-124}$$

This demonstrates that the two trace terms are equal, as claimed. It is clear that they contribute with the opposite signs if there are three γ's, corresponding to $\pi^0 \to 3\gamma$. In general two trace terms, corresponding to the same closed loop but with the directions reversed, cancel one another when the process under consideration is forbidden by charge conjugation.

The S-matrix element we now have to evaluate has the form

$$\begin{aligned}\langle k', k'' | S | q \rangle = & -2e^2 G \int d^4x' \int d^4x'' \int d^4x''' \\ & \times \langle k' | A_\mu(x') | 0 \rangle \langle k'' | A_\nu(x'') \, 0 \rangle \langle 0 | \varphi_3(x''') | q \rangle \\ & \times \mathrm{Tr}[\gamma_5 S_F(x''' - x') \gamma_\mu S_F(x' - x'') \\ & \times \gamma_\nu S_F(x'' - x''')] \end{aligned} \tag{5-125}$$

In evaluating this matrix element we choose the rest frame of the π^0, then the two photons in the final state are emitted in the opposite directions and carry the same amount of energy, namely, $\mu/2$ each.

The single-particle wave functions are

$$\langle 0 | \varphi_3(x''') | q \rangle = (2\mu V)^{-1/2} e^{iqx'''} \tag{5-126}$$

$$\langle k' | A_\mu(x') | 0 \rangle = (\mu V)^{-1/2} e_\mu^{(\lambda')} e^{-ik'x'} \tag{5-127a}$$

$$\langle k'' | A_\nu(x'') | 0 \rangle = (\mu V)^{-1/2} e_\nu^{(\lambda'')} e^{-ik''x''} \tag{5-127b}$$

From here on, we shall refer to $e_\mu^{(\lambda')}$ as e'_μ and to $e_\nu^{(\lambda'')}$ as e''_ν.[3] We also recall the Fourier representation of $S_F(x)$:

$$S_F(x) = \frac{i}{(2\pi)^4} \int d^4p \, e^{ipx} \frac{ip\gamma - m}{p^2 + m^2 - i\epsilon} \tag{5-128}$$

Inserting these expressions into the S-matrix element, we find

$$\langle k', k'' | S | q \rangle = -2e^2 G \frac{e'_\mu e''_\nu}{\mu V} \cdot (2\mu V)^{-1/2} \int d^4x' \int d^4x''$$

$$\times \int d^4x''' \exp(-ik'x' - ik''x'' + iqx''')$$

$$\times \frac{i^3}{(2\pi)^{12}} \int d^4a d^4b d^4c \, \mathrm{Tr} \left[\gamma_5 \frac{ia\gamma - m}{a^2 + m^2 - i\epsilon} \right.$$

$$\left. \times \gamma_\mu \frac{ib\gamma - m}{b^2 + m^2 - i\epsilon} \gamma_\nu \frac{ic\gamma - m}{c^2 + m^2 - i\epsilon} \right]$$

$$\times \exp[ia(x''' - x') + ib(x' - x'') + ic(x'' - x''')]$$

(5-129)

where a, b, and c are the momenta of the virtual protons in the closed loop. We perform x integrations to obtain the three δ functions that express the conservation of energy momentum at the vertices.

$$\langle k'k'' | S | q \rangle = -\sqrt{2}\, e^2 G \frac{e'_\mu e''_\nu}{(\mu V)^{3/2}} \cdot i^3 (2\pi)^{12-12} \int d^4a d^4b d^4c$$

$$\times \delta^4(b - a - k') \delta^4(c - b - k'') \delta^4(q + a - c)$$

$$\times \mathrm{Tr} \left[\gamma_5 \frac{ia\gamma - m}{a^2 + m^2 - i\epsilon} \gamma_\mu \frac{ib\gamma - m}{b^2 + m^2 - i\epsilon} \right.$$

$$\left. \times \gamma_\nu \frac{ic\gamma - m}{c^2 + m^2 - i\epsilon} \right] \tag{5-130}$$

We first perform integrations over a and c by means of two of the δ functions which require that

$$a = b - k' \qquad \text{and} \qquad c = b + k''$$

leaving the remaining δ function expressing the over-all conservation of energy momentum:

$$\int d^4a d^4c \, \delta^4(b - a - k') \delta^4(c - b - k'') \delta^4(q + a - c) = \delta^4(q - k' - k'')$$

(5-131)

Expressing a and c in terms of b, we find the S-matrix element corresponding to Fig. 5-6.

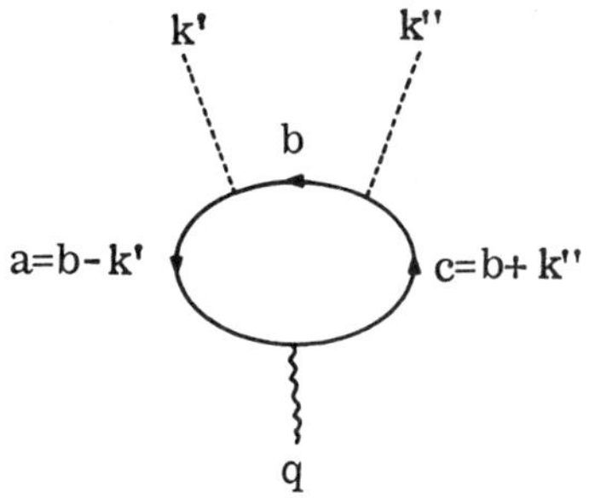

Figure 5-6. A diagram for $\pi^0 \to 2\gamma$ labeled with momenta.

$$\langle k', k'' | S | q \rangle = i\sqrt{2}\, e^2 G \frac{e'_\mu e''_\nu}{(\mu V)^{3/2}} \delta^4(q - k' - k'')$$
$$\times \int d^4 b \; \mathrm{Tr}\left[\gamma_5 \frac{i(b-k')\gamma - m}{(b-k')^2 + m^2 - i\epsilon}\right.$$
$$\left. \times \gamma_\mu \frac{ib\gamma - m}{b^2 + m^2 - i\epsilon} \gamma_\nu \frac{i(b+k'')\gamma - m}{(b+k'')^2 + m^2 - i\epsilon}\right] \tag{5-132}$$

Clearly, not all the momenta of the process are determined uniquely; there is a freedom in the choice of the momenta of the virtual protons over which an integration has to be carried out.

Thus the problem is reduced to the evaluation of the four-dimensional integration, (5-132). The general method to integrate such expressions is called Feynman's method.

In order to simplify the integrand, let us first evaluate the trace.

$$\mathrm{Tr} = \mathrm{Tr}\,[\gamma_5(i(b-k')\gamma - m)\gamma_\mu(ib\gamma - m)\gamma_\nu(i(b+k'')\gamma - m)] \tag{5-133}$$

using the following facts:

$$\mathrm{Tr}(\gamma_5) = \mathrm{Tr}(\gamma_5\gamma_\alpha) = \mathrm{Tr}(\gamma_5\gamma_\alpha\gamma_\beta) = \mathrm{Tr}(\gamma_5\gamma_\alpha\gamma_\beta\gamma_\gamma) = 0$$

$$\mathrm{Tr}(\gamma_5\gamma_\alpha\gamma_\beta\gamma_\gamma\gamma_\delta\gamma_\epsilon) = 0$$

$$\mathrm{Tr}(\gamma_5\gamma_\alpha\gamma_\beta\gamma_\gamma\gamma_\delta) = 4\,\epsilon_{\alpha\beta\gamma\delta}$$

where

$$\epsilon_{\alpha\beta\gamma\delta} = \begin{cases} 1 & \text{if } (\alpha\beta\gamma\delta) \text{ is an even permutation of } (1234) \\ -1 & \text{if } (\alpha\beta\gamma\delta) \text{ is an odd permutation of } (1234) \\ 0 & \text{otherwise} \end{cases}$$

Then

$$\begin{aligned} \mathrm{Tr} &= -4m\, \epsilon_{\mu\nu\alpha\beta} [b_\alpha (b + k'')_\beta - (b - k')_\alpha (b + k'')_\beta + (b - k')_\alpha b_\beta] \\ &= -4m\, \epsilon_{\mu\nu\alpha\beta} k'_\alpha k''_\beta \end{aligned} \qquad (5\text{-}134)$$

since $\epsilon_{\mu\nu\alpha\beta} b_\alpha b_\beta = 0$. Since Tr is independent of b, it can be factored out of the integral. Hence

$$\begin{aligned} \langle k', k'' |S| q\rangle &= -i4m \sqrt{2}\, e^2 G \frac{1}{(\mu V)^{3/2}} (\epsilon_{\mu\nu\alpha\beta} e'_\mu e''_\nu k'_\alpha k''_\beta) \\ &\times \delta^4(q - k' - k'') \int d^4 b \\ &\times \frac{1}{[(b-k')^2 + m^2 - i\epsilon][b^2 + m^2 - i\epsilon][(b+k'')^2 + m^2 - i\epsilon]} \end{aligned} \qquad (5\text{-}135)$$

Now we shall discuss Feynman's method. He found the following formula.

$$\int d^4 p \frac{1}{[p^2 + \Lambda - i\epsilon]^3} = \frac{i\pi^2}{2(\Lambda - i\epsilon)} \qquad (5\text{-}136)$$

Since this is one of the fundamental formulas in the Feynman-Dyson theory we shall prove it here.

From the expression for the integral

$$\int_0^\infty e^{-i\alpha x}\, dx = \lim_{\epsilon\to 0} \int_0^\infty e^{-i\alpha x - \epsilon x}\, dx = \lim_{\epsilon \to 0} \frac{-i}{\alpha - i\epsilon} \qquad (5\text{-}137)$$

we have

$$\int_0^\infty \exp\, [-i(p^2 + \Lambda)x]\, dx = \frac{-i}{p^2 + \Lambda - i\epsilon} \qquad (5\text{-}138)$$

or, differentiating twice with respect to Λ

$$\int_0^\infty x^2 \exp[-i(p^2 + \Lambda)x]\, dx = \frac{2i}{(p^2 + \Lambda - i\epsilon)^3} \qquad (5\text{-}139)$$

Then we integrate the above equation over p, using the Gaussian integral formula

$$\int \exp(-ip^2 x)\, d^4 p = -\frac{i\pi^2}{x^2} \qquad (x > 0) \qquad (5\text{-}140)$$

The left-hand side can be expressed as a product of four Gaussian integrals so that the right-hand side immediately follows, namely,

$$\int \exp(-ip^2 x)\, d^4 p = \left(\int \exp(-ip_1^{\,2} x)\, dp_1\right)\left(\int \exp(-ip_2^{\,2} x)\, dp_2\right)$$

$$\left(\int \exp(-ip_3^{\,2} x) dp_3\right)\left(\int \exp(ip_0^{\,2} x)\, dp_0\right)$$

The integration of the integral representation of $(p^2 + \Lambda - i\epsilon)^{-3}$ with respect to p, with the help of the Gaussian integral formula, yields

$$\int d^4 p \int_0^\infty dx\, x^2 \exp[-i(p^2 + \Lambda)x] = -i\pi^2 \int_0^\infty dx\, \exp(-i\Lambda x)$$

$$= -\frac{\pi^2}{\Lambda - i\epsilon} = 2i \int \frac{d^4 p}{(p^2 + \Lambda - i\epsilon)^3}$$

Therefore

$$\int \frac{d^4 p}{(p^2 + \Lambda - i\epsilon)^3} = \frac{i\pi^2}{2(\Lambda - i\epsilon)} \qquad \text{Q.E.D.} \qquad (5\text{-}136)$$

Before we can apply this formula, however, we must put the integrand into the form $[p^2 + \Lambda - i\epsilon]^{-3}$. For this purpose the following prescription is used:

$$\frac{1}{AB} = \int_0^1 \frac{dx}{[xA + (1 - x)B]^2} \qquad (5\text{-}141)$$

which is valid provided that the denominator does not vanish. From (5-141) we can derive many similar formulas such as

$$\frac{1}{ABC} = 2\int_0^1 dx \int_0^1 x\,dy \, \frac{1}{[A(1-x) + Bxy + Cx(1-y)]^3} \tag{5-142}$$

We go back to the integral

$$I = \int d^4b \, \frac{1}{ABC} \tag{5-143}$$

where

$$\begin{aligned} A &= (b-k')^2 + m^2 - i\epsilon = b^2 - 2bk' + m^2 - i\epsilon \\ B &= b^2 + m^2 - i\epsilon \\ C &= (b+k'')^2 + m^2 - i\epsilon = b^2 + 2bk'' + m^2 - i\epsilon \end{aligned} \tag{5-144}$$

and we have used $k'^2 = k''^2 = 0$ to simplify these expressions. Then

$$\begin{aligned} I &= \int d^4b \int_0^1 2dx \int_0^1 x\,dy \\ &\quad \times \frac{1}{[b^2 + m^2 + (1-x)(-2bk') + x(1-y)(2bk'') - i\epsilon]^3} \\ &= \int_0^1 2dx \int_0^1 x\,dy \int d^4b \\ &\quad \times \frac{1}{[(b-(1-x)k' + x(1-y)k'')^2 + m^2 - ((1-x)k' - x(1-y)k'')^2 - i\epsilon]^3} \\ &= 2\int_0^1 dx \int_0^1 x\,dy \int d^4p \\ &\quad \times \frac{1}{[p^2 + m^2 + 2x(1-x)(1-y)\,k'k'' - i\epsilon]^3} \end{aligned} \tag{5-145}$$

where we have introduced a transformation of the variable of integration by

$$p = b - (1-x)\,k' + x(1-y)\,k'' \tag{5-146}$$

The p integration can be performed immediately since it is now of the standard form.

$$I = i\pi^2 \int_0^1 dx \int_0^1 x dy \frac{1}{m^2 + 2(1-x)\, x(1-y)\, k'k'' - i\epsilon} \tag{5-147}$$

From $q = k' + k''$, we get $-\mu^2 = q^2 = (k' + k'')^2 = 2k'k''$, so that

$$I = i\pi^2 \int_0^1 dx \int_0^1 x dy \frac{1}{m^2 - \mu^2 x(1-x)(1-y)} \tag{5-148}$$

Inasmuch as the denominator does not vanish, we drop the $i\epsilon$. We further notice that

$$m^2 \gg \frac{1}{4}\mu^2 \geq \mu^2 x(1-x)(1-y) \tag{5-149}$$

so that the integral is well approximated by

$$I \approx \frac{i\pi^2}{m^2} \int_0^1 dx \int_0^1 x dy = \frac{i\pi^2}{2m^2} \tag{5-150}$$

The S-matrix element reduces to

$$\langle k', k'' | S | q \rangle = \frac{2}{m}^{3/2} \pi^2 e^2 G \frac{1}{(\mu V)^{3/2}}$$
$$\times (\epsilon_{\mu\nu\alpha\beta}\, e'_\mu\, e''_\nu\, k'_\alpha\, k''_\beta)\, \delta^4(q - k' - k'') \tag{5-151}$$

The resulting transition probability or the decay width is given by

$$\Gamma(\pi^0 \to 2\gamma) = \left(\frac{1}{2}\right) \frac{V}{(2\pi)^4} \cdot \frac{V}{(2\pi)^3} \int d^3k' \frac{V}{(2\pi)^3} \int d^3k''$$
$$\times \delta^4(q - k' - k'') \left| t_{fi} \right|^2 \tag{5-152}$$

where the factor $\frac{1}{2}$ is introduced to avoid counting the same state twice because of the presence of identical particles in the final state. The phase space integral can be carried out immediately, as done in Λ decay, yielding

$$\Gamma(\pi^0 \to 2\gamma) = \frac{V^3}{(2\pi)^9} \cdot \frac{\mu^2}{8} \left| t_{fi} \right|^2 \tag{5-153}$$

or

$$1/\tau = \frac{V^3}{(2\pi)^9}\cdot\frac{\mu^2}{8}\cdot\frac{(2\pi)^4}{2m^2}\quad e^4G^2\frac{1}{(\mu V)^3}\left|\epsilon_{\mu\nu\alpha\beta}e'_\mu e''_\nu k'_\alpha k''_\beta\right|^2 \tag{5-154}$$

$$= \frac{1}{(2\pi)^5}\cdot\frac{1}{16m^2{}_\mu}\cdot\; e^4G^2\left|\epsilon_{\mu\nu\alpha\beta}e'_\mu e''_\nu k'_\alpha k''_\beta\right|^2$$

This is the transition probability for specific polarization states e' and e'' of the two photons. To get the total transition probability we must sum over all these polarization states:

$$\sum_{\lambda',\lambda''}\left|\epsilon_{\mu\nu\alpha\beta}\,e'_\mu e''_\nu k'_\alpha k''_\beta\right|^2$$

$$= -\sum_{\lambda',\lambda''}\epsilon_{\mu\nu\alpha\beta}\,e'_\mu e''_\nu k'_\alpha k''_\beta\,\epsilon_{\mu'\nu'\alpha'\beta'}e'_{\mu'}e''_{\nu'}k'_{\alpha'}k''_{\beta'} \tag{5-155}$$

where the minus sign results from the fact that ϵ is imaginary, since it always has the fourth component of a vector as a factor. Then, using the completeness conditions

$$\sum_{\lambda'} e'_\mu e'_{\mu'} = \delta_{\mu\mu'} \qquad \sum_{\lambda''} e''_\nu e''_{\nu'} = \delta_{\nu\nu'} \tag{5-156}$$

(5-155) becomes

$$-\epsilon_{\mu\nu\alpha\beta}\,\epsilon_{\mu\nu\alpha'\beta'}\,k'_\alpha k'_{\alpha'}\,k''_\beta k''_{\beta'}$$

$$= -2(\delta_{\alpha\alpha'}\,\delta_{\beta\beta'} - \delta_{\alpha\beta'}\,\delta_{\alpha'\beta})\,k'_\alpha k'_{\alpha'}\,k''_\beta k''_{\beta'}$$

$$= -2(k'^2k''^2 - (k'k'')^2)$$

$$= 2\,(k'k'')^2$$

$$= \frac{1}{2}\,\mu^4 \tag{5-157}$$

With this result, the total decay width is

$$1/\tau = \Gamma(\pi^0\to 2\gamma) = \frac{1}{(2\pi)^5}\cdot\frac{1}{16m^2{}_\mu}\cdot e^4G^2\cdot\frac{\mu^4}{2}$$

$$= \frac{\alpha^2}{64\pi^3}\;G^2\left(\frac{\mu}{m}\right)^2\mu \tag{5-158}$$

Converting this result into the CGS units, we get[4]

$$1/\tau = \frac{\alpha^2}{64\pi^3} \cdot \frac{G^2}{\hbar c}\left(\frac{\mu}{m}\right)^2 \left(\frac{\mu c^2}{\hbar}\right)$$

$$\tau = 0.7 \times 10^{-15} \left(\frac{G^2}{4\pi \hbar c}\right)^{-1} \text{ sec}$$

$$\approx 0.5 \times 10^{-16} \text{ sec} \tag{5-159}$$

This should be compared to the experimentally observed π^0 lifetime of $\sim 2 \times 10^{-16}$ sec. In this case the agreement between theory and experiment is much better than in pion-nucleon scattering.

Finally it is worthwhile to mention that we used two distinct ways of summing over the photon polarization states in handling the problems of Compton scattering and of π^0 decay. In the former problem summation was carried out only over the two transversely-polarized photon states, whereas in the latter problem it was done over all four polarization states including the longitudinal and scalar photons. In other words, we employed in the former problem the so-called Coulomb gauge in which the longitudinal and the scalar photons have been eliminated, while in the latter problem we employed the Lorentz gauge in which all four polarizations have been retained.

As long as we are dealing with gauge invariant quantities the final results should not depend on whether we employ the Coulomb or the Lorentz gauge.

5-5 DERIVATIVE COUPLINGS AND LAGRANGIAN METHOD

So far in this chapter we have discussed only simple interactions for which we may assume

$$\mathcal{H}_{int}(x) = -\mathcal{L}_{int}(x) \tag{5-160}$$

This is, however, no longer the case when $\mathcal{L}_{int}(x)$ involves the derivatives of field operators.[5] The total Lagrangian density is defined as

$$\mathcal{L}(x) = \mathcal{L}_f(x) + \mathcal{L}_{int}(x) \tag{5-161}$$

Let us find the interaction Hamiltonian density $\mathcal{H}_{int}(x)$ *in the interaction representation* from the given Lagrangian density.

We assume that $\mathcal{L}_{int}(x)$ involves the first order derivatives of spinless real fields φ_α. For complex fields we can always employ the real field representations introduced in Section 2-4.

The operator that is canonically conjugate to $\varphi_\alpha(x)$ is defined, for a free field, as

$$\pi_\alpha(x) = \frac{\partial \mathcal{L}_f}{\partial \dot{\varphi}_\alpha(x)} = \dot{\varphi}_\alpha(x) \tag{5-162}$$

and for an interacting field, as

$$\pi'_\alpha(x) = \frac{\partial \mathcal{L}}{\partial \dot{\varphi}_\alpha(x)} = \dot{\varphi}_\alpha(x) + \frac{\partial \mathcal{L}_{int}}{\partial \dot{\varphi}_\alpha(x)} \tag{5-163}$$

We now shall study the relationship between the interaction and Heisenberg representations. As has been shown in connection with the Yang-Feldman formalism, the relationship is given by

$$\varphi_\alpha(x) = U(x_0, -\infty)^{-1} \varphi_\alpha^{in}(x)\, U(x_0, -\infty) \tag{5-164}$$

For the operator that is canonically conjugate to φ_α we have

$$\pi'_\alpha(x) = U(x_0, -\infty)^{-1} \pi_\alpha^{in}(x)\, U(x_0, -\infty) \tag{5-165}$$

The operators in the interaction representation are designated by a superscript, in. The Hamiltonian density in the Heisenberg representation is given by

$$\mathcal{H}(x) = \sum_\alpha \pi'_\alpha(x)\, \dot{\varphi}_\alpha(x) - \mathcal{L}(\varphi_\alpha(x), \dot{\varphi}_\alpha(x)) \tag{5-166}$$

where we have suppressed the field operators irrelevant to the present problem. The $\mathcal{L}$ is a function of $\varphi_\alpha(x)$, $\nabla\varphi_\alpha(x)$, and $\dot{\varphi}_\alpha(x)$, but again with suppression of the space derivatives. The Hamiltonian density in the interaction representation is then

$$\mathcal{H}^{in}(x) = U(x_0, -\infty)\, \mathcal{H}(x)\, U(x_0, -\infty)^{-1} \tag{5-167}$$

Using the inverse transformation formulas

$$U(x_0, -\infty)\, \varphi_\alpha(x)\, U(x_0, -\infty)^{-1} = \varphi_\alpha^{in}(x) \tag{5-168}$$

and

$$U(x_0, -\infty)\, \pi'_\alpha(x)\, U(x_0, -\infty)^{-1} = \pi_\alpha^{in}(x) \tag{5-169}$$

we find that $\dot{\varphi}_\alpha$ is transformed as

$$U(x_0, -\infty)\, \dot{\varphi}_\alpha(x)\, U(x_0, -\infty)^{-1}$$

$$= U(x_0, -\infty) \left(\pi'_\alpha(x) - \frac{\partial \mathcal{L}_{int}}{\partial \dot{\varphi}_\alpha(x)} \right) U(x_0, -\infty)^{-1}$$

$$= \pi_\alpha^{in}(x) - \frac{\partial \mathcal{L}_{int}}{\partial \dot{\varphi}_\alpha^{in}(x)}$$

$$= \dot{\varphi}_\alpha^{in}(x) - \frac{\partial \mathcal{L}_{int}}{\partial \dot{\varphi}^{in}(x)} \tag{5-170}$$

Hence

$$\mathcal{H}^{in}(x) = \sum_\alpha \pi_\alpha^{in}(x) \left(\pi_\alpha^{in}(x) - \frac{\partial \mathcal{L}_{int}}{\partial \dot{\varphi}_\alpha^{in}(x)} \right)$$

$$- \mathcal{L} \left(\varphi_\alpha^{in}(x),\ \dot{\varphi}_\alpha^{in}(x) - \frac{\partial \mathcal{L}_{int}}{\partial \dot{\varphi}_\alpha^{in}(x)} \right) \tag{5-171}$$

We expand the second term into the appropriate Taylor series:

$$\mathcal{L} \left(\varphi_\alpha^{in}(x),\ \dot{\varphi}_\alpha^{in}(x) - \frac{\partial \mathcal{L}_{int}}{\partial \dot{\varphi}_\alpha^{in}(x)} \right) = \mathcal{L}\, (\varphi_\alpha^{in}(x),\ \dot{\varphi}_\alpha^{in}(x))$$

$$- \sum_\alpha \frac{\partial \mathcal{L}_{int}}{\partial \dot{\varphi}_\alpha^{in}(x)}$$

$$\times \frac{\partial \mathcal{L}\, (\varphi_\alpha^{in},\ \dot{\varphi}_\alpha^{in})}{\partial \dot{\varphi}_\alpha^{in}(x)}$$

$$+ \frac{1}{2} \sum_{\alpha, \beta} \frac{\partial \mathcal{L}_{int}}{\partial \dot{\varphi}_\alpha^{in}(x)} \frac{\partial \mathcal{L}_{int}}{\partial \dot{\varphi}_\beta^{in}(x)}$$

$$\times \frac{\partial^2 \mathcal{L}\, (\varphi_\alpha^{in},\ \dot{\varphi}_\alpha^{in})}{\partial \dot{\varphi}_\alpha^{in}(x)\, \partial \dot{\varphi}_\beta^{in}(x)}$$

We then use

$$\frac{\partial^2 \mathcal{L}}{\partial\dot{\varphi}_\alpha^{in}\,\partial\dot{\varphi}_\beta^{in}} = \delta_{\alpha\beta} \qquad \frac{\partial \mathcal{L}}{\partial\dot{\varphi}_\alpha^{in}} = \dot{\varphi}_\alpha^{in} + \frac{\partial \mathcal{L}_{int}}{\partial\dot{\varphi}_\alpha^{in}}$$

$$= \pi_\alpha^{in} + \frac{\partial \mathcal{L}_{int}}{\partial\dot{\varphi}_\alpha^{in}} \tag{5-172}$$

Substitution of these results into $\mathcal{L}$ yields

$$\mathcal{L}\left(\varphi_\alpha^{in}(x),\ \dot{\varphi}_\alpha^{in}(x) - \frac{\partial \mathcal{L}_{int}}{\partial\dot{\varphi}_\alpha^{in}(x)}\right)$$

$$= \mathcal{L}(\varphi_\alpha^{in}(x),\ \dot{\varphi}_\alpha^{in}(x)) - \sum_\alpha \pi_\alpha^{in}(x)\,\frac{\partial \mathcal{L}_{int}}{\partial\dot{\varphi}_\alpha^{in}(x)}$$

$$- \frac{1}{2}\sum_\alpha \frac{\partial \mathcal{L}_{int}}{\partial\dot{\varphi}_\alpha^{in}(x)} \cdot \frac{\partial \mathcal{L}_{int}}{\partial\dot{\varphi}_\alpha^{in}(x)} \tag{5-173}$$

and hence

$$\mathcal{H}^{in}(x) = \sum_\alpha \pi_\alpha^{in}(x)\,\pi_\alpha^{in}(x) - \mathcal{L}(\varphi_\alpha^{in}(x),\ \varphi_\alpha^{in}(x))$$

$$+ \frac{1}{2}\sum_\alpha \frac{\partial \mathcal{L}_{int}}{\partial\dot{\varphi}_\alpha^{in}(x)} \cdot \frac{\partial \mathcal{L}_{int}}{\partial\dot{\varphi}_\alpha^{in}(x)}$$

$$= \mathcal{H}_f^{in}(x) + \mathcal{H}_{int}^{in}(x) \tag{5-174}$$

where

$$\mathcal{H}_{int}^{in}(x) = -\mathcal{L}_{int}^{in}(x) + \frac{1}{2}\sum_\alpha \left(\frac{\partial \mathcal{L}_{int}}{\partial\dot{\varphi}_\alpha^{in}(x)}\right)^2 \tag{5-175}$$

Example (5-1) $\mathcal{L}_{\text{int}} = -iF\,\bar{\psi}\gamma_\mu\psi(\partial\varphi/\partial x_\mu)$

In this case the general formula (5-175) immediately gives the result

$$\mathcal{H}_{\text{int}} = -\mathcal{L}_{\text{int}} + \frac{1}{2}F^2(\bar{\psi}\gamma_4\psi)^2 \tag{5-176}$$

When we use the Hamiltonian density derived in the T-S equation, however, we have to be careful about the definition of the time axis. At any point x the time axis should be chosen in the direction of the normal to the spacelike hypersurface $\sigma(x)$. We choose a unit vector n which is in the direction of the future normal to σ at x. Then at this point, $n = (0, 0, 0, i)$.

Now we can write $\mathcal{H}_{\text{int}}$ in the covariant form.

$$\mathcal{H}_{\text{int}}(x) = -\mathcal{L}_{\text{int}}(x) + \frac{1}{2}n_\mu n_\nu \frac{\partial\mathcal{L}_{\text{int}}}{\partial\varphi_{\alpha:\mu}(x)} \cdot \frac{\partial\mathcal{L}_{\text{int}}}{\partial\varphi_{\alpha:\nu}(x)} \tag{5-177}$$

Hence the Hamiltonian density turns out to be dependent on the choice of the family of spacelike hypersurfaces $\{\sigma\}$. However, the definition of the T product also is dependent on $\{\sigma\}$, which can be shown in the following method.

When the interaction involves derivatives of field operators, there are expressions like

$$\langle 0|\,T\left[\frac{\partial\varphi(x)}{\partial x_\mu}, \frac{\partial\varphi(y)}{\partial y_\nu}\right]|0\rangle \tag{5-178}$$

in the S-matrix elements, but this quantity is also σ dependent. In order to see this let us first prove the following formula:

$$\langle 0|\,T\left[\frac{\partial\varphi(x)}{\partial x_\mu}, \varphi(y)\right]|0\rangle = \frac{\partial}{\partial x_\mu}\langle 0|\,T[\varphi(x), \varphi(y)]|0\rangle \tag{5-179}$$

We start from the definition

$$T[\varphi(x), \varphi(y)] = \frac{1}{2}\epsilon(x_0 - y_0)[\varphi(x), \varphi(y)] + \frac{1}{2}\{\varphi(x), \varphi(y)\} \tag{5-180}$$

Hence

$$\frac{\partial}{\partial x_\mu} T[\varphi(x), \varphi(y)] = \frac{1}{2} \epsilon(x_0 - y_0) \left[\frac{\partial \varphi(x)}{\partial x_\mu}, \varphi(y)\right]$$

$$+ \frac{1}{2} \left\{ \frac{\partial \varphi(x)}{\partial x_\mu}, \varphi(y) \right\}$$

$$+ \frac{\partial}{\partial x_\mu} \left(\frac{1}{2} \epsilon(x_0 - y_0) \right) \cdot [\varphi(x), \varphi(y)]$$

$$= T \left[\frac{\partial \varphi(x)}{\partial x_\mu}, \varphi(y)\right]$$

$$+ \frac{1}{i} \delta_{\mu 4} \delta(x_0 - y_0) [\varphi(x), \varphi(y)]$$

$$= T \left[\frac{\partial \varphi(x)}{\partial x_\mu}, \varphi(y)\right]$$

since $\delta(x_0 - y_0)[\varphi(x), \varphi(y)]$ vanishes as a consequence of their equal-time commutator. Next we evaluate

$$\frac{\partial}{\partial y_\nu} T \left[\frac{\partial \varphi(x)}{\partial x_\mu}, \varphi(y)\right] = T \left[\frac{\partial \varphi(x)}{\partial x_\mu}, \frac{\partial \varphi(y)}{\partial y_\nu}\right]$$

$$+ i\, \delta_{\nu 4} \delta(x_0 - y_0) \left[\frac{\partial \varphi(x)}{\partial x_\mu}, \varphi(y)\right] \quad (5\text{-}181)$$

Now the second term does not vanish, but it is equal to

$$\delta_{\mu 4} \delta_{\nu 4} \delta(x_0 - y_0) [\dot{\varphi}(x), \varphi(y)]$$

$$= -i \delta_{\mu 4} \delta_{\nu 4} \delta(x_0 - y_0) \delta^3 (x - y)$$

$$= -i \delta_{\mu 4} \delta_{\nu 4} \delta^4 (x - y)$$

$$= i\, n_\mu n_\nu \delta^4 (x - y) \quad (5\text{-}182)$$

Thus the vacuum expectation value is given by

$$\langle 0 | T \left[\frac{\partial \varphi(x)}{\partial x_\mu}, \frac{\partial \varphi(y)}{\partial y_\nu}\right] | 0 \rangle = \frac{\partial^2}{\partial x_\mu \partial y_\nu} \Delta_F(x - y)$$

$$- i n_\mu n_\nu \delta^4 (x - y) \quad (5\text{-}183)$$

which depends on the normal, n to σ at x.

Thus in the S-matrix formula both $\mathcal{H}_{int}(x)$ and T are dependent on the choice of the family of spacelike hypersurfaces $\{\sigma\}$. The microscopic causality condition implies, however, that the S matrix should be independent of the choice of $\{\sigma\}$. In fact, we can prove that these two kinds of family dependences cancel each other, and the resultant S matrix is again independent of the choice of the family of spacelike surfaces. We further can prove that

$$S = 1 + \sum_{n=1}^{\infty} \frac{(-i)^n}{n!} \int_{-\infty}^{\infty} d^4x_1 \cdots \int_{-\infty}^{\infty} d^4x_n \, T[\mathcal{H}_{int}(x_1) \cdots \mathcal{H}_{int}(x_n)]$$

$$= 1 + \sum_{n=1}^{\infty} \frac{i^n}{n!} \int_{-\infty}^{\infty} d^4x_1 \cdots \int_{-\infty}^{\infty} d^4x_n \, T^*[\mathcal{L}_{int}(x_1) \cdots \mathcal{L}_{int}(x_n)] \tag{5-184}$$

where T* is the invariant modification of T. The T* satisfies

$$T^*\left(\frac{\partial\varphi(x)}{\partial x_\mu}, \frac{\partial\varphi(y)}{\partial y_\nu}\right) = \frac{\partial^2}{\partial x_\mu \, \partial y_\nu} T[\varphi(x), \varphi(y)] \neq T\left(\frac{\partial\varphi(x)}{\partial x_\mu}, \frac{\partial\varphi(y)}{\partial y_\nu}\right) \tag{5-185}$$

Roughly speaking, the ordering operator T is appropriate to the Hamiltonian formalism, whereas T* is suitable to the Lagrangian formalism.

The difference between T and T* consists in the treatment of time derivatives of operators. For simplicity, let us consider particle mechanics and assume

$$p(t) = \dot{q}(t) \tag{5-186}$$

Under the T symbol, p(t) is regarded as an operator defined at a sharp time t—the so-called monopole quantity—so that

$$T[p(t), a(t')] = \begin{cases} p(t)a(t') & \text{for} \quad t > t' \\ a(t')p(t) & \text{for} \quad t' > t \end{cases} \tag{5-187}$$

Under the T* symbol, on the contrary, $p(t) = \dot{q}(t)$ is regarded as an operator defined at two adjacent time points, $t + \epsilon$ and t—the so-called dipole quantity—so that

$$T^*[\dot{q}(t), a(t')] = \lim_{\epsilon \to 0} T\left[\frac{q(t+\epsilon) - q(t)}{\epsilon}, a(t')\right]$$

$$= \frac{d}{dt} T(q(t), a(t')) \tag{5-188}$$

By using this definition of the T* symbol we can prove the S-matrix formula expressed in terms of the interaction Lagrangian density.

When the interaction Lagrangian density involves derivatives of field operators, the Lagrangian method is much simpler than the Hamiltonian method. Examples of interactions involving derivatives of field operators are the pion-nucleon interaction with a derivative coupling and the electromagnetic interactions of charged bosons.

PROBLEMS

5-1. The lifetime τ of a moving unstable particle is given by

$$\tau = \tau_0 (1 - \beta^2)^{-1/2}$$

where τ_0 is the lifetime of the same particle at rest, and β the velocity of the particle divided by the light velocity.

Verify this relation by directly calculating the decay probability for $\Lambda^0 \to p + \pi^-$ in a moving system.

5-2. Find the lowest-order Feynman diagrams for $\pi \to e^+ + e^- + \gamma$ and $\pi^0 \to e^+ + e^-$. State the reason why the former is more probable than the latter.

Note: You are supposed to know the presence of the strong pion-nucleon interaction as well as the electromagnetic interactions of charged particles. Particles appearing in the diagrams are limited to the pion, nucleon, electron, and photon.

5-3. Calculate the differential and total cross sections of the process $\gamma + \pi^+ \to \pi^+ + \gamma$ in the laboratory system. Find the approximate forms of the total cross section at very high and low energies.

5-4. Calculate the total cross section for

$$\pi^- + p \to p + \pi^-$$

in the center-of-mass system by using perturbation theory.

5-5. Express the lifetime of Λ for the decay $\Lambda \to p + \pi^-$ in terms of the coupling constant f when the decay interaction is given by

$$-\mathcal{L}_{int} = i(f/\mu)\,\bar{\psi}_p \gamma_\lambda \psi \;\; (\partial\varphi/\partial x_\lambda) + \text{H. c.}$$

where μ is the pion rest mass.

5-6. Given the interaction

$$\mathcal{H}_{int} = g_a \Phi_c^\dagger \Phi_c \varphi_a + g_b \Phi_c^\dagger \Phi_c \varphi_b$$

calculate the lifetime of the particle a for the process

$$a \to 2b$$

All the particles in this problem are scalar. The a and b are neutral and c is charged; assume

$$m_c \gg m_a > 2m_b$$

This is an analog of $\pi^0 \to 2\gamma$

5-7. Suppose that we evaluate the S-matrix element for the nucleon-nucleon scattering in the second order corresponding to the Feynman diagram in Figure 5-7.

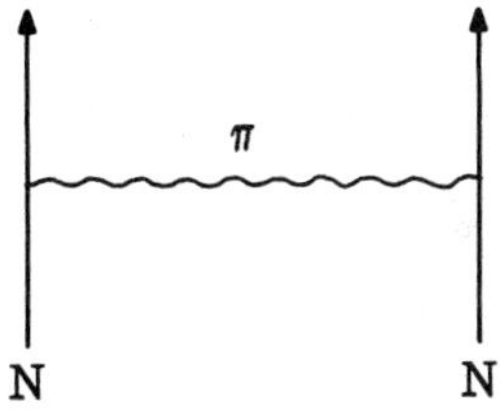

Figure 5-7. A diagram for the nucleon-nucleon scattering.

Prove that the following two interactions lead to the same S-matrix element in this order:

$$\mathcal{L}^{(1)}_{int} = -iG\bar{\psi}\gamma_5 \tau\psi \cdot \varphi$$

$$\mathcal{L}^{(2)}_{int} = -i\,\frac{G}{2M}\,\bar{\psi}\gamma_5\gamma_\lambda \tau\psi(\partial\varphi/\partial x_\lambda)$$

where M denotes the nucleon mass. This statement is called the *equivalence theorem.*

5-8. Suppose that neutral scalar field φ is coupled to the nucleon field through the interaction

$$\mathcal{L}_{int} = -i \frac{G}{2M} \bar{\psi}\gamma_\lambda \psi (\partial\varphi/\partial x_\lambda)$$

Show in the lowest order that the cross section for the scattering of this neutral scalar meson on a nucleon vanishes *(divergence theorem).*

REFERENCES

1. For examples in quantum electrodynamics, refer to G. Källén, *Encyclopedia of Physics* (Julius Springer, Berlin, 1958), Vol. 5, Part 1.
2. O. Klein and Y. Nishina, Z. f. Physik 52, 853 (1929).
3. The value of λ has not yet been established. For example, $\lambda = -0.18 \pm 0.05$ in J. Hamilton, P. Menotti, G. C. Oades, and L. L. J. Vick, Phys. Rev. **128**, 1881 (1962) and $\lambda = -0.01$ in S. Weinberg, Phys. Rev. Letters, **17**, 616 (1966).
4. H. Fukuda and Y. Miyamoto, Progr. Theor. Phys. (Kyoto) **4**, 347 (1949). J. Steinberger, Phys. Rev. **76**, 1180 (1949).
5. S. Kanesawa and S. Tomonaga, Progr. Theor. Phys. (Kyoto) **3**, 1 and 101 (1948).
6. F. J. Dyson, Phys. Rev. **73**, 929 (1948). K. M. Case, Phys. Rev. **76**, 1 and 14 (1949). P. Moldauer and K. M. Case, Phys. Rev. **91**, 459A (1953). L. L. Foldy, Phys. Rev. **84**, 168 (1951).

CHAPTER 6
RENORMALIZATION

We have investigated, so far, applications of perturbation theory in the lowest order and avoided serious difficulties. However, we sometimes have to calculate higher-order corrections to account for small deviations of the measured quantities from theoretical values. In particular, perturbation theory seems to be meaningless in meson theory.

As we proceed to higher-order corrections, however, we encounter the so-called divergence difficulties; that is, each term of the perturbation expansion higher than the lowest order diverges. Another difficulty is that we are not sure about the correctness of the decomposition of the Hamiltonian into free and interaction parts. These two difficulties are not unrelated, and in some cases such as quantum electrodynamics and pion theory we can eliminate divergences from the theory by altering the decomposition of the Hamiltonian. This redecomposition procedure is called renormalization.[1]

In this connection we refer to the theory of rearrangement collisions, in which the total Hamiltonian is decomposed into free and interaction parts in various ways:

$$H = H_a + V_a = H_b + V_b \tag{6-1}$$

The decomposition is fixed by the appropriate boundary conditions.

6-1 CONDITIONS FOR THE PROPER DECOMPOSITION OF THE HAMILTONIAN

In order to give proper boundary conditions we make three intuitively plausible assumptions.

Basic Assumptions

(1) There are two complete sets of state vectors, sometimes called *in* and *out* states, which are eigenstates of the total Hamiltonian.

$$H_{total}\ \Psi_a^{(+)} = E_a\ \Psi_a^{(+)} \tag{6-2a}$$

$$H_{total}\ \Psi_a^{(-)} = E_a\ \Psi_a^{(-)} \tag{6-2b}$$

The in and out states are related to the free particle states by the Lippmann-Schwinger equations[2]:

$$\Psi_a^{(\pm)} = \Phi_a + \frac{1}{E_a - H_0 \pm i\epsilon}\ V\ \Psi_a^{(\pm)} \tag{6-3}$$

In relativistic field theory we generalize these relations by requiring the in and out states to be eigenstates of the four-momentum operator P_μ with eigenvalue p_μ:

$$P_\mu\ \Psi_a^{(+)} = (p_\mu)_a \Psi_a^{(+)} \tag{6-4a}$$

$$P_\mu\ \Psi_a^{(-)} = (p_\mu)_a \Psi_a^{(-)} \tag{6-4b}$$

As is shown in the formal theory of scattering, the S-matrix elements are given by

$$S_{ba} = (\Psi_b^{(-)},\ \Psi_a^{(+)}) \tag{6-5}$$

(2) There exist the vacuum state and stable single particle states satisfying

$$\Psi_0^{(+)} = \Psi_0^{(-)} \qquad \Psi_a^{(+)} = \Psi_a^{(-)} \tag{6-6}$$

Notice that Ψ_0 is an eigenstate of the total Hamiltonian, whereas the vacuum state Φ_0 used in the preceding chapters is an eigenstate of the free Hamiltonian. In terms of the S matrix these conditions are expressed as

$$S_{00} = 1 \qquad S_{aa} = 1 \tag{6-7}$$

As we have discussed in Section 4-7, the first condition implies neglect of all the bubble diagrams.

(3) We assume the hypothesis of adiabatic switching, that is, we can turn off interactions in such a way that eigenstates of the total Hamiltonian approach those of the free Hamiltonian adiabatically. If we use the ordinary interaction representation this condition is not satisfied, and consequently Dyson's formula for the S matrix becomes

invalid in higher orders. Hence it is imperative to alter the decomposition.

The principal physical effects that require renormalization are mass renormalization and charge renormalization.

Mass Renormalization

In field theory an electron does not represent a closed system but is always interacting with its own field, called the self-field. That is, an electron interacts with the Coulomb field generated by and surrounding that same electron. Since the electron is observed to be stable, the matrix element for (electron) → (electron), S_{ee}, should be equal to 1; but the possibility of processes like the emission and absorption of virtual photons, that is, the interaction of the electron with its self-field, makes S_{ee} unequal to 1.

In order to make it equal to 1 we can modify the decomposition by adding a term of the form $\delta m\bar{\psi}\psi$ to H_f and subtracting it from H_{int}. This, in turn, may be interpreted as meaning that the mass term in the free Hamiltonian is changed from $m\bar{\psi}\psi$ to $(m + \delta m)\bar{\psi}\psi$, so that the effective electron mass is $m + \delta m$. Then if H_{total} is properly decomposed, namely, if δm is chosen correctly, the electron with the rest mass $m + \delta m$ is a stable state in the sense that $S_{ee} = 1$, and the observed electron mass would be $m + \delta m$. This observed mass can be determined by experiment, but we can never determine m experimentally since the S matrix in the proper interaction representation involves m always in this special combination $m + \delta m$, so that $S = S(m + \delta m)$. Therefore, in order to compare the theory with experiment we substitute the experimental mass for the expression $m + \delta m$. In fact the calculated value of δm is divergent but the divergence difficulty is avoided by this interpretation.

Charge Renormalization

We know that the vacuum is a dielectric medium. Therefore, we may write Coulomb's law as

$$V = \frac{ee'}{\epsilon}\left(\frac{1}{4\pi r}\right) \tag{6-8}$$

We assume r to be large compared with the electron Compton wave length, since otherwise the self-fields of two interacting charges would be distorted and we would expect more involved effects. The ϵ is the dielectric constant of the vacuum, and if it is calculated in perturbation theory the value of ϵ is also divergent. We interpret $e/\epsilon^{1/2}$ to be the observed charge although $\epsilon^{1/2}$ is divergent, inasmuch as S is a function of e only in the combination $e/\epsilon^{1/2}$. Therefore, if we put $e/\epsilon^{1/2} = e_{exp}$ we can avoid this divergence difficulty. Then Coulomb's law is given by

$$V = (e)_{exp}\,(e')_{exp}\left(\frac{1}{4\pi r}\right) \tag{6-9}$$

This renormalization is called the charge renormalization.

6-2 MASS RENORMALIZATION

A Feynman diagram starting with an electron line and ending with another electron line is called a self-energy diagram. When this diagram cannot be divided in two by cutting just one internal electron line, it is called an irreducible self-energy diagram; otherwise it is called a reducible self-energy diagram. The simplest examples of each type of these diagrams are shown in Fig. 6-1. The self-energy δm is determined from the irreducible self-energy diagrams. In the S-matrix expansion of the process $a \to a$,

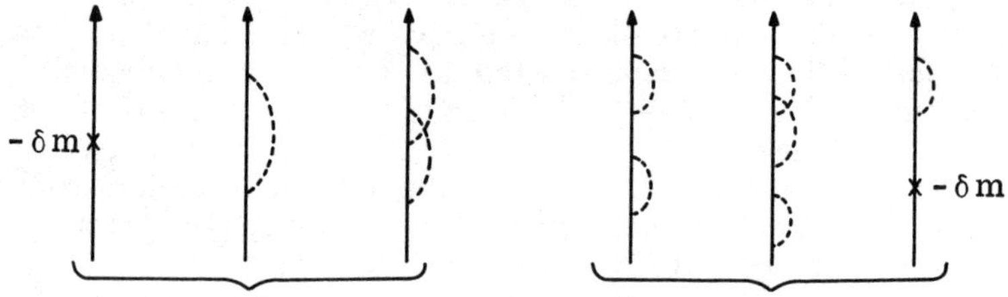

Figure 6-1. (a) Examples of irreducible self-energy diagrams.
(b) Examples of reducible self-energy diagrams.

$$S_{aa} = 1 + \sum_{n=1}^{\infty} \frac{(-i)^n}{n!} \int d^4x_1 \ldots d^4x_n$$

$$\times \langle a | T_c [\mathcal{H}_{int}(x_1) \ldots \mathcal{H}_{int}(x_n)] | a \rangle \qquad (6\text{-}10)$$

the condition $S_{aa} = 1$ implies that all terms with $n \geq 1$ should vanish. This condition fixes the value of δm in the effective interaction Hamiltonian.

$$\mathcal{H}_{int} = -ie\, \bar{\psi}\gamma_\mu \psi \cdot A_\mu - \delta m\, \bar{\psi}\psi \qquad (6\text{-}11)$$

Let us first evaluate the lowest-order S-matrix in order to find δm in the second order.

$$S^{(2)}_{a'a} = \frac{e^2}{2} \int d^4x_1 d^4x_2 \langle a' | T[\bar{\psi}(x_1)\gamma_\mu \psi(x_1), \bar{\psi}(x_2)\gamma_\nu \psi(x_2)] | a \rangle$$

$$\times \delta_{\mu\nu} D_F(x_1 - x_2) - i \int d^4x\, (-\delta m^{(2)}) \langle a' | \bar{\psi}(x)\psi(x) | a \rangle$$

$$= e^2 \int d^4x_1 d^4x_2 \langle a' | \bar{\psi}(x_1) | 0 \rangle \gamma_\mu S_F(x_1 - x_2) \gamma_\mu \langle 0 | \psi(x_2) | a \rangle$$

$$\times D_F(x_1 - x_2) + i\, \delta m^{(2)} \int d^4x \langle a' | \bar{\psi}(x) | 0 \rangle \langle 0 | \psi(x) | a \rangle$$

$$(6\text{-}12)$$

We unify this expression by using

$$\langle 0|\psi(x_2)|a\rangle = \exp[ip_a(x_2 - x_1)]\langle 0|\psi(x_1)|a\rangle$$

which follows from the fact that

$$\langle 0|\psi(x_2)|a\rangle \sim \exp(ip_a x_2) = \exp[ip_a(x_2 - x_1) + ip_a x_1]$$

Then

$$S^{(2)}_{a'a} = \int d^4x_1 \langle a'|\bar{\psi}(x_1)|0\rangle$$

$$\times \Big(i\,\delta m^{(2)} + e^2 \int d^4x_2\, \gamma_\mu S_F(x_1 - x_2)\gamma_\mu D_F(x_1 - x_2)$$

$$\times \exp[ip(x_2 - x_1)]\Big)\langle 0|\psi(x_1)|a\rangle$$

$$= \int d^4x \langle a'|\bar{\psi}(x)|0\rangle \left(i\delta m^{(2)} + e^2 \int d^4y\, \gamma_\mu S_F(y)\gamma_\mu D_F(y)e^{-ipy}\right)$$

$$\times \langle 0|\psi(x)|a\rangle \qquad (6\text{-}13)$$

If we define

$$\Sigma^{*(2)}(p) = i\,e^2 \int d^4y\, \gamma_\mu S_F(y)\gamma_\mu D_F(y)\, e^{-ipy} \qquad (6\text{-}14)$$

we may write

$$S^{(2)}_{a'a} = i\int d^4x \langle a'|\bar{\psi}(x)|0\rangle \left(\delta m^{(2)} - \Sigma^{*(2)}(p)\right)\langle 0|\psi(x)|a\rangle \qquad (6\text{-}15)$$

The generalization of $\Sigma^*(p)$ is obtained by summing the contributions to the self-energy from all possible irreducible diagrams, and is called the proper self-energy operator or mass operator. The superscript star is a reminder that only irreducible (proper) processes are considered. Let us now evaluate $\Sigma^{*(2)}(p)$, the lowest-order approximation to the mass operator, using

$$S_F(x) = \frac{i}{(2\pi)^4} \int d^4q\, e^{iqx} \frac{iq\gamma - m}{q^2 + m^2 - i\epsilon}$$

$$D_F(x) = \frac{-i}{(2\pi)^4} \int d^4k\, e^{ikx} \frac{1}{k^2 - i\epsilon}$$

Inserting these expressions into the definition of $\Sigma^{*(2)}(p)$, we find

$$\Sigma^{*(2)}(p) = \frac{ie^2}{(2\pi)^4} \int d^4k\, \gamma_\mu \frac{i(p-k)\gamma - m}{(p-k)^2 + m^2 - i\epsilon} \gamma_\mu \frac{1}{k^2 - i\epsilon} \tag{6-16}$$

We first sum over μ using

$$\gamma_\mu \gamma_\mu = 4 \qquad \gamma_\mu \gamma_\lambda \gamma_\mu = -2\gamma_\lambda$$

and obtain

$$\Sigma^{*(2)}(p) = \frac{ie^2}{(2\pi)^4} \int d^4k \frac{-2i(p-k)\gamma - 4m}{(p-k)^2 + m^2 - i\epsilon} \cdot \frac{1}{k^2 - i\epsilon}$$

$$= \frac{ie^2}{(2\pi)^4} \int d^4k \int_0^1 dx \frac{-2i(p-k)\gamma - 4m}{(k^2 + xp^2 - 2xpk + xm^2 - i\epsilon)^2} \tag{6-17}$$

where we have introduced a Feynman parameter x to unify the denominator. We further make the substitution

$$k \to k' = k - xp \tag{6-18}$$

to get

$$\Sigma^{*(2)}(p) = \frac{ie^2}{(2\pi)^4} \int_0^1 dx \int d^4k' \frac{-2i(p - k' - xp)\gamma - 4m}{(k'^2 + xp^2 - x^2p^2 + xm^2 - i\epsilon)^2}$$

$$= \frac{-ie^2}{(2\pi)^4} \int_0^1 dx \int d^4k' \frac{2i(1-x)p\gamma + 4m}{(k'^2 + x(1-x)p^2 + xm^2 - i\epsilon)^2} \tag{6-19}$$

The term linear in k' vanishes upon integration because of the odd symmetry of each component.

Since $p^2 = -(ip\gamma)^2$, $\Sigma^*(p)$ can be regarded as a function of $ip\gamma$; so we expand $\Sigma^{*(2)}(p)$ in powers of $(ip\gamma + m)$:

$$\Sigma^{*(2)}(p) = A + B(ip\gamma + m) + C(p) \tag{6-20}$$

where C(p) is of the second and higher orders in $(ip\gamma + m)$. In order to evaluate A we put $ip\gamma = -m$ in $\Sigma^{*(2)}(p)$ and get

$$A = \frac{-ie^2}{(2\pi)^4}\int_0^1 dx \int d^4k' \frac{-2m(1-x) + 4m}{(k'^2 - x(1-x)m^2 + xm^2 - i\epsilon)^2}$$

$$= \frac{-ie^2}{(2\pi)^4}\int_0^1 dx\ 2m(1+x)\int d^4k' \frac{1}{(k'^2 + x^2m^2 - i\epsilon)^2} \tag{6-21}$$

To carry out the integration we use

$$\frac{1}{(k'^2 + x^2m^2 - i\epsilon)^2} = 2\int_{m^2}^{\infty} \frac{x^2 dM^2}{(k'^2 + x^2M^2 - i\epsilon)^3} \tag{6-22}$$

then we find, by utilizing Feynman's formula, the following result:

$$A = \frac{-4ie^2 m}{(2\pi)^4}\int_0^1 dx\, x^2(1+x) \int_{m^2}^{\infty} dM^2 \int d^4k'$$

$$\times \frac{1}{(k'^2 + x^2M^2 - i\epsilon)^3}$$

$$= \frac{2\pi^2 e^2 m}{(2\pi)^4}\int_0^1 dx\,(1+x)\int_{m^2}^{\infty}\frac{dM^2}{M^2}$$

$$= (3/4\pi)\,\alpha m\ \ln(\infty/m^2) \tag{6-23}$$

This is a divergent expression and, therefore, the physical content is obscure; however, the important points are that A is logarithmically divergent and that it is positive.

Returning to our expression for $S^{(2)}_{a'a}$, we have

$$S^{(2)}_{a'a} = \frac{i(2\pi)^4}{V}\, \bar{u}(p') \left(\delta m - \Sigma^{*(2)}(p)\right) u(p)\, \delta^4(p' - p) \tag{6-24}$$

In this expression $\Sigma^{*(2)}(p)$ can be replaced by A because of the Dirac equation

$$(ip\gamma + m)\, u(p) = 0$$

Higher-order terms in $(ip\gamma + m)$ vanish leaving only the constant term A in $\Sigma^{*(2)}(p)$. Therefore, for $S_{aa} = 1$ we must have

$$\delta m^{(2)} = A \tag{6-25}$$

That A or δm is positive is reasonable because if we consider only the Coulomb field, the self-energy of an electron is given by

$$\lim_{r \to 0} \frac{e^2}{4\pi r} > 0 \tag{6-26}$$

6-3 FIELD-OPERATOR RENORMALIZATION

Generally the proper self-energy operator may be written as

$$\Sigma^*(p) = \delta m + (ip\gamma + m)B + C(p) \tag{6-27}$$

The first term is canceled by the counterterm $-\delta m\, \bar{\psi}\psi$ in the interaction Hamiltonian. This is mass renormalization. It is not sufficient, however, to properly interpret the theory.

A second condition arises when we consider the contributions from reducible self-energy diagrams in Fig. 6-2.

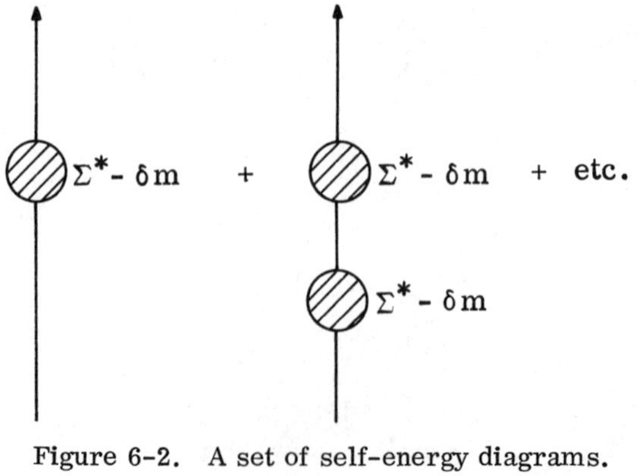

Figure 6-2. A set of self-energy diagrams.

Then we have to replace $\Sigma^* - \delta m$ by the sum of contributions from the above set of diagrams. It is not difficult to show that $\Sigma^* - \delta m$ must be replaced by

$$\left(\Sigma^*(p) - \delta m\right) - \left(\Sigma^*(p) - \delta m\right) \frac{1}{ip\gamma + m} \left(\Sigma^*(p) - \delta m\right) + \cdots = \frac{\Sigma^*(p) - \delta m}{1 + \dfrac{\Sigma^*(p) - \delta m}{ip\gamma + m}} \tag{6-28}$$

In the limit $ip\gamma + m \rightarrow 0$ this expression certainly vanishes as

$$(ip\gamma + m) \frac{B}{1 + B} \sim 0 \tag{6-29}$$

In the case of an isolated electron line nothing peculiar happens. When we consider, however, a slightly more complicated process we encounter a strange situation. The electron-one photon interaction, for example, involves a set of diagrams in Fig. 6-3, which gives contributions to the S matrix like

$$e\,\bar{u}(p')\,\gamma_\mu u(p) + e\,\bar{u}(p')\gamma_\mu \frac{-\left(\Sigma^*(p) - \delta m\right)}{ip\gamma + m} u(p) + \cdots$$
$$= e\,\bar{u}(p')\gamma_\mu \left(1 + \frac{\Sigma^*(p) - \delta m}{ip\gamma + m}\right)^{-1} u(p) = \frac{e}{1 + B} \quad \bar{u}(p')\gamma_\mu u(p) \tag{6-30}$$

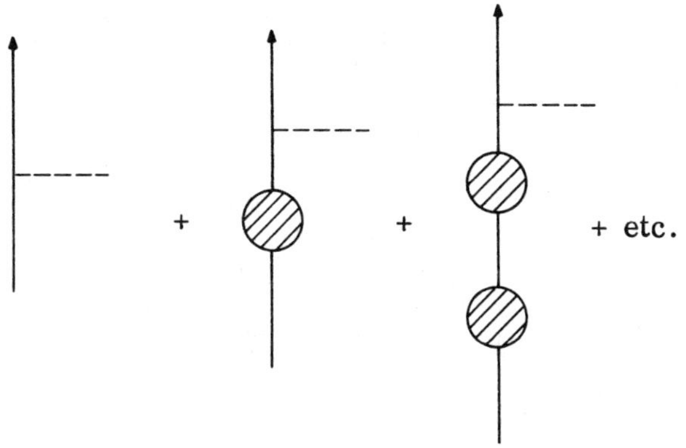

Figure 6-3. A set of diagrams representing self-energy corrections to a vertex.

This equation shows that in the course of the interaction the normalization of either the electron wave function or the charge is changed. This in turn suggests that the hypothesis of the adiabatic switching of the interaction would certainly not be justified from the point of view of the conservation of probability. Therefore, it seems necessary that B, in the proper decomposition of the Hamiltonian, vanish. Then the problem is, how to split the Lagrangian or Hamiltonian correctly so as to get B = 0. This is accomplished by so-called wave-function renormalization, or by charge renormalization.

There are many different ways to formulate this renormalization prescription, but in order to understand it in the most elementary but unified way we shall consider the problem as one of the correct decomposition of the Lagrangian.[3]

Let us perform the transformations

$$\psi(x) = Z_2^{1/2}\,\psi(x)_r \qquad \bar{\psi}(x) = Z_2^{1/2}\,\bar{\psi}(x)_r \tag{6-31}$$

and

$$A_\mu(x) = Z_3^{1/2}\,A_\mu(x)_r \tag{6-32}$$

where ψ_r, $\bar{\psi}_r$, and $A_\mu(x)_r$ are called renormalized field operators. These transformations are defined in the Heisenberg representation. Rewriting the total Lagrangian density in terms of renormalized field operators, we get

$$\begin{aligned}\mathcal{L}_{\text{total}} = & -Z_2\,\bar{\psi}_r\,[\gamma_\mu(\partial_\mu - ieZ_3^{1/2}A_{\mu r}) + m_{\text{exp}} - \delta m]\,\psi_r \\ & -\frac{1}{4}Z_3\,F_{\mu\nu r}\,F_{\mu\nu r} \\ & -\frac{1}{2}Z_3\left(\frac{\partial A_{\mu r}}{\partial x_\mu}\right)^2\end{aligned} \tag{6-33}$$

The Z transformations are related to the wave function renormalization. In the case of the electromagnetic field, however, a peculiar situation exists, although this is the only exceptional case. The longitudinal and scalar fields practically do not interact since their state is fixed or frozen by the Lorentz condition, or we can write an equation

$$\frac{\partial A_\mu}{\partial x_\mu} = \frac{\partial A_\mu^{\text{in}}}{\partial x_\mu} \tag{6-34}$$

which follows from the Yang-Feldman equation for A_μ and the conservation law $\partial_\mu j_\mu = 0$. Consequently, we apply the Z transformation only to the transverse fields, that is

$$F_{\mu\nu}(x) = Z_3^{1/2}\, F_{\mu\nu}(x)_r \qquad \text{(transverse)} \tag{6-35}$$

$$\frac{\partial A_\mu(x)}{\partial x_\mu} = \frac{\partial A_\mu(x)_r}{\partial x_\mu} \qquad \text{(longitudinal and scalar)} \tag{6-36}$$

In view of this interpretation the Lagrangian density is given by

$$\begin{aligned}\mathcal{L}_{\text{total}} &= -Z_2\, \bar{\psi}_r\, [\gamma_\mu(\partial_\mu - ieZ_3^{1/2}\, A_{\mu r}) + m_{\text{exp}} - \delta m]\, \psi_r \\ &\quad - \frac{1}{4}\, Z_3\, F_{\mu\nu r}\, F_{\mu\nu r} \\ &\quad - \frac{1}{2}\left(\frac{\partial A_{\mu r}}{\partial x_\mu}\right)^2\end{aligned} \tag{6-37}$$

We decompose the total Lagrangian into free and interaction parts:

$$\begin{aligned}\mathcal{L}_f &= -\bar{\psi}_r[\gamma_\mu\, \partial_\mu + m_{\text{exp}}]\,\psi_r - \frac{1}{4} F_{\mu\nu r}\, F_{\mu\nu r} \\ &\quad - \frac{1}{2}\left(\frac{\partial A_{\mu r}}{\partial x_\mu}\right)^2\end{aligned} \tag{6-38}$$

$$\begin{aligned}\mathcal{L}_{\text{int}} &= (1 - Z_2)\,\bar{\psi}_r\, [\,\gamma_\mu \partial_\mu + m_{\text{exp}}\,]\,\psi_r \\ &\quad + (1 - Z_3)\,\frac{1}{4}\, F_{\mu\nu r}\, F_{\mu\nu r} \\ &\quad + ieZ_2 Z_3^{1/2} A_{\mu r} \cdot \bar{\psi}_r \gamma_\mu \psi_r + Z_2\, \delta m\, \bar{\psi}_r \psi_r\end{aligned} \tag{6-39}$$

Now we introduce the interaction representation corresponding to the above decomposition, which we call the renormalized interaction representation. From here on we simply shall denote the operators in the renormalized interaction representation by ψ, $\bar{\psi}$, and A_μ.

The constants Z_2 and Z_3 are determined so as to make $S_{ee} = S_{\gamma\gamma} = 1$ as well as to conserve the wave-function normalization.

In order to carry out this program within the framework of perturbation theory, we first assume that renormalization constants can be expanded into power series in the coupling constant:

$$\delta m = \delta m^{(2)} + \delta m^{(4)} + \cdots$$

$$Z_2 = 1 + Z_2^{(2)} + Z_2^{(4)} + \cdots \tag{6-40}$$

$$Z_3 = 1 + Z_3^{(2)} + Z_3^{(4)} + \cdots$$

Next we calculate the proper self-energy operator Σ^* in the lowest order. This requires caution, since the interaction Lagrangian involves derivatives of the field operators. In such a case we have to use the Lagrangian method introduced in Section 5-5:

$$S = 1 + \sum_{n=1}^{\infty} \frac{i^n}{n!} \int d^4x_1 \cdots d^4x_n$$

$$\times\, T^*[\mathcal{L}_{int}(x_1) \cdots \mathcal{L}_{int}(x_n)] \tag{6-41}$$

In evaluating the S-matrix elements to second order, we can utilize our previous calculation.

$$S^{(2)}_{a'a} = \int d^4x \,\langle a' |\, \bar{\psi}(x) | \,0\rangle \;\Delta\; \langle 0 |\, \psi(x) |\, a\rangle \tag{6-42}$$

where

$$\Delta = iZ_2\,\delta m + i(1 - Z_2)(ip\gamma + m_{exp})$$

$$+\, e^2 Z_2{}^2 Z_3 \int d^4y\, \gamma_\mu S_F(y) \gamma_\mu D_F(y) e^{-ipy} \tag{6-43}$$

The expression Δ is a function of $ip\gamma$ and should satisfy the following conditions:

(1) $\Delta = 0$ for $ip\gamma + m_{exp} = 0$, so that $S_{ee} = 1$.

(2) $\Delta/(ip\gamma + m_{exp}) = 0$ for $ip\gamma + m_{exp} = 0$, so that the S-matrix elements are properly normalized.

In general, if one takes contributions from all irreducible self-energy diagrams the result is

$$\Delta = i Z_2 \delta m + i(1 - Z_2)(ip\gamma + m_{exp}) - i \sum^*(p) \tag{6-44}$$

where $\Sigma^*(p)$ can be expanded in powers of $(ip\gamma + m_{exp})$:

$$\sum^*(p) = A + B(ip\gamma + m_{exp}) + C(p) \tag{6-45}$$

with

$$C(p)/(ip\gamma + m_{exp}) \to 0 \qquad \text{for} \quad ip\gamma + m_{exp} \to 0 \tag{6-46}$$

Then the boundary conditions (1) and (2) imply

$$Z_2 \delta m = A \qquad 1 - Z_2 = B \tag{6-47}$$

To second order we have

$$\begin{aligned} (Z_2 \delta m)^{(2)} &= Z_2^{(0)} \delta m^{(2)} = \delta m^{(2)} = A^{(2)} \\ (1 - Z_2)^{(2)} &= -Z_2^{(2)} = B^{(2)} \end{aligned} \tag{6-48}$$

and we can determine δm and Z_2 to this order. Thus we have

$$\Delta = -i C(p) = -i \sum_{ren}^*(p) \tag{6-49}$$

and

$$\frac{\sum_{ren}^*(p)}{ip\gamma + m_{exp}} \to 0 \qquad \text{for} \quad ip\gamma + m_{exp} \to 0 \tag{6-50}$$

In second order we can replace $e^2 Z_2^2 Z_3$ by e^2 and $\Sigma_{ren}^{*(2)}(p)$ from our previous expression for $\Sigma^{*(2)}(p)$ by expanding it in powers of $(ip\gamma + m_{exp})$ and retaining only terms of second and higher orders. Then we find that both A and B are logarithmically divergent, but C(p) or Σ^*_{ren} (p) is convergent.

6-4 RENORMALIZED PROPAGATION FUNCTIONS

We have defined $S_F(x)$ by

$$S_F(x-y) = \langle 0|T[\psi(x)\bar{\psi}(y)]|0\rangle$$
$$= \frac{-i}{(2\pi)^4}\int d^4p \frac{e^{ipx}}{ip\gamma + m - i\epsilon} \qquad (6\text{-}51)$$

in the interaction representation. This function corresponds to an internal line in a Feynman diagram: Let us consider an internal electron line with all possible higher-order corrections as shown in Fig. 6-4.

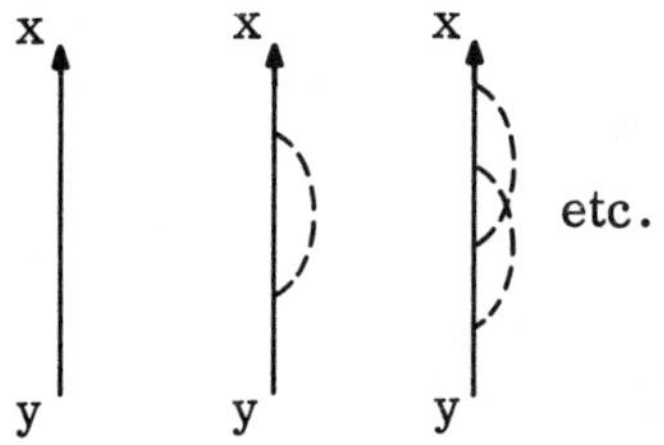

Figure 6-4. A set of diagrams corresponding to the electron propagation function.

Their sum gives us the complete electron propagation function including higher-order corrections. Let us define the electron propagation function in momentum space by

$$S_F(p) = \frac{1}{ip\gamma + m - i\epsilon} \qquad (6\text{-}52)$$

so that

$$S_F(x) = \frac{-i}{(2\pi)^4}\int d^4p\, e^{ipx}\, S_F(p) \qquad (6\text{-}53)$$

The propagation function including all higher-order corrections is given by

$$S'_F(x-y) = \langle 0|T[\psi(x)\ \bar{\psi}(y)]|0\rangle$$

$$+ \sum_{n=1}^{\infty} \frac{i^n}{n!} \int d^4x_1 \cdots d^4x_n$$

$$\times \langle 0|T_c^*[\psi(x)\bar{\psi}(y)\mathcal{L}_{int}(x_1)\cdots\ \mathcal{L}_{int}(x_n)]|0\rangle \qquad (6\text{-}54)$$

Since the S matrix is expressed by

$$S = 1 + \sum_{n=1}^{\infty} \frac{i^n}{n!} \int d^4x_1 \cdots d^4x_n$$

$$\times\ T^*[\mathcal{L}_{int}(x_1) \cdots \mathcal{L}_{int}(x_n)] \qquad (6\text{-}55)$$

we can formally write $S'_F(x-y)$ as

$$S'_F(x-y) = \langle 0|\ T_c^*[\psi(x),\ \bar{\psi}(y),\ S]|0\rangle$$

$$= \frac{\langle 0|T^*[\psi(x),\ \bar{\psi}(y),\ S]|0\rangle}{\langle 0|S|0\rangle} \qquad (6\text{-}56)$$

From graphical considerations one can see that

$$S'_F(p) = S_F(p) - S_F(p)\sum\nolimits^*_{ren}(p)\ S_F(p) + \cdots$$

$$= S_F(p) - S_F(p)\sum\nolimits^*_{ren}(p)\ S'_F(p)$$

$$= S_F(p) - S'_F(p)\sum\nolimits^*_{ren}(p)\ S_F(p) \qquad (6\text{-}57)$$

The last relationship is called Dyson's equation, and its solution is given by

$$S'_F(p) = \frac{S_F(p)}{1 + \Sigma^*_{ren}(p)\, S_F(p)}$$

$$= \left[ip\gamma + m + \Sigma^*_{ren}(p) \right]^{-1} \tag{6-58}$$

In this section m represents the experimental mass m_{exp}. As we have seen in the previous section $\Sigma^*_{ren}(p) \sim (ip\gamma + m)^2$; so

$$(ip\gamma + m)\, S'_F(p) \to 1 \qquad \text{as} \qquad ip\gamma + m \to 0 \tag{6-59}$$

This is a very important property of $S'_F(p)$ in the renormalized interaction representation. In a Feynman diagram an internal electron line always should be represented by $S'_F(p)$; the function $S_F(p)$ is only an approximation.

To second-order $S'_F(p)$ is given by

$$S'_F(p) = S_F(p) - S_F(p) \quad \Sigma^{*(2)}_{ren}(p)\, S_F(p)$$

$$= \frac{1}{ip\gamma + m - i\epsilon} + \frac{e^2}{16\pi^2} \int_m^\infty \frac{dM}{M^3(M^2 - m^2)}$$

$$\times \left[\frac{(M + m)^2 (M^2 + m^2 - 4mM)}{ip\gamma + M - i\epsilon} + \frac{(M - m)^2 (M^2 + m^2 + 4mM)}{ip\gamma - M + i\epsilon} \right] \tag{6-60}$$

where M is a parameter of integration and related to the Feynman parameter x in (6-19) by $m^2 = (1 - x)M^2$. Incidentally, this integral diverges as $M \to m$, giving rise to the so-called infrared divergence problem.

So far we have discussed the electron propagator; now we turn our attention to the photon propagator. The free-photon propagator is

$$\delta_{\mu\nu} D_F(x-y) = \langle 0|T[A_\mu(x)A_\nu(y)]|0\rangle$$

$$= \delta_{\mu\nu} \frac{-i}{(2\pi)^4} \int d^4k \frac{e^{ik(x-y)}}{k^2 - i\epsilon} \tag{6-61}$$

while the propagator including higher-order corrections is given by

$$D'_F(x-y)_{\mu\nu} = D'_{\mu\nu}(x-y) = \frac{\langle 0|T^*[A_\mu(x)A_\nu(y)S]|0\rangle}{\langle 0|S|0\rangle} \tag{6-62}$$

in complete analogy with the electron case.

Let us calculate $D'_{\mu\nu}(x-y)$ to second order in the coupling constant. The second order diagrams for this quantity are shown in Fig. 6-5.

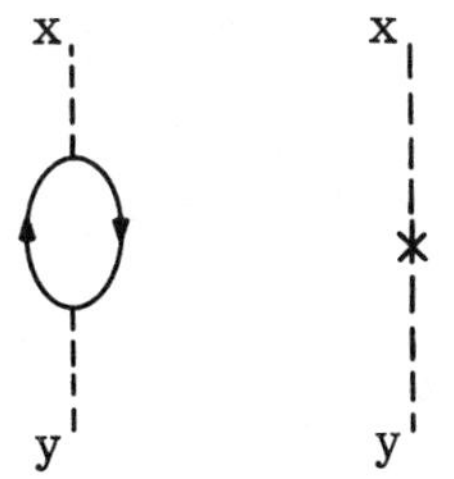

Figure 6-5. Diagrams corresponding to the photon propagation function to second order.

The interaction Lagrangian responsible for these diagrams is

$$\frac{1}{4}(1 - Z_3) F_{\mu\nu}(x)_r F_{\mu\nu}(x)_r$$

$$+ ieZ_2 Z_3^{1/2} A_\mu(x)_r \cdot \bar{\psi}_r(x) \gamma_\mu \psi_r(x) \tag{6-63}$$

Then to second order in e we find

$$D'_{\mu\nu}(x-y) = \delta_{\mu\nu} D_F(x-y) + \frac{i^2}{2!}\int d^4x' d^4x''$$

$$\times \langle 0 | T_c^*[A_\mu(x) A_\nu(y) \mathcal{L}_{int}(x') \mathcal{L}_{int}(x'')] | 0 \rangle$$

$$+ \frac{i}{1!}\int d^4x' \ \langle 0 | T^*[A_\mu(x) A_\nu(y) \mathcal{L}_{int}(x')] | 0 \rangle$$

$$= \delta_{\mu\nu} D_F(x-y) + \frac{e^2}{2} \ Z_2{}^2 Z_3 \int d^4x' d^4x''$$

$$\times \ \langle 0 | T_c^*[A_\mu(x) A_\nu(y) A_\rho(x') A_\sigma(x'')] | 0 \rangle$$

$$\times \ \langle 0 | T^*[\bar{\psi}(x')\gamma_\rho \psi(x'), \ \bar{\psi}(x'')\gamma_\sigma \psi(x'')] | 0 \rangle$$

$$+ \ i(1 - Z_3)\int d^4x'$$

$$\times \langle 0 | T^*[A_\mu(x), \ A_\nu(y), \ \frac{1}{4} \ F_{\rho\sigma}(x') \ F_{\rho\sigma}(x')] | 0 \rangle$$

$$= \delta_{\mu\nu} D_F(x-y) + e^2 Z_2{}^2 Z_3 \int d^4x' d^4x'' \ D_F(x-x') D_F(y-x'')$$

$$\times \ \langle 0 | T[\bar{\psi}(x')\gamma_\mu \psi(x'), \ \bar{\psi}(x'')\gamma_\nu \psi(x'')] | 0 \rangle$$

$$+ \ \frac{i}{2} \ (1 - Z_3)\int d^4x' \ \langle 0 | T[A_\mu(x), \ F_{\rho\sigma}(x')] | 0 \rangle$$

$$\times \langle 0 | \ T[A_\nu(y), \ F_{\rho\sigma}(x')] \ 0 \rangle$$

$$= \delta_{\mu\nu} D_F(x-y) - e^2 Z_2{}^2 Z_3 \int d^4x' d^4x'' \ D_F(x-x') \ D_F(y-x'')$$

$$\times \ \mathrm{Tr}[\gamma_\mu S_F(x'-x'')\gamma_\nu \ S_F(x''-x')]$$

$$+ \ \frac{i}{2} \ (1 - Z_3)\int d^4x' \ \left(\delta_{\mu\sigma} \frac{\partial}{\partial x'_\rho} - \delta_{\mu\rho} \frac{\partial}{\partial x'_\sigma}\right)$$

$$\times \ D_F(x-x') \ \left(\delta_{\nu\sigma} \frac{\partial}{\partial x'_\rho} - \delta_{\nu\rho} \frac{\partial}{\partial x'_\sigma}\right) \ D_F(y-x') \qquad (6\text{-}64)$$

The integrand in the last term is equal to

$$\left(\delta_{\mu\sigma}\frac{\partial}{\partial x_\rho}-\delta_{\mu\rho}\frac{\partial}{\partial x_\sigma}\right)\left(\delta_{\nu\sigma}\frac{\partial}{\partial y_\rho}-\delta_{\nu\rho}\frac{\partial}{\partial y_\sigma}\right)$$

$$\times\ D_F(x-x')\ D_F(y-x')$$

$$=2\left(\delta_{\mu\nu}\frac{\partial^2}{\partial x_\rho\,\partial y_\rho}-\frac{\partial^2}{\partial x_\nu\,\partial y_\mu}\right)$$

$$\times\ D_F(x-x')\ D_F(y-x') \tag{6-65}$$

Hence we obtain

$$\begin{aligned}
D'_{\mu\nu}(x-y) &= \delta_{\mu\nu}D_F(x-y)\\
&\quad - e^2Z_2{}^2Z_3\int d^4x'd^4x''\quad D_F(x-x')\,D_F(y-x'')\\
&\quad \times\ \mathrm{Tr}\,[\gamma_\mu S_F(x'-x'')\,\gamma_\nu\,S_F(x''-x')]\\
&\quad + i(1-Z_3)\int d^4x'\left(\delta_{\mu\nu}\frac{\partial^2}{\partial x_\rho\,\partial y_\rho}-\frac{\partial^2}{\partial x_\nu\,\partial y_\mu}\right)\\
&\quad \times\ D_F(x-x')\,D_F(y-x')\\
&= \delta_{\mu\nu}D_F(x-y)\\
&\quad - e^2Z_2{}^2Z_3\int d^4x'd^4x''\quad D_F(x-x')\,D_F(x''-y)\\
&\quad \times\ \mathrm{Tr}\,[\gamma_\mu S_F(x'-x'')\,\gamma_\nu\,S_F(x''-x')]\\
&\quad + i(1-Z_3)\left[\delta_{\mu\nu}(-\Box_x)+\frac{\partial^2}{\partial x_\mu\,\partial x_\nu}\right]\int d^4x'\\
&\quad \times\ D_F(x-x')\,D_F(x'-y)
\end{aligned} \tag{6-66}$$

By using $\Box_x D_F(x-x') = i\delta^4(x-x')$, the last term can be expressed as

$$(1-Z_3)\left(\delta_{\mu\nu}-\frac{1}{\Box_x}\frac{\partial^2}{\partial x_\mu\,\partial x_\nu}\right)D_F(x-y) \tag{6-67}$$

We set

$$D'_{\mu\nu}(x) = \frac{-i}{(2\pi)^4} \int d^4k \; e^{ikx} \; D'_{\mu\nu}(k) \tag{6-68}$$

then in momentum space we get

$$D'_{\mu\nu}(k) = \frac{\delta_{\mu\nu}}{k^2 - i\epsilon} - \frac{ie^2 Z_2^{\,2} Z_3}{(2\pi)^4} \; \frac{1}{(k^2 - i\epsilon)^2} \int d^4p$$

$$\times \; \mathrm{Tr}\left[\gamma_\mu \frac{1}{ip\gamma + m - i\epsilon} \gamma_\nu \frac{1}{i(p-k)\gamma + m - i\epsilon}\right]$$

$$+ \; (1 - Z_3) \frac{1}{k^2 - i\epsilon} \left(\delta_{\mu\nu} - \frac{k_\mu k_\nu}{k^2} \right) \tag{6-69}$$

The second term is represented in Fig. 6-6.

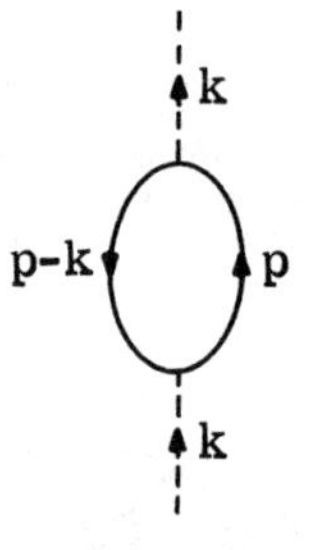

Figure 6-6. A vacuum polarization diagram labelled with momenta.

Before carrying out the p integration let us discuss a consequence of the conservation of charge. The trace term comes from the vacuum polarization contribution

$$\langle 0 | T\,[j_\mu(x),\; j_\nu(y)] | 0 \rangle \tag{6-70}$$

One can *formally* prove that the divergence of (6-70) vanishes:

$$\partial_\mu T[j_\mu(x),\; j_\nu(y)] = T[\partial_\mu j_\mu(x),\; j_\nu(y)]$$

$$+ \; \partial_\mu \left(\frac{1}{2}\, \epsilon(x_0 - y_0) \right) \cdot [j_\mu(x),\; j_\nu(y)]$$

$$= \delta(x_0 - y_0)\, [j_0(x),\; j_\nu(y)] = 0 \tag{6-71}$$

In momentum space this condition is expressed by

$$k_\mu \int d^4p \ \mathrm{Tr}\left[\gamma_\mu \frac{1}{ip\gamma + m - i\epsilon}\gamma_\nu \frac{1}{i(p-k)\gamma + m - i\epsilon}\right] = 0 \qquad (6\text{-}72)$$

Let us denote the result of the p integration by $F_{\mu\nu}(k)$, then from Lorentz invariance the most general form of $F_{\mu\nu}(k)$ is given by

$$F_{\mu\nu}(k) = \delta_{\mu\nu}\, f(k^2) - k_\mu k_\nu\, g(k^2) \qquad (6\text{-}73)$$

The condition of the vanishing divergence requires that

$$k_\mu F_{\mu\nu}(k) = k_\nu\,[f(k^2) - k^2\, g(k^2)] = 0 \qquad (6\text{-}74)$$

so that

$$F_{\mu\nu}(k) = (k^2\,\delta_{\mu\nu} - k_\mu k_\nu)\, g(k^2) \qquad (6\text{-}75)$$

Now we carry out the p integration of the trace term. The trace is calculated in the usual manner:

$$\mathrm{Tr}\left[\gamma_\mu \frac{ip\gamma - m}{p^2 + m^2 - i\epsilon}\gamma_\nu \frac{i(p-k)\gamma - m}{(p-k)^2 + m^2 - i\epsilon}\right]$$

$$= \frac{4[\delta_{\mu\nu}(p^2 - pk + m^2) - 2p_\mu p_\nu + p_\mu k_\nu + p_\nu k_\mu}{(p^2 + m^2 - i\epsilon)\,[(p-k)^2 + m^2 - i\epsilon]}$$

The integration is then performed by parametrizing as we have done for similar integrals.

$$F_{\mu\nu}(k) = 4\int d^4p\ \frac{\delta_{\mu\nu}(p^2 - pk + m^2) - 2p_\mu p_\nu + p_\mu k_\nu + p_\nu k_\mu}{(p^2 + m^2 - i\epsilon)\,[(p-k)^2 + m^2 - i\epsilon]}$$

$$= 4\int_0^1 dx \int d^4p$$

$$\times \frac{\delta_{\mu\nu}(p^2 - pk + m^2) - 2p_\mu p_\nu + p_\mu k_\nu + p_\nu k_\mu}{[p^2 + m^2 + x(k^2 - 2pk) - i\epsilon]^2} \qquad (6\text{-}76)$$

The change of variable $p \to p' = p - xk$ makes the denominator even in p'.

$$F_{\mu\nu}(k) = 4 \int_0^1 dx \int d^4p' \frac{N}{[p'^2 + m^2 + x(1-x)k^2 - i\epsilon]^2} \tag{6-77}$$

where

$$N = \delta_{\mu\nu} [(p' + xk)(p' + xk - k) + m^2] - 2(p' + xk)_\mu (p' + xk)_\nu$$
$$+ (p' + xk)_\mu k_\nu + (p' + xk)_\nu k_\mu \tag{6-78}$$

Since the denominator is an even function of p', the part of the numerator which is odd in p' vanishes upon integration; the even part of N is

$$\delta_{\mu\nu} [p'^2 - x(1-x)k^2 + m^2] - 2p'_\mu p'_\nu - 2x^2 k_\mu k_\nu + 2xk_\mu k_\nu$$
$$= \delta_{\mu\nu} p'^2 - 2p'_\mu p'_\nu + \delta_{\mu\nu} [m^2 + x(1-x)k^2]$$
$$- 2x(1-x)(\delta_{\mu\nu} k^2 - k_\mu k_\nu) \tag{6-79}$$

We know that the final result should be proportional to $(\delta_{\mu\nu}k^2 - k_\mu k_\nu)$; therefore we retain only the last term, since the other terms should vanish upon integration in order to obtain the right form. This argument, however, is not correct, and we shall come back to this problem at the end of this section. For the time being, we keep only the last term

$$F_{\mu\nu}(k) = -8(\delta_{\mu\nu}k^2 - k_\mu k_\nu) \int_0^1 dx\, x(1 - x)$$
$$\times \int d^4p \frac{1}{[p^2 + m^2 + x(1-x)k^2 - i\epsilon]^2} \tag{6-80}$$

We split the integrand in two:

$$\frac{1}{[p^2 + m^2 + x(1-x)k^2 - i\epsilon]^2} = \frac{1}{(p^2 + m^2 - i\epsilon)^2}$$
$$+ \left(\frac{1}{[p^2 + m^2 + x(1-x)k^2 - i\epsilon]^2} - \frac{1}{(p^2 + m^2 - i\epsilon)^2} \right)$$

then we substitute this expression into $D'_{\mu\nu}(k)$.

$$D'_{\mu\nu}(k) = \frac{\delta_{\mu\nu}}{k^2 - i\epsilon} - \frac{ie^2 Z_2{}^2 Z_3}{(2\pi)^4} \frac{1}{(k^2 - i\epsilon)^2}$$

$$\times \; (-8)\,(\delta_{\mu\nu} k^2 - k_\mu k_\nu) \left[\int_0^1 dx\, x(1-x) \right.$$

$$\times \int \frac{d^4p}{(p^2 + m^2 - i\epsilon)^2} + \int_0^1 dx\, x(1-x)$$

$$\left. \times \int d^4p \left(\frac{1}{[p^2 + m^2 + x(1-x)k^2 - i\epsilon]^2} - \frac{1}{(p^2 + m^2 - i\epsilon)^2} \right) \right]$$

$$+ \; (1 - Z_3) \frac{1}{(k^2 - i\epsilon)^2} (k^2 \delta_{\mu\nu} - k_\mu k_\nu)$$

$$= \frac{\delta_{\mu\nu}}{k^2 - i\epsilon} + \frac{8ie^2 Z_2{}^2 Z_3}{(2\pi)^4} \frac{1}{(k^2 - i\epsilon)^2} (\delta_{\mu\nu} k^2 - k_\mu k_\nu)$$

$$\times \int_0^1 dx\, x(1-x)$$

$$\times \int d^4p \left(\frac{1}{[p^2 + m^2 + x(1-x)k^2 - i\epsilon]^2} - \frac{1}{(p^2 + m^2 - i\epsilon)^2} \right)$$

$$+ \frac{k^2 \delta_{\mu\nu} - k_\mu k_\nu}{(k^2 - i\epsilon)^2} \left[1 - Z_3 + \frac{8ie^2 Z_2{}^2 Z_3}{(2\pi)^4} \right.$$

$$\left. \times \int_0^1 dx\, x(1-x) \int d^4p \, \frac{1}{(p^2 + m^2 - i\epsilon)^2} \right] \tag{6-81}$$

In order to determine Z_3 to second order we can use

$$D_F(k)\; \Pi^*_{\text{ren}}(k) \Big|\, \text{one photon, } k \rangle = 0 \tag{6-82}$$

corresponding to the similar statement

$$S_F(p)\, \Sigma^*_{\overline{\text{ren}}}(p) \Big| \text{ one electron, p} \Big\rangle = 0 \tag{6-83}$$

The $\Pi^*(k)$ is the proper self-energy operator for the photon and is usually called the polarization operator because of its connection with vacuum polarization. As we soon shall see this condition forbids the presence of the double pole $(k^2 - i\epsilon)^{-2}$ in $D'_{\mu\nu}(k)$, so that we are led to the following relation:

$$1 - Z_3 + \frac{8ie^2 Z_2{}^2 Z_3}{(2\pi)^4} \int_0^1 dx\, x(1-x) \int d^4p \frac{1}{(p^2 + m^2 - i\epsilon)^2} = 0 \tag{6-84}$$

To second order $Z_2{}^2 Z_3$ in front of the integral can be replaced by 1.

$$Z_3 = 1 + \frac{8ie^2}{(2\pi)^4} \int_0^1 dx\, x(1-x) \int d^4p \frac{1}{(p^2 + m^2 - i\epsilon)^2} \tag{6-85}$$

The integral on the right-hand side is divergent so that Z_3 is also a divergent quantity, but no divergent expression enters in $D'_{\mu\nu}(k)$:

$$D'_{\mu\nu}(k) = \frac{\delta_{\mu\nu}}{k^2 - i\epsilon} + \frac{8ie^2}{(2\pi)^4} \frac{\delta_{\mu\nu} k^2 - k_\mu k_\nu}{(k^2 - i\epsilon)^2} \times \int_0^1 dx\, x(1-x) \int d^4p \left(\frac{1}{[p^2 + m^2 + x(1-x)k^2 - i\epsilon]^2} - \frac{1}{(p^2 + m^2 - i\epsilon)^2} \right) \tag{6-86}$$

Since $D_F(k) = (k^2 - i\epsilon)^{-1}$ we get, for the second-order photon propagator,

$$D'_{\mu\nu}(k) = \delta_{\mu\nu} D_F(k) - D_F(k)\, \Pi^*_{\mu\nu}(k)\, D_F(k) \tag{6-87}$$

where $\Pi^*_{\mu\nu}(k)$ is the renormalized polarization operator given by

$$\Pi^*_{\mu\nu}(k) = -\frac{8ie^2}{(2\pi)^4}(\delta_{\mu\nu}k^2 - k_\mu k_\nu)$$

$$\times \int_0^1 dx\, x(1-x) \int d^4p$$

$$\times \left(\frac{1}{[p^2 + m^2 + x(1-x)k^2 - i\epsilon]^2} - \frac{1}{(p^2 + m^2 - i\epsilon)^2} \right) \tag{6-88}$$

The integral over momentum p may be carried out by transformation into the following form:

$$\int d^4p\ (\cdots) = 2\int d^4p \int_{m^2}^{\infty} dM^2$$

$$\times \left(\frac{1}{[p^2 + M^2 + x(1-x)k^2 - i\epsilon]^3} - \frac{1}{(p^2 + M^2 - i\epsilon)^3} \right)$$

$$= i\pi^2 \int_{m^2}^{\infty} dM^2 \left(\frac{1}{M^2 + x(1-x)k^2 - i\epsilon} - \frac{1}{M^2 - i\epsilon} \right)$$

So we find to second order

$$\Pi^*_{\mu\nu}(k) = \frac{e^2}{2\pi^2}(\delta_{\mu\nu}k^2 - k_\mu k_\nu)$$

$$\times \int_0^1 dx\, x(1-x)\, \ln\left(\frac{m^2}{m^2 + x(1-x)k^2 - i\epsilon} \right) \tag{6-89}$$

It is clear that the integrand vanishes for $k^2 = 0$, and in fact $\Pi^*_{\mu\nu}$ given for small values of k^2 by the following approximate formula:

$$\Pi^*_{\mu\nu}(k) \approx \frac{e^2}{60\pi^2}(\delta_{\mu\nu}k^2 - k_\mu k_\nu)\left(-\frac{k^2}{m^2}\right) \tag{6-90}$$

Therefore, for small values of k^2 the expression $D_F(k)\ \Pi^*_{\mu\nu}(k)$ is proportional to $(\delta_{\mu\nu} k^2 - k_\mu k_\nu)$ so that it vanishes when applied to a photon state, that is, $k^2 = 0$, $k_\nu e_\nu = 0$.

In general we can write the polarization operator $\Pi^*_{\mu\nu}(k)$ as

$$\Pi^*_{\mu\nu}(k) = (\delta_{\mu\nu} k^2 - k_\mu k_\nu) k^2 f(k^2) \tag{6-91}$$

The inclusion of higher order corrections makes Dyson's equation for the photon propagator read

$$\begin{aligned} D'_{\mu\nu}(k) &= \delta_{\mu\nu} D_F(k) - D_F(k)\, \Pi^*_{\mu\lambda}(k)\, D'_{\lambda\nu}(k) \\ &= \delta_{\mu\nu} D_F(k) - D'_{\mu\lambda}(k)\, \Pi^*_{\lambda\nu}(k)\, D_F(k) \end{aligned} \tag{6-92}$$

Addendum

In the derivation of the expression for $D'_{\mu\nu}(k)$ we have neglected a term in the integral of a trace on the ground that it does not fit the form required by current conservation. By actual calculation this term diverges; however, we can show that its second derivative with respect to $s = -k^2$ vanishes.

The integral we have discarded is

$$\begin{aligned} &\int_0^1 dx \int d^4p\, \frac{\delta_{\mu\nu} p^2 - 2p_\mu p_\nu + \delta_{\mu\nu}(m^2 - x(1-x)s)}{[p^2 + m^2 - x(1-x)s - i\epsilon]^2} \\ &= \int_0^1 dx \int d^4p\, \frac{\delta_{\mu\nu} p^2 - \frac{2}{4}\delta_{\mu\nu} p^2 + \delta_{\mu\nu}(m^2 - x(1-x)s)}{[p^2 + m^2 - x(1-x)s - i\epsilon]^2} \\ &= \delta_{\mu\nu}\, I(s) \end{aligned} \tag{6-93}$$

The $p_\mu p_\nu$ term survives upon integration only when $\mu = \nu$ so that it has been replaced by its invariant average value:

$$p_\mu p_\nu \to \delta_{\mu\nu} \frac{1}{4}(p_1 p_1 + p_2 p_2 + p_3 p_3 + p_4 p_4) = \frac{1}{4}\delta_{\mu\nu} p^2 \tag{6-94}$$

The integral I(s) can be written as

$$I(s) = \int_0^1 dx \int d^4p \times \left(\frac{1}{p^2 + m^2 - x(1-x)s - i\epsilon} - \frac{\frac{1}{2}p^2}{[p^2 + m^2 - x(1-x)s - i\epsilon]^2} \right)$$

If a cutoff is to be introduced to avoid divergences, this form suggests $I(s) \to 0$ as $s \to \infty$. We shall come back to this alternative approach later.

Differentiating I(s) twice with respect to s, we find

$$\begin{aligned}
I''(s) &= \int_0^1 dx \int d^4p \left(\frac{2x^2(1-x)^2}{[p^2 + m^2 - x(1-x)s - i\epsilon]^3} - \frac{3x^2(1-x)^2 p^2}{[p^2 + m^2 - x(1-x)s - i\epsilon]^4} \right) \\
&= \int_0^1 dx\, x^2(1-x)^2 \int d^4p \left(\frac{-1}{[p^2 + m^2 - x(1-x)s - i\epsilon]^3} + \frac{3(m^2 - x(1-x)s)}{[p^2 + m^2 - x(1-x)s - i\epsilon]^4} \right) \\
&= \int_0^1 dx\, x^2(1-x)^2 \left(\frac{-i\pi^2}{2[m^2 - x(1-x)s - i\epsilon]} + \frac{i\pi^2(m^2 - x(1-x)s)}{2[m^2 - x(1-x)s - i\epsilon]^2} \right) \\
&= 0
\end{aligned} \qquad (6\text{-}95)$$

Therefore the integral I(s) must be linear in s, namely,

$$I(s) = A + Bs \qquad (6\text{-}96)$$

The constants may be found easily:

$$A = I(0) = \frac{1}{2} \int d^4p \left(\frac{1}{p^2 + m^2 - i\epsilon} + \frac{m^2}{(p^2 + m^2 - i\epsilon)^2} \right) \tag{6-97}$$

This expression is quadratically divergent.

$$B = I'(0) = \int_0^1 dx\, x(1-x) \int d^4p$$

$$\times \left(\frac{1}{(p^2 + m^2 - i\epsilon)^2} - \frac{p^2}{(p^2 + m^2 - i\epsilon)^3} \right)$$

$$= \frac{1}{6} \int d^4p \, \frac{m^2}{(p^2 + m^2 - i\epsilon)^3}$$

$$= \frac{i\pi^2}{12} \tag{6-98}$$

Alternatively, one could adopt the point of view that

$$I(s) \to 0 \qquad \text{as} \qquad s \to \infty \tag{6-99}$$

and combine this condition with $I''(s) = 0$ to get the result

$$I(s) = 0 \tag{6-100}$$

This is clearly a specious argument and is presented here only for the convenience of its result. The correct argument for dropping this integral can be given, however, only after the introduction of the dispersion approach in Chapter 8.

6-5 VERTEX RENORMALIZATION

We have discussed the renormalization of the electron propagator as well as of the photon propagator. This is not the whole story, however. We must also consider renormalization of the vertex, that is, corrections owing to higher-order processes which take place around the vertex, such as those in Fig. 6-7.

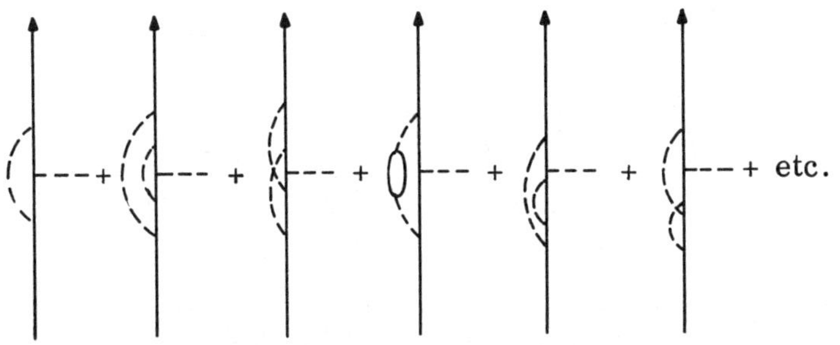

Figure 6-7. Diagrams representing vertex corrections.

This leads to the so-called vertex operator Γ_μ which takes the place of γ_μ, its first-order approximation, in the same way that S_F' and D_F' should take the place of the approximations $(ip\gamma + m - i\epsilon)^{-1}$ and $(k^2 - i\epsilon)^{-1}$, respectively. Thus the total picture for the electron-photon interaction looks like the Feynman diagram in Fig. 6-8.

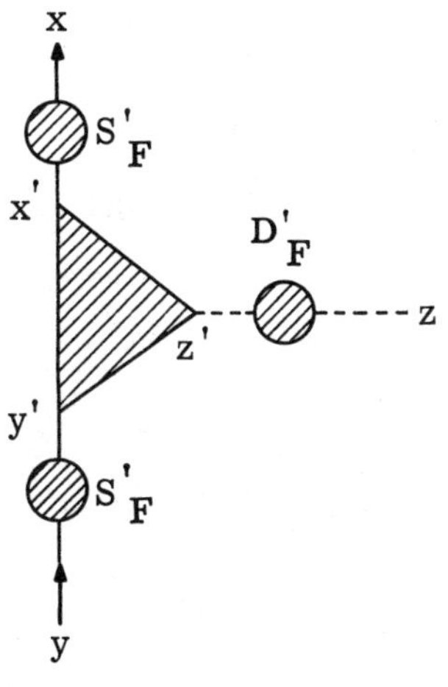

Figure 6-8. The diagram representing the most general three-point function.

We consider corrections owing to all Feynman diagrams with one incoming electron line, one outgoing electron line, and one photon line as indicated in Fig. 6-8; then the corresponding analytic expression is given by an expression similar to those for the propagators mentioned in the preceding section.

$$\frac{\langle 0 | T^* [\psi(x), \ \bar{\psi}(y), \ A_\nu(z), \ S] | 0 \rangle}{\langle 0 | S | 0 \rangle} \tag{6-101}$$

The vertex function Γ_μ is defined by equating (6-101) to

$$-e Z_2 Z_3^{1/2} \int d^4x' d^4y' d^4z' S_F'(x-x') \Gamma_\mu(x'y'; z') S_F'(y'-y) \times D_{\mu\nu}'(z'-z) \tag{6-102}$$

in conformity with the structure indicated in Figure 6-8. The propagators S_F' and D_F' represent renormalized ones, but Γ_μ is not yet renormalized. It is the object of this section to discuss how to renormalize this vertex function.

As we have seen, in all expressions the coupling constant e^2 occurs in the form $e^2 Z_2{}^2 Z_3$, where Z_2 and Z_3 are divergent constants. Furthermore, if we introduce the vertex renormalization we find that the correct combination is $e^2 Z_1^{-2} Z_2{}^2 Z_3$, where Z_1 is the vertex renormalization constant.

In lowest-order perturbation theory Γ_μ is given by

$$\Gamma_\mu(xy ; z) = \gamma_\mu \, \delta^4(x-z) \, \delta^4(y-z) \tag{6-103}$$

where Γ_μ is a function of two independent vectors $x - z$ and $z - y$, and the corresponding vertex function in momentum space is defined by

$$\Gamma_\mu(xy ; z) = \frac{1}{(2\pi)^8} \int d^4p \, d^4q \, e^{ip(x-z)+iq(z-y)} \, \Gamma_\mu(p, q) \tag{6-104}$$

Then, in the lowest order, $\Gamma_\mu(p, q)$ is given by

$$\Gamma_\mu(p, q) = \gamma_\mu \tag{6-105}$$

Before continuing with discussions of higher-order corrections let us give the definition of the observed charge. If Ψ_e is an eigenstate of the total Hamiltonian corresponding to a one-electron state, and $j_\mu(x)$ is the current operator in the Heisenberg representation, we take its diagonal matrix element and find the following relation:

$$\begin{aligned}(\Psi_e, j_0(x)\Psi_e) &= V^{-1} \int (\Psi_e, j_0(x)\Psi_e) \, d^3x \\ &= (\Psi_e, Q\Psi_e)/V \\ &= e_{obs} / V \end{aligned} \tag{6-106}$$

where Q is the charge operator and e_{obs} denotes the observed charge of the electron. The 0th component of the charge density in the renormalized interaction representation is related to the same quantity in the Heisenberg representation by a transformation,

$$j_0(x) = U(x_0, 0)^{-1} j_0^{in}(x) U(x_0, 0) \tag{6-107}$$

and Ψ_e is related to the corresponding state Φ_e in the interaction representation, as we shall see in Section 7-5, by

$$\Psi_e = U(0, -\infty) \Phi_e \tag{6-108}$$

Hence

$$\begin{aligned}(\Psi_e, j_0(x) \Psi_e) &= (\Phi_e, U(x_0, -\infty)^{-1} j_0^{in}(x) U(x_0, -\infty) \Phi_e) \\ &= (\Phi_e, U(\infty, -\infty)^{-1} \Phi_e) \\ &\quad \times (\Phi_e, U(\infty, x_0) j_0^{in}(x) U(x_0, -\infty) \Phi_e)\end{aligned} \tag{6-109}$$

In the renormalized interaction representation we may use

$$\begin{aligned}(\Phi_e, U(\infty, -\infty)^{-1} \Phi_e) &= (\Phi_e, U(\infty, -\infty) \Phi_e)^{-1} \\ &= (\Phi_0, U(\infty, -\infty) \Phi_0)^{-1}\end{aligned} \tag{6-110}$$

so that

$$(\Psi_e, j_0(x) \Psi_e) = \frac{(\Phi_e, T^* [j_0^{in}(x) S] \Phi_e)}{\langle 0 | S | 0 \rangle} = e_{obs}/V \tag{6-111}$$

An inspired scrutiny of the Feynman diagrams reveals that the middle expression in the above equation is given by

$$e Z_2 Z_3^{1/2} \bar{u}(p) \Gamma_4(p, p) u(p)/V \tag{6-112}$$

If Z_1 is now defined by

$$\bar{u}(p) \Gamma_\mu(p, p) u(p) = Z_1^{-1} \bar{u}(p) \gamma_\mu u(p) \tag{6-113}$$

we can simplify the above expression by using

$$\bar{u}(p) \Gamma_4(p, p) u(p) = Z_1^{-1} u^*(p) u(p) = Z_1^{-1} \tag{6-114}$$

Then we finally obtain

$$e_{obs} = Z_1^{-1} Z_2 Z_3^{1/2} e \qquad (6\text{-}115)$$

We define the renormalized vertex function $\Gamma_{\mu r}$ by

$$\Gamma_\mu(p, q)_r = Z_1 \Gamma_\mu(p, q) \qquad (6\text{-}116)$$

so that

$$u(p) \Gamma_\mu(p, p)_r u(p) = \bar{u}(p) \gamma_\mu u(p) \qquad (6\text{-}117)$$

The charge $e Z_2 Z_3^{1/2}$ may be written as

$$e Z_2 Z_3^{1/2} = e Z_1^{-1} Z_2 Z_3^{1/2} - (Z_1^{-1} - 1) e Z_2 Z_3^{1/2}$$

$$\equiv e_{obs} - \delta e \qquad (6\text{-}118)$$

and the interaction can be written as

$$\mathcal{L}_{int}(x) = i(e_{obs} - \delta e) \bar{\psi}\gamma_\mu \psi \cdot A_\mu + \text{bilinear terms} \qquad (6\text{-}119)$$

in the renormalized interaction representation.

From now on we shall expand everything in powers of e_{obs} rather than of e, then it will be shown that

$$\delta e = (1 - Z_1) e_{obs} \sim \mathcal{O}(e_{obs}^3) \qquad (6\text{-}120)$$

thereby justifying our expansions introduced so far.

The product $(e_{obs} - \delta e) \Gamma_\mu(xy; z) = e_{obs} \Gamma_\mu(xy; z)_r$ satisfies the following equation in third order:

$$e_{obs} \Gamma_\mu(xy; z)_r = (e_{obs} - \delta e) \gamma_\mu \delta^4(x - z) \delta^4(y - z)$$

$$+ e_{obs}^3 \gamma_\lambda S_F(x - z) \gamma_\mu S_F(z - y) \gamma_\lambda \qquad (6\text{-}121)$$

$$\times D_F(x - y)$$

where we have replaced $(e_{obs} - \delta e)^3$ by e_{obs}^3 in the last term since the difference represents further higher-order corrections.

We convert Eq. (6-121) into an equation in momentum space (Fig. 6-9).

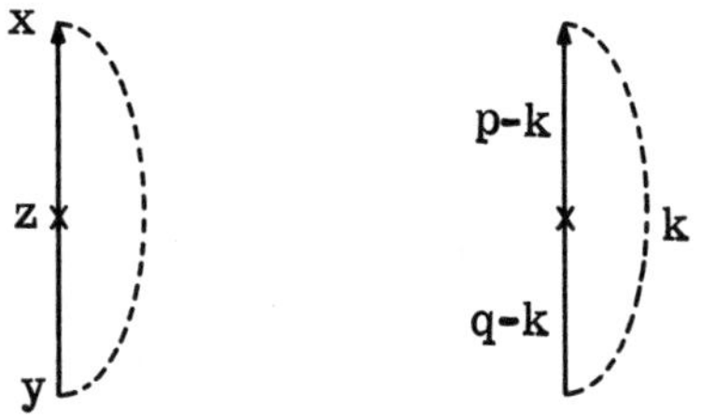

Figure 6-9 The third-order vertex diagram in configuration and momentum spaces.

$$e_{obs}\,\Gamma_\mu(p, q)_r = (e_{obs} - \delta e)\,\gamma_\mu + \frac{(-i)^3}{(2\pi)^4}\,e_{obs}^3 \int d^4k\,\gamma_\lambda \frac{1}{i(p-k)\gamma + m - i\epsilon} \times \gamma_\mu \frac{1}{i(q-k)\gamma + m - i\epsilon}\,\gamma_\lambda \cdot \frac{1}{k^2 - i\epsilon} \qquad (6\text{-}122)$$

In order to fix δe we use the definition of $\Gamma_{\mu r}$:

$$\bar{u}(p)\,\Gamma_\mu(p, p)_r\,u(p) = \bar{u}(p)\,\gamma_\mu u(p) \qquad (6\text{-}123)$$

After the denominator of the integrand is rationalized, the numerator is given by

$$\gamma_\lambda\,[i(p-k)\gamma - m]\,\gamma_\mu\,[i(q-k)\gamma - m]\,\gamma_\lambda \qquad (6\text{-}124)$$

We sum over λ by using

(1) $\gamma_\lambda \gamma_a \gamma_\lambda = -2\gamma_a$

(2) $\gamma_\lambda \gamma_a \gamma_b \gamma_\lambda = 4\delta_{ab}$

(3) $\gamma_\lambda \gamma_a \gamma_b \gamma_c \gamma_\lambda = -2\gamma_c \gamma_b \gamma_a$

Thus the numerator becomes

$$N = -2i(q-k)\gamma \cdot \gamma_\mu \cdot i(p-k)\gamma$$
$$-4mi((p-k)_\mu + (q-k)_\mu) - 2m^2\gamma_\mu \tag{6-125}$$

and our equation for $\Gamma_{\mu r}$ is

$$e_{obs}\,\Gamma_\mu(p, q)_r = (e_{obs} - \delta e)\,\gamma_\mu + \frac{(-ie_{obs})^3}{(2\pi)^4}$$
$$\times \int d^4k \, \frac{N}{[(p-k)^2 + m^2 - i\epsilon][(q-k)^2 + m^2 - i\epsilon](k^2 - i\epsilon)} \tag{6-126}$$

In order to unify the denominator we recall Feynman's formula:

$$\frac{1}{abc} = 2\int_0^1 dx_1 \int_0^1 dx_2 \int_0^1 dx_3\, \delta(1 - \Sigma x_i)$$
$$\times \frac{1}{(x_1 a + x_2 b + x_3 c)^3}$$

It is also convenient to introduce the following set of momenta:

$$P = \frac{1}{2}(p+q) \qquad \Delta = p - q \tag{6-127}$$

so that

$$p = P + \frac{\Delta}{2} \qquad q = P - \frac{\Delta}{2} \tag{6-128}$$

Then

$$\frac{1}{[(p-k)^2+m^2-i\epsilon][(q-k)^2+m^2-i\epsilon](k^2-i\epsilon)}$$

$$= 2\int_0^1 dx_1 \int_0^1 dx_2 \int_0^1 dx_3\, \delta(1-\Sigma x_i) \tag{6-129}$$

$$\times\ [(P^2+\frac{\Delta^2}{4}+m^2)(x_1+x_2) + P\Delta(x_1-x_2)$$

$$-\ 2(x_1+x_2)\,kP - (x_1-x_2)\,k\Delta + k^2 - i\epsilon]^{-3}$$

The variable x_3 does not enter in the integrand at all, so we can easily perform the x_3 integration, using

$$\int_0^1 dx_1 \int_0^1 dx_2 \int_0^1 dx_3\, \delta(1-\Sigma x_i)\, f(x_1, x_2)$$
$$= \int_\Delta dx_1\, dx_2\, f(x_1, x_2) \tag{6-130}$$

where Δ is a triangular domain, illustrated in Fig. 6-10.

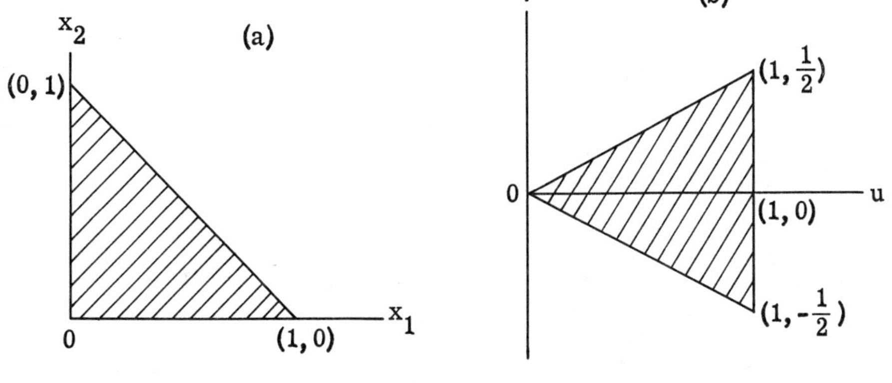

Figure 6-10. (a) The triangular domain Δ.
(b) The triangular domain Δ'.

Furthermore, the domain Δ is transformed in the domain Δ' by the following transformation:

$$x_1 + x_2 = u \quad \left(x_1 - x_2\right)/2 = v \quad \text{with } u/2 \geq v \geq -u/2 \quad 0 \leq u \leq 1 \tag{6-131}$$

then

$$\int_{\Delta} dx_1 dx_2 f(x_1, x_2) = \int_{\Delta'} du\, dv\, f'(u, v) \tag{6-132}$$

where

$$f'(u, v) = f(u/2 + v, u/2 - v) \tag{6-133}$$

Hence, from (6-129),

$$\frac{1}{[\ ][\ ][\ ]} = 2 \int_{\Delta'} dudv \left[(P^2 + \Delta^2/4 + m^2)\, u + 2P\Delta v - 2kPu - 2k\Delta v + k^2 - i\epsilon\right]^{-3}$$

$$= 2 \int_{\Delta'} dudv \left[k'^2 + (P^2 + \Delta^2/4 + m^2)\, u + 2P\Delta v - (uP + v\Delta)^2 - i\epsilon\right]^{-3} \tag{6-134}$$

where $k' = k - uP - v\Delta$.

The numerator N is now

$$-\frac{1}{2}N = i(P - \Delta/2 - k)\gamma \cdot \gamma_\mu \cdot i(P + \Delta/2 - k)\gamma + 4im(P - k)_\mu + m^2\gamma_\mu$$

$$= i(P - \Delta/2 - uP - v\Delta - k')\gamma \cdot \gamma_\mu \cdot i(P + \Delta/2 - uP - v\Delta - k')\gamma + 4im(P - uP - v\Delta - k')_\mu + m^2\gamma_\mu \tag{6-135}$$

Retaining only the terms of even order in k', we get

$$i((1-u)P - (\tfrac{1}{2}+v)\Delta)\gamma\cdot\gamma_\mu\cdot i((1-u)P - (-\tfrac{1}{2}+v)\Delta)\gamma$$

$$- k'\gamma\cdot\gamma_\mu\cdot k'\gamma + 4im((1-u)P - v\Delta)_\mu + m^2\gamma_\mu$$

In order to simplify the calculation let us take the matrix element of $\Gamma_{\mu r}$ between two single-electron states, so that we may use

$$p^2 + m^2 = q^2 + m^2 = P^2 + \Delta^2/4 + m^2 = 0 \qquad P\Delta = 0 \tag{6-136}$$

and

$$\begin{aligned} \bar{u}(p)(ip\gamma + m) &= \bar{u}(p)(iP\gamma + \tfrac{1}{2}i\Delta\gamma + m) = 0 \\ (iq\gamma + m)u(q) &= (iP\gamma - \tfrac{1}{2}i\Delta\gamma + m)u(q) = 0 \end{aligned} \tag{6-137}$$

When $\bar{u}(p)Au(q) = \bar{u}(p)Bu(q)$, we use the equivalence notation:

$$A \sim B \tag{6-138}$$

Then we can eliminate $iP\gamma$ from the numerator, and the even part of the numerator is

$$\begin{aligned} -\tfrac{1}{2}N_{\text{even}} &\sim [-(1-u)(\tfrac{1}{2}i\Delta\gamma + m) - (\tfrac{1}{2}+v)\,i\Delta\gamma]\,\gamma_\mu \\ &\quad \times [(1-u)(\tfrac{1}{2}i\Delta\gamma - m) + (\tfrac{1}{2}-v)\,i\Delta\gamma] \\ &\quad + k'^2\gamma_\mu - 2k'_\mu(k'\gamma) + 4im((1-u)P_\mu - v\Delta_\mu) + m^2\gamma_\mu \\ &= [-i\Delta\gamma(1-u/2+v) - m(1-u)]\,\gamma_\mu\,[i\Delta\gamma(1-u/2-v) - m(1-u)] \\ &\quad + k'^2\gamma_\mu - 2k_\mu'(k'\gamma) + 4im((1-u)P_\mu - v\Delta_\mu) + m^2\gamma_\mu \end{aligned} \tag{6-139}$$

Furthermore, the term in the denominator is

$$\begin{aligned} k'^2 - (uP + v\Delta)^2 - i\epsilon &= k'^2 - u^2P^2 - v^2\Delta^2 - i\epsilon \\ &= k'^2 + u^2m^2 + \Delta^2(u^2/4 - v^2) - i\epsilon \end{aligned} \tag{6-140}$$

This is even in v, so we can drop odd-order terms of v in the numerator, since Δ' is symmetric under $v \to -v$. Furthermore, the following relation is useful:

$$
\begin{aligned}
2iP_\mu &= \gamma_\mu (iP\gamma) + (iP\gamma)\gamma_\mu \\
&\sim \gamma_\mu \left(\frac{1}{2} i\gamma\Delta - m\right) - \left(\frac{1}{2} i\gamma\Delta + m\right)\gamma_\mu \\
&= \frac{1}{2} i [\gamma_\mu, \gamma_\nu] \Delta_\nu - 2m\gamma_\mu \\
&= -\sigma_{\mu\nu}\Delta_\nu - 2m\gamma_\mu
\end{aligned}
\tag{6-141}
$$

where the tensor operator $\sigma_{\mu\nu}$ is defined as

$$\sigma_{\mu\nu} = (\gamma_\mu\gamma_\nu - \gamma_\nu\gamma_\mu)/2i$$

Then the remaining part of the numerator is equivalent to

$$
\begin{aligned}
&\gamma\Delta \cdot \gamma_\mu \cdot \gamma\Delta[(1-u/2)^2 - v^2] - m(1-u)(1-u/2)\, i[\gamma_\mu, \gamma_\nu]\Delta_\nu \\
&+ m^2(1-u)^2\gamma_\mu + (k'^2/2)\gamma_\mu - 2m(\sigma_{\mu\nu}\Delta_\nu + 2m\gamma_\mu)(1-u) + m^2\gamma_\mu \\
&= 2\Delta_\mu \cdot \gamma\Delta \left[\left(1-\frac{u}{2}\right)^2 - v^2\right] - m\sigma_{\mu\nu}\Delta_\nu u(1-u) \\
&+ \gamma_\mu[-\Delta^2((1-u/2)^2 - v^2) + m^2(1-u)^2 - 4m^2(1-u) + m^2 \\
&+ k'^2/2]
\end{aligned}
$$

where we have made the replacement $k'_\mu \cdot k'\gamma \to \frac{1}{4} k'^2 \gamma_\mu$.

We introduce the assumption that Δ^2 is much smaller than m^2, and neglect Δ^2, thereby keeping only linear terms in Δ. The denominator becomes approximately

$$k'^2 + u^2m^2 - i\epsilon$$

and the numerator becomes

$$-u(1-u)\, m\sigma_{\mu\nu}\Delta_\nu + m^2\gamma_\mu[(1-u)^2 - 4(1-u) + 1] + \frac{1}{2}k'^2\gamma_\mu$$

Hence for small values of Δ^2 and with both the incoming and outgoing electrons on the mass shell,

$$e_{obs}\,\Gamma_\mu(p, q)_r \sim (e_{obs} - \delta e)\,\gamma_\mu - 4\frac{ie^3_{obs}}{(2\pi)^4}\int_{\Delta'} du\,dv$$

$$\times \int d^4k' \frac{[-u(1-u)\,m\sigma_{\mu\nu}\Delta_\nu + m^2\gamma_\mu(u^2 + 2u - 2) + \frac{1}{2}k'^2\gamma_\mu]}{(k'^2 + m^2u^2 - i\epsilon)^3}$$

$$+ \mathcal{O}(\Delta^2)$$

(6-142)

In the limit $\Delta \to 0$, the right-hand side must be equal to $e_{obs}\gamma_\mu$, so that the integral must be equal to $\delta e\,\gamma_\mu$ in this limit. Hence

$$\delta e = -4\frac{ie^3_{obs}}{(2\pi)^4}\int_{\Delta'} du\,dv \times \int d^4k' \frac{m^2(u^2 + 2u - 2) + \frac{1}{2}k'^2}{(k'^2 + m^2u^2 - i\epsilon)^3} \qquad (6\text{-}143)$$

which is logarithmically divergent.

After the above identification of δe we are left with

$$e_{obs}\,\Gamma_\mu(p, q)_r \sim e_{obs}\,\gamma_\mu + 4\frac{ie^3_{obs}}{(2\pi)^4}\int_{\Delta'} du\,dv \times \int d^4k' \frac{u(1 - u)\,m\sigma_{\mu\nu}\Delta_\nu}{(k'^2 + m^2u^2 - i\epsilon)^3} + \mathcal{O}(\Delta^2) \qquad (6\text{-}144)$$

The second term on the right-hand side is readily evaluated:

$$4\,\frac{ime^3}{(2\pi)^4}\,\sigma_{\mu\nu}\Delta_\nu \int_{\Delta'} du\,dv \int d^4k'\;\frac{u(1-u)}{(k'^2+m^2u^2-i\epsilon)^3}$$

$$= \frac{4ime^3}{(2\pi)^4}\,\sigma_{\mu\nu}\Delta_\nu \int_0^1 du\;\frac{i\pi^2u^2(1-u)}{2m^2u^2}$$

$$= -\left(\frac{e^2}{4\pi}\right)\frac{e}{4\pi m}\,\sigma_{\mu\nu}\Delta_\nu \tag{6-145}$$

where we denoted e_{obs} by e for simplicity. Hence

$$e\Gamma_{\mu r} \sim e\gamma_\mu - \frac{e}{4\pi m}\,\alpha\,\sigma_{\mu\nu}\Delta_\nu + \mathcal{O}(\Delta^2) \tag{6-146}$$

This implies a change in the magnetic moment of the electron. From this result we shall deduce in the next section that the magnetic moment of the electron in the Dirac theory

$$\boldsymbol{\mu} = \frac{e}{2m}\,\boldsymbol{\sigma} \tag{6-147}$$

is modified to

$$\boldsymbol{\mu} = \left(1+\frac{\alpha}{2\pi}\right)\frac{e}{2m}\,\boldsymbol{\sigma} \tag{6-148}$$

The accuracy of this result is demonstrated by comparison of the theoretical value of the g factor with the experimental result.

Schwinger[4]: $g = 2\left(1+\frac{\alpha}{2\pi}\right) = 2\times 1.00116$

Sommerfield[5]: $g = 2\left(1+\frac{\alpha}{2\pi} - 0.328\,\frac{\alpha^2}{\pi^2}\right) = 2\times 1.0011596$

The experimental value by Wilkinson and Crane[6] can be described as

$$g = 2\left(1+\frac{\alpha}{2\pi} - (0.327\pm 0.005)\,\frac{\alpha^2}{\pi^2}\right)$$

$$= 2(1.001159622 \pm 0.000000027)$$

6-6 ELECTRON IN AN EXTERNAL FIELD

In this section we shall discuss problems concerning one electron in the presence of an external field. In order to incorporate the external field we replace

$$-i(e_{obs} - \delta e)\, \bar{\psi}\gamma_\mu \psi \cdot A_\mu$$

by

$$-i(e_{obs} - \delta e)\, \bar{\psi}\gamma_\mu \psi (A_\mu + A_\mu^{ext}) \tag{6-149}$$

in the interaction Hamiltonian.

To simplify the problem we assume that A_μ^{ext} is weak and static, and expand the S matrix in powers of the external field:

$$S = S_0 + S_1 + S_2 + \cdots \tag{6-150}$$

We study the first-order term which is given by

$$S_1 = i \int d^4x\; T^*[S_0\,,\, j_\mu(x)]\, A_\mu^{ext}(x) \tag{6-151}$$

or

$$\langle b|S_1|a\rangle = i \sum_{n=0}^{\infty} \frac{i^n}{n!} \int d^4x\, d^4x_1 \ldots d^4x_n$$

$$\times \langle b|T^*[j_\mu(x)\, \mathcal{L}_{int}(x_1) \ldots \mathcal{L}_{int}(x_n)]|a\rangle A_\mu^{ext}(x) \tag{6-152}$$

Anomalous Magnetic Moment of the Electron

If we include the second-order radiative corrections to the interaction of an electron with an external field, the Feynman diagrams are those given in Fig. 6-11.

The process in the first diagram is proportional to e, the others

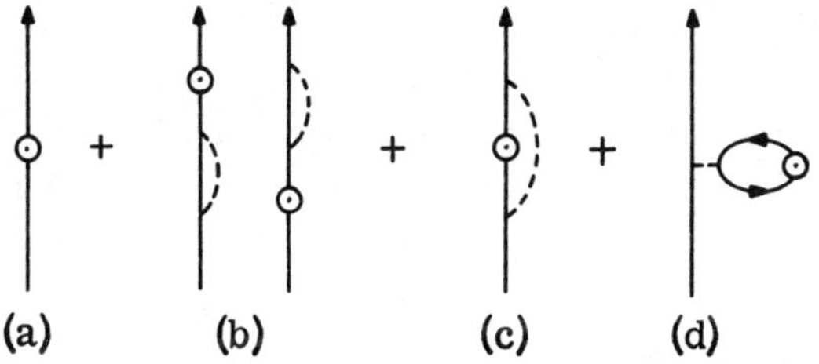

Figure 6-11. Diagrams representing the interaction of an electron with an external field to third order.
(a) The lowest order diagram.
(b) The self-energy type corrections.
(c) The vertex type corrections.
(d) The vacuum polarization type corrections.

being proportional to e^3. The lowest-order process has been discussed in Section 5-1, and the matrix element is

$$\langle q'|S_1|q\rangle = -\frac{e}{V}\bar{u}^{(r')}(q')\gamma_\mu u^{(r)}(q)A_\mu^{ext}(q-q') \times 2\pi\delta(q_0 - q_0') \tag{6-153}$$

Consider the third-order correction and, in so doing, retain only the first-order term in q - q'. Then in the light of the result of the previous section, the matrix element to third order is

$$\langle q'|S_1|q\rangle = -\frac{e}{V}\bar{u}^{(r')}(q')\left[\gamma_\mu - \frac{\alpha}{4\pi m}\sigma_{\mu\nu}(q'-q)_\nu\right] \times u^{(r)}(q)A_\mu^{ext}(q-q') \circ 2\pi\delta(q_0 - q_0') \tag{6-154}$$

or, on reverting to coordinate space,

$$\begin{aligned}\langle q'|S_1|q\rangle &= -e\int d^4x\left[\langle q'|\bar{\psi}(x)|0\rangle\gamma_\mu\langle 0|\psi(x)|q\rangle - \frac{\alpha}{4\pi m}i\frac{\partial}{\partial x_\nu}\langle q'|\bar{\psi}(x)|0\rangle\sigma_{\mu\nu}\langle 0|\psi(x)|q\rangle\right]A_\mu^{ext}(x)\\ &= -e\int d^4x\left[\langle q'|\bar{\psi}(x)|0\rangle\gamma_\mu\langle 0|\psi(x)|q\rangle A_\mu^{ext}(x) + \frac{i\alpha}{4\pi m}\langle q'|\bar{\psi}(x)|0\rangle\sigma_{\mu\nu}\langle 0|\psi(x)|q\rangle\frac{\partial A_\mu^{ext}(x)}{\partial x_\nu}\right]\end{aligned} \tag{6-155}$$

The second term is equivalent to an effective Hamiltonian

$$\mathcal{H}_{eff}(x) = -\frac{e\alpha}{8\pi m}\bar{\psi}(x)\sigma_{\mu\nu}\psi(x)F_{\mu\nu}^{ext}(x) \tag{6-156}$$

Suppose that the external field is a uniform magnetic field, then this Hamiltonian reduces, in the nonrelativistic approximation, to

$$H_{eff} = -\frac{e\alpha}{4\pi m}\sigma\cdot H \tag{6-157}$$

This means that the electron receives an additional, so-called anomalous, magnetic moment from radiative corrections, which is of magnitude

$$\left(\frac{\alpha}{2\pi}\right)\left(\frac{e}{2m}\right)\sigma \tag{6-158}$$

Adding this correction to the Dirac moment $\left(\frac{e}{2m}\right)\sigma$, we get

$$\mu = \left(1 + \frac{\alpha}{2\pi}\right)\left(\frac{e}{2m}\right)\sigma$$

This value agrees very well with the observed magnetic moment of the electron.

Lamb Shift

In the hydrogen atom the electron is not free, but is bound to the Coulomb field of the proton. This complicates the problem considerably. The shift of an energy level is given by

$$\Delta E_a = - \int d^3x \, \langle a | \Delta j_\mu(x) | a \rangle \, A_\mu^{ext}(x) \tag{6-159}$$

where $\Delta j_\mu(x)$ denotes the radiative correction to the operator $j_\mu(x)$. In the present problem one has to evaluate all third-order diagrams correctly. Since the electron is not free the renormalized self-energy operator gives a nonvanishing contribution. We know, in principle, how to evaluate the required matrix elements, but the exact calculation is much too complicated to carry out here.

Therefore, we reproduce Bethe's nonrelativistic but intuitive version of the Lamb shift calculation. Lamb and Retherford[7] found, experimentally, an energy difference in the 2s and 2p levels of hydrogen,

$$E(2s) - E(2p_{1/2}) \sim 0.033 \text{ cm}^{-1}$$

$$\text{or} \qquad 1000 \text{ megacycles} \tag{6-160}$$

The Dirac theory predicts that these levels are degenerate,

$$E(2s) = E(2p_{1/2}) \tag{6-161}$$

and Bethe[8] explained this energy difference in terms of radiative corrections to the energy levels of the bound electron in hydrogen. He calculated the electron self-energy in bound states and attributed the level shift to the difference between free and bound electron self-energies.

$$\Delta E = W_{bound} - W_{free} \tag{6-162}$$

where W_a denotes the electron self-energy in the state a. In nonrelativistic second-order perturbation theory this difference diverges logarithmically. Because the relativistic theory is less singular, it may be hoped that the relativistic calculation gives a convergent result, and such is in fact the case. Hence it is necessary to introduce a cutoff momentum K in nonrelativistic theory at about m, the rest energy of the electron.

The second-order self-energy of the electron in a bound state m is

$$W_m = -\frac{2\alpha}{3\pi}\int_0^K kdk \sum_n |\mathbf{v}_{mn}|^2/(E_n - E_m + k) \qquad (6\text{-}163)$$

where **v** denotes the electron velocity and is identified as

$$v = p/m \qquad (6\text{-}164)$$

For a free electron, using the same cutoff, we have

$$W_0 = -\frac{2\alpha}{3\pi}\int_0^K kdk \frac{(v^2)_{mm}}{k} \qquad (6\text{-}165)$$

Hence the level shift is given by

$$W' = W_m - W_0 = \frac{2\alpha}{3\pi}\int_0^K dk \sum_n \frac{|v_{mn}|^2 (E_n - E_m)}{E_n - E_m + k} \qquad (6\text{-}166)$$

Assuming that $K >> E_n - E_m$ we can approximate this expression by

$$W' = \frac{2\alpha}{3\pi}\sum_n |v_{mn}|^2 (E_n - E_m)\, \ell n \frac{K}{E_n - E_m} \qquad (6\text{-}167)$$

In order to carry out the summation we replace the argument of the logarithm by its average value:

$$W' = \frac{2\alpha}{3\pi}\ell n \frac{K}{\langle E_n - E_m\rangle_{av}} \sum_n |v_{mn}|^2 (E_n - E_m) \qquad (6\text{-}168)$$

The summation is then readily carried out:

$$\sum_n |p_{nm}|^2(E_n - E_m) = -\int \psi_m^* \nabla V \cdot \nabla \psi_m d^3x$$

$$= \frac{1}{2}\int \psi_m^* \psi_m \nabla^2 V \, d\, x$$

$$= 2\pi\, \alpha Z\, (\psi_m(0))^2 \qquad (6\text{-}169)$$

Since $\psi_m(0) = 0$ except for an S state, W'_m survives only for an S state in which

$$\left(\psi_m(0)\right)^2 = \frac{1}{\pi}\left(\frac{Z}{na}\right)^3 \qquad (6\text{-}170)$$

where a is the Bohr radius and n the principal quantum number. The result is

$$W'_{n,\,\ell=0} = \frac{8}{3\pi}\,\alpha^3\ \text{Ry}\,\frac{Z^4}{n^3}\,\ell n\,\frac{K}{\langle E_n - E_m\rangle_{av}} \qquad (6\text{-}171)$$

where Ry is the ionization energy of the ground state of hydrogen. In this way, with $K \approx m$, Bethe[8] found an energy shift of about 1040 megacycles for $E(2s) - E(2P_{1/2})$. More recent values for the Lamb shift are listed below:

Theoretical
- 1057.64 ± 0.21 Mc/sec[9]
- 1057.50 ± 0.11 Mc/sec[10]

Experimental
- 1057.77 ± 0.10 Mc/sec[11]
- 1058.05 ± 0.10 Mc/sec[12]

In this connection we also give the measured values of α^{-1}.

$\alpha^{-1} =$
- 137.0388 ± 0.0006[11]
- 137.0359 ± 0.0004[12]

6-7 DOUBLE MEANING OF THE CONSERVATION OF CHARGE AND WARD'S IDENTITY

Ward's Identity

Let us recall the relationship between the bare and observed charges:

$$e_{obs} = Z_1^{-1}\,Z_2 Z_3^{1/2}\,e \qquad (6\text{-}172)$$

where Z_1 is the vertex renormalization constant, Z_2 the electron wave function renormalization constant, and Z_3 the photon wave function renormalization constant, or Z_3^{-1} is the dielectric constant of the vacuum in the sense mentioned in Section 6-1.

When there are different kinds of particles, we distinguish these renormalization constants by particle subscripts a, b, and so on; thus, for particle a the above relationship is written as

$$(e_a)_{obs} = Z_{1a}^{-1} Z_{2a} Z_3^{1/2} e_a \qquad \text{etc.} \tag{6-173}$$

Let us consider a reaction

$$a + b \to c + d \tag{6-174}$$

then the conservation of charge, which follows from field equations, implies

$$e_a + e_b = e_c + e_d \tag{6-175}$$

The experimental conservation of charge, however, is expressed by

$$(e_a)_{obs} + (e_b)_{obs} = (e_c)_{obs} + (e_d)_{obs} \tag{6-176}$$

Equations(6-175) and (6-176) both must hold; substitution of (6-176) for e_{obs} in terms of e leads to the conclusion that the two equations as well as all other sets of equations expressing the conservation of charge are consistent only when $Z_1^{-1} Z_2$ is independent of the particle species:

$$Z_{1a}^{-1} Z_{2a} = Z_{1b}^{-1} Z_{2b} = Z_{1c}^{-1} Z_{2c} = Z_{1d}^{-1} Z_{2d} \qquad \text{etc.} \tag{6-177}$$

Ward [13, 14] has proved that the above relation is true by showing that

$$Z_{1a} = Z_{2a} \qquad Z_{1b} = Z_{2b} \qquad \text{etc.} \tag{6-178}$$

This equality between Z_1 and Z_2 is called Ward's identity, and in the discussion that follows we shall outline the proof.

We recall the definitions of Z_1 and Z_2 in quantum electrodynamics.

$$S_F'(p)_{unr} = Z_2 S_F'(p)_r \tag{6-179}$$

$$\Gamma_\mu(p, q)_{unr} = Z_1^{-1} \Gamma_\mu(p, q)_r \tag{6-180}$$

In order to study this problem we also recall the definition of the vertex functions:

$$\langle 0 | T^*[\psi(x), \bar{\psi}(y), A_\nu(z), S] | 0\rangle / \langle 0 | S | 0\rangle$$

$$= -e \int d^4x' d^4y' d^4z' S_F'(x - x') \Gamma_\mu(x'y':z')$$

$$S_F'(y' - y) D_{\mu\nu}'(z' - z) \tag{6-181}$$

We define both the renormalized and unrenormalized vertex functions by Eq. (6-181); in the former case the left-hand side is defined in the renormalized interaction representation and all the operators and propagators are renormalized, with e representing e_{obs}; in the latter case everything is unrenormalized with e representing the bare charge e. Therefore these two definitions are not independent, since renormalized and unrenormalized quantities are related to each other through simple scale transformations; it is obvious that the two definitions are equivalent provided that

$$e_{obs} = Z_1^{-1} Z_2 Z_3^{1/2} e \tag{6-182}$$

Since Ward's identity is expressed in terms of divergent renormalization constants, we have to convert the statement to a more tractable form. Ward has shown that the equality $Z_1 = Z_2$ follows from the relation, called the Ward relation[13],

$$\Gamma_\mu(p, p)_{unr} = -i \frac{\partial}{\partial p_\mu} S_F'(p)^{-1}_{unr} \tag{6-183}$$

Leaving the proof of this relation for later, we now shall derive Ward's identity from it. First, we replace the unrenormalized Γ and S_F' by corresponding functions multiplied by the appropriate renormalization constants.

$$\Gamma_\mu(p, p)_r = -i \frac{\partial}{\partial p_\mu} Z_1 Z_2^{-1} S_F'(p)^{-1}_r \tag{6-184}$$

If we take the matrix element of this equation between the same single-electron states represented by a Dirac spinor $u(p)$, we have

$$\bar{u}(p)\Gamma_\mu(p, p)_r u(p) = Z_1 Z_2^{-1} \bar{u}(p) \left(-i \frac{\partial}{\partial p_\mu} S_F'(p)_r^{-1}\right) u(p) \tag{6-185}$$

From the definition of the renormalized vertex function it follows that the left-hand side reduces to $\bar{u}(p)\,\gamma_\mu\,u(p)$. The renormalized propagator also satisfies the condition

$$(ip\gamma + m)\,S'_F(p)_r \to 1 \qquad \text{as} \qquad ip\gamma + m \to 0 \tag{6-186}$$

so that

$$\bar{u}(p)\left(-i\frac{\partial}{\partial p_\mu} S'_F(p)_r^{-1}\right)u(p) = \bar{u}(p)\,\gamma_\mu u(p) \tag{6-187}$$

since

$$\bar{u}(p)\,(ip\gamma + m) = (ip\gamma + m)\,u(p) = 0$$

Hence

$$\bar{u}(p)\,\gamma_\mu u(p) = Z_1 Z_2^{-1}\,\bar{u}(p)\,\gamma_\mu u(p) \tag{6-188}$$

or Ward's identity

$$Z_1 Z_2^{-1} = 1 \tag{6-189}$$

Thus our problem is reduced to proving the Ward relation between S'_F and Γ. From Ward's identity it is obvious that the Ward relation should hold for renormalized quantities too, namely,

$$\Gamma_\mu(p, p)_r = -i\frac{\partial}{\partial p_\mu}\,S'_F\,(p)_r^{-1} \tag{6-190}$$

provided that it holds for unrenormalized quantities.

Ward-Takahashi Relations

The original Ward relation was generalized by Takahashi[15] to the case of finite momentum transfer

$$-i(p - q)_\nu\,S'_F(p)\,\Gamma_\nu\,(p, q)\,S'_F(q) = S'_F(p) - S'_F(q) \tag{6-191}$$

If we make p - q infinitesimal, this relation reduces to that of Ward. The Ward-Takahashi relation is valid for both renormalized and unrenormalized quantities, just as charge conservation holds for both bare and observed charges. In the discussion that follows we shall give a set of relationships that is a generalization of the Ward-Takahashi relation.

We shall discuss this problem for *unrenormalized* expressions in the Heisenberg representation. The Lagrangian density expressing the interaction between matter fields and the electromagnetic field is introduced by Dirac's substitution discussed in Section 3-4.

$$\mathcal{L}_{\text{total}} = \mathcal{L}_{\text{matter}}(\varphi_a(x),\ (\partial_\mu - ie_a A_\mu(x))\,\varphi_a(x)) + \mathcal{L}_{\text{rad}} \tag{6-192}$$

where e_a denotes the charge of the quantum of the field $\varphi_a(x)$. We also assume that there are several charged fields.

Thus we define the current operator j_μ as

$$\Box A_\mu(x) = -j_\mu(x) \tag{6-193}$$

of which the explicit form follows from the Euler equation:

$$j_\mu(x) = \frac{\partial}{\partial A_\mu(x)} \mathcal{L}_{\text{matter}}(\varphi_a(x),\ \partial_\mu \varphi_a(x) - ie_a A_\mu(x)\,\varphi_a(x)) \tag{6-194}$$

Let us also define an operator $D_{a\mu}(x)$ as

$$D_{a\mu}(x) = \partial_\mu \varphi_a(x) - ie_a A_\mu(x)\,\varphi_a(x) \tag{6-195}$$

then the current $j_\mu(x)$ is

$$\begin{aligned} j_\mu(x) &= \sum_a \frac{\partial D_{a\nu}(x)}{\partial A_\mu(x)} \cdot \frac{\partial \mathcal{L}_{\text{matter}}(x)}{\partial D_{a\nu}(x)} \\ &= -i \sum_a e_a \varphi_a(x)\, \frac{\partial \mathcal{L}_{\text{matter}}(x)}{\partial \varphi_{a:\mu}(x)} \\ &= -i \sum_a e_a \varphi_a(x)\, \frac{\partial \mathcal{L}_{\text{total}}(x)}{\partial \varphi_{a:\mu}(x)} \end{aligned} \tag{6-196}$$

In the summation over a, we always include the field operator $\varphi_{\bar{a}}(x) = \varphi_a^\dagger(x)$ corresponding to the antiparticle $\bar{a}$. For $\mu = 0$,

$$\begin{aligned} j_0(x) = -ij_4(x) &= -i \sum_a e_a \varphi_a(x)\, \frac{\partial \mathcal{L}_{\text{total}}(x)}{\partial \dot{\varphi}_a(x)} \\ &= -i \sum_a e_a \varphi_a(x)\, \pi_a^\dagger(x) \end{aligned} \tag{6-197}$$

where $\pi^\dagger_a(x)$ is the canonical conjugate of $\varphi_a(x)$.

From the canonical commutation relation

$$[\varphi_a(x), \pi^\dagger_b(y)]_\pm = i\delta_{ab}\,\delta^3(x - y) \qquad \text{for} \quad x_0 = y_0 \tag{6-198}$$

we immediately get

$$[\varphi_a(x), j_0(y)] = e_a\varphi_a(x)\,\delta^3(x - y) \qquad \text{for} \quad x_0 = y_0 \tag{6-199}$$

The commutator (-) is used for boson fields and the anticommutator (+) for fermion fields. From the last commutation relation we can derive the integrated relation

$$[\varphi_a(x), Q] = e_a\,\varphi_a(x) \tag{6-200}$$

where Q is the total charge of the system defined as

$$Q = \int d^3x\, j_0(x) \tag{6-201}$$

Then we derive a generalization[16] of the Ward-Takahashi relation; recalling the definition of the T product, we write the following equation in the Heisenberg representation:

$$\begin{aligned}
\Box_x\, & T\,[A_\mu(x)\, A_\nu(x')\, A_\sigma(x'') \ldots \varphi_a(x_a)\, \varphi_b(x_b) \ldots] \\
&= -T\,[j_\mu(x)\, A_\nu(x')\, A_\sigma(x'') \ldots \varphi_a(x_a)\, \varphi_b(x_b) \ldots] \\
&\quad - \delta(x_0 - x_0')\, T\Big[[A_\mu(x), A_\nu(x')]A_\sigma(x'') \ldots \\
&\quad \varphi_a(x_a)\, \varphi_b(x_b) \ldots\Big] - \ldots \\
&= -T\,[j_\mu(x)\, A_\nu(x')\, A_\sigma(x'') \ldots \varphi_a(x_a)\, \varphi_b(x_b) \ldots] \\
&\quad + i\delta^4(x - x')\, \delta_{\mu\nu}\, T\Big[A_\sigma(x'') \ldots \varphi_a(x_a)\, \varphi_b(x_b) \ldots\Big] \\
&\quad + \ldots
\end{aligned} \tag{6-202}$$

where use has been made of the following canonical commutation relation for the electromagnetic field.

$$\delta(x_0 - x_0')\,[\dot{A}_\mu(x), A_\nu(x')] = -i\delta_{\mu\nu}\delta^4(x - x') \tag{6-203}$$

Equation (6-202) may be written symbolically in a more transparent form,

$$\Box_x T[A_\mu(x) \ldots] = -T[j_\mu(x) \ldots] + i\frac{\delta}{\delta A_\mu(x)} T\,[\ldots] \tag{6-204}$$

with the stipulation that

$$\frac{\delta A_\nu(x')}{\delta A_\mu(x)} = \delta_{\mu\nu}\delta^4(x - x') \tag{6-205}$$

Equation (6-204) is not useful, however, unless $j_\mu(x)$ is expressed explicitly in terms of field operators; so we try to eliminate the current operator $j_\mu(x)$ from Eq. (6-204). It satisfies the conservation law

$$\partial_\mu j_\mu(x) = 0 \tag{6-206}$$

Differentiating (6-204) with respect to x_μ, we get

$$\begin{aligned}\Box_x \frac{\partial}{\partial x_\mu} T[A_\mu(x)\ldots] &= \delta(x_0 - x_{a0})\, T\,[A_\nu(x')\, A_\sigma(x'')\ldots[\varphi_a(x_a), \\ &\qquad j_0(x)]\varphi_b(x_b)\ldots] + \ldots \\ &\quad + i\delta_{\mu\nu}\frac{\partial}{\partial x_\mu}\delta^4(x - x')\, T[A_\sigma(x'')\ldots\varphi_a(x_a)\,\varphi_b(x_b)\ldots] \\ &\quad + \ldots \\ &= \left[\sum_a e_a\delta^4(x - x_a) + i\frac{\partial}{\partial x_\mu}\cdot\frac{\delta}{\delta A_\mu(x)}\right] \\ &\quad \times T[A_\nu(x')\, A_\sigma(x'')\ldots\varphi_a(x_a)\,\varphi_b(x_b)\ldots]\end{aligned} \tag{6-207}$$

where we have used the conservation law $\partial_\mu j_\mu(x) = 0$ as well as the equal-time commutation relations between field operators and $j_0(x)$ derived at the beginning of this discussion. The relation (6-207) can be written in a formal but more transparent form as

$$\Box_x \frac{\partial}{\partial x_\mu} T[A_\mu(x)\ldots] = \left[\sum_a e_a \delta^4(x - x_a) + i \frac{\partial}{\partial x_\mu}\left(\frac{\delta}{\delta A_\mu(x)}\right)\right] T[\ldots] \tag{6-208}$$

which represents a generalization of the Ward-Takahasi relation, as we shall see below.

Since (6-208) has been derived in the Heisenberg representation, we have to relate it to the interaction representation. For this purpose we use the following relation:

$$(\Psi_0, T[ABC\ldots]\Psi_0) = \frac{(\Phi_0, T[A^{in}B^{in}C^{in}\ldots S]\Phi_0)}{(\Phi_0, S\Phi_0)} \tag{6-209}$$

where Ψ_0 is the vacuum state in the Heisenberg representation, and operators with the superscript *in* are defined in the interaction representation. The discussion of the relation (6-209), derived by M. Gell-Mann and F.E. Low, is postponed until we cover that topic in Section 7-5.

The use of the relationship of Eq. (6-209) enables us to express propagation functions in terms of Heisenberg operators:

$$S'_F(x - y) = \langle 0|T[\psi(x), \bar{\psi}(y)]|0\rangle \tag{6-210}$$

$$D'_{\mu\nu}(x - y) = \langle 0|T[A_\mu(x), A_\nu(y)]|0\rangle \tag{6-211}$$

were $|0\rangle$ represents Ψ_0. All the Heisenberg operators in (6-210) and (6-211) are still unrenormalized. The vertex function is then defined, in the Heisenberg representation, by

$$\langle 0|T[\psi(x), \bar{\psi}(y), A_\nu(z)]|0\rangle = -e\int d^4x'd^4y'd^4z'S'_F(x - x')$$
$$\times\ \Gamma_\mu(x'y' : z')S'_F(y' - y)D'_{\mu\nu}(z' - z) \tag{6-212}$$

Application of the generalized Ward-Takahashi relation (6-208) to these functions leads to

$$\Box_x \frac{\partial}{\partial x_\mu}\langle 0|T[A_\mu(x), A_\nu(y)]|0\rangle = i\frac{\partial}{\partial x_\nu}\delta^4(x - y) \tag{6-213}$$

and

$$\square_z \frac{\partial}{\partial z_\mu} \langle 0|T[\psi(x), \bar{\psi}(y), A_\mu(z)]|0\rangle$$

$$= -e\left\{\langle 0|T[\psi(x), \bar{\psi}(y)]|0\rangle \delta^4(y-z) - \langle 0|T[\psi(x), \bar{\psi}(y)]|0\rangle \delta^4(x-z)\right\}$$

(6-214)

Expressing the left-hand side of Eq. (6-214) in terms of the propagation functions and the vertex function and using Eq. (6-213), we obtain

$$-ie\int d^4x' d^4y'\, S'_F(x-x') \frac{\partial}{\partial z_\nu} \Gamma_\nu(x'y' : z)\, S'_F(y'-y)$$

$$= -e\, S'_F(x-y)[\delta^4(y-z) - \delta^4(x-z)] \qquad (6\text{-}215)$$

Converting this relation to momentum space, we get the Takahashi relation:

$$-i(p-q)_\nu S'_F(p)\,\Gamma_\nu(p, q)\, S'_F(q) = S'_F(p) - S'_F(q) \qquad (6\text{-}216)$$

The generalized Ward-Takahashi relation holds not only in the unrenormalized form but also in the renormalized form[16], as a consequence of the Ward identity $Z_1 = Z_2$.

PROBLEM

6-1. Show that the renormalized second-order propagation function for the nucleon interacting with the pion field through the pseudoscalar coupling (6-217) is given, for $m \gg \mu$, by

$$S'_F(p) = \frac{1}{ip\gamma + m} + \frac{3G^2}{16\pi^2} \int_{m+\mu}^{\infty} \frac{dM}{2M^3} (M^2 - m^2)$$

$$\times \left[\frac{1}{ip\gamma + M - i\epsilon} + \frac{1}{ip\gamma - M + i\epsilon}\right] \qquad (6\text{-}217)$$

REFERENCES

1. J. Schwinger (Ed.), *Quantum Electrodynamics* (Dover Publications, Inc., New York, 1958).

2. B. A. Lippmann and J. Schwinger, Phys. Rev. **79**, 469 (1950).

3. P. T. Matthews and A. Salam, Phys. Rev. **94**, 185 (1954).

4. J. Schwinger, Phys. Rev. **75**, 898 (1949).

5. C. M. Sommerfield, Phys. Rev. **107**, 328 (1957).

6. D. T. Wilkinson and H. R. Crane, Phys. Rev. **130**, 852 (1963). For an older result, refer to H. M. Foley and P. Kusch, Phys. Rev. **73**, 412 (1948).

7. W. E. Lamb, Jr. and R. C. Retherford, Phys. Rev. **72**, 241 (1947).

8. H. A. Bethe, Phys. Rev. **73**, 416 (1948).

9. G. W. Erickson and D. R. Yennie, Ann. Phys. (N. Y.) **35**, 271 (1965).

10. M. F. Soto, Jr., Phys. Rev. Letters, **17**, 1153 (1966).

11. S. Triebwasser, E. S. Dayhoff and W. E. Lamb, Jr., Phys. Rev. **89**, 98 (1953).

12. W. H. Parker, B. N. Taylor and D. N. Langenberg, Phys. Rev. Letters **18**, 287 (1967).

13. J. C. Ward, Phys. Rev. **78**, 182 (1950).

14. G. Källén, Helv. Phys. Acta **26**, 755 (1953).

15. Y. Takahashi, Nuovo Cimento **6**, 370 (1957).

16. K. Nishijima, Phys. Rev. **119**, 485 (1960). See also E. Kazes, Nuovo Cimento **13**, 1226 (1959).

CHAPTER 7

GREEN'S FUNCTIONS AND BOUND STATES

We have been studying general aspects of the Feynman-Dyson theory and have learned how to do computations involving the S matrix. The Dyson formula for the S matrix,

$$S = \sum_{n=0}^{\infty} \frac{(-i)^n}{n!} \int d^4x_1 \ldots d^4x_1 \, T \, [\mathcal{H}(x_1) \ldots \mathcal{H}(x_n)]$$

$$= \sum_{n=0}^{\infty} \frac{i^n}{n!} \int d^4x_1 \ldots d^4x_n \, T^* \, [\mathcal{L}(x_1) \ldots \mathcal{L}(x_n)]$$

has been extensively used to compute the amplitudes for various reactions. A question which naturally arises is how one can generalize this approach so as to include bound states, for example, how one can compute the transition amplitude for

$$p + p \rightarrow d + \pi^+$$

It is clear that the S-matrix formula above is intrinsically bound to perturbation theory and cannot be used for this problem; instead it is necessary to formulate field theory in terms of Green's functions.

7-1 LADDER APPROXIMATION AND THE BETHE-SALPETER EQUATION

In the problem of bound states it is known that perturbation theory never generates bound states, so that we must consider an approximation which is better than the power series expansion.

Let us consider a scattering problem and try to sum contributions from a particular set of diagrams, the so-called ladder diagrams shown in Fig. 7-1.

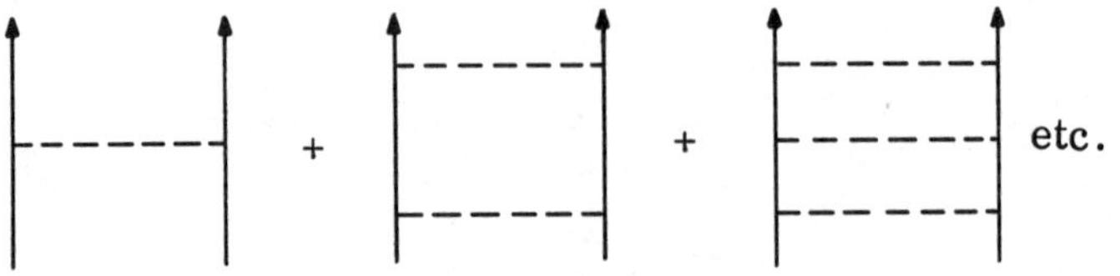

Figure 7-1. A set of ladder diagrams.

Let us assume a model in which spinless particles a and b are interacting by exchanging other spinless mesons c. The interaction Hamiltonian is then

$$\mathcal{H}_{int} = g_a \Phi_a^\dagger \Phi_a \varphi_c + g_b \Phi_b^\dagger \Phi_b \varphi_c \tag{7-1}$$

with g_a and g_b representing the relevant coupling constants. Then the scattering amplitude between a and b is given by

$$\begin{aligned}
&\langle a' \, b' | S | ab \rangle \\
&= (-i)^2 g_a g_b \int d^4x_1 \, d^4x_2 \, \langle a' | \Phi_a^\dagger(x_1) | 0 \rangle \langle b' | \Phi_b^\dagger(x_2) | 0 \rangle \, D_F(x_1 - x_2) \\
&\quad \times \langle 0 | \Phi_a(x_1) | a \rangle \, \langle 0 | \Phi_b(x_2) | b \rangle \\
&\quad + (-i)^4 g_a^2 g_b^2 \int d^4x_1 \, d^4x_2 \, d^4x_3 \, d^4x_4 \, \langle a' | \Phi_a^\dagger(x_1) | 0 \rangle \, \langle b' | \Phi_b^\dagger(x_2) | 0 \rangle \\
&\quad \times D_F(x_1 - x_2) \, \Delta_{Fa}(x_1 - x_3) \, \Delta_{Fb}(x_2 - x_4) \\
&\quad \times D_F(x_3 - x_4) \, \langle 0 | \Phi_a(x_3) | a \rangle \, \langle 0 | \Phi_b(x_4) | b \rangle + \cdots
\end{aligned} \tag{7-2}$$

Where D_F is used for the propagator of c to distinguish it from those of a and b, even though the mass of c is not necessarily zero.

We define a function $\mathcal{F}(x_1 x_2 : ab)$ as

$$\mathcal{F}(x_1 x_2 : ab) = \langle 0 | \Phi_a(x_1) | a \rangle \quad \langle 0 | \Phi_b(x_2) | b \rangle$$

$$- g_a g_b \int d^4x_3 \, d^4x_4 \, \Delta_{Fa}(x_1 - x_3) \, \Delta_{Fb}(x_2 - x_4)$$

$$\times D_F(x_3 - x_4) \quad \langle 0 | \Phi_a(x_3) | a \rangle \quad \langle 0 | \Phi_b(x_4) | b \rangle$$

$$+ \ldots \tag{7-3}$$

This series is akin to a geometric series in which the multiplicative parameter is an integral over the functions $\Delta_{Fa} \Delta_{Fb} D_F$; as a result we get an integral equation for the unknown function $\mathcal{F}$.

$$\mathcal{F}(x_1 x_2 : ab) = \mathcal{F}_0(x_1 x_2 : ab)$$

$$+ \int d^4x_3 d^4x_4 \, \Delta_{Fa}(x_1 - x_3) \, \Delta_{Fb}(x_2 - x_4)$$

$$\times (-g_a g_b) \, D_F(x_3 - x_4) \, \mathcal{F}(x_3 x_4 : ab) \tag{7-4}$$

where $\mathcal{F}_0$ is the inhomogeneous term

$$\mathcal{F}_0(x_1 x_2 : ab) = \mathcal{F}(x_1 : a) \, \mathcal{F}(x_2 : b)$$

$$= \langle 0 | \Phi_a(x_1) | a \rangle \langle 0 | \Phi_b(x_2) | b \rangle \tag{7-5}$$

Equation (7-4) is called the Bethe-Salpeter equation and $\mathcal{F}$ the Feyman amplitude. The scattering amplitude is then given by

$$\langle a' b' | S | ab \rangle = \int d^4x_1 \, d^4x_2 \, \mathcal{F}^*_0(x_1 x_2 : a' b')$$

$$\times (-g_a g_b D_F(x_1 - x_2)) \, \mathcal{F}(x_1 x_2 : ab) \tag{7-6}$$

There is a strong formal resemblance between the Bethe-Salpeter formalism and the Lippmann-Schwinger formalism; the analog of formula (7-6) in the Lippmann-Schwinger formalism is

$$T_{BA} = (\Phi_B, V \Psi_A^{(+)}) \tag{7-7}$$

and the analog of the Bethe-Salpeter (B-S) equation (7-4), is the Lippmann-Schwinger (L-S) equation

$$\Psi_A^{(+)} = \Phi_A + \frac{1}{E_a - H_0 + i\epsilon} V \Psi_A^{(+)} \tag{7-8}$$

In the Bethe-Salpeter formalism $\mathcal{F}$ takes the place of $\Psi^{(+)}$, $\mathcal{F}_0$ of Φ, $-g_a g_b D_F$ of V, and $\Delta_{Fa} \Delta_{Fb}$ of $(E - H_0 + i\epsilon)^{-1}$. Making use of the formal resemblance between the B-S and L-S formalism, we can find a relativistic equation for bound states. In the L-S formalism the equation for bound states is

$$\Psi_B = \frac{1}{E_B - H_0} V \Psi_B \tag{7-9}$$

This equation is obtained from the L-S equation for scattering states (7-8), by dropping the inhomogeneous term; so that by dropping the inhomogeneous term $\mathcal{F}_0$ in the B-S equation (7-4), we get a corresponding relativistic equation for bound states:

$$\mathcal{F}(x_1 x_2 : B) = \int d^4x_3 d^4x_4 \, \Delta_{Fa}(x_1 - x_3) \Delta_{Fb}(x_2 - x_4) \times (-g_a g_b) D_F(x_3 - x_4) \, \mathcal{F}(x_3 x_4 : B) \tag{7-10}$$

The letter B stands for a bound state. We also can write the B-S equation in a differential form by applying Klein-Gordon operators; then

$$(\Box_1 - m_a^2)(\Box_2 - m_b^2) \, \mathcal{F}(x_1 x_2 : \quad) = g_a g_b D_F(x_1 - x_2) \, \mathcal{F}(x_1 x_2 : \quad) \tag{7-11}$$

This equation is valid not only for scattering states but also for bound states, and was first proposed by Nambu.[2]

7-2 WICK'S SOLUTION

In the preceding section we derived a relativistic equation for bound states (7-10); we now shall show, by means of a solvable example, how this equation determines the binding energy. The following is, in fact, the only case that has been solved in closed form.

Wick[3] considered the case of two spinless bosons of equal mass exchanging a spinless and massless boson.

$$m_a = m_b = \mu \qquad m_c = 0 \qquad g_a = g_b = f \tag{7-12}$$

We transform (7-12) to center-of-mass and relative coordinates by setting

$$X = \frac{1}{2}(x_1 + x_2) \qquad x = x_1 - x_2 \tag{7-13}$$

and correspondingly

$$P = p_1 + p_2 \qquad p = \frac{1}{2}(p_1 - p_2) \tag{7-14}$$

Then we can define $\Psi(p)$, the Fourier transform of the Feynman amplitude, by

$$\mathcal{F}(x_1 x_2 : B) = e^{iPX} \int d^4p \; e^{ipx}\, \Psi(p) \tag{7-15}$$

The B-S equation in momentum space then assumes the form

$$\left[\left(\frac{P}{2} + p\right)^2 + \mu^2\right]\left[\left(\frac{P}{2} - p\right)^2 + \mu^2\right] \Psi(p)$$

$$= -i\frac{f^2}{(2\pi)^4} \int d^4q \; \frac{\Psi(q)}{(p-q)^2 - i\epsilon} \tag{7-16}$$

Equation (7-16) is very difficult to solve, but Wick succeeded in finding a transformation which reduces this equation to an ordinary differential equation. He assumed that $\Psi(p)$ could be expressed as

$$\Psi(p) = \int_{-1}^{1} \frac{g(z)dz}{\left(p^2 + zpP + \mu^2 + \frac{P^2}{4} - i\epsilon\right)^3} \tag{7-17}$$

Substituting this integral representation into (7-16) for $\Psi(p)$, we now carry out the q integration by making use of

$$\int d^4q \; \frac{1}{(p-q)^2 - i\epsilon} \cdot \frac{1}{\left(q^2 + zqP + \mu^2 + \frac{P^2}{4} - i\epsilon\right)^3}$$

$$= \frac{i\pi^2}{2\left(\mu^2 + \frac{P^2}{4} - z^2\frac{P^2}{4}\right)} \cdot \frac{1}{\left(p^2 + zpP + \mu^2 + \frac{P^2}{4} - i\epsilon\right)} \tag{7-18}$$

and the Equation (7-16) reduces to

$$\Psi(p) = \left(p^2 + pP + \mu^2 + \frac{P^2}{4}\right)^{-1} \left(p^2 - pP + \mu^2 + \frac{P^2}{4}\right)^{-1} \frac{f^2}{(4\pi)^2}$$

$$\times \int_{-1}^{1} \frac{dz}{2\left(\mu^2 + \frac{P^2}{4} - z^2 \frac{P^2}{4}\right)} \frac{g(z)}{p^2 + zpP + \mu^2 + \frac{P^2}{4} - i\epsilon} \tag{7-19}$$

For convenience, we change to dimensionless parameters by setting the coupling constant equal to $\lambda = (f/4\pi\mu)^2$ and the squared mass of the bound state $M^2 = -P^2 = 4\mu^2\eta^2$; the possible values of η are from 0 to 1, with small values corresponding to strong binding and values near 1 to weak binding.

$$\Psi(p) = \frac{\lambda}{\left(p^2 + pP + \mu^2 + \frac{P^2}{4}\right)\left(p^2 - pP + \mu^2 + \frac{P^2}{4}\right)}$$

$$\times \int_{-1}^{1} \frac{dx\, g(x)}{2(1-\eta^2+\eta^2x^2)} \cdot \frac{1}{\left(p^2 + xpP + \mu^2 + \frac{P^2}{4} - i\epsilon\right)}$$

$$= \frac{\lambda}{2} \int_{-1}^{1} \frac{dy}{\left(p^2 + ypP + \mu^2 + \frac{P^2}{4}\right)^2}$$

$$\times \int_{-1}^{1} \frac{dx\, g(x)}{2\left(1-\eta^2+\eta^2x^2\right)\left(p^2 + xpP + \mu^2 + \frac{P^2}{4} - i\epsilon\right)}$$

$$= \int_0^1 \zeta d\zeta \int_{-1}^{1} dy \int_{-1}^{1} dx \frac{\lambda}{2(1-\eta^2+\eta^2x^2)}$$

$$\times \frac{g(x)}{\left[p^2 + (\zeta y+(1-\zeta)x)pP + \mu^2 + \frac{P^2}{4} - i\epsilon\right]^3}$$

Comparing this expression for $\Psi(p)$ with (7-17) we see that the original equation for $\Psi(p)$ reduces to

$$g(z) = \int_0^1 \zeta d\zeta \int_{-1}^{1} dy \int_{-1}^{1} dx \frac{\lambda\, g(x)}{2(1-\eta^2+\eta^2x^2)} \delta[z-(\zeta y+(1-\zeta)x)] \tag{7-20}$$

First we carry out the ζ integration. The argument of the δ function vanishes when

$$\zeta = \frac{z - x}{y - x} \tag{7-21}$$

which happens if and only if

$$\begin{aligned} & x > z > y \\ \text{or} \qquad & y > z > x \end{aligned} \tag{7-22}$$

since ζ changes from 0 to 1. Thus

$$\begin{aligned} g(z) &= \int_{\substack{x>z>y \\ x<z<y}} dx\, dy\, \frac{z - x}{y - x} \cdot \frac{1}{|x - y|} \cdot \frac{\lambda\, g(x)}{2(1 - \eta^2 + \eta^2 x^2)} \\ &= \int_z^1 dx \int_{-1}^z dy\, \frac{x - z}{(x - y)^2} \cdot \frac{\lambda\, g(x)}{2(1 - \eta^2 + \eta^2 x^2)} \\ &\quad + \int_z^1 dy \int_{-1}^z dx\, \frac{z - x}{(x - y)^2} \cdot \frac{\lambda\, g(x)}{2(1 - \eta^2 + \eta^2 x^2)} \\ &= \int_z^1 dx\, \frac{1 + z}{1 + x} \cdot \frac{\lambda\, g(x)}{2(1 - \eta^2 + \eta^2 x^2)} \\ &\quad + \int_{-1}^z dx\, \frac{1 - z}{1 - x} \cdot \frac{\lambda\, g(x)}{2(1 - \eta^2 + \eta^2 x^2)} \end{aligned} \tag{7-23}$$

We see that $g(1) = g(-1) = 0$. If we differentiate this equation twice we obtain a second-order differential equation for $g(z)$.

$$g''(z) = -\frac{\lambda}{1-z^2} \cdot \frac{g(z)}{1 - \eta^2 + \eta^2 z^2} \tag{7-24}$$

We have to solve this equation with the boundary condition

$$g(\pm 1) = 0 \tag{7-25}$$

This is clearly an eigenvalue problem.

In order to solve this problem we shall use an intuitive approximation. The ladder approximation is supposed to be good only when λ is small, so that the binding energy is also supposed to be small, namely,

$$1 \gg 1 - \eta > 0 \tag{7-26}$$

If this is the case, the function $(1 - \eta^2 + \eta^2 z^2)^{-1}$ has a sharp peak at $z = 0$; Wick approximated it by a δ function

$$(1 - \eta^2 + \eta^2 z^2)^{-1} \to \delta(z) \int_{-1}^{1} \frac{dz}{(1 - \eta^2 + \eta^2 z^2)} \approx \frac{\pi}{(1 - \eta^2)^{1/2}}\, \delta(z) \tag{7-27}$$

Then the differential equation reduces to

$$g''(z) \approx - \lambda \pi (1 - \eta^2)^{-1/2}\, g(0)\, \delta(z) \tag{7-28}$$

and the solution satisfying the prescribed boundary conditions is easily formed:

$$g(z) \approx \frac{\pi}{2} (1 - \eta^2)^{-1/2}\, g(0)\, \lambda\, (1 - |z|) \tag{7-29}$$

Since g(z) has no nodes, it represents the lowest eigenfunction; hence the lowest eigenvalue must be given approximately by

$$\lambda \approx \frac{2}{\pi} (1 - \eta^2)^{1/2} \tag{7-30}$$

which follows from the solution (7-29) by setting $z = 0$. Thus the mass of the bound state $2\mu\eta$ is determined in terms of the coupling constant λ.

The solutions corresponding to excited bound states have been obtained by Cutkosky.[4]

7-3 SCHWINGER'S THEORY OF GREEN'S FUNCTIONS, I

In preceding sections the Bethe-Salpeter equation was derived on the basis of graphical considerations. It is desirable, however, to develop a method which does not depend on intuition; that is the object of this section. We also shall summarize here what we have learned and what we have to learn.

It is possible to formulate the eigenvalue problem for bound states in a completely Lorentz-invariant fashion by means of the Bethe-Salpeter equation. The unknown functions, the Feynman amplitudes, are, however, not directly related to the probability amplitudes, so that we do not know how to normalize them, nor do we know how the Feynman amplitudes for bound states enter into the amplitudes for reactions involving bound states. These problems will be studied later.

Schwinger's formulation[5] of Green's functions will be discussed here for scalar fields, and extended to the case of two-particle systems in the next section. In the preceding section we have solved the Bethe-Salpeter equation for a simple model, but since it is the only case that can be solved in a closed form we shall adhere to it.

We shall consider here the interaction between a charged scalar field φ and a scalar photon field A. In the interaction representation their propagators are given by

$$\langle 0|T[\varphi(1),\ \varphi^\dagger(2)]|0\rangle = \Delta_F(1-2)$$

$$\langle 0|T[A(1),\ A(2)|0\rangle = D_F(1-2) \tag{7-31}$$

These functions are Green's functions for the Klein-Gordon operators:

$$(\Box - \mu^2)\Delta_F(x) = i\delta^4(x) \qquad \Box D_F(x) = i\delta^4(x) \tag{7-32}$$

and have the integral representations introduced in Section 4-5.

The interaction Hamiltonian density is given by

$$\mathcal{H}_{int} = f\,\varphi^\dagger\varphi A + JA - \mathcal{H}_{self} \tag{7-33}$$

where J is an external c-number source and $\mathcal{H}_{self}$ is the counter term introduced to cancel divergences. The coupling constant f is of the dimension of mass in natural units.

Functional Differentiation

By using (7-33) we can write Dyson's formula

$$U[\infty,\ -\infty] = T\exp\left(-i\int_{-\infty}^{\infty} d^4x\,\mathcal{H}_{int}(x)\right) \tag{7-34}$$

If we set $U = U[\infty,\ -\infty]$ for simplicity, U is a functional of the external source J. We define the functional derivative of U with respect to J as

$$\frac{\delta U[J]}{\delta J(x)} = \lim_{\epsilon\to 0}\frac{U[J(y)+\epsilon\,\delta^4(y-x)] - U[J(y)]}{\epsilon} \tag{7-35}$$

In the present case the functional derivative of U with respect to J is clearly

$$\begin{aligned} i\frac{\delta U}{\delta J(x)} &= i\frac{\delta}{\delta J(x)}\,T\exp\left(-i\int d^4y\,\mathcal{H}_{int}(y)\right) \\ &= T\left[\exp\left(-i\int d^4y\,\mathcal{H}_{int}(y)\right),\ \frac{\delta}{\delta J(x)}\int d^4z\,\mathcal{H}_{int}(z)\right] \\ &= T\left[\exp\left(-i\int d^4y\,\mathcal{H}_{int}(y)\right),\ A(x)\right] \\ &= T[U,\ A(x)] \end{aligned} \tag{7-36}$$

This result easily is extended to the general case:

$$i \frac{\delta}{\delta J(x)} T[U, a, b, \ldots] = T[U, A(x), a, b, \ldots] \tag{7-37}$$

Green's Functions

The n-point Green's function is defined to be

$$\langle A(1), B(2), \ldots, Z(n) \rangle = \frac{\langle 0 | T[U, A(1), B(2), \ldots, Z(n)] | 0 \rangle}{\langle 0 | U | 0 \rangle} \tag{7-38}$$

We have already encountered the two- and three-point Green's functions in the chapter on renormalization; here we define propagation functions in the presence of an external source as

$$\Delta'_F(1, 2) = \langle \varphi(1), \varphi^\dagger(2) \rangle \tag{7-39}$$

In the limit of no external source, $\Delta'_F(1, 2)$ reduces to $\Delta'_F(1 - 2)$. For the scalar photon field the propagation function is defined as

$$D'_F(1, 2) = \langle A(1), A(2) \rangle - \langle A(1) \rangle \langle A(2) \rangle \tag{7-40}$$

In the presence of an external source the second term does not vanish; it subtracts the unwanted contributions from the disconnected processes in which a virtual photon created at point 1 (or 2) by the external source is destroyed and absorbed by the vacuum without reaching the point 2 (or 1). From the definitions of $\langle A \rangle$ and of the functional derivative we easily can verify the following relation:

$$D'_F(1, 2) = i \frac{\delta \langle A(1) \rangle}{\delta J(2)} = i \frac{\delta \langle A(2) \rangle}{\delta J(1)} \tag{7-41}$$

Reduction Formula in the Interaction Representation

Now let us recapitulate some of the formalism leading to the so-called reduction formulas for the time-ordered product of field operators. We recall from Wick's theorem of Section 4-5 that the T product can be expressed as a sum over the possible operator contractions in the normal product. Suppose we apply the Klein-Gordon operator $K_x = \Box_x - \mu^2$ to the field operator $\varphi(x)$ inside the T product; when $\varphi(x)$ appears in the normal product, the result is zero, since the operator in the normal product creates or destroys a real particle on the mass shell, whereas when $\varphi(x)$ is contracted the result is not zero as seen from

$$K_x \, \Delta_F(x - y) = i\delta^4(x - y)$$

Thus we are led, with the help of Wick's theorem, to the result

$$K_x \, T[\varphi(x), A, B, \ldots] = i \frac{\delta}{\delta\varphi^\dagger(x)} T[A, B, \ldots] \tag{7-42}$$

where as before we have

$$\frac{\delta\varphi^\dagger(y)}{\delta\varphi^\dagger(x)} = \delta^4(x - y) \tag{7-43}$$

Formula (7-42) is called the reduction formula *in the interaction representation* and is distinguished from (8-114) in the Heisenberg representation which is usually referred to as the LSZ reduction formula. We postpone a detailed discussion of the reduction formulas until Section 7-6, where we shall discuss how the reduction formulas determine the commutation relations and how they are useful for deriving a new formula for the S matrix. Because of their importance, we list here the reduction formulas for scalar, spinor, and photon fields:

$$(\Box_x - \mu^2) \, T[\varphi(x) \ldots] = i \frac{\delta}{\delta\varphi^\dagger(x)} T[\ldots] \tag{7-44}$$

$$(\gamma\partial + m) \, T[\psi(x) \ldots] = i \frac{\delta}{\delta\bar{\psi}(x)} T[\ldots] \tag{7-45}$$

$$\Box_x \, T[A_\mu(x) \ldots] = i \frac{\delta}{\delta A_\mu(x)} T[\ldots] \tag{7-46}$$

Schwinger's Equations for Green's Functions

The foregoing formalism can be used to set up a closed set of equations for Green's functions. In the discussion that follows we shall omit the contributions from the counter term $\mathcal{H}_{\text{self}}$ for simplicity. We start from the following manipulations:

$$K_1 \, T[U, \varphi(1), \varphi^\dagger(2)] = i \frac{\delta}{\delta\varphi^\dagger(1)} T[U, \varphi^\dagger(2)]$$

$$= i\delta^4(1 - 2) \, U$$

$$+ \; T[U, \varphi^\dagger(2), \frac{\delta}{\delta\varphi^\dagger(1)} \int d^4x\, \mathcal{H}_{int}(x)]$$

$$= i\delta^4(1-2)\,U + f\,T[U, \varphi(1), A(1), \varphi^\dagger(2)]$$

$$= i\delta^4(1-2)\,U$$

$$+ \; if \frac{\delta}{\delta J(1)}\, T[U, \varphi(1), \varphi^\dagger(2)] \qquad (7\text{-}47)$$

We take the vacuum expectation value of (7-47) and obtain

$$\left(K_1 - f\,\langle A(1)\rangle - if \frac{\delta}{\delta J(1)}\right) \Delta_F'(1, 2) = i\delta^4(1-2) \qquad (7\text{-}48)$$

where use has been made of the relation

$$i \frac{\delta}{\delta J(1)} \langle 0|U|0\rangle = \langle 0|T[UA(1)]|0\rangle$$

$$= \langle 0|U|0\rangle\ \langle A(1)\rangle \qquad (7\text{-}49)$$

The above equation is written symbolically as

$$D_J(1)\,\Delta_F'(1, 2) = i\delta^4(1-2) \qquad (7\text{-}50)$$

In order to eliminate the functional derivative we define the proper self-energy operator Σ^* by

$$if \frac{\delta}{\delta J(1)} \Delta_F'(1, 2) = \int d^4x_3\ \Sigma^*(1, 3)\,\Delta_F'(3, 2)$$

$$\equiv \Sigma^*(1)\ \Delta_F'(1, 2) \qquad (7\text{-}51)$$

Then Eq. (7-50) can be rewritten as

$$D_N(1)\Delta_F'(1, 2) = i\delta^4(1-2) \qquad (7\text{-}52)$$

where

$$D_N(1) = K_1 - f\,\langle A(1)\rangle - \Sigma^*(1) \qquad (7\text{-}53)$$

Differentiation of (7-52) with respect to J(3) gives

$$D_N(1)\left(i\frac{\delta}{\delta J(3)}\Delta'_F(1, 2)\right) = -\left(-f\,D'_F(1, 3) - i\frac{\delta\sum^*(1)}{\delta J(3)}\right) \times \Delta'_F(1, 2) \qquad (7\text{-}54)$$

This relation can be regarded as an equation for the functional derivative of Δ'_F, and it readily can be solved with the help of Green's function Δ'_F, to yield the relation

$$i\frac{\delta}{\delta J(3)}\Delta'_F(1, 2) = \int d^4x_4\,\Delta'_F(1, 4)\left(-if\,D'_F(4, 3) + \frac{\delta\sum^*(4)}{\delta J(3)}\right) \times \Delta'_F(4, 2) \qquad (7\text{-}55)$$

If we regard $\langle A\rangle$ as the independent external variable instead of J, then the formulas involving functional derivatives eventually are simplified. We use

$$i\frac{\delta}{\delta J(1)} = \int d^4x_2\, i\frac{\delta\langle A(2)\rangle}{\delta J(1)}\cdot\frac{\delta}{\delta\langle A(2)\rangle} = \int d^4x_2\,D'_F(1, 2)\frac{\delta}{\delta\langle A(2)\rangle} \qquad (7\text{-}56)$$

and define the vertex function Γ by

$$i\frac{\delta}{\delta\langle A(3)\rangle}\Delta'_F(1, 2) = f\int d^4x_4\,d^4x_5\,\Delta'_F(1, 4)\,\Gamma(45{:}3) \times \Delta'_F(5, 2) \qquad (7\text{-}57)$$

By comparing this definition with (7-55) we immediately find the connection between Γ and Σ^*:

$$f\,\Gamma(12; 3) = f\,\delta^4(1 - 2)\,\delta^4(1 - 3) + \frac{\delta\sum^*(1, 2)}{\delta\langle A(3)\rangle} \qquad (7\text{-}58)$$

Furthermore, by rewriting (7-51) we have a relation between it and (7-57).

$$\Sigma^*(1, 2) = -if^2\int d^4x_3\,d^4x_4\,\Delta'_F(1, 3)\,\Gamma(32{:}4)\,D'_F(4, 1) \qquad (7\text{-}59)$$

This equation may be represented graphically in Fig. 7-2.

We next have to derive an equation satisfied by D'_F; for this purpose, we derive an equation for $\langle A \rangle$ first by applying $\Box_1$ on $T[U, A(1)]$ and then by taking the vacuum expectation value.

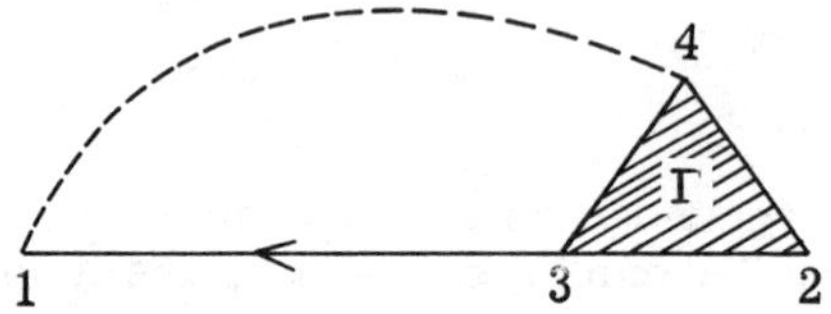

Figure 7-2. The diagram representing the proper self-energy operator.

$$\Box_1 \langle A(1) \rangle = f\, \Delta'_F(1, 1) + J(1) \tag{7-60}$$

By differentiating this equation with respect to J(2) we get

$$\Box_1 D'_F(1, 2) = i\delta^4(1 - 2) + if \frac{\delta}{\delta J(2)} \Delta'_F(1, 1) \tag{7-61}$$

Recalling the formula (7-57) to express the functional derivative of Δ'_F in terms of propagators and the vertex function we get

$$\Box_1 D'_F(1, 2) = i\delta^4(1-2) - if^2 \int d^4x_3\, d^4x_4\, d^4x_5\, \Delta'_F(1, 3) \times \Gamma(34:5)\, \Delta'_F(4, 1)\, D'_F(5, 2) \tag{7-62}$$

Thus if we define the photon polarization operator Π^* as

$$\left(\Box_1 - \Pi^*(1)\right) D'_F(1, 2) = i\delta^4(1 - 2) \tag{7-63}$$

where

$$\Pi^*(1)\, D'_F(1, 2) = \int d^4x_3\ \Pi^*(1, 3)\, D'_F(3, 2) \tag{7-64}$$

it has the following representation:

$$\Pi^*(1, 2) = -if^2 \int d^4x_3\, d^4x_4\, \Delta'_F(1, 3) \times \Gamma(34:2)\, \Delta'_F(4, 1) \tag{7-65}$$

which corresponds graphically to the diagram in Fig. 7-3.

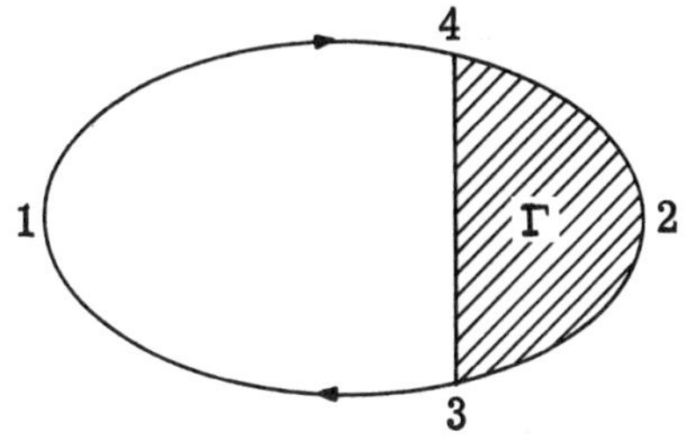

Figure 7-3. The diagram representing the photon polarization operator.

To summarize, we have obtained the following set of equations:

$$\left(K_1 - f\,\langle A(1)\rangle - \Sigma^*(1)\right)\,\Delta'_F(1, 2) = i\delta^4(1-2)$$

$$\left(\Box_1 - \Pi^*(1)\right)\,D'_F(1, 2) = i\delta^4(1-2)$$

$$\Sigma^*(1, 2) = -if^2 \int d^4x_3\, d^4x_4\, \Delta'_F(1, 3)\,\Gamma(32:4)\,D'_F(4, 1)$$

$$\Pi^*(1, 2) = -if^2 \int d^4x_3\, d^4x_4\, \Delta'_F(1, 3)\,\Gamma(34:2)\,\Delta'_F(4, 1)$$

$$f\Gamma(12:3) = f\delta^4(1-2)\,\delta^4(1-3) + \frac{\delta\,\Sigma^*(1, 2)}{\delta\,\langle A(3)\rangle}$$

In order to establish the apparatus necessary to handle the Bethe-Salpeter equation we must discuss two-body Green's functions, which constitute the subject of the next section.

7-4 SCHWINGER'S THEORY OF GREEN'S FUNCTIONS, II

For processes involving bound states, the formalism of two-body Green functions is indispensable. The two-body Green's function is defined as

$$K(12; 34) = \langle \varphi(1)\,\varphi(2)\,\varphi^\dagger(3)\,\varphi^\dagger(4)\rangle \qquad (7\text{-}66)$$

We assume that it satisfies a Bethe-Salpeter equation of the form

$$K(12;\,34) = \Delta_F'(12;\,34) + \int d^4x_5 \cdots d^4x_8 \; \Delta_F'(1,\,5) \times \Delta_F'(2,\,6)\; G(56;\,78)\; K(78;\,34) \qquad (7\text{-}67)$$

where, in analogy with the two possible ways of contracting the four free-field operators, the inhomogeneous term is given by

$$\Delta_F'(12;\,34) = \Delta_F'(1,\,3)\,\Delta_F'(2,\,4) + \Delta_F'(1,\,4)\,\Delta_F'(2,\,3) \qquad (7\text{-}68)$$

The Bethe-Salpeter equation for K is clearly a generalization of the ladder approximation, and is represented by the diagrams in Fig. 7-4.

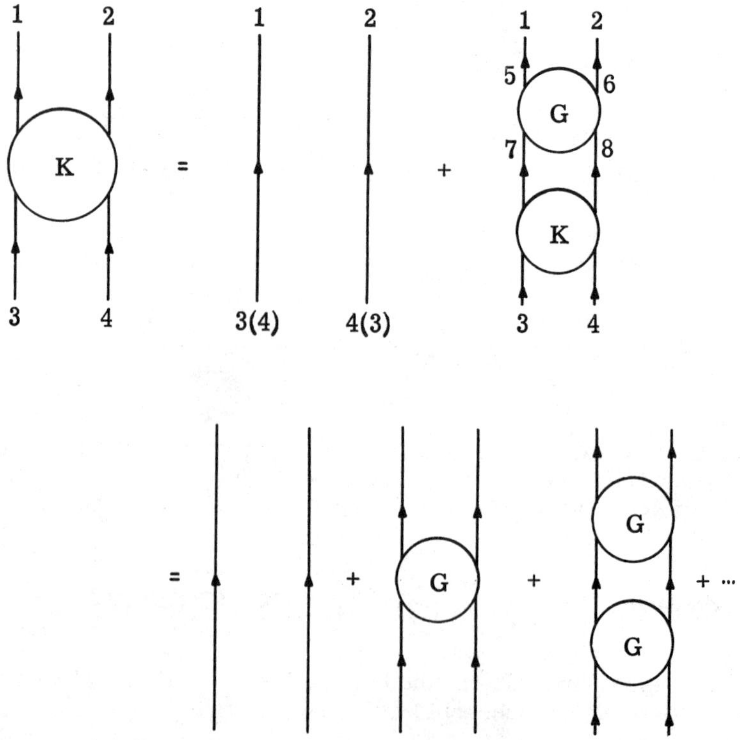

Figure 7-4. A graphical representation of the Bethe - Salpeter equation for the two-body Green's function K.

We have introduced two unknown quantities, K and G, but have only one equation; so that to determine these quantities we have to get one more equation. For this purpose we start with the reduction formula

$$(\Box_1 - \mu^2)\, T[U\, \varphi(1)\, \varphi(2)\, \varphi^\dagger(3)\, \varphi^\dagger(4)]$$

$$= i \frac{\delta}{\delta\varphi^\dagger(1)} T[U\, \varphi(2)\, \varphi^\dagger(3)\, \varphi^\dagger(4)]$$

$$= i\,\delta^4(1-3)\, T[U\, \varphi(2)\, \varphi^\dagger(4)] + i\,\delta^4(1-4)\, T[U\, \varphi(2)\, \varphi^\dagger(3)]$$

$$+ T\left[U, \left(\frac{\delta}{\delta\varphi^\dagger(1)} \int d^4x\, \mathcal{H}(x)\right), \varphi(2), \varphi^\dagger(3), \varphi^\dagger(4)\right]$$

$$= i\,\delta^4(1-3)\, T[U\, \varphi(2)\, \varphi^\dagger(4)] + i\,\delta^4(1-4)\, T[U\, \varphi(2)\, \varphi^\dagger(3)]$$

$$+ f\, T[U\, \varphi(1)\, A(1)\, \varphi(2)\, \varphi^\dagger(3)\, \varphi^\dagger(4)]$$

$$= i\,\delta^4(1-3)\, T[U\, \varphi(2)\, \varphi^\dagger(4)] + i\,\delta^4(1-4)\, T[U\, \varphi(2)\, \varphi^\dagger(3)]$$

$$+ if \frac{\delta}{\delta J(1)} T[U\, \varphi(1)\, \varphi(2)\, \varphi^\dagger(3)\, \varphi^\dagger(4)] \qquad (7\text{-}69)$$

Taking the vacuum expectation value of Eq. (7-69) and dividing it by $\langle 0|U|0\rangle$, we get

$$\left(K_1 - f\,\langle A(1)\rangle - if \frac{\delta}{\delta J(1)}\right) K(12;34) = i\,\delta^4(1-3)\,\Delta_F'(2,4)$$

$$+ i\,\delta^4(1-4)\,\Delta_F'(2,3)$$

or

$$D_J(1)\, K(12;34) = i\,\delta^4(1-3)\,\Delta_F'(2,4) + i\,\delta^4(1-4)\,\Delta_F'(2,3) \qquad (7\text{-}70)$$

This yields another equation to determine K. We also can apply D_J to the Bethe-Salpeter equation, utilizing the procedure

$$D_J(1)\,[\Delta_F'(1,2)\, A] = i\,\delta^4(1-2)\, A - if \frac{\delta A}{\delta J(1)}\, \Delta_F'(1,2) \qquad (7\text{-}71)$$

valid for any quantity A which is not a function of "1", to obtain

$$
\begin{aligned}
& i\delta^4(1-3)\,\Delta_F'(2,4) + i\delta^4(1-4)\,\Delta_F'(2,3) \\
&\quad = i\delta^4(1-3)\,\Delta_F'(2,4) + i\delta^4(1-4)\,\Delta_F'(2,3) \\
&\quad - \Delta_F'(1,3)\, if\, \frac{\delta}{\delta J(1)}\,\Delta_F'(2,4) \\
&\quad - \Delta_F'(1,4)\, if\, \frac{\delta}{\delta J(1)}\,\Delta_F'(2,3) \\
&\quad + i\int d^4x_5 \cdots d^4x_8\, \delta^4(1-5)\,\Delta_F'(2,6)\,G(56;78)\,K(78;34) \\
&\quad - if\int d^4x_5 \cdots d^4x_8\, \Delta_F'(1,5) \\
&\quad \times \frac{\delta}{\delta J(1)}\,(\Delta_F'(2,6)\,G(56;78)\,K(78;34))
\end{aligned} \tag{7-72}
$$

Carrying out the functional differentiations we find

$$
\begin{aligned}
& \int d^4x_5\, d^4x_6\, G(12;56)\,K(56;34) \\
&\quad + f^2\int d^4x_5\, d^4x_6\, \Gamma(26;5)\,D_F'(5,1)\,\Delta_F'(16;34) \\
&\quad + f^2\int d^4x_1'\, d^4x_2'\, d^4x_5 \cdots d^4x_8\, \Gamma(26;5)\,D_F'(5,1) \\
&\quad \times \Delta_F'(1,1')\,\Delta_F'(6,2')\,G(1'2';78)\,K(78;34) \\
&\quad - f\int d^4x_1'\, d^4x_5\, d^4x_6\, \Delta_F'(1,1') \\
&\quad \times \frac{\delta}{\delta J(1)}\,(G(1'2;56)\,K(56;34)) = 0
\end{aligned} \tag{7-73}
$$

This equation is too complicated to disentangle, so it is used to get the lowest-order expression for G; we keep only the first two terms and replace K by Δ_F', D_F' by D_F, and Γ by $\delta^4 \times \delta^4$, and we obtain

$$
G(12;34) = -f^2 D_F(1-2)\,\delta^4(1-3)\,\delta^4(2-4) \tag{7-74}
$$

Inserting this result in Eq. (7-67) for K, we get to lowest order

$$
\begin{aligned}
K(12;34) &= \Delta_F'(12;34) - f^2\int d^4x_5\, d^4x_6\, \Delta_F'(2,6) \\
&\quad \times \Delta_F'(1,5)\,D_F(5-6)\,K(56;34)
\end{aligned} \tag{7-75}
$$

This is the Bethe-Salpeter equation for the two-body Green's function; its structure is the same as that of the Feynman amplitude:

$$\mathcal{F}(12; ab) = \mathcal{F}(1, a)\,\mathcal{F}(2, b) + \mathcal{F}(1, b)\,\mathcal{F}(2, a) - f^2 \int d^4x_5\, d^4x_6\, \Delta_F'(1, 5)\, \Delta_F'(2, 6) \times D_F(5 - 6)\,\mathcal{F}(56; ab) \tag{7-76}$$

Schwinger's theory is useful so far as the derivation of equations is concerned, but for the physical interpretation of the results we have to study the relation of Green's functions and of Feynman amplitudes to the Heisenberg representation.

7-5 THE GELL-MANN-LOW RELATION

In this section we shall study the relation of Green's functions to the Heisenberg representation. This relation was studied and clarified by Gell-Mann and Low[6] and forms the basis for subsequent sections. In deriving this relation, however, we have to use the so-called *hypothesis of adiabatic switching*.

We shall discuss various relationships in the interaction representation first. Let us recall the following formula derived in Section 2-2:

$$[P_\mu^0, \varphi(x)] = i\, \frac{\partial \varphi(x)}{\partial x_\mu} \tag{7-77}$$

where P_μ^0 is the total energy-momentum of the free fields and $\varphi(x)$ an arbitrary free-field operator. In particular we find

$$[P_0^0, \varphi(x)] = -i\, \frac{\partial \varphi(x)}{\partial t} \tag{7-78}$$

and also, for the interaction Hamiltonian $H_{int}(t) = \int_{x_0=t} d^3x\, \mathcal{H}_{int}(x)$,

$$[P_0^0, H_{int}(t)] = -i\, \frac{\partial H_{int}(t)}{\partial t} \tag{7-79}$$

Defining $U(t, t_0)$ by use of Dyson's formula:

$$U(t, t_0) = \sum_{n=0}^{\infty} \frac{(-i)^n}{n!} \int_{t_0}^{t} dt_1 \cdots \int_{t_0}^{t} dt_n \; T[H_{int}(t_1) \cdots H_{int}(t_n)] \tag{7-80}$$

and taking the commutator between $P_0^{\;0}$ and U, we thereby obtain time derivatives of H_{int} in the T product:

$$[P_0^{\;0}, U(t, t_0)] = \sum_{n=0}^{\infty} \frac{(-i)^n}{n!} \int_{t_0}^{t} dt_1 \cdots \int_{t_0}^{t} dt_n \times [P_0^{\;0}, T[H_{int}(t_1) \cdots H_{int}(t_n)]]$$

$$= -i \sum_{n=0}^{\infty} \frac{(-i)^n}{n!} \int_{t_0}^{t} dt_1 \cdots \int_{t_0}^{t} dt_n \times \left(\frac{\partial}{\partial t_1} + \cdots + \frac{\partial}{\partial t_n}\right) T[H_{int}(t_1) \cdots H_{int}(t_n)]$$

$$= - \sum_{n=1}^{\infty} \frac{(-i)^{n-1}}{(n-1)!} \int_{t_0}^{t} dt_1 \cdots \int_{t_0}^{t} dt_{n-1} \times \Big(H_{int}(t)\, T[H_{int}(t_1) \cdots H_{int}(t_{n-1})] - T[H_{int}(t_1) \cdots H_{int}(t_{n-1})]\, H_{int}(t_0) \Big)$$

$$= -H_{int}(t)\, U(t, t_0) + U(t, t_0)\, H_{int}(t_0) \qquad (7\text{-}81)$$

We set $t = 0$, $t_0 = -\infty$, and $P_0^{\;0} + H_{int}(0) = H_{total}$ to get

$$H_{total}\, U(0, -\infty) = U(0, -\infty)\,(P_0^{\;0} + H_{int}(-\infty)) \qquad (7\text{-}82)$$

Now we shall introduce the hypothesis of adiabatic switching of interactions, which here means simply

$$H_{int}(-\infty) = 0 \qquad (7\text{-}83)$$

This is an incomplete form of the more rigorous asymptotic condition. Then we find

$$H_{total}\, U(0, -\infty) = U(0, -\infty)\, P_0^{\;0} \qquad (7\text{-}84)$$

Although this relationship itself is physically meaningless we shall see later that some results derived on the basis of this relation are meaningful. Let us introduce the free vacuum state Φ_0, the lowest eigenstate of $P_0{}^0$, by

$$P_0{}^0 \, \Phi_0 = 0 \tag{7-85}$$

Hence

$$H_{\text{total}} \, U(0, -\infty) \, \Phi_0 = U(0, -\infty) \, P_0{}^0 \, \Phi_0 = 0 \tag{7-86}$$

This means that

$$\Psi_0 = U(0, -\infty) \, \Phi_0 \tag{7-87}$$

is an eigenstate of the total Hamiltonian, called the true vacuum state. Actually, no matter how adiabatically the interactions are switched on, the energy level of the vacuum state is shifted; but since this shift is common to all the states we ignore it. In other words, we give up any attempt to make the derivation look more rigorous at the present stage.

The Gell-Mann-Low relation referred to in Section 6-7 is given by

$$(\Psi_0, \, T[A^{(H)} \, B^{(H)} \cdots] \, \Psi_0)$$

$$= \frac{(\Phi_0, \, T[U(\infty, -\infty) \, AB \cdots] \, \Phi_0)}{(\Phi_0, \, U(\infty, -\infty) \, \Phi_0)} \tag{7-88}$$

where $A^{(H)} \, B^{(H)} \cdots$ stands for Heisenberg operators. The relation between Φ_0 and Ψ_0 is now known; and we can give the relation between A and $A^{(H)}$, which already has been discussed in Section 3-3:

$$\varphi_\alpha{}^{(H)} \, (\mathbf{x}, t) = U(t, t_0)^{-1} \, \varphi_\alpha(\mathbf{x}, t) \, U(t, t_0) \tag{7-89}$$

Depending on the choice of t_0, we can introduce various interaction representations, such as

$$\varphi_\alpha = \varphi_\alpha^{\text{in}} \qquad \text{for} \quad t_0 = -\infty \tag{7-90}$$

and

$$\varphi_\alpha = \varphi_\alpha^{\text{out}} \qquad \text{for} \quad t_0 = \infty \tag{7-91}$$

Here we shall choose $t_0 = 0$, so that

$$\varphi^{(H)}(\mathbf{x}, t) = U(t, 0)^{-1} \varphi(\mathbf{x}, t) U(t, 0) \tag{7-92}$$

Without loss of generality we shall consider, with $t_1 > t_2$,

$$(\Phi_0, T[U(\infty, -\infty) A(t_1) B(t_2)] \Phi_0)$$

$$= (\Phi_0, U(\infty, t_1) A(t_1) U(t_1 t_2) B(t_2) U(t_2, -\infty) \Phi_0)$$

$$= (\Phi_0, U(\infty, t_1) U(t_1, 0) A^{(H)}(t_1) U(t_1, 0)^{-1}$$

$$\times U(t_1, t_2) U(t_2, 0) B^{(H)}(t_2) U(t_2, 0)^{-1} U(t_2, -\infty) \Phi_0)$$

$$= (\Phi_0, U(\infty, 0) A^{(H)}(t_1) B^{(H)}(t_2) U(0, -\infty) \Phi_0)$$

$$= (\Phi_0, U(\infty, -\infty) \Phi_0)$$

$$\times (\Phi_0, U(0, -\infty)^{-1} A^{(H)}(t_1) B^{(H)}(t_2) U(0, -\infty) \Phi_0)$$

$$= (\Phi_0, U(\infty, -\infty) \Phi_0)(\Psi_0, A^{(H)}(t_1) B^{(H)}(t_2) \Psi_0) \tag{7-93}$$

using the properties of the U matrix that $U(t_1, t_2) U(t_2, t_3) = U(t_1, t_3)$ and the fact that $(\Phi_0\ U(\infty, -\infty) \Phi_\alpha) = (\Phi_0, U(\infty, -\infty) \Phi_0) (\Phi_0, \Phi_\alpha)$.

Thus we have established the relation

$$(\Psi_0, T[A^{(H)}(t_1) B^{(H)}(t_2)] \Psi_0)$$

$$= \frac{(\Phi_0, T[U(\infty, -\infty) A(t_1) B(t_2)] \Phi_0)}{(\Phi_0, U(\infty, -\infty) \Phi_0)} \tag{7-94}$$

By generalizing this formula we get the Gell-Mann-Low relation. This relation is important in giving the relation between Green's functions and Heisenberg operators; namely, in the absence of an external source we have

$$\langle AB\cdots\rangle = \frac{(\Phi_0,\ T[U(\infty,\ -\infty)\ AB\cdots]\ \Phi_0)}{(\Phi_0,\ U(\infty,\ -\infty)\ \Phi_0)}$$

$$= (\Psi_0,\ T[A^{(H)}\ B^{(H)}\cdots]\ \Psi_0) \tag{7-95}$$

In the next section we shall derive a formula expressing the S-matrix elements in terms of Heisenberg operators based on this relation.

7-6 THE REDUCTION FORMULA AND THE S-MATRIX ELEMENTS IN TERMS OF GREEN'S FUNCTIONS

The ideas that have been developed in the preceding section can be incorporated into a new S-matrix formalism. By means of the reduction formulas the S-matrix elements are expressible in terms of Green's functions.

In this section we first shall show that the reduction formula contains the prescription for quantizing the fields and then shall give a discussion establishing the role of Green's functions in the S-matrix.

Derivation of Commutation Relations from the Reduction Formulas

The general form of the reduction formula in the interaction representation is

$$D_{\alpha\beta}(\partial_x)\ T[\varphi_\beta(x)\cdots] = i\frac{\delta}{\delta\varphi^\dagger_\alpha(x)}\ T[\cdots] \tag{7-96}$$

where $D_{\alpha\beta}(\partial)$ is the differential operator introduced in the discussion of Peierls' method of quantization:

$$[\mathfrak{L}]_{\varphi^\dagger_\alpha(x)} = D_{\alpha\beta}(\partial)\ \varphi_\beta(x) = 0 \tag{7-97}$$

In particular, the formula

$$D_{\alpha\beta}(\partial_x)\ T[\varphi_\beta(x),\ \varphi^\dagger_\gamma(y)]$$

$$= i\frac{\delta\varphi^\dagger_\gamma(y)}{\delta\varphi^\dagger_\alpha(x)} = i\delta_{\alpha\gamma}\ \delta^4(x-y) \tag{7-98}$$

alone is sufficient to fix the commutation relations. In order to see this we express the T product in two different forms:

$$\begin{aligned} T[\varphi_\beta(x), \varphi^\dagger_\gamma(y)] &= \theta(x_0 - y_0)\, \varphi_\beta(x)\, \varphi^\dagger_\gamma(y) \\ &\quad + \theta(y_0 - x_0)\, \varphi^\dagger_\gamma(y)\, \varphi_\beta(x) \\ &= \theta(x_0 - y_0)\, [\varphi_\beta(x), \varphi^\dagger_\gamma(y)] + \varphi^\dagger_\gamma(y)\, \varphi_\beta(x) \\ &= \theta(y_0 - x_0)\, [\varphi^\dagger_\gamma(y), \varphi_\beta(x)] + \varphi_\beta(x)\, \varphi^\dagger_\gamma(y) \end{aligned}$$

Substituting these expressions in the reduction formula and using the free-field equation for $\varphi_\beta(x)$ we obtain

$$\begin{aligned} D_{\alpha\beta}(\partial_x)\,(\theta(x_0 - y_0)\,[\varphi_\beta(x), \varphi^\dagger_\gamma(y)]) &= i\delta^4(x - y) \\ D_{\alpha\beta}(\partial_x)\,(\theta(y_0 - x_0)\,[\varphi^\dagger_\gamma(y), \varphi_\beta(x)]) &= i\delta^4(x - y) \end{aligned} \tag{7-99}$$

From this, follows

$$\theta(x_0 - y_0)\,[\varphi_\beta(x), \varphi^\dagger_\gamma(y)] = -i\Delta^R_{\beta\gamma}(x - y) \tag{7-100}$$

because the left-hand size obviously possesses the retarded character, so that it should be equal to the right-hand side, which satisfies the same equation and the same boundary condition. Similarly we get

$$\begin{aligned} \theta(y_0 - x_0)\,[\varphi^\dagger_\gamma(y), \varphi_\beta(x)] &= -\theta(y_0 - x_0)\,[\varphi_\beta(x), \varphi^\dagger_\gamma(y)] \\ &= -i\Delta^A_{\beta\gamma}(x - y) \end{aligned} \tag{7-101}$$

Hence

$$\begin{aligned} [\varphi_\beta(x), \varphi^\dagger_\gamma(y)] &= \theta(x_0 - y_0)[\varphi_\beta(x), \varphi^\dagger_\gamma(y)] \\ &\quad + \theta(y_0 - x_0)[\varphi_\beta(x), \varphi^\dagger_\gamma(y)] \\ &= i\,(\Delta^A_{\beta\gamma}(x - y) - \Delta^R_{\beta\gamma}(x - y)) \\ &= i\,\Delta_{\beta\gamma}(x - y) \\ &= iC_{\beta\gamma}(\partial_x)\,\Delta(x - y) \end{aligned} \tag{7-102}$$

Thus the commutation relations are determined directly from the reduction formula.

The S-Matrix Elements in Terms of Green's Functions

In Feynman-Dyson theory we know how to express the S matrix in terms of the interaction Hamiltonian density, but in order to include bound states as well as to discuss dispersion relations it is important to express the S-matrix elements in terms of Green's functions, since Dyson's formula is closely related to perturbation theory.

In this connection we shall introduce two alternative formulations; the first one is related to the Bethe-Salpeter equation, and the second one is suited to an axiomatic formulation of field theory. We shall illustrate these formulations by the elastic scattering of two scalar particles:

$$a + b \to b + a \tag{7-103}$$

Let φ_a and φ_b be the field operators for particles a and b, respectively. In the *renormalized interaction representation* the corresponding S-matrix element is given, with J = 0, by

$$\begin{aligned}\langle a'b'|S|ab\rangle &= \langle a'b'|U(\infty, -\infty)|ab\rangle / \langle 0|U(\infty, -\infty)|0\rangle \\ &= \langle a'b'|U(\infty, -\infty)_{\text{conn}}|ab\rangle\end{aligned} \tag{7-104}$$

where $U(\infty, -\infty)$ is given by Dyson's formula. In order to find the matrix element for the scattering process (7-103), we have to select one of each of the operators $\varphi_a, \varphi_b, \varphi_a^\dagger$, and $\varphi_b^\dagger$, describing the destruction of the incident particles and creation of the outgoing particles; the rest of the operators in the T product describe virtual processes. The selected field operators must be replaced by the single-particle wave functions:

$$\begin{aligned}\varphi_a &\to \langle 0|\varphi_a|a\rangle \qquad & \varphi_b &\to \langle 0|\varphi_b|b\rangle \\ \varphi_a^\dagger &\to \langle a'|\varphi_a^\dagger|0\rangle \qquad & \varphi_b^\dagger &\to \langle b'|\varphi_b^\dagger|0\rangle\end{aligned} \tag{7-105}$$

We can express this operation in a compact form by using functional derivatives with respect to field operators to pick out those operators from U:

$$\begin{aligned}\langle a'b'|S - 1|ab\rangle = \int & d^4x_1\ d^4x_2\ d^4x_1'\ d^4x_2' \\ & \times \langle a'|\varphi_a^\dagger(x_1')|0\rangle\ \langle b'|\varphi_b^\dagger(x_2')|0\rangle \\ & \times \langle 0|\frac{\delta}{\delta\varphi_a^\dagger(x_1')}\ \frac{\delta}{\delta\varphi_b^\dagger(x_2')}\ \frac{\delta}{\delta\varphi_a(x_1)}\ \frac{\delta}{\delta\varphi_b(x_2)} \\ & \times U(\infty, -\infty)|0\rangle \\ & \times \langle 0|\varphi_a(x_1)|a\rangle\ \langle 0|\varphi_b(x_2)|b\rangle \\ & \times \langle 0|U(\infty, -\infty)|0\rangle^{-1}\end{aligned} \tag{7-106}$$

It is obvious how to generalize this formula for more complicated processes. The next step consists of eliminating the functional derivatives since they are rather symbolic. In order to do this we utilize the reduction formula

$$(\Box_1 - \mu^2)\, T[\varphi(1)\cdots] = i\frac{\delta}{\delta\varphi^\dagger(1)} T[\cdots] \qquad (7\text{-}107)$$

We then find

$$\begin{aligned}
\langle a'b'|\, S-1\, |ab\rangle = & \int d^4x_1\, d^4x_2\, d^4x_1'\, d^4x_2' \\
& \times \langle a'|\varphi_a^\dagger(x_1')|0\rangle \langle b'|\varphi_b^\dagger(x_2')|0\rangle \\
& \times (-i)(\Box_{1'} - \mu_a^2)(-i)(\Box_{2'} - \mu_b^2) \\
& \times (-i)(\Box_1 - \mu_a^2)(-i)(\Box_2 - \mu_b^2) \\
& \times \langle 0|\, T[\varphi_a(x_1')\varphi_b(x_2')\varphi_a^\dagger(x_1)\varphi_b^\dagger(x_2)\, U(\infty, -\infty)]|0\rangle \\
& \div \langle 0|\, U(\infty, -\infty)\, |0\rangle \\
& \times \langle 0|\varphi_a(x_1)|a\rangle \langle 0|\varphi_b(x_2)|b\rangle \\
= & \int d^4x_1\, d^4x_2\, d^4x_1'\, d^4x_2' \\
& \times \langle a'|\varphi_a^\dagger(x_1')|0\rangle \langle b'|\varphi_b^\dagger(x_2')|0\rangle \\
& \times (-i)K_{1'}^a(-i)K_{2'}^b(-i)K_1^a(-i)K_2^b \\
& \times \langle \varphi_a(x_1')\, \varphi_b(x_2')\, \varphi_a^\dagger(x_1)\, \varphi_b^\dagger(x_2)\rangle \\
& \times \langle 0|\varphi_a(x_1)|a\rangle \langle 0|\varphi_b(x_2)|b\rangle
\end{aligned} \qquad (7\text{-}108)$$

Generalization of this formula to more complicated cases is straightforward. In the discussion that follows we shall study various modifications of this formula.

(1) Heisenberg representation

By making use of the Gell-Mann-Low relation, we can write the S-matrix element in terms of Heisenberg operators:

$$\langle a'b' | S - 1 | ab \rangle = \int d^4x_1\, d^4x_2\, d^4x_1'\, d^4x_2'$$

$$\times\ (\Psi_{a'},\ \varphi_a^{\dagger(H)}(x_1')\ \Psi_0)\ \ (\Psi_{b'},\ \varphi_b^{\dagger(H)}(x_2')\ \Psi_0)$$

$$\times\ (-i)\, K_{1'}^a\, (-i)\, K_{2'}^b\, (-i)\, K_1^a\, (-i)\, K_2^b$$

$$\times\ (\Psi_0,\ T[\varphi_a^{(H)}(x_1')\, \varphi_b^{(H)}(x_2')\, \varphi_a^{\dagger(H)}(x_1)\, \varphi_b^{\dagger(H)}(x_2)]\, \Psi_0)$$

$$\times\ (\Psi_0,\ \varphi_a^{(H)}(x_1)\ \Psi_a)\ (\Psi_0,\ \varphi_b^{(H)}(x_2)\ \Psi_b) \qquad (7\text{-}109)$$

In this formula only Heisenberg operators and Heisenberg state vectors appear. In obtaining (7-109) we have used the relation

$$(\Psi_0,\ \varphi_a^{(H)}(x)\ \Psi_a) = \langle 0 | \varphi_a(x) | a \rangle = (2p_0V)^{-1/2}\ e^{ipx} \qquad (7\text{-}110)$$

This relation can be understood as follows: The x dependence of the left-hand side is given by e^{ipx} and the presence of the normalization factor $(2p_0V)^{-1/2}$ is understandable, so that the general form must be

$$(\Psi_0,\ \varphi_a^{(H)}(x)\ \Psi_a) = c(2p_0V)^{-1/2}\ e^{ipx} \qquad (7\text{-}111)$$

Thus, introducing an adequate scale transformation of the field operator we can make c equal to unity. This scale transformation is exactly the wave function renormalization, and in fact c = 1 for the renormalized Heisenberg operator.

The S-matrix formula in the Heisenberg representation has an advantage over the corresponding one in the interaction representation in that we need not split the total Hamiltonian artificially into free and interaction parts. As mentioned in connection with renormalization, the correct way of splitting the total Hamiltonian is by no means trivial; furthermore it is indispensable to introduce divergent renormalization constants into the interaction part. The S-matrix formula (7-109), however, does not explicitly depend on the way in which the Hamiltonian is decomposed into two parts. The recent trend in field theory is toward formulation of the theory without the introduction of an explicit Hamiltonian, thus avoiding the treatment of divergent constants. We readily can see that this form of the S-matrix elements fits the purpose of the axiomatic field theory.

(2) Decomposition of Green's functions

The preceding method is general enough to accommodate bound states in the theory and is simpler than the one that follows. The latter method, however, is more intuitive and helps introduce bound states into the theory in a rather natural way.

The method essentially consists in factorization of the Green's functions under consideration. First, we have to introduce the concept of connected and disconnected diagrams and then study the relation between them. Let us take arbitrary Green's function (with J = 0)

$$\langle AB \cdots Z \rangle \tag{7-112}$$

and draw all the corresponding diagrams. For the sake of definiteness we choose a neutral scalar field $\varphi(x)$ and consider

$$\langle \varphi(x_1) \cdots \varphi(x_n) \rangle = \langle 0 | T[U(\infty, -\infty)\, \varphi(x_1) \cdots \varphi(x_n)] | 0 \rangle \div \langle 0 | U(\infty, -\infty) | 0 \rangle \tag{7-113}$$

in the interaction representation. The corresponding diagrams in Fig. 7-5 may be classified into two groups according to whether the points $x_1, \ldots, x_n$ are all connected together by internal lines or not.

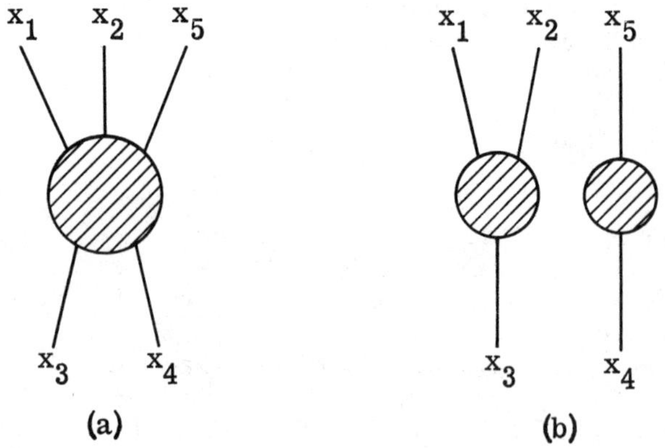

Figure 7-5. (a) An example of connected diagrams.
(b) An example of disconnected diagrams.

The former is called *connected* and the latter *disconnected*. For simplicity let us introduce the notation

$$\tau(x_1, \ldots, x_n) = \langle \varphi(x_1), \ldots, \varphi(x_n) \rangle \tag{7-114}$$

and denote the sum of contributions from connected diagrams alone by

$$\rho(x_1, \ldots, x_n) = \langle \varphi(x_1), \ldots, \varphi(x_n) \rangle_{\text{conn}} \tag{7-115}$$

The relations between the τ's and ρ's can be derived easily by means of the following recursion relation:

$$\tau(x\, x_1 \cdots x_n) = \rho(x\, x_1 \cdots x_n) + \sum_{\text{comb}} \rho(x\, x_1' \cdots x_k')\, \tau(x_{k+1}' \cdots x_n') \tag{7-116}$$

This relation is proved as follows. Consider a point x in a τ function; then for connected diagrams the sum of contributions is given by $\rho(x\, x_1 \cdots x_n)$, and for disconnected diagrams we choose a factor which represents the contributions from the partial diagrams directly connected to the point x, $\rho(x\, x_1' \cdots x_k')$, and the rest is represented by a factor $\tau(x_{k+1}' \cdots x_n')$, where $(x_1' \cdots x_n')$ is a permutation of $(x_1 \cdots x_n)$. The relation between the τ's and ρ's is expressed in a far simpler way with the help of generating functionals.

Let us use Dyson's formula for the transformation functional

$$U(\infty, -\infty) = \sum_{n=0}^{\infty} \frac{(-i)^n}{n!} \int d^4x_1 \cdots d^4x_n\, T[\mathcal{H}(x_1) \cdots \mathcal{H}(x_n)] \tag{7-117}$$

with

$$\mathcal{H}(x) = \mathcal{H}_{\text{int}}(x) + \varphi(x)\, J(x) \tag{7-118}$$

This U is a functional of the external c-number source J, so that we shall denote it by U[J] hereafter. Differentiation of U[J] with respect to J gives, as we have already seen in Section 7-3, the formula

$$i \frac{\delta}{\delta J(x)}\, T\left[U[J]\, abc \cdots\right] = T\left[U[J]\, \varphi(x)\, abc \cdots\right] \tag{7-119}$$

Define $\mathcal{T}[J]$ as

$$\mathcal{T}[J] = \langle 0 | U[J] | 0 \rangle / \langle 0 | U[0] | 0 \rangle \tag{7-120}$$

then $\mathcal{T}[J]$ is the generating functional of a set of Green's functions

$$\begin{pmatrix} 1 \\ \tau(x_1) \\ \tau(x_1 x_2) \\ \tau(x_1 x_2 x_3) \\ \cdot \\ \cdot \\ \cdot \end{pmatrix} \tag{7-121}$$

in the sense that

$$\left(i \frac{\delta}{\delta J(x_1)} \cdots i \frac{\delta}{\delta J(x_n)} \mathcal{T}[J] \right)_{J=0} = \tau(x_1 \cdots x_n) \tag{7-122}$$

where $\mathcal{T}[J]$ has the Taylor expansion

$$\mathcal{T}[J] = 1 + \sum_{n=1}^{\infty} \frac{(-i)^n}{n!} \int d^4x_1 \cdots d^4x_n \; \tau(x_1 \cdots x_n) \, J(x_1) \cdots J(x_n) \tag{7-123}$$

Now let us ask a question: What is the generating functional of the ρ's? Define $\mathcal{R}[J]$ as

$$\mathcal{R}[J] = \sum_{n=1}^{\infty} \frac{(-i)^n}{n!} \int d^4x_1 \cdots d^4x_n \; \rho(x_1 \cdots x_n) \, J(X_1) \cdots J(x_n) \tag{7-124}$$

so that

$$\left(i \frac{\delta}{\delta J(x_1)} \cdots i \frac{\delta}{\delta J(x_n)} \mathcal{R}[J] \right)_{J=0} = \rho(x_1 \cdots x_n) \tag{7-125}$$

Then we can see that the recursion relation connecting the τ's to the ρ's is obtained by differentiating the following functional equation with respect to $J(x_1), \ldots, J(x_n)$:

$$i \frac{\delta \mathcal{T}[J]}{\delta J(x)} = i \frac{\delta \mathcal{R}[J]}{\delta J(x)} \mathcal{T}[J] \tag{7-126}$$

This functional differential equation readily can be integrated to yield

$$\mathcal{T}[J] = \exp(\mathcal{R}[J]) \tag{7-127}$$

The formalism here is essentially that of linked cluster expansions in statistical mechanics. Therefore, since we can express an arbitrary τ function in terms of the ρ functions, we shall confine ourselves to connected diagrams corresponding to the ρ functions in studying the S-matrix elements. The S-matrix elements corresponding to disconnected diagrams are given by the products of connected ones, as we shall see in the next chapter.

(3) Introduction of the interaction kernel σ

We now shall come back to the scattering problem

$$a + b \rightarrow b' + a'$$

The generalization of the previous argument to two- or three-field cases is obvious. Then what we have to evaluate is an expression like

$$\int d^4x_1\ d^4x_2\ d^4x_1'\ \ d^4x_2'\ \ \langle a'\left|\varphi_a{}^{\dagger}(x_1')\right|0\rangle\ \langle b'\left|\varphi_b^{\dagger}(x_2')\right|0\rangle$$

$$\times\ K_1^{a'}\ K_2^{b'}\ K_1^{a}\ K_2^{b}\ \rho(x_1'\ x_2'\ x_1\ x_2) \tag{7-128}$$

$$\times\ \langle 0\left|\varphi_a(x_1)\right|a\rangle\langle 0\left|\varphi_b(x_2)\right|b\rangle$$

If we draw connected Feynman diagrams corresponding to ρ (see Fig. 7-6), we find that we always can write it as a product of external propagators and the interaction kernel σ, that is,

$$\rho(x_1'\ x_2'\ x_1x_2) = \int d^4y_1'\ d^4y_2'\ d^4y_1\ d^4y_2\ \ \Delta_{Fa}'(x_1' - y_1')$$

$$\times\ \Delta_{Fb}'(x_2' - y_2')\quad \sigma(y_1' : y_2' : y_1 : y_2) \tag{7-129}$$

$$\times\ \Delta_{Fa}'(y_1 - x_1)\ \Delta_{Fb}'(y_2 - x_2)$$

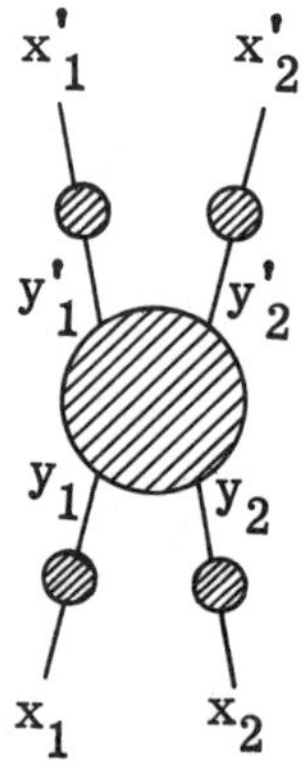

Figure 7-6. Decomposition of a ρ function into a σ function and propagation functions.

This defines σ; the factorization is, intuitively, always possible. Let us substitute this expression in the S-matrix element, then the x integration of (7-128) reduces to the form

$$\int d^4x(-i)\ K_x^a\ \Delta_{Fa}'(y - x)\ \langle 0\left|\varphi_a(x)\right|a\rangle \tag{7-130}$$

This integral is equal to $\langle 0|\varphi_a(y)|a\rangle$ as the following will prove.

In carrying out the integral (7-130) we assume that Δ'_{Fa} is the renormalized propagator and that it has the integral representation

$$\Delta'_{Fa}(y-x) = \frac{-i}{(2\pi)^4}\int d^4q \frac{e^{iq(y-x)}}{q^2+\mu_a^2+\Pi^*_{ren}(q^2)-i\epsilon} \tag{7-131}$$

where $\Pi^*_{ren}(q^2)/(q^2+\mu_a^2) \to 0$ as $q^2+\mu_a^2 \to 0$. The x-dependence of the single-particle wave function is given by

$$\langle 0|\varphi_a(x)|a\rangle = e^{ipx}\langle 0|\varphi_a(0)|a\rangle \tag{7-132}$$

Substituting these relationships into the integral we obtain

$$\int d^4x(-i)\,K_x^a\Delta'_{Fa}(y-x)\,\langle 0|\varphi_a(x)|a\rangle$$

$$= \int d^4q \frac{q^2+\mu_a^2}{q^2+\mu_a^2+\Pi^*_{ren}(q^2)}\,\delta^4(p-q)$$

$$\times \langle 0|\varphi_a(0)|a\rangle\, e^{iqy} \tag{7-133}$$

$$= \frac{p^2+\mu_a^2}{p^2+\mu_a^2+\Pi^*_{ren}(p^2)}\langle 0|\varphi_a(0)|a\rangle\, e^{ipy}$$

Using the mass-shell condition $p^2+\mu_a^2 = 0$ we get

(7-133) $= \langle 0|\varphi_a(y)|a\rangle$

Thus we have proved the desired relation, and now we know that

$K^a_{1'}$, $K^b_{2'}$, $K^a_1 K^b_2\, \rho(x'_1\, x'_2\, x_1\, x_2)$ can be replaced by $\sigma(x'_1\, x'_2\, x_1\, x_2)$.

So

$$\langle a'b'|S-1|ab\rangle = \int d^4y_1\, d^4y_2\, d^4y'_1\, d^4y'_2$$

$$\times \langle a'|\varphi_a^\dagger(y'_1)|0\rangle\langle b'|\varphi^\dagger(y'_2)|0\rangle$$

$$\times \sigma(y'_1 : y'_2 : y_1 : y_2)$$

$$\times \langle 0|\varphi_a(y_1)|a\rangle\langle 0|\varphi_b(y_2)|b\rangle \tag{7-134}$$

(4) Grouping of particles in the interaction kernel

Let us next study how to generalize the prescriptions for writing the S-matrix elements for more complicated reactions involving composite particles. In order to avoid inessential complications let us assume that all the particles are scalar. The functions σ have been derived from the ρ's by amputating the external propagators; we also shall introduce other kinds of amputations.

We shall revert to the neutral scalar field; the σ function is defined in this case as

$$\rho(x_1 \cdots x_n) = \int d^4y_1 \cdots d^4y_n \; \Delta_F'(x_1 - y_1) \cdots \Delta_F'(x_n - y_n) \times \sigma(y_1 : \cdots : y_n) \tag{7-135}$$

which corresponds to the diagrammatical representation similar to Fig. 7-6.

However, it also is possible to group some of the particles; for instance, we can combine two particles x_1 and x_2 as in Fig. 7-7 and write ρ as

$$\rho(x_1 \cdots x_n) = \int d^4y_1 \, d^4y_2 \cdots d^4y_n \; K(x_1x_2; y_1y_2) \times \Delta_F'(x_3 - y_3) \cdots \Delta_F'(x_n - y_n) \times \sigma(y_1y_2 : y_3 : \cdots : y_n) \tag{7-136}$$

Here we have used the two-particle propagator K as defined in Section 7-4 instead of the product of two single-particle propagators.

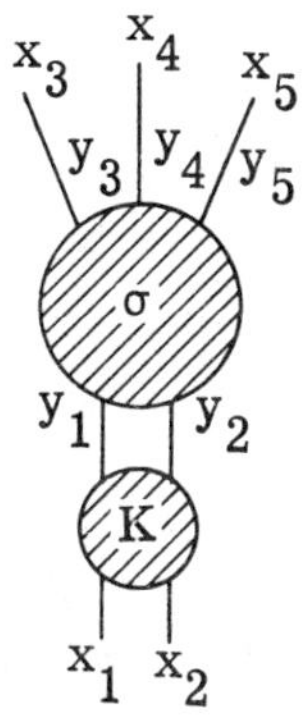

Figure 7-7. Decomposition of a ρ function into a σ function, a two-body Green's function, and propagation functions.

When such groupings are introduced we separate the variables in σ belonging to different groups by colons, for example, $\sigma(x_1 x_2 : x_3 x_4 : x_5 : x_6 : x_7)$, and so on. We can obtain a relationship between σ's of different groupings. Take, for instance, the case in which two of the particles are grouped together as opposed to that in which they are not, $\sigma(x_1 x_2 : \cdots)$ and $\sigma(x_1 : x_2 : \cdots)$, respectively, then

$$\int d^4y_1\, d^4y_2\, K(x_1x_2; y_1y_2)\, \sigma(y_1 y_2 : \cdots)$$
$$= \int d^4y_1\, d^4y_2\, \Delta_F'(x_1 - y_1)\, \Delta_F'(x_2 - y_2)\, \sigma(y_1 : y_2 : \cdots) \tag{7-137}$$

We now assume that particles "1" and "2" are not identical, to avoid symmetrization complications; as a matter of fact we have already made this assumption in defining $\sigma(y_1 y_2 : \cdots)$. Then K satisfies the Bethe-Salpeter equation of Section 7-4:

$$K(12; 34) = \Delta_F'(1 - 3)\, \Delta_F'(2 - 4) + \int d^4x_5 \cdots d^4x_8$$
$$\times\, \Delta_F'(1 - 5)\, \Delta_F'(2 - 6)\, G(56; 78)\, K(78; 34) \tag{7-138}$$

Substituting this equation into (7-137) we find the relation

$$\sigma(1:2:\cdots) = \sigma(12 : \cdots) + \int d^4x_3 \cdots d^4x_6$$
$$\times\, G(12; 34)\, K(34; 56)\, \sigma(56 : \cdots) \tag{7-139}$$

This formula is particularly important in considering composite particle reactions. This subject will be discussed in the next section.

7-7 THE S-MATRIX ELEMENTS FOR COMPOSITE-PARTICLE REACTIONS

Before writing the S-matrix elements for composite-particle reactions we shall review the Bethe-Salpeter formalism.

Feynman Amplitudes

The Feynman amplitudes satisfy the following equation for scattering states:

$$\mathfrak{F}(12; ab) = \langle 0|\varphi_a(1)|a\rangle\langle 0|\varphi_b(2)|b\rangle + \int d^4x_3 \cdots d^4x_6\ \Delta'_{Fa}(1-3)\,\Delta'_{Fb}(2-4) \times G(34; 56)\,\mathfrak{F}(56; ab) \qquad (7\text{-}140)$$

This has the formal solution

$$\mathfrak{F}(12; ab) = \langle 0|\varphi_a(1)|a\rangle\langle 0|\varphi_b(2)|b\rangle + \int d^4x_3 \cdots d^4x_6\, K(12; 34)\, G(34; 56) \times \langle 0|\varphi_a(5)|a\rangle\langle 0|\varphi_b(6)|b\rangle \qquad (7\text{-}141)$$

When particles a and b are bound to form a bound state B, the inhomogeneous term is absent.

$$\mathfrak{F}(12; B) = \int d^4x_3 \cdots d^4x_6\ \Delta'_{Fa}(1-3)\,\Delta'_{Fb}(2-4) \times G(34; 56)\,\mathfrak{F}(56; B) \qquad (7\text{-}142)$$

Reciprocal Feynman Amplitudes

The reciprocal Feynman amplitude $\widetilde{\mathfrak{F}}$ is defined as

$$\widetilde{\mathfrak{F}}(12; ab) = \langle a|\varphi_a^\dagger(1)|0\rangle\langle b|\varphi_b^\dagger(2)|0\rangle + \int d^4x_3 \cdots d^4x_6\ \widetilde{\mathfrak{F}}(56; ab)\, G(56; 34) \times \Delta'_{Fa}(3-1)\,\Delta'_{Fb}(4-2) \qquad (7\text{-}143)$$

which has the formal solution

$$\widetilde{\mathfrak{F}}(12; ab) = \langle a|\varphi_a^\dagger(1)|0\rangle\langle b|\varphi_b^\dagger(2)|0\rangle + \int d^4x_3 \cdots d^4x_6\ \langle a|\varphi_a^\dagger(5)|0\rangle \times \langle b|\varphi_b^\dagger(6)|0\rangle\, G(56; 34)\, K(34; 12) \qquad (7\text{-}144)$$

With the help of the reciprocal Feynman amplitude the S-matrix element for the scattering process a + b → b' + a' can be written in two alternative forms:

$$\langle a'b' | S - 1 | ab \rangle = \int d^4x_1 \cdots d^4x_4 \, \tilde{\mathcal{F}}_0(12; a'b') \, G(12; 34) \times \mathcal{F}(34; ab)$$

$$= \int d^4x_1 \cdots d^4x_4 \, \tilde{\mathcal{F}}(12; a'b') \, G(12; 34) \times \mathcal{F}_0(34; ab) \tag{7-145}$$

where

$$\mathcal{F}_0(34; ab) = \langle 0 | \varphi_a(3) | a \rangle \langle 0 | \varphi_b(4) | b \rangle \tag{7-146}$$

and

$$\tilde{\mathcal{F}}_0(12; a'b') = \langle a' | \varphi_a{}^\dagger(1) | 0 \rangle \langle b' | \varphi_b{}^\dagger(2) | 0 \rangle \tag{7-147}$$

Formulas (7-145) correspond to the following formulas in the Lippmann-Schwinger formalism, respectively:

$$T_{BA} = (\Phi_B, V \Psi_A{}^{(+)}) = (\Psi_B{}^{(-)}, V \Phi_A) \tag{7-148}$$

Green's Functions

The two-body Green's function K(12; 34) for two distinguishable scalar particles a and b is given by the equation

$$K(12; 34) = \Delta'_{Fa}(1 - 3) \, \Delta'_{Fb}(2 - 4) + \int d^4x_5 \cdots d^4x_8 \times \Delta'_{Fa}(1 - 5) \, \Delta'_{Fb}(2 - 6) G(56; 78) K(78; 34) \tag{7-149}$$

The graphical representation of this equation has been shown in Fig. 7-4. Successive approximation gives the solution of this equation in terms of a set of generalized ladder diagrams, shown in Fig. 7-4, which also can be expressed by a graphical equation as in Fig. 7-8.

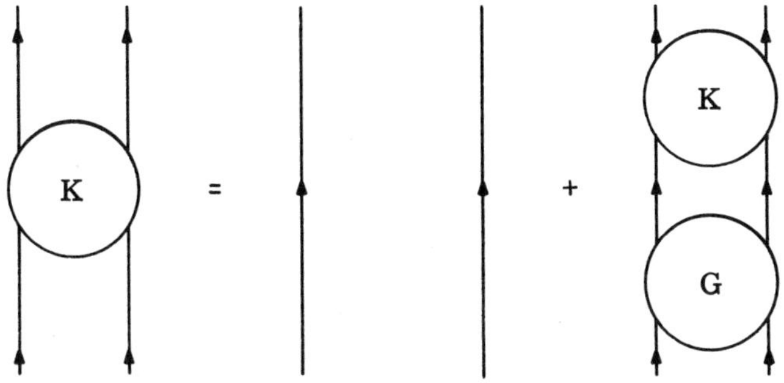

Figure 7-8. An alternative, reciprocal form of the equation in Fig. 7-4.

Therefore, we get another Bethe-Salpeter equation for K which is the reciprocal of the original equation:

$$
\begin{aligned}
K(12;\,34) = {} & \Delta'_{Fa}(1-3)\,\Delta'_{Fb}(2-4) \\
& + \int d^4x_5 \cdots d^4x_8\, K(12;\,56)\; G(56;\,78) \\
& \times \Delta'_{Fa}(7-3)\,\Delta'_{Fb}(8-4)
\end{aligned}
\qquad (7\text{-}150)
$$

Composite Particle Reactions

With these preliminaries we shall study the reaction

$$p + p \to n + p + \pi^+ \qquad (7\text{-}151)$$

and then switch to the special case in which the neutron and proton in the final state are bound to form a deuteron:

$$p + p \to d + \pi^+ \qquad (7\text{-}152)$$

For simplicity we are assuming that all the particles are scalar; therefore the S-matrix element for the first reaction is given by

$$
\begin{aligned}
S(p' + p'' \to n + p + \pi^+) = {} & \int d^4x\, d^4y\, d^4z\, d^4x'\, d^4x'' \\
& \times \langle p|\,\varphi_p(x)\,|0\rangle \langle n\,|\varphi_n(y)|\,0\rangle \\
& \times \langle \pi^+\,|\,\varphi(z)\,|\,0\rangle\, \sigma(x{:}y{:}z{:}x'{:}x'') \\
& \times \langle 0|\,\varphi_p(x')\,|\,p'\rangle \\
& \times \langle 0\,|\varphi_p(x'')\,|p''\rangle
\end{aligned}
\qquad (7\text{-}153)
$$

Then let us use the relationship between $\sigma(xy: \cdots)$ and $\sigma(x:y: \cdots)$ derived in the preceding section.

$$\int d^4x\, d^4y \langle p \left| \varphi_p^\dagger(x) \right| 0 \rangle \langle n \left| \varphi_n^\dagger(y) \right| 0 \rangle \sigma(x:y: \cdots)$$

$$= \int d^4x\, d^4y \langle p \left| \varphi_p^\dagger(x) \right| 0 \rangle \langle n \left| \varphi_n^\dagger(y) \right| 0 \rangle \sigma(xy: \cdots)$$

$$+ \int d^4x\, d^4y\, d^4x_1 \cdots d^4x_4 \langle p \left| \varphi_p^\dagger(x) \right| 0 \rangle \langle n \left| \varphi_n^\dagger(y) \right| 0 \rangle$$

$$\times\, G(xy;\, 12) K(12;\, 34) \sigma(34: \cdots)$$

$$= \int d^4x\, d^4y\ \tilde{\mathfrak{F}}(xy;\, pn)\, \sigma(xy: \cdots) \qquad (7\text{-}154)$$

This result derives from the formal solution of the Bethe-Salpeter equation for the reciprocal Feynman amplitude. Thus

$$S(p' + p'' \to n + p + \pi^+) = \int d^4x\, d^4y\, d^4z\, d^4x'\, d^4x''$$

$$\times\, \tilde{\mathfrak{F}}(xy:pn) \langle \pi^+ \left| \varphi^\dagger(z) \right| 0 \rangle$$

$$\times\, \sigma(xy:z:x':x'')$$

$$\times\, \langle 0 \left| \varphi_p(x') \right| p' \rangle \langle 0 \left| \varphi_p(x'') \right| p'' \rangle \qquad (7\text{-}155)$$

In this form it is obvious how to obtain the S-matrix element for the second reaction, with d instead of pn. The reciprocal Feynman amplitude $\tilde{\mathfrak{F}}$ can be considered as a distorted wave function, and satisfies the same Bethe-Salpeter equation for both states, except for the inhomogeneous term which is absent for $\tilde{\mathfrak{F}}(xy;\, d)$.

Therefore, on the basis of this intuitive argument we get the matrix element for the second reaction: [7]

$$S(p' + p'' \to d + \pi^+) = \int d^4x\, d^4y\, d^4z\, d^4x'\, d^4x''$$

$$\times\, \tilde{\mathfrak{F}}(xy;\, d) \langle \pi^+ \left| \varphi^\dagger(z) \right| 0 \rangle$$

$$\times\, \sigma(xy:z:x':x'') \langle 0 \left| \varphi_p(x') \right| p' \rangle \langle 0 \left| \varphi_p(x'') \right| p'' \rangle$$

$$(7\text{-}156)$$

Generalization of this method to other processes is straightforward, so we shall not discuss it. After obtaining the result (7-156) we notice a serious problem; in order to compute (7-156) it is necessary to find a procedure to normalize the Feynman amplitude for the bound state. This problem, *the normalization problem,* is one of the most interesting in the theory of bound states. For an elementary particle this turns out to be the renormalization problem, that is,

$$(\Psi_o,\ \varphi^{(H)}(x)\ \Psi_a) = \langle 0 \,|\, \varphi(x) \,|\, a \rangle \tag{7-157}$$

which is precisely the condition to fix the renormalization constant. Since this is an important problem we shall devote the next section to it.

7-8 THE NORMALIZATION PROBLEM

In order to discuss the normalization problem we need some preparation; the best way to obtain it is to study the Lippmann-Schwinger formalism. We shall quote here only the necessary results (refer to the original articles for details [8,9]).

The Heisenberg state vectors, being simultaneous eigenvectors of the four components of P_μ form a complete set. There are two important choices of the state vectors as discussed in Section 6-1:

$$\left\{\Psi^{(+)}\right\} \qquad \text{and} \qquad \left\{\Psi^{(-)}\right\}$$

A state vector belonging to the first set is considered to consist of incident plane waves and outgoing spherical waves, while a state vector in the second set consists of plane waves and converging spherical waves. The scattering-state wave function occurring in elementary scattering theory is an example of the first set. Then the S-matrix element is

$$S_{\beta\alpha} = (\Psi_\beta^{(-)},\ \Psi_\alpha^{(+)}) = \langle\ \beta(-) \,|\, \alpha(+) \rangle \tag{7-158}$$

The S matrix represents transitions between two complete orthonormal sets of states, and is therefore unitary. This formula also fits our present purpose since it does not involve the explicit form of the interaction.

Suppose that an external source J is introduced, then the Heisenberg state vectors are no longer constant and the S matrix is a functional of the external source:

$$S_{\beta\alpha}[J] = \langle\ \beta(-) \,|\, U(\infty,\ -\infty) \,|\, \alpha(+)\ \rangle \tag{7-159}$$

where

$$U(\infty, -\infty) = \sum_{n=0} \frac{(-i)^n}{n!} \int d^4x_1 \cdots d^4x_n \, T[\mathcal{H}_{ext}(x_1) \cdots \mathcal{H}_{ext}(x_n)] \tag{7-160}$$

with

$$\mathcal{H}_{ext}(x) = \varphi^{(H)}(x) J(x) \tag{7-161}$$

in the Heisenberg representation. This is essentially a result from the formal Lippmann-Schwinger theory.

Hence the first-order correction in the S-matrix element owing to the action of the external source is

$$S_{\beta\alpha}^{(1)}[J] = (-i) \int d^4x \, \langle \beta(-) | \varphi^{(H)}(x) | \alpha(+) \rangle J(x) \tag{7-162}$$

We can compute this same quantity starting from the S-matrix elements of the previous section in the interaction representation. Using the single-particle wave functions

$$\mathcal{F}_a(x) = \langle 0 | \varphi_a(x) | a \rangle \qquad \tilde{\mathcal{F}}_b(y) = \langle b | \varphi_b^\dagger(y) | 0 \rangle \tag{7-163}$$

we can obtain the result

$$\begin{aligned} \frac{\delta}{\delta J(x)} S_{\beta\alpha}^{(1)}[J] = (-i) \int & \prod_{\alpha} d^4x_a \, \mathcal{F}_a(x_a)(-i) K_{x_a} \\ & \times \prod_{\beta} d^4y_b \, \tilde{\mathcal{F}}_b(y_b)(-i) K_{y_b} \\ & \times \tau(x_a, \cdots, y_b, \cdots, x) \end{aligned} \tag{7-164}$$

Hence we conclude

$$\begin{aligned} \langle \beta(-) | \varphi^{(H)}(x) | \alpha(+) \rangle = \int & \prod_{\alpha} d^4x_a \, \mathcal{F}_a(x_a)(-i) K_{x_a} \\ & \times \prod_{\beta} d^4y_b \, \tilde{\mathcal{F}}(y_b)(-i) K_{y_b} \\ & \times (\Psi_0, T[\varphi_a^{\dagger(H)}(x_a) \cdots \varphi_b^{(H)}(y_b) \cdots \\ & \times \varphi^{(H)}(x)] \Psi_0) \end{aligned} \tag{7-165}$$

This result can be further generalized for an n-point matrix element.

$$\langle\beta(-)\left|T[A^{(H)}B^{(H)}\cdots]\right|\alpha(+)\rangle = \int \prod_{\alpha} d^4x_a\, \mathcal{F}(x_a)(-i)K_{x_a}$$

$$\times \prod_{\beta} d^4y_b\, \tilde{\mathcal{F}}_b(y_b)(-i)K_{y_b}$$

$$\times (\Psi_0, T[\varphi_a^{\dagger(H)}(x_a)\cdots$$

$$\times \varphi_b^{(H)}(y_b)\cdots A^{(H)}B^{(H)}\cdots]\Psi_0) \tag{7-166}$$

Starting from these results one can derive the so-called LSZ reduction formula in the Heisenberg representation, but our purposes are much more limited.

The Feynman Amplitudes

We have defined the Feynman amplitude $\mathcal{F}$ as the solution of a Bethe-Salpeter equation but we can also express it in terms of Heisenberg operators. As special cases of (7-165) and (7-166) we get

$$(\Psi_0, \varphi_n^{(H)}(x)\Psi_n) = \int d^4y\, \mathcal{F}_n(y)(-i)K_y$$

$$\times (\Psi_0, T[\varphi_n^{(H)}(x)\ \varphi_n^{\dagger(H)}(y)]\Psi_0) \tag{7-167}$$

and

$$(\Psi_0, T[\varphi_n^{(H)}(x_1), \varphi_p^{(H)}(x_2)]\Psi_{np}^{(+)})$$

$$= \int d^4y_1 d^4y_2\, \mathcal{F}_n(y_1)\mathcal{F}_p(y_2)(-i)K_{y_1}(-i)K_{y_2}$$

$$\times (\Psi_0, T[\varphi_n^{(H)}(x_1)\varphi_p^{(H)}(x_2)\varphi_n^{\dagger(H)}(y_1)\varphi_p^{\dagger(H)}(y_2)]\Psi_0) \tag{7-168}$$

Or, in terms of the two- and four-body Green's functions we may write

$$(\Psi_0, \varphi_n^{(H)}(x)\Psi_n) = \int d^4y\, \mathcal{F}_n(y)(-i)K_y\Delta_F'(x-y) = \mathcal{F}_n(x) \tag{7-169}$$

and

$$(\Psi_0, T[\varphi_n^{(H)}(x_1), \varphi_p^{(H)}(x_2)]\Psi_{np}^{(+)}) = \int d^4y_1 d^4y_2$$

$$\times \mathcal{F}_n(y_1)\mathcal{F}_p(y_2)(-i)K_{y_1}(-i)K_{y_2}K(x_1x_2; y_1y_2)$$

$$= \mathcal{F}(x_1x_2; np) \tag{7-170}$$

In obtaining the last result we substitute the Bethe-Salpeter equation for K and use Eq. (7-169) to eliminate the Klein-Gordon operators; we then recognize that the formal solution of the Bethe-Salpeter equation for $\mathcal{F}$ is reproduced.

Similarly we can deduce

$$(\Psi_{np}^{(-)}, T[\varphi_n^{\dagger(H)}(x_1)\varphi_p^{\dagger(H)}(x_2)]\Psi_o) = \widetilde{\mathcal{F}}(x_1x_2; np) \tag{7-171}$$

These formulas also are true with np replaced by d.

Matrix Elements

Formula (7-165) can be rewritten as follows:

$$\langle \beta(-)|\varphi^{(H)}(z)|\alpha(+)\rangle = \int \prod_\alpha d^4x_a\, \mathcal{F}_a(x_a)(-i)K_{x_a}$$

$$\times \prod_\beta d^4y_b\, \widetilde{\mathcal{F}}_b(y_b)(-i)K_{y_b}$$

$$\times \tau(x_a \cdots y_b \cdots z)$$

$$= \int \prod_\alpha d^4x_a\, \mathcal{F}_a(x_a) \prod_\beta d^4y_b\, \widetilde{\mathcal{F}}_b(y_b)\, \sigma(x_a : \cdots : y_b : \cdots \| z) \tag{7-172}$$

where $\| z$ means that the leg or the propagator originating from z has not been amputated from τ. If we replace φ by the electromagnetic current j_μ, we obtain a similar expression

$$\langle \beta(-)|j_\mu^{(H)}(z)|\alpha(+)\rangle = \int \prod_\alpha d^4x_a\, \mathcal{F}_a(x_a) \prod_\beta d^4y_b\, \widetilde{\mathcal{F}}_b(y_b)$$

$$\times \sigma_\mu(x_a : \cdots : y_b : \cdots \| z) \tag{7-173}$$

In particular, if we use deuteron states in the matrix element, then with the help of the relation derived in the preceding section we get

$$(\Psi_{d'}, j_\mu^{(H)}(z)\,\Psi_d) = \int d^4x_1 d^4x_2 d^4y_1 d^4y_2\; \tilde{\mathcal{F}}(y_1y_2, d')$$

$$\times\, \sigma_\mu(y_1y_2 : x_1x_2 \| z)\; \mathcal{F}(x_1x_2, d) \qquad (7\text{-}174)$$

This is the formula to be utilized in normalizing the Feynman amplitude for the deuteron state.

We take a conserved current j_μ, where

$$\frac{\partial j_\mu^{(H)}(x)}{\partial x_\mu} = 0 \qquad (7\text{-}175)$$

then $Q = \int j_0^{(H)}(x)\, d^3x$ is a constant of motion; Q is such a quantity as *charge*, baryon number, isospin, *energy momentum*, and so on. Suppose that Ψ_d is an eigenstate of Q with an eigenvalue q, that is,

$$Q\Psi_d = q\Psi_d \qquad (7\text{-}176)$$

then the diagonal matrix element of j_0 must be given by

$$(\Psi_d, j_0^{(H)}(x)\,\Psi_d) = V^{-1}(\Psi_d, Q\Psi_d) = qV^{-1} \qquad (7\text{-}177)$$

Thus this matrix element is a known quantity. In Eq. (7-174), we have learned how to express the left-hand side of the relation (7-177) in terms of the Feynman amplitudes; consequently the normalization of the Feynman amplitudes is, in principle, determined.

7-9 EXAMPLES OF THE BOUND-STATE PROBLEMS

In the preceding sections we have developed a general theory of bound states; now we shall give some simple illustrations of the methods.[7]

Normalization of Wick's Solution

The only known analytical solution of the Bethe-Salpeter equation is that of Wick, so to begin with we shall study the normalization of this solution. This solution was obtained in a model in which a charged scalar field interacts with a scalar photon field, and the bound state consists of two charged particles.

The current operator is given by

$$j_\mu^{(H)} = i\left(\varphi^{(H)} \frac{\partial \varphi^{\dagger(H)}}{\partial x_\mu} - \frac{\partial \varphi^{(H)}}{\partial x_\mu} \varphi^{\dagger(H)}\right)$$

$$+ \text{(term linear in } A_\mu) \tag{7-178}$$

in units of the elementary charge e. The second term now will be discarded. Then the total charge of the system in these units is

$$Q^{(H)} = \int d^3x\, j_0^{(H)}(x) \tag{7-179}$$

In order to calculate the expectation value of $Q^{(H)}$ or $j_\mu^{(H)}$ we shall introduce the following Lagrangian:

$$-\mathcal{L} = f\varphi^\dagger \varphi A + JA + q_\mu j_\mu \tag{7-180}$$

where J and q_μ are external c-number sources. Because of the derivatives in j_μ, $\mathcal{H}$ is slightly different from $-\mathcal{L}$.

In order to reproduce Schwinger's formulation in this case we define

$$U(\infty, -\infty) = T^*\left[\exp\left(i \int_{-\infty}^{\infty} d^4x\, \mathcal{L}(x)\right)\right] \tag{7-181}$$

Then the only modification of the preceding formulation is the addition of the formula

$$i\frac{\delta}{\delta q_\mu(z)}\, T[\,U(\infty, -\infty)\, abc\cdots] = T[\,U(\infty, -\infty)\, j_\mu(z)\, abc\cdots] \tag{7-182}$$

The equations for Green's functions are also slightly modified:

$$\left[D_J(1) + i\left(q_\mu(1) \frac{\partial}{\partial x_{1\mu}} + \frac{\partial}{\partial x_{1\mu}} q_\mu(1)\right)\right] \Delta_F'(1, 2) = i\delta^4(1 - 2) \tag{7-183}$$

with the operator $D_J(1)$ defined in Section 7-3; the equation for the photon propagator remains unchanged:

$$\Box_1 \langle A(1)\rangle = f\Delta_F'(1, 1) + J(1) \tag{7-184}$$

From the equation for the q derivative and the Gell-Mann—Low relation we obtain

$$\left(i \frac{\delta}{\delta q_\mu(z)} \Delta'_F (1, 2)\right)_{\substack{J=0\\q=0}} = \langle \varphi(1)\, \varphi_2^\dagger(2)\, j_\mu(z)\rangle_0 - \langle j_\mu(z)\rangle_0 \, \Delta'_F (1 - 2)$$
$$= (\Psi_0, T[\varphi^{(H)}(1)\, \varphi^{\dagger(H)}(2) \times j_\mu^{(H)}(z)]\, \Psi_0) \qquad (7\text{-}185)$$

since $\langle j_\mu(z)\rangle_0 = 0$, where the subscript 0 implies J = 0 and q = 0. From the equation for $\Delta'_F\ (1, 2)$ we find

$$\left(i \frac{\delta}{\delta q_\mu(z)} \Delta'_F (1, 2)\right)_0 = i\left(\frac{\partial \Delta'_F (1 - z)}{\partial z_\mu} \Delta'_F (z - 2) - \Delta'_F (1 - z)\frac{\partial \Delta'_F (z-2)}{\partial z_\mu}\right) + \int d^4x_3\, d^4x_4\, \Delta'_F (1 - 3)\left(\frac{\delta \Sigma^*(3, 4)}{\delta q_\mu (z)}\right)_0 \times \Delta'_F (4 - 2)$$
$$\equiv i \int d^4x_3\, d^4x_4\, \Delta'_F (1 - 3)\, D_\mu(34; z) \times \Delta'_F (4 - 2) \qquad (7\text{-}186)$$

where the unknown structure Σ^* has to be defined corresponding to the new modified equation for Δ'_F, and the vector vertex function D_μ is defined by the above equation and reduces, to lowest order, to

$$D_\mu(34; z) = \frac{\partial}{\partial z_\mu} \delta^4(3 - z) \cdot \delta^4(4 - z) - \delta^4(3 - z) \cdot \frac{\partial}{\partial z_\mu} \delta^4(4 - z) \qquad (7\text{-}187)$$

In order to study bound states, we need the derivative of the two-body Green's function

$$\left(i \frac{\delta}{\delta q_\mu(z)} K(12; 34)\right)_0 = (\Psi_0, T[j_\mu^{(H)}(z)\, \varphi^{(H)}(1)\, \varphi^{(H)}(2) \times \varphi^{\dagger(H)}(3)\, \varphi^{\dagger(H)}(4)]\, \Psi_0) \qquad (7\text{-}188)$$

To find this functional derivative we write the Bethe-Salpeter equation for K symbolically by using an integral operator $\mathcal{I}$ (12)

$$(1 - \mathcal{Q}(12))\,K(12;34) = \Delta'_F(12;34) \tag{7-189}$$

where

$$\mathcal{Q}(12)\,K(12;34) = \int d^4x_5\, d^4x_6\, I(12;56)\, K(56;34) \tag{7-190}$$

Differentiating this equation with respect to $q_\mu(z)$ we get

$$(1 - \mathcal{Q}(12))\left(i\frac{\delta}{\delta q_\mu(z)} K(12;34)\right) = i\frac{\delta}{\delta q_\mu(z)} \Delta'_F(12;34) + i\frac{\delta\mathcal{Q}(12)}{\delta q_\mu(z)}\cdot K(12;34) \tag{7-191}$$

The explicit form of $\delta\mathcal{Q}/\delta q_\mu(z)$ is

$$\begin{aligned}\left(i\frac{\delta I(12;78)}{\delta q_\mu(z)}\right)_0 = {} & i\int d^4x_3 \ldots d^4x_6\,[\Delta'_F(1-3) \\ & \times D_\mu(34;z)\,\Delta'_F(4-5)\,.\,\Delta'_F(2-6) \\ & + \Delta'_F(1-5)\,\Delta'_F(2-3)\,D_\mu(34;z) \\ & \times \Delta'_F(4-6)]\,G(56;78) \\ & + \int d^4x_5\, d^4x_6\, \Delta'_F(1-5)\,\Delta'_F(2-6) \\ & \times \left(i\frac{\delta}{\delta q_\mu(z)} G(56;78)\right)_0\end{aligned} \tag{7-192}$$

We know that the first term on the right-hand side of (7-191) for $(\delta K/\delta q_\mu)$ is expressible in terms of Δ'_F; therefore it is possible to find the formal solution of this equation with the help of Green's function K. Without explicitly showing the calculation we shall give the final result. It should be mentioned that for bound states the first term on the right-hand side of this equation does not give any contribution, since in this case the inhomogeneous term should be excluded. In terms of the Feynman amplitude (7-191) becomes

$$\begin{aligned}(1 - \mathcal{Q}(12))_0\,(\Psi_0,\ & T[j_\mu^{(H)}(z)\,\varphi^{(H)}(1)\,\varphi^{(H)}(2)]\,\Psi_B) \\ & = i\left(\frac{\delta\mathcal{Q}(12)}{\delta q_\mu(z)}\right)_0 (\Psi_0,\ T[\varphi^{(H)}(1)\,\varphi^{(H)}(2)]\,\Psi_B) \\ & = i\left(\frac{\delta\mathcal{Q}(12)}{\delta q_\mu(z)}\right)_0 \mathcal{F}(12;B)\end{aligned} \tag{7-193}$$

The formal solution, with the help of the Green's function K, is

$$
\begin{aligned}
&(\Psi_0, T[j_\mu^{(H)}(z)\,\varphi^{(H)}(1)\,\varphi^{(H)}(2)]\,\Psi_B) \\
&\quad = i\int d^4x_3 \cdots d^4x_8\; K(12;36)\,D_\mu(34;z)\,\Delta_F'(4-5) \\
&\quad \times G(56;78)\,\mathcal{F}(78;B) + \frac{1}{2}\int d^4x_3 \cdots d^4x_6\; K(12;34) \\
&\quad \times \left(i\,\frac{\delta}{\delta q_\mu(z)}\,G(34;56)\right)_0 \mathcal{F}(56;B)
\end{aligned}
\tag{7-194}
$$

from which

$$
\begin{aligned}
(\Psi_{B'}, j_\mu^{(H)}(z)\,\Psi_B) &= i\int d^4x_3 \cdots d^4x_8\; \tilde{\mathcal{F}}(36;B')\,D_\mu(34;z) \\
&\quad \times \Delta_F'(4-5)\,G(56;78)\,\mathcal{F}(78;B) \\
&\quad + \frac{1}{2}\int d^4x_3 \cdots d^4x_6\; \tilde{\mathcal{F}}(34;B') \\
&\quad \times \left(i\,\frac{\delta}{\delta q_\mu(z)}\,G(34;56)\right)_0 \mathcal{F}(56;B)
\end{aligned}
\tag{7-195}
$$

In the lowest-order approximation the second term is neglected and the lowest-order approximations to D_μ and G obtained in (7-187) and in Section 7-4, respectively, are used so that

$$
\begin{aligned}
(\Psi_{B'}, j_\mu^{(H)}(z)\,\Psi_B) &\approx -if^2\int d^4x_1\,d^4x_2 \left(\frac{\partial\,\tilde{\mathcal{F}}(z2;B')}{\partial z_\mu}\,\Delta_F(z-1)\right. \\
&\quad \left. - \tilde{\mathcal{F}}(z2;B')\,\frac{\partial\,\Delta_F(z-1)}{\partial z_\mu}\right) \\
&\quad \times D_F(1-2)\,\mathcal{F}(12;B)
\end{aligned}
\tag{7-196}
$$

By making use of (7-196) we can normalize the Feynman amplitude $\mathcal{F}(12;B)$ in Wick's solution. However, there is one more problem to solve before carrying out the explicit calculation for normalization; namely, given $\mathcal{F}(12;B)$, to calculate $\tilde{\mathcal{F}}(12;B)$. By definition,

$$
\mathcal{F}(12;B) = (\Psi_0, T[\varphi^{(H)}(1)\,\varphi^{(H)}(2)]\,\Psi_B) \tag{7-197}
$$

and

$$
\tilde{\mathcal{F}}(12;B) = (\Psi_B, T[\varphi^{\dagger(H)}(1)\,\varphi^{\dagger(H)}(2)]\,\Psi_0) \tag{7-198}
$$

respectively. In ordinary quantum mechanics these two amplitudes are related to one another through complex conjugation, but such is not the case for the Feynman amplitudes:

$$
\begin{aligned}
\mathcal{F}(12; B)^* &= (\Psi_0, T[\varphi^{(H)}(1)\,\varphi^{(H)}(2)]\,\Psi_B)^* \\
&= (\Psi_B, T[\varphi^{(H)}(1)\,\varphi^{(H)}(2)]^\dagger\,\Psi_0) \\
&= (\Psi_B, \tilde{T}[\varphi^{\dagger(H)}(1)\,\varphi^{\dagger(H)}(2)]\,\Psi_0) \\
&\neq \tilde{\mathcal{F}}(12; B)
\end{aligned}
$$

since the antitime-ordering operator $\tilde{T}$ has to be used. The desired relation is obtained with the help of the time-reversal operation; the result is

$$\tilde{\mathcal{F}}(1, 2; B) = \mathcal{F}(-1, -2; B) \tag{7-199}$$

Now let us substitute Wick's solution in the form

$$\mathcal{F}(12; B) = i\mu^4 C\, e^{iPX} \int_{-1}^{1} dz\, g(z) \int d^4p \times \frac{e^{ipx}}{[p^2 + zpP + \mu^2 + P^2/4 - i\epsilon]^3} \tag{7-200}$$

where $g(z) = 1 - |z|$ in the weak coupling limit, and C is the normalization constant to be determined. Then $\tilde{\mathcal{F}}$ is given by

$$\tilde{\mathcal{F}}(12; B) = i\mu^4 C\, e^{-iPX} \int_{-1}^{1} dz\, g(z) \times \int \frac{d^4p\; e^{ipx}}{[p^2 + zpP + \mu^2 + P^2/4 - i\epsilon]^3} \tag{7-201}$$

where use has been made of the evenness of the function g(z). Returning to the matrix element of the current, we may write, for the diagonal element, the expression

$$(\Psi_B, j_\mu^{(H)}(z)\,\Psi_B) = -2if^2 \int d^4x_1\, d^4x_2\, \frac{\partial \tilde{\mathcal{F}}(z2; B)}{\partial z_\mu} \times \Delta_F(z - 1)\, D_F(1 - 2)\, \mathcal{F}(12; B) \tag{7-202}$$

Substituting the integral representations for the various expressions in the integrand and using the Feynman integral formula derived in Section 7-2 we obtain

$$(\Psi_B, j_\mu^{(H)}(z)\, \Psi_B)$$

$$= 2if^2 C^2 \int d^4x_1\, d^4x_2\, \frac{\partial}{\partial z_\mu} \left(\mu^4 \exp\left[\frac{-iP(z + x_2)}{2}\right]\right.$$

$$\times \int_{-1}^{1} d\xi\, g(\xi) \int d^4p\, \frac{\exp[ip(z - x_2)]}{[p^2 + \xi pP + \mu^2 + P^2/4 - i\epsilon]^3}$$

$$\times \frac{-i}{(2\pi)^4} \int d^4k\, \frac{\exp[ik(z - x_1)]}{k^2 + \mu^2 - i\epsilon} \cdot \frac{\mu^4 \pi^2}{2(2\pi)^4}$$

$$\times \exp\left[\frac{iP(x_1 + x_2)}{2}\right] \int_{-1}^{1} d\zeta\, \frac{g(\zeta)}{\mu^2 + (1 - \zeta^2)P^2/4}$$

$$\times \int d^4q\, \frac{\exp[iq(x_1 - x_2)]}{q^2 + \zeta qP + \mu^2 + P^2/4 - i\epsilon} \tag{7-203}$$

This integral is not as complex as it looks, since the x integrations give δ functions of momenta, which make two of the momentum integrations trivial, leaving

$$f^2 \mu^8 C^2 \pi^2 \int_{-1}^{1} d\xi\, g(\xi) \int_{-1}^{1} d\zeta\, g(\zeta) \left(\mu^2 + (1 - \zeta^2)\frac{P^2}{4}\right)^{-1}$$

$$\times \int d^4p\, \frac{i(p - P/2)_\mu}{[p^2 - \xi pP + \mu^2 + P^2/4 - i\epsilon]^3}$$

$$\times \frac{i(p - P/2)_\mu}{[(p - P/2)^2 + \mu^2 - i\epsilon]\,[p^2 + \zeta pP + \mu^2 + P^2/4 - i\epsilon]}$$

$$= f^2 \pi^4 C^2 I \cdot \frac{P_\mu}{2} \tag{7-204}$$

where

$$I = \int_0^1 dx \int_0^1 dy \int_{-1}^1 d\xi \int_{-1}^1 d\zeta$$

$$\frac{y^2(1-y)(1-z)\, g(\xi)\, g(\zeta)}{(1-\eta^2+\eta^2 z^2)^3\,(1-\eta^2+\eta^2\zeta^2)} \tag{7-205}$$

with $z = (1-y)(1-x+\zeta x) + \xi y$. Hence the normalization condition is

$$\frac{2}{V} = f^2 \pi^4 C^2 (P_0/2) \cdot I \tag{7-206}$$

since the charge of the bound state is 2; the normalization constant is thereby fixed:

$$C = \frac{2}{\pi^2 f} \cdot (IP_0 V)^{-1/2} \tag{7-207}$$

The factor $(P_0 V)^{-1/2}$ is typical of Bose particles. In evaluating the integral I we can employ the approximations used in Section 7-2, assuming that the binding is weak ($\eta^2 \approx 1$),

$$(1-\eta^2+\eta^2\zeta^2)^{-1} \approx \pi(1-\eta^2)^{-1/2}\,\delta(\zeta)$$

$$(1-\eta^2+\eta^2 z^2)^{-3} \approx \frac{3}{8}\pi(1-\eta^2)^{-5/2}\,\delta(z) \tag{7-208}$$

and find

$$I = \frac{1}{32}\pi^2 (1-\eta^2)^{-3} \tag{7-209}$$

and

$$C = \frac{16}{\pi^3 f}(1-\eta^2)^{3/2}(2P_0 V)^{-1/2} \tag{7-210}$$

Thus the normalization problem has been solved for this case.

Electromagnetic Form Factor of the Bound State B

Once the normalization constant is determined we can compute various cross sections. In the normalization problem just discussed, we have computed the diagonal matrix element

$$(\Psi_P, j_0^{(H)}(z)\,\Psi_P) = \frac{2}{V} \tag{7-211}$$

where P denotes the energy momentum of the bound state B. In general the expression

$$\left(P_0'\right)^{1/2} (\Psi_{P'}, j_\mu^{(H)}(0)\,\Psi_P)\left(P_0\right)^{1/2} \tag{7-212}$$

transforms as the μth component of a vector, so that we can write

$$(\Psi_P, j_\mu^{(H)}(0)\,\Psi_P) = \left(\frac{2}{P_0 V}\right) \cdot P_\mu \tag{7-213}$$

or, more generally, we get

$$\left(P_0'\right)^{1/2} (\Psi_{P'}, j_\mu^{(H)}(0)\,\Psi_P)\left(P_0\right)^{1/2} = V^{-1}\,(P_\mu + P_\mu')\,F[(P' - P)^2] \tag{7-214}$$

The function F is called the electromagnetic form factor of the composite particle B and is normalized to

$$F(0) = 1 \tag{7-215}$$

as can be seen by setting $P' = P$ in (7-214). This form factor can be obtained by evaluating the nondiagonal element of $j_\mu^{(H)}$. Let us discuss the physical meaning of the form factor. An immediate extension of the normalization problem is the elastic scattering of the B particle by a Coulomb field:

$$\langle B'|S|B\rangle = ie\int d^4x(\Psi_{B'}, j_\mu^{(H)}(x)\,\Psi_B)\,A_\mu^{ext}(x) \tag{7-216}$$

If we use a constant form factor this matrix element becomes identical with that of an elementary particle and the cross section is given by the Rutherford formula

$$\left(\frac{d\sigma}{d\Omega}\right)_{\mathcal{L}} = \left(\frac{2eQE}{8\pi P^2 \sin^2\frac{\theta}{2}}\right)^2 \tag{7-217}$$

where E and P are the energy and momentum of the B particle, Q the charge of the source of the Coulomb field, namely, the nuclear charge, and 2e the charge of the B particle. If we take account of the form factor the cross section is given by

$$\frac{d\sigma}{d\Omega} = \left(\frac{d\sigma}{d\Omega}\right)_{\mathcal{L}} \left[F((P' - P)^2)\right]^2 \tag{7-218}$$

and the square of the momentum transfer is

$$(P' - P)^2 = 4P^2 \sin^2\frac{\theta}{2} \tag{7-219}$$

The expression for the form factor is very complicated; for small values of the momentum transfer F can be expanded:

$$F = 1 - \frac{35}{96}\left(\frac{P^2}{\mu B}\right)\sin^2\frac{\theta}{2} + \cdots \tag{7-220}$$

The second term clearly expresses the structure correction as is clear from the appearance of the binding energy B. In fact, $(\mu B)^{1/2}$ is inversely proportional to the radius of the bound state.

Photo Disintegration of the Bound State B

We next shall discuss the disintegration of the bound state B by an incoming scalar photon "γ", namely

$$"\gamma" + B \rightarrow M + M \tag{7-221}$$

For this purpose we need to know the Feynman amplitude

$$\mathcal{F}\,(12;\, 0;\, B) = (\Psi_0,\ T[\varphi^{(H)}(1)\varphi^{(H)}(2)\,A^{(H)}(0)]\,\Psi_B) \tag{7-222}$$

This amplitude is given, again with the help of the two-body Green's function K, by

$$\begin{aligned} \mathcal{F}(12;\, 0;\, B) = {} & -if\int d^4x_3 \cdots d^4x_9\ K(12;\, 36)\ \Gamma(34;\, 9)\ \Delta'_F(4 - 5) \\ & \times\ G(56;\, 78)\ \mathcal{F}(78;\, B)\ D'_F(9 - 0) \\ & -\frac{f}{2}\int d^4x_3 \cdots d^4x_7\ K(12;\, 34)\ \Gamma(34;\, 56;\, 7) \\ & \times\ \mathcal{F}(56;\, B)\ D'_F(7 - 0) \end{aligned} \tag{7-223}$$

where Γ is defined by

$$\left(i \frac{\delta}{\delta J(0)} G(34;\,56)\right)_0 = -f\int d^4x_7\, \Gamma(34;\,56;\,7)\, D'_F(7-0) \qquad (7\text{-}224)$$

Then the S-matrix element immediately is given by

$$\begin{aligned} S(``\gamma" + B \to 2M) &= -if\int d^4x_3 \cdots d^4x_9\, \tilde{\mathcal{F}}(36;\,f)\, \Gamma(34;\,9)\, \Delta'_F(4-5) \\ &\quad\times G(56;\,78)\,\mathcal{F}(78;\,B)\,\mathcal{F}(9;\,\gamma) \\ &\quad -\frac{f}{2}\int d^4x_3 \cdots d^4x_7\, \tilde{\mathcal{F}}(34;\,f)\, \Gamma(34;\,56;\,7)\,\mathcal{F}(56;\,B) \\ &\quad\times \mathcal{F}(7;\,\gamma) \end{aligned} \qquad (7\text{-}225)$$

In the lowest-order approximation the second integral is neglected, the vertex function is a delta function, and $\tilde{\mathcal{F}}$ is replaced by the Born expression $\tilde{\mathcal{F}}_0$, so that

$$\begin{aligned} S_{fi} &= S(``\gamma" + B \to 2M) \\ &\approx if^3\int d^4x_1\, d^4x_2\, d^4x_3\, \tilde{\mathcal{F}}_0(23;\,f)\, \Delta_F(1-3)\, D_F(1-2) \\ &\quad\times \mathcal{F}(12;\,B)\,\mathcal{F}(3;\,\gamma) \end{aligned} \qquad (7\text{-}226)$$

where

$$\tilde{\mathcal{F}}_0(23;\,f) = \tilde{\mathcal{F}}(2;\,a)\,\tilde{\mathcal{F}}(3;\,b) + \tilde{\mathcal{F}}(3;\,a)\,\tilde{\mathcal{F}}(2;\,b) \qquad (7\text{-}227)$$

and a and b denote the energy momenta of the mesons in the final state, respectively. If the known integral representations are substituted, the calculation is straightforward; in the center-of-mass system we obtain

$$\frac{d\sigma}{d\Omega} = \pi^2 \left(\frac{2f\mu}{\pi}\right)^4 (1-\eta^2)^2 \left(\frac{P}{k^5E^2}\right) \frac{[E^2 + (E^2 - 4\mu^2)\cos^2\theta]^2}{[E^2 - (E^2 - 4\mu^2)\cos^2\theta]^4} \qquad (7\text{-}228)$$

where E is the total energy of the system, and p and k are the magnitudes of the relative momenta in the final and initial states, respectively;

$$E = 2\left(\mu^2 + p^2\right)^{1/2} = k + \left(M^2 + k^2\right)^{1/2} \qquad M = 2\mu\eta \qquad (7\text{-}229)$$

In this way we can compute various cross sections, expectation values related to bound states, at least in principle. We shall not discuss these problems further, however, since this approach offers various difficulties.

(1) Existence of the solution of the Bethe-Salpeter equation for two spinor particles in the ladder approximation seems to be denied.

(2) Because of the presence of extra variables, such as the relative time, it is generally difficult to solve the Bethe-Salpeter equation.

(3) Very often we are confronted with difficulties in renormalizing divergences in the Bethe-Salpeter equation. This difficulty derives essentially from the fact that the Bethe-Salpeter equation is not based on the perturbative approach, so that the techniques which are used to remove divergences in perturbation theory cannot be used here, and the proper method of doing so within the framework of the Bethe-Salpeter formalism is not known. For these reasons we shall proceed to dispersion theory.

PROBLEMS

7.1 Reproduce Schwinger's formulation for electrodynamics, namely for the system described by the interaction.

$$\mathcal{H}_{\text{int}} = -ie\,\bar{\psi}\,\gamma_\mu \psi\, A_\mu + J_\mu A_\mu$$

where J_μ denotes an external c-number source. Derive the following set of equations:

$$\left(\gamma \frac{\partial}{\partial x_1} + m - ie\,\gamma(1)\,\langle A(1)\rangle + \Sigma^*(1)\right) S_F'(1, 2) = -i\delta^4(1-2)$$

$$\left(\Box_1 - \Pi^*(1)\right) D_F'(1, 2) = i\delta^4(1-2),$$

$$\Sigma^*(1, 2) = ie^2 \int d^4x_3\, d^4x_4\, \gamma(1)\, S'(1, 3)\, \Gamma(32{:}4)\, D_F'(4, 1)$$

$$\Pi^*(1, 2) = -ie^2 \int d^4x_3\, d^4x_4\, \mathrm{Tr}\left[\gamma(1)\, S_F'(1, 3)\, \Gamma(34{:}2) S_F'(4, 1)\right]$$

$$e\Gamma(12; 3) = e\gamma(3)\,\delta^4(1-2)\delta^4(1-3) - i\,\frac{\delta\,\Sigma^*(1, 2)}{\delta\,\langle A(3)\rangle}$$

where

$$S'_F(1, 2) = \langle \psi(1), \bar{\psi}(2) \rangle \qquad \text{etc.}$$

It should be mentioned that the arguments 1, 2, ··· represent not only the space-time variables but also the vector or spinor indices.

7-2 From the definition

$$T[U(\infty, -\infty)\, A(t_1)\, B(t_2) \cdots]$$

$$= \sum_{n=0}^{\infty} \frac{(-i)^n}{n!} \int_{-\infty}^{\infty} dt_1' \cdots \int_{-\infty}^{\infty} dt_n'$$

$$\times\, T\,[H_{int}(t_1') \cdots H_{int}(t_n')\, A(t_1)\, B(t_2) \cdots]$$

derive the following relationship:

$$T[U(\infty, -\infty)\, A(t_1)\, B(t_2)] = U(\infty, t_1)\, A(t_1)\, U(t_1, t_2)\, B(t_2)\, U(t_2, -\infty)$$

$$\text{for} \quad t_1 > t_2$$

REFERENCES

1. E. E. Salpeter and H. A. Bethe, Phys. Rev. **84**, 1232 (1951).
2. Y. Nambu, Progr. Theor. Phys. (Kyoto) **5**, 614 (1950).
3. G. C. Wick, Phys. Rev. **96**, 1124 (1954).
4. R. E. Cutkosky, Phys. Rev. **96**, 1135 (1954).
5. J. Schwinger, Proc. Natl. Acad. Sci. **37**, 452 and 455 (1951).
6. M. Gell-Mann and F. E. Low, Phys. Rev. **84**, 350 (1951).
7. K. Nishijima, Progr. Theor. Phys. (Kyoto) **10**, 549 (1953); **12**, 279 (1954); **13**, 305 (1955).
8. B. A. Lippmann and J. Schwinger, Phys. Rev. **79**, 469 (1950).
9. M. Gell-Mann and M. L. Goldberger, Phys. Rev. **91**, 398 (1953).

CHAPTER 8
DISPERSION THEORY

In lecture notes covering a variety of subjects it is difficult to keep the same set of notations throughout the notes. For this reason we have used the symbols φ and $\varphi^{(H)}$ to denote the field operators in the interaction and Heisenberg representations, respectively, in sections in which the interaction representation dominates. Similarly, we have used Φ or the bracket notation for state vectors in the interaction representation, while Ψ has been used to represent a Heisenberg state. In the present chapter, however, we shall use exclusively the bracket notation for Heisenberg state vectors and the symbol φ for the Heisenberg operators. Whenever it will be necessary to refer to the interaction representation we shall use the superscript in. The superscript in denotes two different concepts, interaction representation and incoming, so that one might wonder how one can clearly distinguish between them. Fortunately it so happens that they are essentially the same and consequently these meanings need not be separated.

Dispersion theory represents an attempt to remedy some of the intrinsic difficulties (divergences, ambiguities, etc.) of the Feynman-Dyson theory and may be considered to be supplementary to it. There are two alternative approaches to dispersion theory. One is a rather sophisticated axiomatic approach, and the other is intuitive, based on perturbation theory. We shall employ the latter approach; however, it is instructive to have some contact with the foundations of the axiomatic approach.

The simplest example of a dispersion relation occurs for the propagation function, so this is where we shall start.

8-1 THE KÄLLÉN-LEHMANN REPRESENTATION [1-4]

We shall introduce a general structure analysis of the propagation function. For simplicity we shall consider a neutral scalar field φ in the Heisenberg representation.

We assume the following postulates.

(1) There exists an operator P_μ expressing the total energy-momentum of the system; it satisfies

$$[P_\mu, P_\nu] = 0 \tag{8-1}$$

and

$$\frac{\partial\varphi(x)}{\partial x_\mu} = i[\varphi(x), P_\mu] \tag{8-2}$$

(2) The simultaneous eigenvectors of P_μ form a complete set

$$P_\mu \Phi_k = k_\mu \Phi_k \tag{8-3}$$

Then we define various functions as vacuum expectation values of products of field operators.

$$\begin{aligned}
(\Phi_0, \varphi(x)\varphi(x')\Phi_0) &= \langle\varphi(x), \varphi(x')\rangle_0 = i\Delta^{(+)\,\prime}(x - x') \\
(\Phi_0, \varphi(x')\varphi(x)\Phi_0) &= \langle\varphi(x')\varphi(x)\rangle_0 = -i\Delta^{(-)\,\prime}(x - x') \\
(\Phi_0[\varphi(x), \varphi(x')]\Phi_0) &= \langle[\varphi(x), \varphi(x')]\rangle_0 = i\Delta'(x - x') \\
(\Phi_0, T[\varphi(x), \varphi(x')]\Phi_0) &= \langle T[\varphi(x), \varphi(x')]\rangle_0 = \Delta_F'(x - x')
\end{aligned} \tag{8-4}$$

First, we study the structure of the function $\Delta^{(+)\prime}$ by inserting a complete set of intermediate states

$$\begin{aligned}
\langle\varphi(x)\varphi(x')\rangle_0 &= \sum_k (\Phi_0, \varphi(x)\Phi_k)(\Phi_k, \varphi(x')\Phi_0) \\
&= \sum_k a_{0k}\, a_{0k}^*\, \exp(ik(x - x'))
\end{aligned} \tag{8-5}$$

where

$$(\Phi_0, \varphi(x)\Phi_k) = a_{0k}\, e^{ikx} \tag{8-6}$$

The x dependence of matrix elements of (8-6) results from postulate (1), discussed first in Section 2-2. On the basis of Lorentz invariance we can define a scalar function ρ as

$$\rho(-k^2) = (2\pi)^3 \sum_{\text{fixed } k_\mu} a_{0k}\, a^*_{0k} \tag{8-7}$$

where the summation must be carried out over all the states belonging to a given eigenvalue k_μ. Then we can write

$$i\Delta^{(+)\,'}(x - x') = \frac{1}{(2\pi)^3} \int d^4k\, \theta(k_0)\, \rho(-k^2) \exp[ik(x - x')] \tag{8-8}$$

The factor of $\theta(k_0)$ results from the requirement that $k_0 > 0$. Let us set

$$\rho(-k^2) = \int d\kappa^2\, \rho(\kappa^2)\, \delta(k^2 + \kappa^2) \tag{8-9}$$

with the assumption $-k^2 > 0$ implied, and get

$$\Delta^{(+)\,'}(x) = \int_0^\infty d\kappa^2\, \rho(\kappa^2)\, \Delta^{(+)}(x, \kappa^2) \tag{8-10}$$

as a weighted integral over free-particle propagators with continuous mass values.

Similarly, we can get for other Green's functions

$$\Delta^{(\;)'}(x) = \int_0^\infty d\kappa^2\, \rho(\kappa^2)\, \Delta^{(\;)}(x, \kappa^2) \tag{8-11}$$

The weight function or the spectral function $\rho(\kappa^2)$ has various important properties. First, from the definition of ρ, it is clear that

$$\rho(\kappa^2) \geq 0 \tag{8-12}$$

Second, it is different from zero only for $\kappa^2 > 0$; this corresponds to the fact that the energy-momentum k_μ must always be timelike.

For the special case of a free field of mass m, the weight function $\rho^{(0)}(\kappa^2)$ is given by

$$\rho^{(0)}(\kappa^2) = \delta(\kappa^2 - m^2) \tag{8-13}$$

In general, stable single-particle intermediate states of mass m give $\rho(\kappa^2)$ a contribution of the form $c\delta(\kappa^2 - m^2)$, so that the general form of $\rho(\kappa^2)$ is

$$\rho(\kappa^2) = c\delta(\kappa^2 - m^2) + \sigma(\kappa^2) \tag{8-14}$$

where $\sigma(\kappa^2)$ represents the contribution from intermediate states with a continuous mass spectrum, such as two-particle states, three-particle states, and so on. Hence, when there is only one kind of neutral scalar meson of mass m, the $\sigma(\kappa^2)$ vanishes for $\kappa^2 \leq (2m)^2$. The constant c is determined by the renormalization condition as we shall see later.

The Fourier transform of the free propagator is given by

$$\Delta_F(-k^2) = \frac{1}{k^2+m^2-i\epsilon}$$

$$\text{with} \quad \Delta_F(x) = \frac{-i}{(2\pi)^4} \int d^4k\, e^{ikx}\, \Delta_F(-k^2) \tag{8-15}$$

Therefore, we get in general

$$\Delta_F'(-k^2) = \int_0^\infty d\kappa^2 \frac{\rho(\kappa^2)}{k^2 + \kappa^2 - i\epsilon} \tag{8-16}$$

If we write $s = -k^2$ and $s' = \kappa^2$,

$$\Delta_F'(s) = \int_0^\infty ds' \frac{\rho(s')}{s' - s - i\epsilon} \tag{8-17}$$

Now we use the formula

$$\frac{1}{x - i\epsilon} = \frac{P}{x} + i\pi\, \delta(x) \tag{8-18}$$

where P denotes Cauchy's principal value, then

$$\Delta_F'(s) = P \int_0^\infty \frac{\rho(s')ds'}{s' - s} + i\pi\, \rho(s) \tag{8-19}$$

Hence, for real values of s we have

$$\operatorname{Re} \Delta_F'(s) = P \int_0^\infty \frac{\rho(s')ds'}{s' - s} \qquad \operatorname{Im} \Delta_F'(s) = \pi\rho(s) \tag{8-20}$$

Thus

$$\operatorname{Re} \Delta_F'(s) = \frac{P}{\pi} \int_0^\infty \frac{ds'}{s' - s} \operatorname{Im} \Delta_F'(s') \tag{8-21}$$

or

$$\Delta_F'(s) = \frac{1}{\pi} \int_0^\infty \frac{ds'}{s' - s - i\epsilon} \operatorname{Im} \Delta_F'(s') \tag{8-22}$$

This is a typical dispersion relation.

In general, when a function f(z) is analytic in the upper half complex z plane we can write it in Cauchy's form (illustrated in Fig. 8-1)

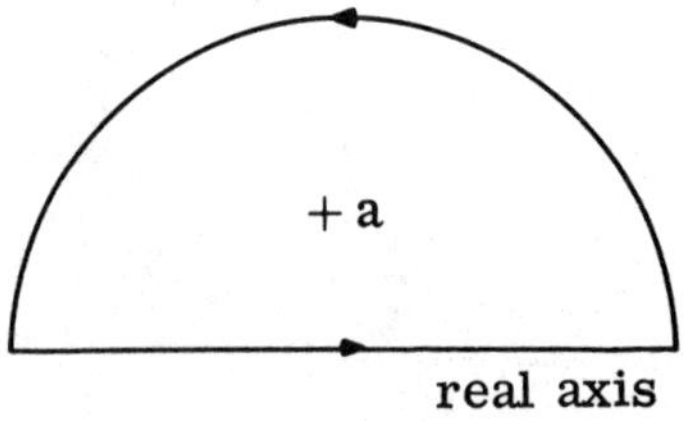

Figure 8-1. The contour for the integral (8-23).

$$f(a) = \frac{1}{2\pi i} \oint \frac{dz}{z - a} f(z) \tag{8-23}$$

provided Im a > 0.

When a is real, f(a) is defined as

$$f(a) = \lim_{\epsilon \to 0} f(a + i\epsilon)$$

Let us assume that f(z) falls off sufficiently rapidly along the semi-circle in the upper half-plane as its radius increases, namely, $f(z) \to 0$ for $|z| \to \infty$, $\text{Im } z > 0$, then the contour integral is reduced to an integral along the real axis:

$$f(a) = \frac{1}{2\pi i} \int_{-\infty}^{\infty} \frac{dx'}{x' - a} f(x') \qquad (\text{Im } a > 0) \tag{8-24}$$

We replace a by $x + i\epsilon$ and take the limit $\epsilon \to 0$:

$$f(x) = \lim_{\epsilon \to 0} f(x + i\epsilon) = \frac{1}{2\pi i} \lim_{\epsilon \to 0} \int_{-\infty}^{\infty} \frac{dx'}{x' - x - i\epsilon} f(x')$$

$$= \frac{1}{2\pi i} \left[P \int_{-\infty}^{\infty} \frac{dx'}{x' - x} f(x') + i\pi f(x) \right]$$

or

$$f(x) = \frac{P}{i\pi} \int_{-\infty}^{\infty} \frac{dx'}{x' - x} f(x') \tag{8-25}$$

Taking the real part we get

$$\text{Re } f(x) = \frac{P}{\pi} \int_{-\infty}^{\infty} \frac{dx'}{x' - x} \text{Im } f(x') \tag{8-26}$$

and for the imaginary part

$$\text{Im } f(x) = -\frac{P}{\pi} \int_{-\infty}^{\infty} \frac{dx'}{x' - x} \text{Re } f(x') \tag{8-27}$$

In physical applications the real and imaginary parts are called the dispersive and absorptive parts, respectively. Writing Re f = F and Im f = G, we can write a pair of conjugate equations

$$F(x) = \frac{P}{\pi} \int_{-\infty}^{\infty} \frac{dx'}{x' - x} G(x') \qquad G(x) = -\frac{P}{\pi} \int_{-\infty}^{\infty} \frac{dx'}{x' - x} F(x') \tag{8-28}$$

The functions F and G are called Hilbert transforms of one another.

We now shall study the connection between the spectral function $\rho(\kappa^2)$ and the renormalization constants. We write the commutator function as

$$\Delta'(x) = \int_0^\infty d\kappa^2 \, \rho(\kappa^2) \Delta(x, \kappa^2) \tag{8-29}$$

and consider the canonical commutation relation for *unrenormalized* field operators

$$\langle [\dot{\varphi}(x, t), \varphi(x', t)] \rangle_0 = i \frac{\partial}{\partial t} \Delta'(x - x', t - t')_{t' = t} \tag{8-30}$$

We know from Section 2-1 that

$$\frac{\partial}{\partial t} \Delta(x, t; \kappa^2)_{t=0} = -\delta^3(x) \tag{8-31}$$

so that

$$\langle [\dot{\varphi}(x, t), \varphi(x', t)] \rangle_0 = -i\delta^3(x - x') \int_0^\infty d\kappa^2 \, \rho(\kappa^2) = -i\delta^3(x - x') \tag{8-32}$$

That is,

$$\int_0^\infty d\kappa^2 \, \rho(\kappa^2) = 1 \tag{8-33}$$

We next introduce the commutator between *renormalized* operators

$$\langle [\varphi(x)_r, \varphi(x')_r] \rangle_0 = i\Delta'(x - x')_r \tag{8-34}$$

where the renormalized field operators are related to the unrenormalized ones by a scale transformation, given by the constants $Z_3^{1/2}$ for the electron field and $Z_3^{1/2}$ for the electromagnetic field. Here we use $Z_3^{1/2}$ for the scalar field, namely,

$$\varphi(x) = Z_3^{1/2} \varphi(x)_r \tag{8-35}$$

and the renormalization condition is

$$\langle 0 | \varphi(x)_r | p \rangle = (2p_0 V)^{-1/2} e^{ipx} \tag{8-36}$$

This means that the discrete part is normalized in the same way as for the free field, that is,

$$\rho_r(\varkappa^2) = \delta(\varkappa^2 - m^2) + \sigma_r(\varkappa^2) \tag{8-37}$$

On the other hand the unrenormalized spectral function $\rho(\varkappa^2) = Z_3 \rho_r(\varkappa^2)$ is given by

$$\rho(\varkappa^2) = c\ \delta(\varkappa^2 - m^2) + \sigma(\varkappa^2) \tag{8-38}$$

and

$$c + \int \sigma(\varkappa^2)\, d\varkappa^2 = 1 \tag{8-39}$$

so that

$$c = Z_3 \qquad \sigma(\varkappa^2) = Z_3\ \sigma_r(\varkappa^2) \tag{8-40}$$

Hence

$$Z_3^{-1} = 1 + \int \sigma_r(\varkappa^2)\, d\varkappa^2 \geq 1 \tag{8-41}$$

since σ_r is positive-definite. Therefore,

$$1 \geq Z_3 \geq 0 \tag{8-42}$$

In the following discussion we use only the renormalized functions and drop the subscript r.

We now shall find an expression for the self-energy in terms of the spectral function σ. For this purpose it is convenient to give a model Lagrangian although the final result appears to be valid independent of of the choice of a model. The model to be used is a mixture of charged and neutral scalar fields:

$$\mathcal{L} = -\left(\frac{\partial \Phi_0{}^\dagger}{\partial x_\lambda} \cdot \frac{\partial \Phi_0}{\partial x_\lambda} + M_0{}^2\ \Phi_0{}^\dagger \Phi_0\right) - \frac{1}{2}\left(\left(\frac{\partial \varphi_0}{\partial x_\lambda}\right)^2 + m_0^2 \varphi_0{}^2\right)$$
$$- g_0\ \Phi_0^\dagger \Phi_0 \varphi_0 \tag{8-43}$$

This time the subscript 0 means *unrenormalized*. These quantities can be expressed in terms of renormalized ones:

$$\Phi_0 = Z_2^{1/2}\,\Phi \qquad \varphi_0 = Z_3^{1/2}\,\varphi \qquad g_0 = Z_1 Z_2^{-1} Z_3^{-1/2}\, g$$

$$M^2 = M_0^2 + \delta M^2 \qquad m^2 = m_0^2 + \delta m^2 \tag{8-44}$$

The field equations for renormalized field operators are

$$(\Box - m^2)\,\varphi = Z_1 Z_3^{-1} g\,\Phi^\dagger\,\Phi - \delta m^2 \varphi$$

$$(\Box - M^2)\,\Phi = Z_1 Z_2^{-1} g\,\Phi\varphi - \delta M^2 \Phi \tag{8-45}$$

Now

$$(\Box_x - m_0^2)\,\langle[\,\varphi(x),\ \varphi(x')\,]\rangle_0 = Z_1 Z_3^{-1} g\,\langle[\,\Phi^\dagger(x)\,\Phi(x),\,\varphi(x')\,]\,\rangle_0 \tag{8-46}$$

Let us differentiate this equation with respect to x_0' and then set $x_0' = x_0$. For both renormalized and unrenormalized operators,

$$\langle[\,\Phi^\dagger(x)\ \Phi(x),\ \dot{\varphi}(x')\,]\rangle_0 = 0 \qquad \text{for } x_0' = x_0 \tag{8-47}$$

Therefore, since

$$(\Box_x - m_0^2)\,\langle[\,\varphi(x)\,,\ \varphi(x')\,]\rangle_0 = i\int d\varkappa^2 \rho(\varkappa^2)\,(\varkappa^2 - m_0^2)\,\Delta(x - x',\,\varkappa^2) \tag{8-48}$$

and

$$\frac{\partial}{\partial x_0'}\ \Delta(x - x',\,\varkappa^2)_{x_0' = x_0} = \delta^3(\mathbf{x} - \mathbf{x}') \tag{8-49}$$

we have

$$0 = (\Box_x - m_0^2)\ \langle[\,\varphi(x),\ \dot{\varphi}(x')\,]\rangle_{0,\ x_0' = x_0}$$

$$= i\,\delta^3(\mathbf{x} - \mathbf{x}')\ \int d\varkappa^2\ (\varkappa^2 - m_0^2)\ \rho(\varkappa^2)$$

or

$$\int \varkappa^2 \rho(\varkappa^2)\, d\varkappa^2 = m_0^2 \int \rho(\varkappa^2)\, d\varkappa^2$$

$$= (m^2 - \delta m^2) \int \rho(\varkappa^2)\, d\varkappa^2 \tag{8-50}$$

so that

$$\delta m^2 = -\frac{\int (\varkappa^2 - m^2)\, \rho(\varkappa^2)\, d\varkappa^2}{\int \rho(\varkappa^2)\, d\varkappa^2}$$

$$= -Z_3 \int (\varkappa^2 - m^2)\, \rho(\varkappa^2)\, d\varkappa^2$$

$$= -Z_3 \int (\varkappa^2 - m^2)\, \sigma(\varkappa^2)\, d\varkappa^2 \leq 0 \tag{8-51}$$

Therefore, the self-energy δm^2 is always negative for a boson provided that the particle under consideration is stable.

We can generalize the above method to spinor fields; but we shall not discuss that problem here.

8-2 CALCULATION OF THE PROPAGATION FUNCTIONS

Various expressions can be computed by using either the Feynman-Dyson theory or the dispersion theory. In the previous section we found the dispersion relation for the propagator, so now we can compute the lowest-order correction to the propagator by using both methods, and compare the results. Again we shall employ the model introduced in the previous section.

First, we use the Feynman-Dyson theory. The propagators for the φ and Φ fields are denoted by D_F and Δ_F, respectively. The interaction Hamiltonian is

$$\mathcal{H}_{int} = g\, \Phi^\dagger \Phi \varphi \tag{8-52}$$

The propagator D'_F of the φ field in the second-order approximation is

$$D'_F(x_1 - x_2)_{unr} = D_F(x_1 - x_2) + (-i)^2 \int d^4y_1\, d^4y_2\, g^2 D_F(x_1 - y_1)$$
$$\times \Delta_F(y_1 - y_2)\, \Delta_F(y_2 - y_1)\, D_F(y_2 - x_2) + \cdots \tag{8-53}$$

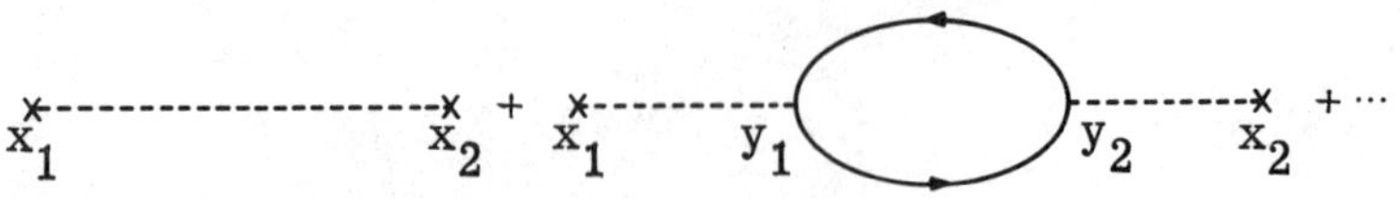

Figure 8-2. Diagrams corresponding to the second-order propagation function.

corresponding to the diagrams in Fig. 8-2, while for the Φ field the propagator is

$$\Delta'_F(x_1 - x_2)_{unr} = \Delta_F(x_1 - x_2) + (-i)^2 \int d^4y_1\, d^4y_2\, g^2\, \Delta_F(x_1 - y_1)$$
$$\times\, \Delta_F(y_1 - y_2)\, D_F(y_1 - y_2)\, \Delta_F(y_2 - x_2) + \cdots \tag{8-54}$$

corresponding to the diagrams in Fig. 8-3. In this section we shall evaluate Δ'_F first by using the Feynman-Dyson theory. We define the Fourier transform of Δ'_F, as in the previous section, as

$$\Delta'_F(x) = \frac{-i}{(2\pi)^4} \int d^4p\, e^{ipx}\, \Delta'_F(p) \tag{8-55}$$

then

$$\Delta'_F(p)_{unr} = \frac{1}{p^2 + M^2 - i\epsilon} + \frac{-i}{(2\pi)^4} g^2 \frac{1}{(p^2 + M^2 - i\epsilon)^2}$$
$$\times \int d^4q\, \frac{1}{[(p-q)^2 + M^2 - i\epsilon](q^2 + m^2 - i\epsilon)} \tag{8-56}$$

The integral corresponds to the self-energy part; let us set

$$I = -i \int \frac{d^4q}{[(p-q)^2 + M^2 - i\epsilon]\,(q^2 + m^2 - i\epsilon)}$$
$$= A + B(p^2 + M^2) + s(p^2)\,(p^2 + M^2)^2 \tag{8-57}$$

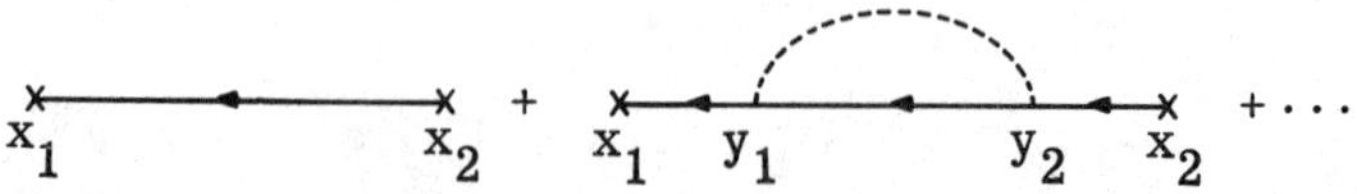

Figure 8-3. Diagrams corresponding to the second-order propagation function.

Then the first term A represents the self-energy of the Φ quantum apart from a trivial factor, and is supposed to be canceled by the counter term expressed diagrammatically in Fig. 8-4.

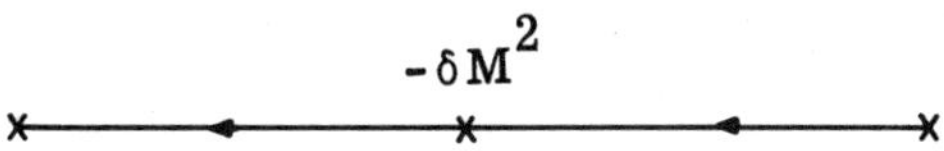

Figure 8-4. Contribution of the counter term introduced to cancel the self-energy.

The B term changes the normalization of the first term, namely,

$$\Delta'_F(p)_{unr} = \frac{1}{p^2 + M^2 - i\epsilon} + \frac{g^2}{(2\pi)^4}\left[\frac{A}{(p^2 + M^2 - i\epsilon)^2} + \frac{B}{p^2 + M^2 - i\epsilon} + s(p^2)\right]$$

$$= \frac{1}{p^2 + M^2 - i\epsilon}\left(1 + \frac{g^2}{(2\pi)^4} B\right) + \cdots \qquad (8\text{-}58)$$

From the definition of the wave-function renormalization constant,

$$Z_2 = 1 + \frac{g^2}{(2\pi)^4} B \qquad (8\text{-}59)$$

Therefore, we can drop the B term by means of wave-function renormalization. Thus, after mass- and wave-function renormalizations we obtain

$$\Delta'_F(p)_{ren} = \frac{1}{p^2 + M^2 - i\epsilon} + \frac{g^2}{(2\pi)^4} s(p^2) \qquad (8\text{-}60)$$

Therefore, we need to calculate only $s(p^2)$.

In order to carry out the integral I we use Feynaman's method:

$$I = -i \int d^4q \int_0^1 dx \, [q^2 - 2xqp + xp^2 + xM^2 + (1 - x)m^2 - i\epsilon]^{-2}$$

$$= -i \int_0^1 dx \int d^4q' \, [q'^2 + x(1 - x)p^2 + xM^2 + (1 - x)m^2 - i\epsilon]^{-2}$$

where $q' = q - xp$. We differentiate and integrate the integrand with respect to M^2 so that we can carry out the q' integration.

$$I = -i\int_0^1 dx \int d^4q' \int_{M^2}^{\infty} dM'^2\, 2x[q'^2 + x(1-x)p^2 + xM'^2 + (1-x)m^2 - i\epsilon]^{-3}$$

$$= \pi^2 \int_0^1 x\,dx \int_{M^2}^{\infty} dM'^2\, [x(1-x)p^2 + xM'^2 + (1-x)m^2 - i\epsilon]^{-1}$$

$$= \pi^2 \int_0^{1-} \frac{dx}{1-x} \int_{M^2}^{\infty} dM'^2 \left[p^2 + M^2 + \left(\frac{M'^2}{1-x} + \frac{m^2}{x} - M^2\right) - i\epsilon\right]^{-1} \tag{8-61}$$

In order to decompose I into the desired form we use

$$\frac{1}{a+b} = \frac{1}{b} - \frac{a}{b^2} + \frac{a^2}{b^2(a+b)} \tag{8-62}$$

partly to expand the integrand in the parameter $a = p^2 + M^2$ using

$$b = \frac{M'^2}{1-x} - M^2 + \frac{m^2}{x} > 0 \tag{8-63}$$

Then

$$A = \pi^2 \int_0^1 \frac{dx}{1-x} \int_{M^2}^{\infty} dM'^2 \left(\frac{M'^2}{1-x} + \frac{m^2}{x} - M^2\right)^{-1} \tag{8-64}$$

$$B = -\pi^2 \int_0^1 \frac{dx}{1-x} \int_{M^2}^{\infty} dM'^2 \left(\frac{M'^2}{1-x} + \frac{m^2}{x} - M^2\right)^{-2} \tag{8-65}$$

and

$$s(p^2) = \pi^2 \int_0^1 \frac{dx}{1-x} \int_{M^2}^{\infty} dM'^2 \left(\frac{M'^2}{1-x} + \frac{m^2}{x} - M^2\right)^{-2} \times \left(p^2 + \frac{M'^2}{1-x} + \frac{m^2}{x} - i\epsilon\right)^{-1} \tag{8-66}$$

Notice that both A and B are real so that these terms contribute only to the real or dispersive part of Δ'_F. This is an important observation: *The renormalization procedure modifies only the dispersive part of the propagation function, leaving its absorptive part unchanged.* It also should be mentioned that $i\epsilon$ in the pole terms $(p^2 + M^2 - i\epsilon)^{-1}$ and $(p^2 + M^2 - i\epsilon)^{-2}$ can be dropped when the absorptive part of Δ'_F is studied for $s = -p^2 > M^2$. Hence

$$\operatorname{Im} \Delta'_F(p)_{unr} = \operatorname{Im} \Delta'_F(p)_{ren} \qquad \text{for} \quad s = -p^2 > M^2 \tag{8-67}$$

Into Eq. (8-66) we introduce a transformation of the variables of integration

$$\varkappa^2 = \frac{M'^2}{1-x} + \frac{m^2}{x} \tag{8-68}$$

and note $\varkappa^2 \geq (M+m)^2$ for $1 \geq x \geq 0$, and $M' \geq M$. We insert

$$\int_{(M+m)^2}^{\infty} d\varkappa^2\, \delta\left[\varkappa^2 - \left(\frac{M'^2}{1-x} + \frac{m^2}{x}\right)\right] \tag{8-69}$$

into the Eq. (8-66) to get

$$s(p^2) = \pi^2 \int_{(M+m)^2}^{\infty} d\varkappa^2 \frac{\tau(\varkappa^2)}{p^2 + \varkappa^2 - i\epsilon} \tag{8-70}$$

where

$$\tau(\varkappa^2) = \frac{1}{(\varkappa^2 - M^2)^2} \int_0^1 \frac{dx}{1-x} \int_{M^2}^{\infty} dM'^2\, \delta\left(\varkappa^2 - \frac{M'^2}{1-x} - \frac{m^2}{x}\right)$$

$$= \frac{1}{(\varkappa^2 - M^2)^2} \frac{[\varkappa^2 - (M-m)^2]^{1/2}[\varkappa^2 - (M+m)^2]^{1/2}}{\varkappa^2}$$

$$\times\, \theta\,[\varkappa^2 - (M+m)^2] \tag{8-71}$$

Hence the renormalized propagator is given by

$$\Delta'_F(p)_{ren} = \frac{1}{p^2 + M^2 - i\epsilon} + \frac{g^2}{(4\pi)^2} \int_{(M+m)^2}^{\infty} \frac{d\varkappa^2\, \tau(\varkappa^2)}{p^2 + \varkappa^2 - i\epsilon}$$

$$= \frac{1}{p^2 + M^2 - i\epsilon} + \int_{(M+m)^2}^{\infty} d\varkappa^2 \frac{\sigma(\varkappa^2)}{p^2 + \varkappa^2 - i\epsilon} \qquad (8\text{-}72)$$

with

$$\sigma(\varkappa^2) = \frac{g^2}{(4\pi)^2}\tau(\varkappa^2)$$

$$= \left(\frac{g}{4\pi}\right)^2 \frac{1}{(\varkappa^2 - M^2)^2} \frac{[\varkappa^2 - (M-m)^2]^{1/2}[\varkappa^2 - (M+m)^2]^{1/2}}{\varkappa^2}$$

$$\times\ \theta[\varkappa^2 - (M+m)^2] \qquad (8\text{-}73)$$

We have verified the Lehmann representation for the renormalized propagation function by a direct calculation. The lower limit $(M + m)^2$ reflects the fact that the least massive state above M is (M + m), where the continuous spectrum starts.

Next we shall compute the propagator on the basis of dispersion theory; the point is that if we know

$$\text{Im}\ \Delta'_F(s) = \pi\, \sigma(s)$$

we can compute $\Delta'_F(s)$ by using the Lehmann representation or the dispersion relation. The *renormalized* propagator can be obtained without recourse to the renormalization procedure, since renormalization does not affect Im $\Delta'_F(s)$.

We compute $\sigma(s)$ or $\rho(s)$ from its definition:

$$\rho(-k^2) = (2\pi)^3 \sum_{P_\alpha = \text{fixed } k} |\langle 0|\Phi(0)|\alpha\rangle|^2$$

$$= (2\pi)^3 \sum_{\alpha} |\langle 0|\Phi(0)|\alpha\rangle|^2\, \delta^4(P_\alpha - k) \qquad (8\text{-}74)$$

First, let us carry out the summation over single-particle states.

$$\rho^{(0)}(-k^2) = (2\pi)^3 \frac{V}{(2\pi)^3} \int d^3p\,(2p_0 V)^{-1}\,\delta^4(p-k)$$

$$= \int d^4p\,\delta(p^2 + M^2)\,\delta^4(p-k)$$

$$= \delta(k^2 + M^2) \tag{8-75}$$

Hence the contribution of the single-particle state is given, as expected, by

$$\rho^{(0)}(\varkappa^2) = \delta(\varkappa^2 - M^2) \tag{8-76}$$

As long as the field operators are renormalized, the discrete part of the spectrum remains unchanged even when interactions are present. The contribution of two-particle states to ρ, or to σ in this case, is

$$\sigma(-k^2) = (2\pi)^3 \sum_{p,q} \left|\langle 0|\Phi(0)|p,\,q\rangle\right|^2 \delta^4(p+q-k) \tag{8-77}$$

where p and q are the energy-momenta of the Φ and φ quanta, respectively. This matrix element (8-77) must be calculated by use of the Feynman-Dyson theory, but as far as the lowest order matrix element is concerned there is no need for renormalization.

$$\langle 0|\Phi(0)|p,\,q\rangle$$

$$\approx \langle 0^{in}|T[U(\infty,-\infty)\Phi^{in}(0)]|p,\,q\ in\rangle$$

$$\approx -i\ \langle 0^{in}|T[\Phi^{in}(0),\ g\int d^4x\,\Phi^{\dagger in}(x)\,\Phi^{in}(x)\,\varphi^{in}(x)]|p,\,q\ in\rangle$$

$$= -ig \int d^4x\ \Delta_F(0-x)\ \langle 0|\Phi(x)|p\rangle\langle 0|\varphi(x)|q\rangle$$

$$= -g\frac{1}{(p+q)^2 + M^2}\quad (2p_0V)^{-1/2}\ (2q_0V)^{-1/2} \tag{8-78}$$

Hence

$$\sigma(-k^2) = (2\pi)^3\left(\frac{V}{(2\pi)^3}\right)^2 \int d^3p\,d^3q\;\delta^4(p + q - k)\left(\frac{g}{(p + q)^2 + M^2}\right)^2$$

$$\times\;(2p_0V)^{-1}\;(2q_0V)^{-1}$$

$$= \frac{1}{(2\pi)^3}\left(\frac{g}{k^2 + M^2}\right)^2 \int\frac{d^3p}{2p_0}\int\frac{d^3q}{2q_0}\;\delta^4(p + q - k) \qquad (8\text{-}79)$$

The integral of (8-79) is already familiar from Sections 5-3 and 5-4; we evaluate it in the center-of-mass system, namely $\mathbf{k} = 0$, $k_0 = \varkappa$, to get

$$\sigma(\varkappa^2) = \frac{1}{(2\pi)^3}\left(\frac{g}{\varkappa^2 - M^2}\right)^2 \pi\frac{q}{\varkappa} \qquad (8\text{-}80)$$

where q is the magnitude of the relative momentum given by

$$\left(M^2 + q^2\right)^{1/2} + \left(m^2 + q^2\right)^{1/2} = \varkappa \qquad (8\text{-}81)$$

or

$$q^2 = \frac{[\varkappa^2 - (M + m)^2]\,[\varkappa^2 - (M - m)^2]}{4\varkappa^2} \qquad (8\text{-}82)$$

Therefore,

$$\sigma(\varkappa^2) = \left(\frac{g}{4\pi}\right)^2\frac{1}{(\varkappa^2 - M^2)^2}\,\frac{[\varkappa^2 - (M + m)^2]^{1/2}\,[\varkappa^2 - (M - m)^2]^{1/2}}{\varkappa^2}$$

$$\times\;\theta[\varkappa^2 - (M + m)^2] \qquad (8\text{-}83)$$

in agreement with the result of the Feynman-Dyson theory. The renormalized Δ'_F is then given by

$$\Delta'_F(s)_{\text{ren}} = \frac{1}{M^2 - s - i\epsilon} + \int_{(M + m)^2}^{\infty} d\varkappa^2\,\frac{\sigma(\varkappa^2)}{\varkappa^2 - s - i\epsilon} \qquad (8\text{-}84)$$

As we have seen Equation (8-84) is already renormalized and no divergence occurs in the course of calculation.

In general the calculation of an amplitude in dispersion theory is divided into two steps: (1) calculation of the absorptive part, and (2) calculation of the dispersive part from the absorptive part by means of a dispersion relation. The process of renormalization corresponds to the proper choice of dispersion relations.

The first step is carried out with the help of the unitarity condition, which is closely related to the asymptotic condition first formulated by Lehmann, Symanzik, and Zimmermann.[5]

8-3 THE ASYMPTOTIC CONDITION

Field theory as it is formulated in the Feynman-Dyson approach is not entirely self-consistent. In order to remedy this situation we need a new approach based on the so-called LSZ asymptotic condition.

In Sections 7-6, 7-7, 7-8, we derived a general formula to express S-matrix elements in terms of Green's functions, that is, for a neutral scalar field

$$\langle p_1, p_2 \cdots |S| q_1, q_2, \cdots \rangle$$

$$= \int \prod_{\alpha} dx_{\alpha} \langle p_{\alpha}|\varphi(x_{\alpha})|0\rangle \; (-i)K_{x_{\alpha}} \prod_{\beta} dy_{\beta} \langle 0|\varphi(y_{\beta})|q_{\beta}\rangle \; (-i)K_{y_{\beta}}$$

$$\times \tau(x_1, \cdots, y_1, \cdots) \tag{8-85}$$

In deriving this formula we have made the rather unrealistic assumption that the interaction completely disappears at $t = \pm\infty$, including the interaction with the self-field. However, since the mass should not change, the interaction Hamiltonian must be chosen so as to cancel the mass shift.

The so-called asymptotic condition introduced by Lehmann, Symanzik, and Zimmermann is an improved version of the hypothesis of the adiabatic switching of interactions. Very crudely speaking it requires that the field operator $\varphi(x)$ behave like a free field at $t = \pm\infty$.

$$\begin{aligned} \varphi(x) &\to \varphi^{in}(x) \qquad \text{for } t \to -\infty \\ &\to \varphi^{out}(x) \qquad \text{for } t \to \infty \end{aligned} \tag{8-86}$$

This is an oversimplified form of the asymptotic condition. It should be recalled that φ^{in} and φ^{out} are the asymptotic fields which first appeared in the Yang-Feldman formalism.

Asymptotic Fields

The φ^{in} and φ^{out} are free fields satisfying the appropriate equations

$$(\Box - m^2)\,\varphi^{in}(x) = 0 \qquad [\varphi^{in}(x),\, \varphi^{in}(y)] = i\Delta(x - y) \tag{8-87}$$

and similar relations for φ^{out}

The assumption that the field operator tends asymptotically to a free-field operator is based on the idea that at $t = \pm\infty$ all the particles are separated so far away from each other that they do not feel any force exerted on them by other particles. In order to formulate this idea more precisely we have to stick to the particle picture and use localized wave packets rather than unnormalizable plane waves.

Suppose f(x) represents the wave function of a wave packet satisfying

$$(\Box - m^2)\, f(x) = 0 \tag{8-88}$$

and a normalization condition

$$-i \int d^3x \left\{ f\frac{\partial f^*}{\partial x_0} - f^* \frac{\partial f}{\partial x_0} \right\} = 1 \tag{8-89}$$

This may appear to be a strange normalization, but in fact it is just the equivalent of the normalization of a boson wave function based on the conserved current discussed in Section 7-8. In the case of a neutral field one has to choose the *energy momentum* tensor for the conserved current.

We define

$$\varphi_f(t) = i \int_{x_0 = t} \left\{ \varphi(x)\frac{\partial f(x)}{\partial x_0} - f(x)\frac{\partial \varphi(x)}{\partial x_0} \right\} d^3x \tag{8-90}$$

and similar operators φ_f^{in} and φ_f^{out} in terms of the asymptotic fields. The $\varphi_f(t)$ is an operator which creates or destroys a wave packet.

The LSZ asymptotic condition is then given by

$$\lim_{\tau \to -\infty} (\Phi,\, \varphi_f(\tau)\Psi) = (\Phi,\, \varphi_f^{in}(t)\Psi)$$

$$\lim_{\tau \to +\infty} (\Phi,\, \varphi_f(\tau)\Psi) = (\Phi,\, \varphi_f^{out}(t)\Psi) \tag{8-91}$$

It should be noted that the right-hand sides actually are independent of t, so that we can drop t from φ_f^{in} and φ_f^{out}.

We further assume that the wave functions of wave packets are chosen to satisfy the orthogonality relation

$$-i\int\left\{f_\alpha\frac{\partial f_\beta^*}{\partial x_0}-f_\beta^*\frac{\partial f_\alpha}{\partial x_0}\right\}d^3x=\delta_{\alpha\beta}\tag{8-92}$$

as well as the completeness condition

$$\sum_\alpha f_\alpha(x)\,f_\alpha^*(x')=i\Delta^{(+)}(x-x')\tag{8-93}$$

Then we can construct two complete sets of state vectors by applying the operators φ^{in} and φ^{out} successively to the vacuum state:

$$\left\{\Phi^{(+)}\right\}\equiv\left\{\Phi^{in}\right\}\tag{8-94}$$

$$\Phi_0$$

$$\Phi_\alpha^{in}=\varphi_\alpha^{in}\,\Phi_0$$

$$\vdots$$

$$\Phi^{in}_{\alpha_1\cdots\alpha_k}=\left(p_{\alpha_1\ldots\alpha_k}\right)^{-1/2}\varphi^{in}_{\alpha}\cdots\varphi^{in}_{\alpha_k}\Phi_0$$

etc.

$$\left\{\Phi^{(-)}\right\}\equiv\left\{\Phi^{out}\right\}\tag{8-95}$$

$$\Phi_0$$

$$\Phi_\alpha^{out}=\varphi_\alpha^{out}\,\Phi_0$$

$$\vdots$$

$$\Phi^{out}_{\alpha_1\cdots\alpha_k}=\left(p_{\alpha_1\ldots\alpha_k}\right)^{-1/2}\varphi^{out}_{\alpha_1}\cdots\varphi^{out}_{\alpha_k}\Phi_0$$

where $p_{\alpha_1 \cdots \alpha_k} = n_1!\, n_2! \cdots n_r!$, with each n representing the number of identical single particle states, is introduced on the basis of the discussion in Section 2-6. We are using φ_α^{in} and φ_α^{out} in the sense of creation operators; their Hermitian conjugates $\varphi_\alpha^{in\,\dagger}$ and $\varphi_\alpha^{out\,\dagger}$ are therefore destruction operators.

The S matrix elements are

$$S_{\beta\alpha} = (\Phi_\beta^{(-)}, \Phi_\alpha^{(+)}) = (\Phi_\beta^{out}, \Phi_\alpha^{in}) \tag{8-96}$$

and φ^{out} is related to φ^{in} by

$$\varphi^{out} = S^{-1}\ \varphi^{in}\, S \tag{8-97}$$

Renormalization Condition

Whenever Φ_α^{in} or Φ_α^{out} represents a one-particle state, then

$$(\Box - m^2)\,(\Phi_0, \varphi(x)\Phi_\alpha^{in}) = (\Box - m^2)\,(\Phi_0, \varphi(x)\Phi_\alpha^{out}) = 0 \tag{8-98}$$

Thus there remains a problem of how to normalize these matrix elements. Incidentally, for single-particle states we have

$$\Phi_\alpha^{in} = \Phi_\alpha^{out} \tag{8-99}$$

and we may choose the normalization

$$(\Phi_0, \varphi(x)\Phi_\alpha) = (\Phi_0, \varphi^{in}(x)\Phi_\alpha) = f_\alpha(x) \tag{8-100}$$

which is exactly the condition that a renormalized operator $\varphi(x)$ should satisfy. Therefore, with this normalization, only renormalized field operators occur in this theory.

8-4 THE LSZ REDUCTION FORMULA

In the preceding section we formulated the asymptotic condition. In order to use it in a practical calculation the so-called LSZ reduction formula is a useful tool.

For convenience we use the notations

$$T(x_1 \cdots x_n) = T[\varphi(x_1) \cdots \varphi(x_n)] \tag{8-101}$$

$$\tau(x_1 \cdots x_n) = (\Phi_0,\ T[\varphi(x_1) \cdots \varphi(x_n)]\,\Phi_0) \tag{8-102}$$

$$K_y = \Box_y - m^2 \qquad f\overleftrightarrow{\frac{\partial}{\partial x}}g = f\,\frac{\partial g}{\partial x} - \frac{\partial f}{\partial x}\,g \tag{8-103}$$

We first shall prove

$$(\Phi_0,\ T(x_1 \cdots x_n)\Phi_\alpha^{\text{in}}) = -i \int d^4y\, f_\alpha(y)\, K_y\, \tau(x_1 \cdots x_n y) \tag{8-104}$$

The left-hand side is

$$(\Phi_0,\ T(x_1 \cdots x_n)\Phi_\alpha^{\text{in}}) = \lim_{y_0 \to -\infty} (\Phi_0,\ T(x_1 \cdots x_n)\,\varphi_\alpha(y_0)\Phi_0)$$

$$= i \lim_{y_0 \to -\infty} \int d^3y\ (\Phi_0,\ T(x_1 \cdots x_n y)\,\Phi_0) \times \overleftrightarrow{\frac{\partial}{\partial y_0}}\ f_\alpha(y)$$

$$= i \lim_{y_0 \to -\infty} \int d^3y\ \tau(x_1 \cdots x_n y)\ \overleftrightarrow{\frac{\partial}{\partial y_0}}\ f_\alpha(y) \tag{8-105}$$

Next we shall show that as y_0 approaches $+\infty$, (8-105) vanishes, namely,

$$i \lim_{y_0 \to +\infty} \int d^3y\ \tau(x_1 \cdots x_n y)\ \overleftrightarrow{\frac{\partial}{\partial y_0}}\ f_\alpha(y) = 0 \tag{8-106}$$

Using a procedure similar to that of Eq. (8-105) we can show that this is equal to

$$(\Phi_0, \varphi_\alpha^{out} T(x_1 \cdots x_n)\Phi_0) = (\varphi_\alpha^{out\dagger} \Phi_0, T(x_1 \cdots x_n)\Phi_0) = 0 \qquad (8\text{-}107)$$

since $\varphi_\alpha^{out\dagger}$ is a destruction operator. Hence

$$(\Phi_0, T(x_1 \cdots x_n)\Phi_\alpha^{in})$$

$$= i\left(\lim_{y_0 \to -\infty} - \lim_{y_0 \to +\infty}\right)\int d^3y\, \tau(x_1 \cdots x_n y) \frac{\overleftrightarrow{\partial}}{\partial y_0} f_\alpha(y)$$

$$= -i\int d^4y \frac{\partial}{\partial y_0}\left[\tau(x_1 \cdots x_n y) \frac{\overleftrightarrow{\partial}}{\partial y_0} f_\alpha(y)\right]$$

$$= -i\int d^4y \left[\tau(x_1 \cdots x_n y) \frac{\partial^2 f_\alpha(y)}{\partial y_0^2} - \frac{\partial^2 \tau(x_1 \cdots x_n y)}{\partial y_0^2} f_\alpha(y)\right] \qquad (8\text{-}108)$$

Now we shall use Green's theorem:

$$\int_V d^3y \left[\tau(x_1 \cdots x_n y)\, \nabla_y^2 f_\alpha(y) - \nabla_y^2 \tau(x_1 \cdots x_n y) \cdot f_\alpha(y)\right]$$

$$= \int_S dS_y \left[\tau(x_1 \cdots x_n y) \frac{\partial}{\partial n_y} f_\alpha(y) - \frac{\partial}{\partial n_y} \tau(x_1 \cdots x_n y) \cdot f_\alpha(y)\right] \to 0 \qquad \text{for} \quad V \to \infty$$

The vanishing of this integral is justified by the assumed localization of the wave packet represented by f_α, and would not be true for a plane wave. Therefore, by adding this integral to (8-108), we get

$$(\Phi_0,\ T(x_1\cdots x_n)\Phi_\alpha^{\text{in}})$$

$$= i\int d^4y\left[\tau(x_1\cdots x_n y)\,\Box_y\, f_\alpha(y) - \Box_y\,\tau(x_1\cdots x_n y)\cdot f_\alpha(y)\right]$$

$$= -i\int d^4y\, f_\alpha(y)\, K_y \tau(x_1\cdots x_n y) \tag{8-109}$$

since $K_y f_\alpha(y) = 0$. Thus the desired formula has been proved. The generalization of this formula is

$$(\Phi_0,\ T(x_1\cdots x_n)\Phi^{\text{in}}_{\alpha_1\cdots\alpha_k})$$

$$= (-i)^k\int d^4y_1\cdots d^4y_k\, f_{\alpha_1}(y_1)\cdots f_{\alpha_k}(y_k)$$

$$\times K_{y_1}\cdots K_{y_k}\,\tau(x_1\cdots x_n y_1\cdots y_k) \tag{8-110}$$

By further generalizing we get

$$(\Phi_\alpha^{\text{out}},\ T(x_1\cdots x_n)\Phi_\beta^{\text{in}}) = -i\int d^4\eta\, f_{\beta_\ell}(\eta) \times K_\eta(\Phi_\alpha^{\text{out}},\ T(x_1\cdots x_n\eta)\Phi^{\text{in}}_{\beta_1\cdots\beta_{\ell-1}})$$

$$(\Phi_\alpha^{\text{out}},\ T(x_1\cdots x_n)\Phi_\beta^{\text{in}}) = -i\int d^4\zeta\, f^*_{\alpha_k}(\zeta) \times K_\zeta(\Phi^{\text{out}}_{\alpha_1\cdots\alpha_{k-1}},\ T(x_1\cdots x_n\zeta)\Phi_\beta^{\text{in}}) \tag{8-111}$$

where $\alpha = \alpha_1\cdots\alpha_k$, $\beta = \beta_1\cdots\beta_\ell$, and we assume that there is no common index between α and β. These reduction formulas give the connection between particle states and field operators, and therefore implicitly contain the prescription for field quantization, just as is the case for the reduction formulas in the interaction representation (Section 7-6).

By repeated application of the reduction formula, the S-matrix elements can be written

$$S_{\alpha\beta} = (\Phi_\alpha^{out}, \Phi_\beta^{in})$$

$$= (-i)^{k+\ell} \int d^4\zeta_1 \cdots d^4\zeta_k\, d^4\eta_1 \cdots d^4\eta_\ell$$

$$\times f^*_{\alpha_1}(\zeta_1) \cdots f^*_{\alpha_k}(\zeta_k)\, f_{\beta_1}(\eta_1) \cdots f_{\beta_\ell}(\eta_\ell)$$

$$\times K_{\zeta_1} \cdots K_{\zeta_k}\, K_{\eta_1} \cdots K_{\eta_\ell}\, \tau(\zeta_1 \cdots \zeta_k\, \eta_1 \cdots \eta_\ell) \qquad (8\text{-}112)$$

This formula, in agreement with the formula (7-109) in Section 7-6, gives the matrix elements of S when none of the single-particle states in α coincides with any in β.

In the Feynman-Dyson theory we have an expression for the S matrix as an operator; we shall try to get a corresponding one in this case. We notice that φ_α^{in} and φ_α^{out} are creation operators, so

$$\varphi^{in}(x) = \sum_\alpha \left\{ \varphi_\alpha^{in} \cdot f_\alpha^*(x) + \varphi_\alpha^{in\dagger} \cdot f_\alpha(x) \right\} \qquad (8\text{-}113)$$

and a similar relation holds for $\varphi^{out}(x)$.

When we try to derive the reduction formula in operator form we easily get

$$-i \int d^4y\, f_\alpha(y)\, K_y\, T(x_1 \cdots x_n\, y)$$

$$= T(x_1 \cdots x_n)\, \varphi_\alpha^{in} - \varphi_\alpha^{out}\, T(x_1 \cdots x_n) \qquad (8\text{-}114)$$

In replacing $f_\alpha(y)$ by $f_\alpha^*(y)$ we get a formula similar to (8-114), with φ_α^{in} and φ_α^{out} replaced by their Hermitian conjugates. Combining these two formulas with (8-113), the expansion formulas for φ^{in} and φ^{out}, and (8-93), the completeness relation for $f_\alpha(x)$, we obtain

$$\int d^4y\, \Delta(y - x)\, K_y\, T(x_1 \cdots x_n\, y)$$

$$= T(x_1 \cdots x_n)\, \varphi^{in}(x) - \varphi^{out}(x)\, T(x_1 \cdots x_n) \qquad (8\text{-}115)$$

We combine this relation with $S\varphi^{\text{out}}(x) = \varphi^{\text{in}}(x)S$ to get

$$\int d^4y\, \Delta(y - x)K_y\, S\, T(x_1 \cdots x_n y)$$

$$= S\, T(x_1 \cdots x_n)\varphi^{\text{in}}(x) - S\, \varphi^{\text{out}}(x)\, T(x_1 \cdots x_n)$$

$$= S\, T(x_1 \cdots x_n)\varphi^{\text{in}}(x) - \varphi^{\text{in}}(x)\, S\, T(x_1 \cdots x_n) \qquad (8\text{-}116)$$

Therefore,

$$\int d^4y\, \Delta(x - y)K_y\, S\, T(x_1 \cdots x_n y)$$

$$= [\varphi^{\text{in}}(x), S\, T(x_1 \cdots x_n)] \qquad (8\text{-}117)$$

This is the basic step in the derivation; by iteration we find

$$\int d^4y_1 \cdots d^4y_\ell\, \Delta(z_1 - y_1) \cdots \Delta(z_\ell - y_\ell)K_{y_1} \cdots K_{y_\ell}$$

$$\times S\, T(x_1 \cdots x_n y_1 \cdots y_\ell)$$

$$= [\varphi^{\text{in}}(z_1), [\varphi^{\text{in}}(z_2), [\cdots \varphi^{\text{in}}(z_\ell), S\, T(x_1 \cdots x_n)] \cdots]$$

$$(8\text{-}118)$$

Then we take the vacuum expectation value of the equation above using $S\Phi_0 = \Phi_0$ to get

$$\int d^4y_1 \cdots d^4y_\ell\, \Delta(z_1 - y_1) \cdots \Delta(z_\ell - y_\ell)$$

$$\times K_{y_1} \cdots K_{y_\ell}\, \tau(x_1 \cdots x_n y_1 \cdots y_\ell)$$

$$= (\Phi_0, [\varphi^{\text{in}}(z_1), [\varphi^{\text{in}}(z_2), \cdots [\Phi^{\text{in}}(z_\ell),$$

$$S\, T(x_1 \cdots x_n)] \cdots]\Phi_0) \qquad (8\text{-}119)$$

If we set n = 0,

$$\int d^4y_1 \cdots d^4y_\ell\, \Delta(z_1 - y_1) \cdots \Delta(z_\ell - y_\ell)$$

$$\times K_{y_1} \cdots K_{y_\ell}\, \tau(y_1 \cdots y_\ell)$$

$$= (\Phi_0, [\varphi^{\text{in}}(z_1), [\varphi^{\text{in}}(z_2), \cdots [\varphi^{\text{in}}(z_\ell),$$

$$S] \cdots]]\Phi_0) \qquad (8\text{-}120)$$

Let us then assume the following expansion of the S matrix:

$$S = \sum_{\ell=0}^{\infty} \frac{1}{\ell!} \int d^4y_1 \cdots d^4y_\ell \, c(y_1 \cdots y_\ell) \times :\varphi^{in}(y_1) \cdots \varphi^{in}(y_\ell): \tag{8-121}$$

where c is totally symmetric in the ℓ variables. The coefficients c are determined by substituting this expansion into the above equation.

Since the only term that survives in the vacuum expectation value of the ℓ-fold commutator is the one involving the normal product of ℓ operators, we get

$$\begin{aligned}(\Phi_0, [\varphi^{in}(z_1), \cdots, [\varphi^{in}(z_\ell), S] \cdots] \Phi_0) &= i^\ell \int d^4y_1 \cdots d^4y_\ell \, \Delta(z_1 - y_1) \cdots \Delta(z_\ell - y_\ell) \, c(y_1 \cdots y_\ell) \\ &= \int d^4y_1 \cdots d^4y_\ell \, \Delta(z_1 - y_1) \cdots \Delta(z_\ell - y_\ell) \\ &\quad \times K_{y_1} \cdots K_{y_\ell} \, \tau(y_1 \cdots y_\ell)\end{aligned} \tag{8-122}$$

This relation determines the coefficient functions c on the mass shell; c must be equal to

$$c(y_1 \cdots y_\ell) = (-i)^\ell K_{y_1} \cdots K_{y_\ell} \, \tau(y_1 \cdots y_\ell) \tag{8-123}$$

Therefore, the S matrix in operator form must be given by

$$S = \sum_{\ell=0}^{\infty} \frac{(-i)^\ell}{\ell!} \int d^4y_1 \cdots d^4y_\ell \, K_{y_1} \cdots K_{y_\ell} \, \tau(y_1 \cdots y_\ell) \times :\varphi^{in}(y_1) \cdots \varphi^{in}(y_\ell): \tag{8-124}$$

By retracing this derivation we can generalize formula (8-124) to

$$S \, T(x_1 \cdots x_n) = \sum_{\ell=0}^{\infty} \frac{(-i)^\ell}{\ell!} \int d^4y_1 \cdots d^4y_\ell \, K_{y_1} \cdots K_{y_\ell} \times \tau(x_1 \cdots x_n y_1 \cdots y_\ell) :\varphi^{in}(y_1) \cdots \varphi^{in}(y_\ell): \tag{8-125}$$

These formulas form the basis of the LSZ theory.

8-5 UNITARITY

One of the most important properties of the τ functions in the LSZ formalism is that one can derive a set of equations for the τ functions which is related to unitarity.

The following expression is unitary, provided J(x) is real:

$$T \exp \left[-i \int_{-\infty}^{\infty} d^4x \, \varphi(x) \, J(x) \right] \tag{8-126}$$

The unitarity condition is expressed by

$$T \exp \left[-i \int_{-\infty}^{\infty} d^4x \, \varphi(x) \, J(x) \right] \cdot \tilde{T} \exp \left[i \int_{-\infty}^{\infty} d^4x \, \varphi(x) \, J(x) \right] = 1 \tag{8-127}$$

By differentiating this relation n times with respect to J(x) and setting J = 0 after that, we obtain

$$\sum_{\text{comb}} (-i)^k \, (i)^{n-k} \, T(x'_1 \cdots x'_k) \, \tilde{T}(x'_{k+1} \cdots x'_n) = 0 \tag{8-128}$$

where summation should be made over all possible ways of dividing the variables $x_1 \ldots x_n$ into two groups $x'_1 \ldots x'_k$ and $x'_{k+1} \ldots x'_n$. We then take the vacuum expectation value to get

$$\sum_{\text{comb}} (-i)^k \, (i)^{n-k} \, (\Phi_0, \, T(x'_1 \cdots x'_k) \, \tilde{T} \, (x'_{k+1} \cdots x'_n) \, \Phi_0) = 0 \tag{8-129}$$

Then we insert a complete set of intermediate states $\{\Phi^{in}\}$ and use the reduction formula

$$(\Phi_0, \, T(x_1 \cdots x_k) \, \Phi^{in}_{\alpha_1 \cdots \alpha_\ell})$$

$$= (-i)^\ell \int d^4u_1 \cdots d^4u_\ell \, f_{\alpha_1}(u_1) \cdots f_{\alpha_\ell}(u_\ell) \, K_{u_1} \cdots K_{u_\ell}$$

$$\times \, \tau(x_1 \cdots x_k u_1 \cdots u_\ell), \tag{8-130}$$

and the conjugate formula

$$(\Phi^{\text{in}}_{\alpha_1\cdots\alpha_\ell}, \tilde{T}(x_{k+1}\cdots x_n)\Phi_0)$$

$$= i^\ell \int d^4v_1 \cdots d^4v_\ell \, f^*_{\alpha_1}(v_1)\cdots f^*_{\alpha_\ell}(v_\ell)\, K_{v_1}\cdots K_{v_\ell}$$

$$\times \tau^*(x_{k+1}\cdots x_n\, v_1\cdots v_\ell) \tag{8-131}$$

The summation over intermediate states can be carried out by use of

$$\sum_\alpha f_\alpha(u)\, f^*_\alpha(v) = i\,\Delta^{(+)}(u - v) \tag{8-132}$$

If we define

$$\bar{\tau}(x_1\cdots x_n) = (-i)^n\, K_{x_1}\cdots K_{x_n}\, \tau(x_1\cdots x_n) \tag{8-133}$$

then the original equation can be written, after applying the operator $K_{x_1}\cdots K_{x_n}$ on it, in the form

$$\bar{\tau}(x_1\cdots x_n) + \bar{\tau}^*(x_1\cdots x_n)$$

$$+ \sum_{\text{comb}}{}' \sum_{\ell=0}^{\infty} \frac{i^\ell}{\ell!} \int d^4u_1\cdots d^4u_\ell\, d^4v_1\cdots d^4v_\ell$$

$$\times \bar{\tau}(x'_1\cdots x'_k\;\; u_1\cdots u_\ell)$$

$$\times \Delta^{(+)}(u_1 - v_1)\cdots\Delta^{(+)}(u_\ell - v_\ell)\, \bar{\tau}^*(x'_{k+1}\cdots x'_n\, v_1\cdots v_\ell) = 0$$

$$\tag{8-134}$$

The factorial $\ell!$ is necessary in order to avoid counting of the same states multiply. The $\sum'$ implies omission of $k = 0$ and $k = n$ in the summation, so that $k > 0$ and $n - k > 0$.

When the particles are on the mass shell (8-134) reduces to the ordinary unitarity condition, as we shall see later; but this relation holds also for particles off the mass shell. For this reason it is called the generalized unitarity condition.

By making use of the generalized unitarity condition we can prove the unitarity of the S matrix.

$$S = 1 + \sum_{\ell=1}^{\infty} \frac{1}{\ell !} \int d^4x_1 \cdots d^4x_\ell \; \tau(x_1 \cdots x_\ell) :\overset{in}{\varphi}(x_1) \cdots \overset{in}{\varphi}(x_\ell): \tag{8-135}$$

and

$$S^\dagger = 1 + \sum_{m=1}^{\infty} \frac{1}{m!} \int d^4y_1 \cdots d^4y_m \; \tau^*(y_1 \cdots y_m) \times :\overset{in}{\varphi}(y_1) \cdots \overset{in}{\varphi}(y_m): \tag{8-136}$$

We take the product $SS^\dagger$ and expand it to normal products, using

$$\begin{aligned} &:\overset{in}{\varphi}(x_1) \cdots \overset{in}{\varphi}(x_\ell): \; :\overset{in}{\varphi}(y_1) \cdots \overset{in}{\varphi}(y_m): \\ &= :\overset{in}{\varphi}(x_1) \cdots \overset{in}{\varphi}(x_\ell) \; \overset{in}{\varphi}(y_1) \cdots \overset{in}{\varphi}(y_m): \\ &+ \sum_{\text{comb}} i\Delta^{(+)}(x_1' - y_1') \; :\overset{in}{\varphi}(x_2') \cdots \overset{in}{\varphi}(x_\ell') \; \overset{in}{\varphi}(y_2') \cdots \overset{in}{\varphi}(y_m'): \\ &+ \sum_{\text{comb}} i\Delta^{(+)}(x_1' - y_1') \, i\Delta^{(+)}(x_2' - y_2') \\ &\times :\overset{in}{\varphi}(x_3') \cdots \overset{in}{\varphi}(x_\ell') \; \overset{in}{\varphi}(y_3') \cdots \overset{in}{\varphi}(y_m'): \\ &+ \cdots \end{aligned} \tag{8-137}$$

Then it can be shown that the coefficient of the normal product in each order reduces to the form of the generalized unitarity condition times

$$\int d^4x_1 \cdots d^4x_n \; :\overset{in}{\varphi}(x_1) \cdots \overset{in}{\varphi}(x_n): \tag{8-138}$$

and hence the product vanishes except for the first term, yielding

$$SS^\dagger = 1 \tag{8-139}$$

Similarly one can derive

$$S^\dagger S = 1 \tag{8-140}$$

8-6 ARBITRARINESS IN THE CHOICE OF FIELD OPERATORS

The formula (7-109) in Section 7-6 was derived originally from the reduction formula in the interaction representation and the hypothesis of adiabatic switching, but as we have seen in Section 8-4 the same formula follows from the LSZ reduction formalism. The same correspondence is true for other matrix elements.

This formula is already very general in that one can write it even if one does not know the explicit form of the interactions. It still reminds us, however, of the origin of the Heisenberg operator φ since it is the field operator that appears in the original Lagrangian, whether or not we know its explicit form. Therefore, let us ask what the necessary and sufficient conditions are for the proper choice of the field operators to make the formula valid. Certainly, one cannot help suspecting that φ and $\varphi^\dagger$ must be the operators present in the original Lagrangian. This is not true, however, as we shall see below.

Since it is instructive to list a set of postulates relevant to formula (70-109) we shall do so here. We require the S matrix to be Lorentz invariant, or more precisely, the vacuum expectation value of the T product of field operators in the integrand to be Lorentz invariant. The independence of the T product on the choice of a frame of reference, as imposed by Lorentz invariance, requires local commutativity; two Heisenberg operators separated by a spacelike distance should commute (or anticommute).

We can conclude that the necessary conditions for formula (7-109) to be valid are (1) the correct transformation property of φ under Lorentz transformations, such as that of being scalar, pseudoscalar, and so on, (2) the locality or local commutativity of φ, and (3) the normalization condition, namely,

$$\langle 0|\varphi(x)|a\rangle = (2p_0 V)^{-1/2} e^{ipx} \tag{8-141}$$

where a is a single-particle state and p the energy-momentum of that state. The last condition is readily satisfied by multiplication of the field operator φ by an appropriate normalization constant if it originally is not properly normalized.

We also can show that these are sufficient conditions. Let $\Phi(x)$ be an operator that satisfies these conditions, then it satisfies

$$\langle 0|\Phi(x)|a\rangle = \langle 0|\varphi(x)|a\rangle \tag{8-142}$$

and

$$\langle a|\Phi^\dagger(x)|0\rangle = \langle a|\varphi^\dagger(x)|0\rangle \tag{8-143}$$

as a consequence of the normalization condition. Furthermore, we can derive the reduction formula for the new operator Φ, that is,

$$\int d^4x \langle a | \Phi^\dagger(x) | 0 \rangle (-i) K_x \langle 0 | T[\Phi(x) A(x_1) \cdots Z(x_n)] | 0 \rangle$$

$$= \langle a | T[A(x_1) \cdots Z(x_n)] | 0 \rangle \tag{8-144}$$

In order to prove this formula let us consider the Fourier transform of the vacuum expectation value of the T product.

$$f(P) = \int d^4x \, e^{-iPx} \langle 0 | T[\Phi(x) A(x_1) \cdots Z(x_n)] | 0 \rangle$$

$$= \left[\int_{T_2}^{\infty} d^4x + \int_{T_1}^{T_2} d^4x + \int_{-\infty}^{T_1} d^4x \right] e^{-iPx}$$

$$\times \langle 0 | T[\Phi(x) A(x_1) \cdots Z(x_n)] | 0 \rangle$$

$$= f_1(P) + f_2(P) + f_3(P) \tag{8-145}$$

where $T_2 = \mathrm{Max}(x_{10}, \ldots, x_{n0})$, and $T_1 = \mathrm{Min}(x_{10}, \ldots, x_{n0})$. Among these three integrals the first and last ones are important.

$$f_1(P) = \int_{T_2}^{\infty} dt \int d^3x \, e^{[-iPx + iP_0 t]} \langle 0 | \Phi(x) \, T[A(x_1) \ldots Z(x_n)] | 0 \rangle$$

$$= iV \sum_\alpha \frac{e^{[i(P_0 - P_{\alpha 0})T_2]}}{P_0 - P_{\alpha 0} + i\epsilon} \delta_{P_\alpha, P} \langle 0 | \Phi(0) | \alpha \rangle$$

$$\times \langle \alpha | T[A(x_1) \ldots Z(x_n)] | 0 \rangle \tag{8-146}$$

Since $P_{\alpha 0} > 0$, (8-146) has discrete and continuous poles along the positive P_0 axis. Similarly,

$$f_3(P) = -iV \sum_\beta \frac{e'^{[i(P_0 + P_{\beta 0})T_1]}}{P_0 + P_{\beta 0} - i\epsilon} \delta_{P_\beta, -P}$$

$$\times \langle 0 | T[A(x_1) \ldots Z(x_n)] | \beta \rangle \langle \beta | \Phi(0) | 0 \rangle \tag{8-147}$$

Since $P_{\beta 0} > 0$, this expression has discrete and continuous poles along the negative P_0 axis. That $f_1(P)$ and $f_3(P)$ have singularities in P_0 results from the fact that the time integrals diverge when the integrands have nonoscillating parts as functions of time; we may infer, therefore, that $f_2(P)$, which is defined as an integral over a finite time interval, is finite for all finite values of P_0. We can write

$$\int d^4x \langle a | \Phi^\dagger(x) | 0 \rangle (-i) K_x \langle 0 | T[\Phi(x) A(x_1) \ldots Z(x_n)] | 0 \rangle$$

$$= (2P_{a0} V)^{-1/2} \lim_{\substack{P_0 \to P_{a0} \\ P \to P_a}} i(P^2 + m_\alpha^2) f(P) \tag{8-148}$$

where $P_{a0} > 0$. Since $P_a^2 + m_a^2 = 0$, there must be a discrete pole in f(P) to cancel this factor if (8-148) is to survive. Such a pole exists in $f_1(P)$ for positive P_0 corresponding to $\alpha = a$, so that (8-148) becomes

$$i (2P_{a0}/V)^{1/2} \lim_{P \to P_\alpha} (P_{a0} - P_0) f_1(P)$$

$$= (2P_{a0} V)^{1/2} \langle 0 | \Phi(0) | a \rangle \langle a | T[A(x_1) \ldots Z(x_n)] | 0 \rangle$$

$$= \langle a | T[A(x_1) \ldots Z(x_n)] | 0 \rangle \tag{8-149}$$

Similarly, we get

$$\int d^4x \langle 0 | \Phi^\dagger(x) | \bar{a} \rangle \quad (-i) K_x \langle 0 | T[\Phi(x) A(x_1) \ldots Z(x_n)] | 0 \rangle$$

$$= \langle 0 | T[A(x_1) \ldots Z(x_n)] | \bar{a} \rangle \tag{8-150}$$

where $\bar{a}$ denotes the antiparticle of a. These results have an extremely important feature; in the reduction formalism any matrix elements of S or of time-ordered products of field operators can be obtained by application of operations of the form

$$\int d^4x \langle 0 | \varphi(x) | a \rangle (-i) K_x \tag{8-151}$$

on the vacuum expectation values of the T products of field operators. In the resulting expressions we can successively replace φ by Φ, and $\varphi^\dagger$ by $\Phi^\dagger$ using the results obtained above. Thus we can completely replace each φ and $\varphi^\dagger$ in the reduction formulas by Φ and $\Phi^\dagger$, respectively.

This leads to the conclusion that as far as reduction formulas are concerned, including the S-matrix formulas like (7-109) as special cases, the correspondence between stable particles and field operators is by no means one-to-one; in fact, we can employ an arbitrary field operator in describing a stable particle, provided the aforementioned conditions are met.

The above conclusion opens a way to introduce field operators even for the so-called composite particles; this problem is discussed briefly in the following paragraph.

Let us assume that a scalar deuteron d consists of a scalar neutron n and a scalar proton p in the sense that

$$\langle 0 | \psi_n(x) \, \psi_p(y) | d \rangle \neq 0 \tag{8-152}$$

then we can introduce an operator $\varphi_d(x)$ by

$$\varphi_d(x) = \lim_{\xi \to 0} \frac{\psi_n(x + \xi) \quad \psi_p(x - \xi)}{(2P_0V)^{1/2} \langle 0 | \psi_n(\xi) \, \psi_p(-\xi) | d \rangle} \tag{8-153}$$

where ξ approaches zero from a spacelike direction. Therefore, this operator is local and satisfies the normalization condition

$$\langle 0 | \varphi_d(x) | d \rangle = \frac{e^{iPx}}{\sqrt{2P_0V}} \tag{8-154}$$

where P denotes the energy-momentum of the deuteron. It is also important to prove that the operator $\varphi_d(x)$ does not depend on the direction of the vector P so that $\varphi_d(x)$ can have the proper transformation property. [6]

Thus we have learned that we can introduce field operators even for composite particles. This construction of the composite particle field is called the HNZ construction. [7-9]

8-7 DISPERSION RELATIONS FOR THE VERTEX FUNCTIONS

We already have studied the dispersion relation for the propagation function; probably the next simplest object is the vertex function. From the point of view of Green's functions we can classify our objects as follows:

2-point function:	propagation function
3-point function:	vertex function or form factor
4-point function:	scattering amplitude
5-point function:	production amplitude

For the last object the correct form of the dispersion relation is not yet known so our study is confined to the first three objects.

In order to study the vertex function we shall employ the simple scalar model already used in preceding sections, with the definitions

$$\tau(xy; z) = (\Phi_0, T[\Phi(x) \Phi^\dagger(y) \varphi(z)] \Phi_0) \tag{8-155}$$

$$\bar{\tau}(xy; z) = (-i)^3 (\Box_x - M^2)(\Box_y - M^2)(\Box_z - m^2) \tau(xy;z) \tag{8-156}$$

and

$$\bar{\tau}(xy; z) = \frac{-i}{(2\pi)^8} \int d^4p_1 \, d^4p_2 \, d^4p_3 \; \delta^4(p_1 + p_2 + p_3) \times e^{(ip_1x + ip_2y + ip_3z)} \mathcal{G}(p_1, p_2, p_3) \tag{8-157}$$

which is represented in the diagram in Fig. 8-5.

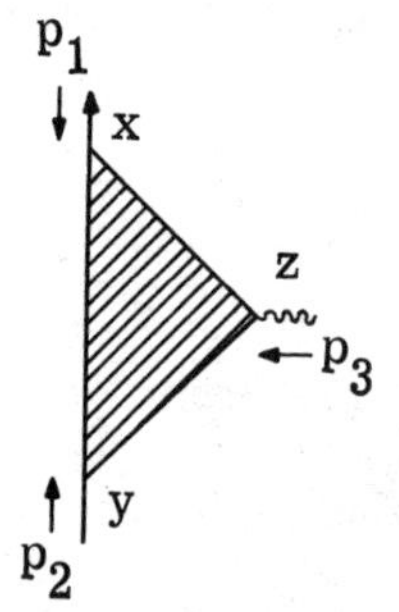

Figure 8-5. The diagram representing the three-point function.

The normalization and sign convention of $\mathcal{G}$ are so chosen as to make it equal to the renormalized coupling constant g when all three particles are on the mass shell, that is, $\mathcal{G} = g$ when $p_1^2 + M^2 = p_2^2 + M^2 = p_3^2 + m^2 = 0$. In the lowest order these functions are given by

$$\tau(xy; z) = (\Phi_0, T[\Phi^{in}(x), \Phi^{\dagger in}(y), \varphi^{in}(z), (-i) \times \int d^4w \, g \, \Phi^{\dagger in}(w) \Phi^{in}(w) \varphi^{in}(w)] \Phi_0)$$

$$= -ig \int d^4w \, \Delta_F(x - w) \Delta_F(w - y) D_F(w - z) \tag{8-158}$$

and

$$\bar{\tau}(xy; z) = -ig\ \delta^4(x - y)\ \delta^4(x - z) \tag{8-159}$$

so

$$\mathcal{G}\ (p_1, p_2, p_3) = g \tag{8-160}$$

The $\mathcal{G}$ is an invariant function of three momenta p_i, and since $\Sigma p_i = 0$ allows scalar products of different momenta to be expressed in terms of the variables p_1^2, p_2^2, and p_3^2, we can use the latter as the independent variables in $\mathcal{G}$:

$$\mathcal{G}\ (p_1\ , p_2, p_3) = F(-p_1^2, -p_2^2, -p_3^2) \tag{8-161}$$

It is too complicated, however, to study the dispersion relation in all three variables, so we shall fix two of them on the mass shell and consider it as a function of the third variable.

$$p_1^2 + M^2 = p_3^2 + m^2 = 0 \qquad\qquad -p_2^2 = s \tag{8-162}$$

The resulting form factor is a function of s alone, $\mathcal{G} = F(s)$. If we use the reduction formula we find that F(s) is equal to the following matrix element for $p_1 = p$, $p_3 = q$:

$$F(s) = \langle p, q(-)\ |\ (\ \Box - M^2)\ \Phi^\dagger(0)\ |0\rangle\ (2p_0 V)^{1/2}\ (2q_0 V)^{1/2} \tag{8-163}$$

with $s = -(p + q)^2$ and $|\ p, q(-)\rangle = \Phi_{p,q}^{out}$.

We now shall study the properties of this function F.

(1) *Mass shell condition*

$$F(M^2) = g \tag{8-164}$$

This is the definition of the renormalized coupling constant g.

(2) *Dispersion relation*

$$F(s) = g + \frac{s - M^2}{\pi} \int_{(M+m)^2}^{\infty} ds' \frac{\operatorname{Im} F(s')}{(s' - M^2)\ (s' - s - i\epsilon)} \tag{8-165}$$

We shall verify this relation to third order in g by computing the contributions from the processes in Fig. 8-6.

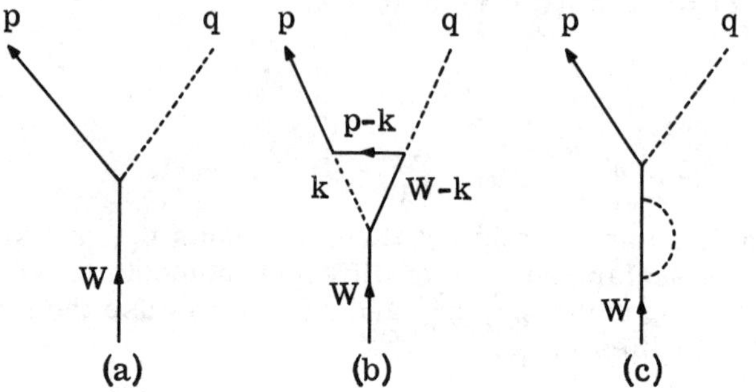

Figure 8-6. Diagrams giving contributions to the form factor F(s).

The contribution from diagram c can be evaluated easily since we already know the renormalized expression for the propagator, so

$$F_a(s) + F_c(s) = g\left(1 + (M^2 - s)\int_{(M+m)^2}^{\infty} ds'\,\frac{\sigma(s')}{s' - s - i\epsilon}\right) \tag{8-166}$$

The remaining contribution is

$$F_b(s) = \frac{g^3}{(2\pi)^4}(-i) \times \int d^4k \frac{1}{(k^2 + m^2 - i\epsilon)[(W-k)^2 + M^2 - i\epsilon][(p-k)^2 + M^2 - i\epsilon]} \tag{8-167}$$

where $W = p + q$. This expression for $F_b(s)$ is not yet properly renormalized; the renormalized expression is

$$F_b(s)_{ren} = F_b(s) - F_b(M^2) \tag{8-168}$$

so that the mass shell condition is satisfied. Let us define J as

$$J = -i\int \frac{d^4k}{(k^2 + m^2 - i\epsilon)[(W-k)^2 + M^2 - i\epsilon][(p-k)^2 + M^2 - i\epsilon]} \tag{8-169}$$

By making use of Feynman's trick we get

$$J = -2i\int_0^1 dx_1 \int_0^{x_1} dx_2$$

$$\times \int d^4k\ [k^2 - 2k((x_1 - x_2)p + (1 - x_1)W) + x_2 m^2$$

$$+ (1 - x_1)(W^2 + M^2) - i\epsilon]^{-3}$$

$$= -2i\int_0^1 dx_1 \int_0^{x_1} dx_2 \int d^4k' \frac{1}{(k'^2 + \Lambda - i\epsilon)^3}$$

with

$$k' = k - ((x_1 - x_2)p + (1 - x_1)W)$$

$$\Lambda = (x_2 - x_1(1 - x_1))m^2 + (1 - x_2)^2 M^2 - x_2(1 - x_1)(s - M^2 - m^2)$$

Carrying out the k' integration we find

$$J = \pi^2 \int_0^1 dx_1 \int_0^{x_1} dx_2 \frac{1}{\Lambda - i\epsilon}$$

$$= \pi^2 \int_a^\infty ds' \frac{\gamma(s')}{s' - s - i\epsilon} \tag{8-170}$$

where

$$\gamma(s') = \int_0^1 dx_1 \int_0^{x_1} dx_2 \frac{1}{x_2(1 - x_1)}\ \delta\left[s' - M^2 - m^2 - \frac{[x_2 - x_1(1 - x_1)]\ m^2 + (1 - x_2)^2 M^2}{x_2(1 - x_1)}\right] \tag{8-171}$$

and a is the threshold value of s' below which $\gamma(s')$ vanishes. This problem reduces to that of finding the minimum value of the function

$$f(x_1, x_2) = \frac{[x_2 - x_1(1 - x_1)]\ m^2 + (1 - x_2)^2 M^2}{x_2(1 - x_1)} \tag{8-172}$$

inside or on the boundary of the right triangle in Fig. 8-7.

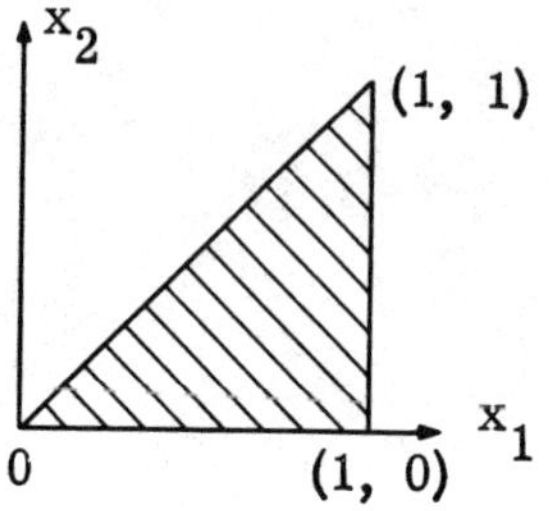

Figure 8-7. Triangular domain for the Feynman parameters x_1 and x_2.

(1) If $f(x_1, x_2)$ has no minimum inside the triangle, its smallest value occurs on the boundary. Since $f = \infty$ for $x_2 = 0$, or for $x_1 = 1$, the smallest value must occur when $x_1 = x_2$; then on this hypotenuse the minimum of $f(x, x)$, namely,

$$f(x, x) = \frac{x^2 m^2 + (1 - x)^2 M^2}{x(1 - x)} \tag{8-173}$$

occurs at $x = M/(M + m)$, and its minimum value is

$$\text{Min } f(x, x) = 2Mm \tag{8-174}$$

Therefore, if there is no minimum inside the triangle we get

$$\text{Min } f = 2Mm, \qquad \text{Min } s' = M^2 + m^2 + 2Mm = (M + m)^2 \tag{8-175}$$

or

$$a = (M + m)^2 \tag{8-176}$$

(2) If $f(x_1, x_2)$ has an absolute minimum inside the triangle, then the coordinates of the minimum are given by

$$\frac{\partial}{\partial x_1} f(x_1, x_2) = \frac{1}{(1 - x_1)^2}\left[m^2 + \frac{(1 - x_2)^2}{x_2} M^2\right] - \frac{m^2}{x_2} = 0 \tag{8-177}$$

and

$$\frac{\partial}{\partial x_2} f(x_1, x_2) = \frac{x_1}{x_2^2} m^2 - \frac{1 - x_2^2}{x_2^2} \frac{M^2}{1 - x_1} = 0 \tag{8-178}$$

Equation (8-177) may be written as

$$(1 - x_1)^2 m^2 = (1 - x_2)^2 M^2 + m^2 x_2$$

then it is clear that there is no solution to this equation for $M > m$ and $1 \geq x_1 \geq x_2 \geq 0$. Therefore,

$$F_b(s)_{ren} = F_b(s) - F_b(M^2)$$

$$= \left(\frac{g}{4\pi}\right)^2 g \int_{(M+m)^2}^{\infty} ds' \, \gamma(s') \left[\frac{1}{s' - s - i\epsilon} - \frac{1}{s - M^2}\right]$$

$$= (s - M^2) \int_{(M+m)^2}^{\infty} \frac{ds'}{(s' - M^2)(s' - s - i\epsilon)} g \left(\frac{g}{4\pi}\right)^2 \gamma(s') \tag{8-179}$$

Hence, in this approximation we can write

$$F(s) = F_a(s) + F_b(s)_{ren} + F_c(s)$$

$$= g + (s - M^2) \int_{(M+m)^2}^{\infty} ds' \frac{\tau(s')}{(s' - M^2)(s' - s - i\epsilon)} \tag{8-180}$$

where

$$\tau(s') = g\sigma(s')(M^2 - s') + g\left(\frac{g}{4\pi}\right)^2 \gamma(s') \tag{8-181}$$

Then taking the absorptive part of F(s) we find

$$\text{Im } F(s) = \pi \, \tau(s) \qquad \text{for} \quad s \geq (M + m)^2 \tag{8-182}$$

Thus we arrive at the desired dispersion relation

$$F(s) = g + \frac{s - M^2}{\pi} \int_{(M+m)^2}^{\infty} ds' \frac{\operatorname{Im} F(s')}{(s' - M^2)(s' - s - i\epsilon)} \tag{8-183}$$

for the *renormalized* vertex function F(s). This form of the dispersion relation is different from the one we obtained previously for the propagator, that is,

$$F(s) = \frac{1}{\pi} \int ds' \frac{\operatorname{Im} F(s')}{s' - s - i\epsilon} \tag{8-184}$$

The latter form is called an unsubtracted dispersion relation, and the other is called a once-subtracted dispersion relation.

There are two cases in which a subtraction is needed, (1) when $F(s_0)$ is given as a boundary condition, and (2) when the dispersion integral diverges in the unsubtracted form. These two conditions are very closely related. In general, a vertex function that requires renormalization (vertex renormalization) needs a subtraction.

It is extremely important to recognize that renormalization or subtraction modifies only the dispersive part leaving the absorptive part unchanged. This is a fundamental reason why dispersion theory is powerful in handling renormalization; namely, when one writes a correct dispersion relation, such as a once-subtracted one, renormalization is already taken into account automatically.

8-8 CALCULATION OF VERTEX FUNCTIONS FROM DISPERSION RELATIONS

In this section we shall obtain the renormalized vertex function by means of dispersion relations. In dispersion theory we first have to evaluate the absorptive or imaginary part of F(s); we shall learn how to compute the imaginary part of F(s) in the discussion that follows.

We go back to the definition:

$$F(s) = \langle p, q(-) | (\Box - M^2) \Phi^\dagger(0) | 0 \rangle \left(2p_0 V\right)^{1/2} \left(2q_0 V\right)^{1/2}$$

We then use the time-reversal operator R, defined as[10]

$$\begin{aligned} \Phi^\dagger(x, t)^R &= \Phi(x, -t) \\ \Phi(x, t)^R &= \Phi^\dagger(x, -t) \end{aligned} \tag{8-185}$$

In particular, we have

$$\Phi^\dagger(0)^R = \Phi(0) \qquad \Phi(0)^R = \Phi^\dagger(0) \tag{8-186}$$

so we can write the vertex function with

$$\langle p, q(-) | (\Box - M^2)\, \Phi^\dagger(0) | 0\rangle = \langle p, q(-) | (\Box - M^2)\, \Phi(0)^R | 0\rangle$$
$$= \langle 0 | (\Box - M^2)\, \Phi(0) | p, q(-)^R\rangle$$
$$= \langle 0 | (\Box - M^2)\, \Phi(0) | -p, -q(+)\rangle$$
$$= \langle -p, -q(+) | (\Box - M^2)\, \Phi^\dagger(0) | 0\rangle^* \tag{8-187}$$

Since this matrix element is a function of only $(p + q)^2 - (p_0 + q_0)^2$,

$$F(s) = \langle p, q(+) | (\Box - M^2)\, \Phi^\dagger(0) | 0\rangle^* (2p_0 V)^{1/2} (2q_0 V)^{1/2} \tag{8-188}$$

and

$$F^*(s) = \langle p, q(+) | (\Box - M^2)\, \Phi^\dagger(0) | 0\rangle \ (2p_0 V)^{1/2} \ (2q_0 V)^{1/2}$$

We use the completeness of the set $\{ \Phi^{(+)} \} = \{ \Phi^{in} \}$.

$$\langle p, q(-) | (\Box - M^2)\, \Phi^\dagger(0) | 0\rangle$$
$$= \sum_\alpha \langle p, q(-) | \alpha(+)\rangle \langle \alpha(+) | (\Box - M^2)\, \Phi^\dagger(0) | 0\rangle \tag{8-189}$$

The one-nucleon intermediate states— where Φ is called the nucleon field—give a vanishing contribution since

$$\langle \text{nucleon} | (\Box - M^2)\, \Phi^\dagger(0) | 0\rangle$$
$$= (\Box - M^2) \langle \text{nucleon} | \Phi^\dagger(x) | 0\rangle_{x=0} = 0 \tag{8-190}$$

Therefore, in the lowest order we shall keep only *(one nucleon + one meson)* intermediate states. This turns out to be exact below the threshold for meson production. The approximate relation is

$$\frac{F(s)}{(2p_0 V)^{1/2} \ (2q_0 V)^{1/2}}$$
$$= \sum_{p', q'} \langle p, q(-) | p', q'(+)\rangle \frac{F^*(s)}{(2p_0 V)^{1/2} \ (2q_0' V)^{1/2}} \tag{8-191}$$

with $s = -W^2 = -(p + q)^2 = -(p' + q')^2$.

We define the transition matrix T from

$$S_{fi} = \delta_{fi} - i\delta^4(P_f - P_i)\, t_{fi}$$

$$= \delta_{fi} - i(2\pi)^4\, \delta^4(P_f - P_i)\, T_{fi}$$

and

$$S^\dagger_{fi} = \delta_{fi} + i(2\pi)^4\, \delta^4(P_f - P_i)\, T^\dagger_{fi}$$

then the relation (8-191) can be written as

$$\frac{\operatorname{Im} F(s)}{(2p_0V)^{1/2}\,(2q_0V)^{1/2}}$$

$$= -\frac{(2\pi)^4}{2} \sum_{p',\, q'} \langle p, q\,|T^\dagger|p', q'\rangle \frac{F(s)}{(2p'_0V)^{1/2}(2q'_0V)^{1/2}}\, \delta^4(p + q - p' - q') \tag{8-192}$$

This is called the unitarity condition for the form factor since it is derivable from the generalized unitary condition in Section 8-5. Those who are not familiar with time reversal may try to derive this result from the generalized unitarity condition.

The Im F(s) is of order g^3, so we shall compute the right-hand side of (8-192) in the lowest order. To the second order,

$$\langle p, q\,|T^{\dagger(2)}|\,p', q'\rangle = \langle p, q\,|T^{(2)}|\,p', q'\rangle$$

$$= -\frac{g^2}{(2p_0V)^{1/2}(2q_0V)^{1/2}(2p'_0V)^{1/2}(2q'_0V)^{1/2}}$$

$$\times \left(\frac{1}{(p + q)^2 + M^2} + \frac{1}{(p - q')^2 + M^2}\right) \tag{8-193}$$

corresponding to the diagrams in Fig. 8-8.

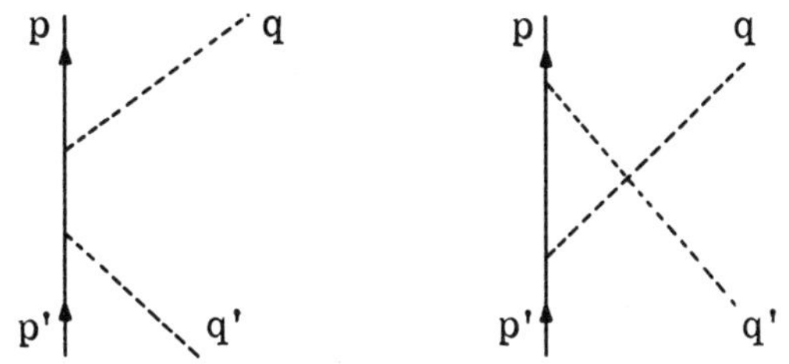

Figure 8-8. Diagrams for meson-nucleon scattering.

Hence

$$\operatorname{Im} F^{(3)}(s) = -\frac{(2\pi)^4}{2}\frac{V}{(2\pi)^3}\int d^3p' \frac{V}{(2\pi)^3}$$

$$\int d^3q' \, \delta^4(p + q - p' - q')$$

$$\times -\frac{g^2}{(2p_0'V)(2q_0'V)}$$

$$\times \left(\frac{1}{(p+q)^2 + M^2} + \frac{1}{(p-q')^2 + M^2}\right) F^{(1)}(s)$$

$$= \frac{g^3}{2(2\pi)^2}\int \frac{d^3p'}{2p_0'}\int \frac{d^3q'}{2q_0'}\, \delta^4(W - p' - q')$$

$$\times \left(\frac{1}{M^2 - s} + \frac{1}{(p-q')^2 + M^2}\right) \tag{8-194}$$

where we have replaced $F^{(1)}(s)$ by g. The two terms in the integrand correspond to the Feynman diagrams in Fig. 8-9; they are identical with those considered in the preceding section.

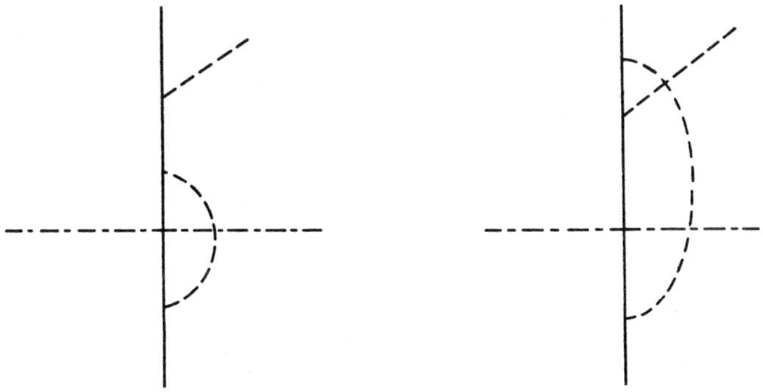

Figure 8-9. Diagrams representing the absorptive part of F(s). The chain lines (- · -) indicate where the intermediate states are inserted.

The integral in (8-194) is of the standard form and we can utilize the familiar result in the center-of-mass system:

$$\int \frac{d^3p'}{2p'_0} \int \frac{d^3q'}{2q'_0} \quad \delta^4(W - p' - q') \ldots = \frac{1}{4} \frac{|q|}{s^{1/2}} \int d\Omega \ldots$$

where

$$|q| = \frac{1}{2} \frac{[s - (M + m)^2]^{1/2} [s - (M - m)^2]^{1/2}}{s^{1/2}} \tag{8-195}$$

In the center-of-mass system,

$$p = \left(-q, \left(M^2 + q^2\right)^{1/2}\right) \qquad q' = \left(q', \left(m^2 + q^2\right)^{1/2}\right) \tag{8-196}$$

so that

$$(p - q')^2 + M^2 = s - M^2 - 2m^2 - 2q^2 (1 - \cos\theta),$$

$$s = M^2 + m^2 + 2q^2 + 2\left(M^2 + q^2\right)^{1/2} \left(m^2 + q^2\right)^{1/2}$$

Hence

$$\operatorname{Im} F^{(3)}(s) = \frac{g}{4} \left(\frac{g}{4\pi}\right)^2 \frac{[s - (M + m)^2]^{1/2} [s - (M - m)^2]^{1/2}}{s}$$

$$\times \int d\Omega \left[\frac{1}{M^2 - s} + \frac{1}{s - M^2 - 2m^2 - 2q^2 + 2q^2 \cos\theta}\right]$$

$$= \frac{g}{4} \left(\frac{g^2}{4\pi}\right) \frac{[s - (M + m)^2]^{1/2} [s - (M - m)^2]^{1/2}}{s}$$

$$\times \left[\frac{1}{M^2 - s} + \frac{1}{4q^2} \ln \left|\frac{s - M^2 - 2m^2}{s - M^2 - 2m^2 - 4q^2}\right|\right] \tag{8-197}$$

We can verify that this expression agrees with the previous result obtained by means of the Feynman-Dyson theory. The first term, corresponding to the diagram representing propagator renormalization, clearly agrees with $\pi g \sigma(s) (M^2 - s)$; and the second term can be shown to be equal to $\pi g (g/4\pi)^2 \gamma(s)$.

8-9 APPLICATION TO ELECTRODYNAMICS

We can repeat the same calculation as the preceding one in quantum electrodynamics. Let us define the current operator j_μ as

$$\Box A_\mu = - j_\mu \tag{8-198}$$

and consider the following matrix element:

$$\langle p, \bar{p}(-) | j_\mu (0) | 0 \rangle \tag{8-199}$$

where $| p, \bar{p}(-) \rangle$ denotes a state in which there are an electron of momentum p and a positron of momentum $\bar{p}$, or more precisely

$$| p, \bar{p}(-) \rangle = | p, \bar{p}(\text{out}) \rangle = a^\dagger (p)^{\text{out}} \, b^\dagger (\bar{p})^{\text{out}} | 0 \rangle \tag{8-200}$$

In the lowest-order perturbation theory we find

$$\langle p, \bar{p}(-) | j_\mu(0) | 0 \rangle = ie\bar{u}(p) \gamma_\mu v(\bar{p})/V \tag{8-201}$$

where e denotes the electronic charge. In general we must have, from the Lorentz transformation properties of the matrix element and the conservation of charge,

$$\langle p, \bar{p}(-) | j_\mu (0) | 0 \rangle$$

$$= \bar{u}(p) \left[i \gamma_\mu F_1 (-W^2) - i\sigma_{\mu\nu} W_\nu F_2 (-W^2) \right] v(\bar{p})/V \tag{8-202}$$

where $W = p + \bar{p}$. In this general case we have two form factors called charge and moment, or Dirac and Pauli, form factors. These factors are normalized by

$$F_1 (0) = e \qquad F_2 (0) = \mu \tag{8-203}$$

where e is the negative electronic charge and μ the anomalous magnetic moment of the electron. The e is a given fixed constant, while μ is not known a priori and must be calculated. Therefore we expect $F_1(s)$ to satisfy a once-subtracted dispersion relation, and $F_2(s)$ to satisfy an unsubtracted dispersion relation. In this section we shall calculate the anomalous magnetic moment μ on this assumption.

From the time-reversal argument of the previous section we can show that

$$\langle p, \bar{p}(+) | j_\mu (0) | 0 \rangle$$

$$= \bar{u}(p) \left[i\gamma_\mu F_1^* (-W^2) - i\sigma_{\mu\nu} W_\nu F_2^* (-W^2) \right] v(\bar{p}) \tag{8-204}$$

where we have set V = 1. We shall combine this relation with the unitarity condition in the form

$$\operatorname{Im}\langle \alpha(-)|j_\mu(0)|0\rangle$$

$$= -\frac{(2\pi)^4}{2}\sum_\beta T^\dagger_{\alpha\beta}\,\delta^4(P_\alpha - P_\beta)\,\langle\beta(-)|j_\mu(0)|0\rangle \qquad (8\text{-}205)$$

Since the lowest-order expression for T is of the second order in e and F is first order in e, we shall compute Im F in the third order.

$$F_1^{(1)}(s) = e \qquad F_2^{(1)}(s) = 0 \qquad (8\text{-}206)$$

Thus we must compute

$$\operatorname{Im}\langle p, \bar{p}(-)|j_\mu^{(3)}(0)|0\rangle$$

$$= -\frac{(2\pi)^4}{2}\sum_{p',\bar{p}'}\langle p, \bar{p}|T^{\dagger(2)}|p', \bar{p}'\rangle\,\delta^4(W - W')\langle p', \bar{p}'|j_\mu^{(1)}(0)|0\rangle \qquad (8\text{-}207)$$

We can compute $T^{\dagger(2)} = T^{(2)}$ using the interaction

$$\mathcal{H}_{\rm int}(x) = -ie\bar{\psi}(x)\,\gamma_\mu\,\psi(x)\,A_\mu(x) \qquad (8\text{-}208)$$

The result is

$$\langle p, \bar{p}|T^{(2)}|p', \bar{p}'\rangle$$

$$= e^2\left[-\frac{1}{(p - p')^2}\bar{u}(p)\,\gamma_\lambda u(p')\cdot\bar{v}(\bar{p}')\,\gamma_\lambda\,v(\bar{p}) + \frac{1}{(p + \bar{p})^2}\bar{u}(p)\,\gamma_\lambda\,v(\bar{p})\cdot\bar{v}(\bar{p}')\,\gamma_\lambda\,u(p')\right] \qquad (8\text{-}209)$$

Now, combining the relationships (8-207) and (8-209), we get

$$\bar{u}(p)\left[i\gamma_\mu \text{ Im } F_1^{(3)}(-W^2) - i\sigma_{\mu\nu} W_\nu \text{ Im } F_2^{(3)}(-W^2)\right]v(\bar{p})$$

$$= -ie^3 \frac{(2\pi)^4}{2} \int \frac{d^3p'}{(2\pi)^3} \int \frac{d^3\bar{p}'}{(2\pi)^3} \delta^4(p + \bar{p} - p' - \bar{p}')$$

$$\times \left[-\frac{1}{(p - p')^2} \bar{u}(p)\gamma_\lambda \frac{-ip'\gamma + m}{2p'_0} \gamma_\mu \frac{-i\bar{p}'\gamma - m}{2\bar{p}'_0} \gamma_\lambda v(\bar{p})\right.$$

$$\left. + \frac{1}{(p + \bar{p})^2} \bar{u}(p)\gamma_\lambda v(\bar{p}) \cdot \text{Tr}\left(\gamma_\mu \frac{-i\bar{p}'\gamma - m}{2\bar{p}'_0} \gamma_\lambda \frac{-ip'\gamma + m}{2p'_0}\right)\right] \tag{8-210}$$

where use has been made of the Casimir operators

$$\sum_{\text{spin}} u(p')\bar{u}(p') = \frac{-ip'\gamma + m}{2p'_0}$$

$$\sum_{\text{spin}} v(\bar{p}')\bar{v}(\bar{p}') = \frac{-i\bar{p}'\gamma - m}{2\bar{p}'_0}$$

The second term, representing the contribution from vacuum polarization, does not contribute to the moment form factor.

Using the Dirac equations

$$\bar{u}(p)(ip\gamma + m) = (-i\bar{p}\gamma + m)v(\bar{p}) = 0$$

and

$$\bar{u}(p)(p - \bar{p})_\mu v(\bar{p}') = i\bar{u}(p)\sigma_{\mu\nu}W_\nu v(\bar{p}') - 2mi\,\bar{u}(p)\gamma_\mu v(\bar{p})$$

and the relations for summing over γ matrices:

$$\gamma_\lambda\gamma_a\gamma_\lambda = -2\gamma_a \qquad \gamma_\lambda\gamma_a\gamma_b\gamma_\lambda = 4\delta_{ab}$$

$$\gamma_\lambda\gamma_a\gamma_b\gamma_c\gamma_\lambda = -2\gamma_c\gamma_b\gamma_a$$

We finally find

$$\text{Im } F_2^{(3)}(s) = \frac{em}{2}\left(\frac{e^2}{4\pi}\right) s^{-1/2}(s - 4m^2)^{-1/2}\,\theta\,(s - 4m^2) \qquad (8\text{-}211)$$

Hence

$$\mu = \text{Re } F_2(0) = \frac{1}{\pi}\int_{4m^2}^{\infty} ds\,\frac{\text{Im } F_2^{(3)}(s)}{s}$$

$$= \frac{em}{2\pi}\,\alpha \int_{4m^2}^{\infty} ds\; s^{-3/2}(s - 4m^2)^{-1/2} = \frac{e}{2m}\left(\frac{\alpha}{2\pi}\right) \qquad (8\text{-}212)$$

The Dirac moment is given by e/2m, so the total magnetic moment of the electron to third order is

$$\left(1 + \frac{\alpha}{2\pi}\right)\frac{e}{2m} \qquad (8\text{-}213)$$

which agrees with Schwinger's result. The benefit of the dispersion method is that one need not be concerned with renormalization.

8-10 GAUGE INVARIANCE AND DISPERSION RELATIONS

In previous sections we learned the connection between renormalization and subtraction in dispersion relations. It was made clear that divergences inherent in the Feynman-Dyson theory occur only in the dispersive parts of Green's functions or of certain transition amplitudes; these divergences can be avoided by the introduction of subtractions in the dispersion relations. The Feynman-Dyson theory is infected, however, not only with the disease of divergences but also with ambiguities. We have encountered a typical example of this kind in evaluating the photon propagator in Section 6-4, that is, the requirement of gauge invariance implies that a certain integral is equal to zero, but an actual evaluation shows that it is not. Therefore, depending on one's attitude, relying either on a formal argument based on gauge invariance or on an actual evaluation of the integral, one arrives at different results.

In this section it will be shown that this disease of ambiguities also can be cured with the help of dispersion relations. This problem will be illustrated by discussion of the process $K_1^0 \to 2\gamma$.[11]

Feynman–Dyson Calculation of the Decay $K_1^0 \to 2\gamma$

The neutral meson K_1^0 is known to decay into two pions:

$$K_1^0 \to \pi^+ + \pi^- \qquad K_1^0 \to 2\pi^0 \tag{8-214}$$

Let us combine the first decay mode with the pair annihilation into two photons $\pi^+ + \pi^- \to 2\gamma$ analogous to the electron-positron annihilation. Consequently, the K_1^0 meson can decay directly in two photons.

$$K_1^0 \to 2\gamma \tag{8-215}$$

First we shall evaluate this decay amplitude by using the Feynman-Dyson technique. Corresponding to the charged-pion mode of decay of the K_0^1 meson, we introduce a phenomenological decay interaction of the form

$$\mathcal{L}_{\text{decay}} = -g\Phi\varphi^\dagger\varphi \tag{8-216}$$

where Φ and φ denote field operators of the K_1 and π^+ mesons, respectively. The electromagnetic interaction of the charged-pion field responsible for the pion annihilation is described by

$$\mathcal{L}_{\text{em}} = -ieA_\lambda(\varphi^\dagger\partial_\lambda\varphi - \partial_\lambda\varphi^\dagger \cdot \varphi) - e^2A_\lambda^2\ \varphi^\dagger\varphi \tag{8-217}$$

The corresponding Feynman diagrams are given in Fig. 8-10.

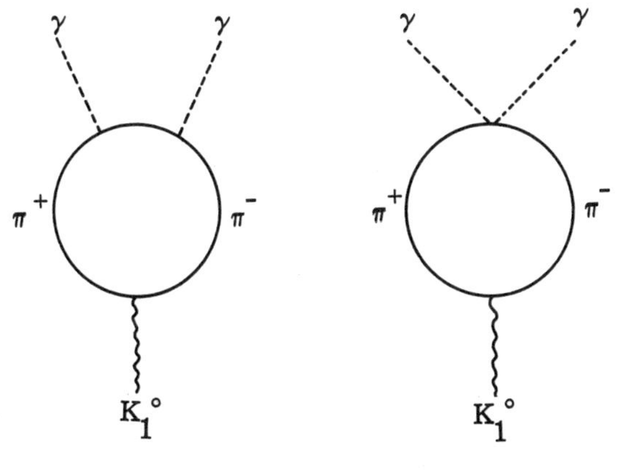

Figure 8-10. The lowest-order diagrams for $K_1^0 \to 2\gamma$.

Since this process is very similar to the process $\pi^0 \rightarrow 2\gamma$ discussed at length in Section 5-4, we immediately may write the S-matrix element. If we denote the four energy-momentum vectors of the K_1^0 meson and of the two photons in the final state by P, k', and k'', respectively, the decay amplitude is given by

$$\langle k', k'' | S | P \rangle = \delta^4 (P_f - P_i) \frac{2ge^2 \epsilon'_\lambda \epsilon''_\mu}{(2k'_0 V)^{1/2} (2k''_0 V)^{1/2} (2P_0 V)^{1/2}}$$

$$\times \left[\int d^4 q \frac{i(2q - k')_\lambda \cdot i(2q + k'')_\mu}{[(q - k')^2 + \mu^2 - i\epsilon] [q^2 + \mu^2 - i\epsilon]} \right.$$

$$\times \frac{1}{[(q + k'')^2 + \mu^2 - i\epsilon]}$$

$$\left. + \delta_{\lambda\mu} \int d^4 q \frac{1}{[(q - k')^2 + \mu^2 - i\epsilon] [(q + k'')^2 + \mu^2 - i\epsilon]} \right] \tag{8-218}$$

The corresponding Feynman diagrams labeled with momenta are shown in Fig. 8-11.

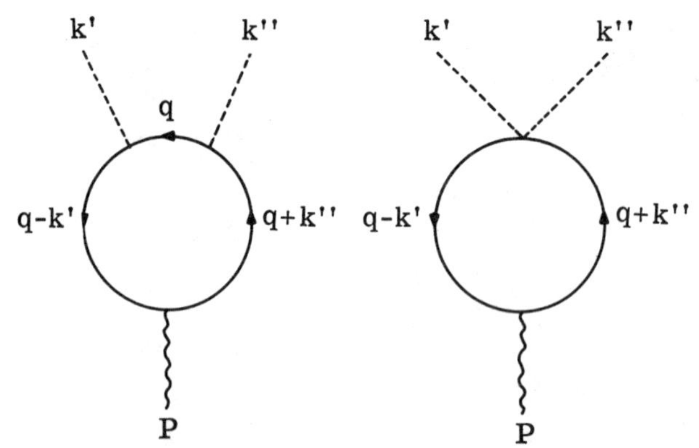

Figure 8-11. Diagrams for $K_1^0 \rightarrow 2\gamma$ labeled with momenta.

The two integrals in the decay amplitude are divergent, but the sum is convergent as a consequence of the cancellation between the two divergences. We may write the sum as

$$\langle k', k''|S|P\rangle = \delta^4(P_f - P_i)\,\frac{2ge^2}{(2k_0'V)^{1/2}\,(2k_0''V)^{1/2}\,(2P_0V)^{1/2}}\,I \tag{8-219}$$

where

$$I = \int d^4q\,\frac{(\epsilon'\epsilon'')(q^2+\mu^2) - 4(\epsilon'q)(\epsilon''q)}{[(q-k')^2+\mu^2 - i\epsilon][q^2+\mu^2 - i\epsilon][(q+k'')^2+\mu^2 - i\epsilon]} \tag{8-220}$$

In deriving this form, use has been made of the Lorentz condition

$$\epsilon_\lambda' k_\lambda' = \epsilon_\lambda'' k_\lambda'' = 0 \tag{8-221}$$

The integral I can be evaluated by use of the standard Feynman technique.

$$I = \int d^4q \int_0^1 2x\,dx \int_0^1 dy$$
$$\times \frac{(\epsilon'\epsilon'')(q^2+\mu^2) - 4(\epsilon'q)(\epsilon''q)}{[(q-(1-x)k' + xyk'')^2 + \mu^2 - (xyk'' - (1-x)k')^2 - i\epsilon]^3} \tag{8-222}$$

We introduce a new variable p by $p = q - (1-x)k' + xyk''$; then the denominator is a function of p^2 alone, and linear terms in p in the numerator vanish upon integration.
Thus we may write

$$I = \int_0^1 2x\,dx \int_0^1 dy \int d^4p\,\frac{N}{D^3} \tag{8-223}$$

where

$$N = (\epsilon'\epsilon'')p^2 - 4(\epsilon'p)(\epsilon''p) + (\epsilon'\epsilon'')\mu^2$$
$$- x(1-x)y[(\epsilon'\epsilon'')P^2 - 4(\epsilon'P)(\epsilon''P)] \tag{8-224}$$

and

$$D = p^2 + \mu^2 + x(1 - x)yP^2 - i\epsilon \tag{8-225}$$

with $P = k' + k''$. In carrying out the p integration we can drop the terms $(\epsilon'\epsilon'')p^2 - 4(\epsilon'p)(\epsilon''p)$ since they cancel each other as discussed at the end of Section 6-4; otherwise the integral would be divergent. After carrying out the p integration and replacing $(\epsilon'\epsilon'')P^2 - 2(\epsilon'P)(\epsilon''P)$ by $2[(\epsilon'\epsilon'')(k'k'') - (\epsilon'k'')(\epsilon''k')]$ on the basis of the Lorentz condition and $k^2 = 0$, we obtain

$$\langle k', k'' | S | P \rangle = \frac{2i\pi^2 ge^2}{(2k'_0 V)^{1/2}(2k''_0 V)^{1/2}(2P_0 V)^{1/2}} \times \delta^4(P_f - P_i) \int_0^1 x dx \int_0^1 dy \; J \tag{8-226}$$

where

$$J = (\epsilon'\epsilon'') - \frac{4x(1 - x)y}{\mu^2 + x(1 - x)yP^2 - i\epsilon}[(\epsilon'\epsilon'')(k'k'') - (\epsilon'k'')(\epsilon''k')] \tag{8-227}$$

The second term, proportional to $(\epsilon'\epsilon'')(k'k'') - (\epsilon'k'')(\epsilon''k')$, is gauge-invariant since it is invariant under the replacements $\epsilon' \to \epsilon' + ak'$, and $\epsilon'' \to \epsilon'' + bk''$, but the first term is *not* gauge-invariant; this is a difficulty similar to the one we have encountered in evaluating the photon propagator. In order to resolve this dilemma let us regard these integrals as functions of the square of the barycentric energy, $s = -P^2$. In the present problem s is the square of the K_1-meson rest mass, and it is regarded as a variable. In the case of the photon propagator the noninvariant term is a linear function of s, and in the present case the coefficient of the nongauge-invariant factor $(\epsilon'\epsilon'')$ is a constant. Therefore, in both cases the coefficients of noninvariant factors have no branch cut in the complex s plane; in other words these functions have *no* absorptive parts. Thus we can jump to the conclusion that not only the divergences but also ambiguities occur only in the dispersive parts; the absorptive parts calculated in the Feynman-Dyson theory are free of both divergences and ambiguities.

Once we take this conclusion for granted we have no difficulties in evaluating the decay amplitude for $K_1^0 \to 2\gamma$. We first evaluate the absorptive part by using

$$\text{Im}\left(\frac{1}{\mu^2 + x(1 - x)yP^2 - i\epsilon}\right) = \pi\,\delta[\mu^2 + x(1 - x)yP^2] \tag{8-228}$$

and the formula

$$\int_0^1 x\,dx \int_0^1 dy\, x(1-x)y\,\delta[\mu^2 - x(1-x)ys]$$

$$= \frac{\mu^2}{s^2} \ln\left(\frac{1+[1-(4\mu^2/s)]^{1/2}}{1-[1-(4\mu^2/s)]^{1/2}}\right) \theta(s-4\mu^2) \qquad (8\text{-}229)$$

We write the decay amplitude as

$$\langle k', k'' | S | P\rangle$$

$$= -i(2\pi)^4 \delta^4(P_f - P_i) \frac{\mathcal{T}_{fi}}{(2k'_0 V)^{1/2}(2k''_0 V)^{1/2}(2P_0 V)^{1/2}} \qquad (8\text{-}230)$$

and define F(s) as

$$\mathcal{T}_{fi} = \frac{ge^2}{2\pi}\mu^2[(\epsilon'\epsilon'')(k'k'') - (\epsilon'k'')(\epsilon''k')]F(s) \qquad (8\text{-}231)$$

then

$$\text{Im } F(s) = \frac{1}{s^2} \ln\left(\frac{1+[1-(4\mu^2/s)]^{1/2}}{1-[1-(4\mu^2/s)]^{1/2}}\right)\theta(s-4\mu^2) \qquad (8\text{-}232)$$

The dispersive part Re F(s) can be calculated from the dispersion relation

$$\text{Re } F(s) = \frac{P}{\pi}\int_{4\mu^2}^{\infty} \frac{ds'}{s'-s}\,\text{Im } F(s') \qquad (8\text{-}233)$$

In this problem there is no primary interaction in the Lagrangian corresponding to the decay $K_1^0 \to 2\gamma$, so that there is no problem of vertex renormalization; therefore, we have to assume an unsubtracted dispersion relation for F(s) just as in the case of the moment form factor in the preceding section.

Dispersion Calculation of the Decay $K_1^0 \to 2\gamma$

We have evaluated the absorptive part of the decay amplitude for $K_1^0 \to 2\gamma$ in the Feynman-Dyson theory. We can derive the same result, moreover, by using the unitarity condition.

In the preceding sections we have derived the unitarity condition for form factors; we now apply it to the present problem in the form

$$\text{Im } T_{fi} = -\frac{(2\pi)^4}{2} \sum_n T^\dagger_{fn}\, \delta^4 (P_f - P_n)\, T_{ni}$$

where $n = |\pi^+\pi^-\rangle$, and

$$T_{ni} = g\,(2P_0 V)^{-1/2} (2q_{+0}V)^{-1/2} (2q_{-0}V)^{-1/2} \tag{8-234}$$

$$T_{fn} = \langle 2\gamma |T| \pi^+\pi^-\rangle$$

$$= (2k'_0 V)^{-1/2} (2k''_0 V)^{-1/2} (2q_{+0}V)^{-1/2} (2q_{-0}V)^{-1/2}$$

$$\times\, 2e^2 \left[(\epsilon'\epsilon'') - \frac{(\epsilon' q_+)(\epsilon'' q_-)}{(q_+ k')} - \frac{(\epsilon' q_-)(\epsilon'' q_+)}{(q_+ k'')} \right] \tag{8-235}$$

where q_+ and q_- denote the energy-momenta of the π^+ and π^- mesons appearing in the intermediate states. Substituting these expressions in the unitarity condition and replacing $\sum_n$ by

$$\sum_n \to \frac{V}{(2\pi)^3} \int d^3 q_+ \frac{V}{(2\pi)^3} \int d^3 q_- \tag{8-236}$$

we can reproduce, after a strightforward but tedious calculation, the result obtained by the Feynman-Dyson theory.

Evaluation of the Dispersion Integral

Since the absorptive part of the decay amplitude has been found we shall proceed to the evaluation of the dispersive part; we have to evaluate the following dispersion integral:

$$I = P \int_{4\mu^2}^{\infty} \frac{ds'}{s'-s} \cdot \frac{1}{s'^2} \ell n \left(\frac{1+[1-(4\mu^2/s')]^{1/2}}{1-[1-(4\mu^2/s')]^{1/2}} \right) \tag{8-237}$$

Introducing new variables x and y by

$$x = 4\mu^2/s \qquad y = 4\mu^2/s' \tag{8-238}$$

we can write I as

$$I = \frac{x}{(4\mu^2)^2} \int_0^1 \frac{y\,dy}{x - y} \ln\left(\frac{1 + (1 - y)^{1/2}}{1 - (1 - y)^{1/2}}\right)$$

$$= -\frac{x}{(4\mu^2)^2} \int_0^1 dy \ln\left(\frac{1 + (1 - y)^{1/2}}{1 - (1 - y)^{1/2}}\right)$$

$$+ \frac{x^2}{(4\mu^2)^2} \int_0^1 \frac{dy}{x - y} \ln\left(\frac{1 + (1 - y)^{1/2}}{1 - (1 - y)^{1/2}}\right)$$

$$= -\frac{2x}{(4\mu^2)^2} + \frac{x^2}{(4\mu^2)^2} \int_0^1 \frac{dy}{x - y} \ln\left(\frac{1 + (1 - y)^{1/2}}{1 - (1 - y)^{1/2}}\right)$$

Let us define J as

$$J = \int_0^1 \frac{dy}{x - y} \ln\left(\frac{1 + (1 - y)^{1/2}}{1 - (1 - y)^{1/2}}\right)$$

then

$$\frac{dJ}{dx} = -\int_0^1 \frac{dy}{(x - y)^2} \ln\left(\frac{1 + (1 - y)^{1/2}}{1 - (1 - y)^{1/2}}\right)$$

$$= \left[\frac{1}{y - x} \ln\left(\frac{1 + (1 - y)^{1/2}}{1 - (1 - y)^{1/2}}\right) + \frac{1}{x} \ln\left(\frac{1 + (1 - y)^{1/2}}{1 - (1 - y)^{1/2}}\right)\right]_0^1$$

$$- \frac{1}{x} \int_0^1 \frac{dy}{(1 - y)^{1/2}} \frac{1}{x - y}$$

$$= -\frac{1}{x} \int_0^1 \frac{dy}{(1 - y)^{1/2}} \frac{1}{x - y}$$

where use has been made of integration by parts and of the relation

$$\frac{d}{dy} \ln\left(\frac{1 + (1 - y)^{1/2}}{1 - (1 - y)^{1/2}}\right) = -\frac{1}{y\,(1 - y)^{1/2}} \tag{8-239}$$

Introducing $1 - y = t^2$ and $1 - x = a^2$, we easily can evaluate dJ/dx.

$$\frac{dJ}{dx} = -\frac{P}{x}\int_0^1 \frac{dy}{(1-y)^{1/2}}\,\frac{1}{x-y}$$

$$= -\frac{2}{x}\,P\int_0^1 \frac{dt}{t^2 - a^2}$$

$$= -\frac{1}{ax}\left[\ln\left|\frac{t-a}{t+a}\right|\right]_0^1$$

$$= \frac{1}{x(1-x)^{1/2}}\,\ln\left|\frac{1+(1-x)^{1/2}}{1-(1-x)^{1/2}}\right|$$

Recalling relation (8-239) and the boundary condition $J \to 0$ for $x \to -\infty$, we readily can integrate the last relation, to get J for $x < 0$.

$$J = -\frac{1}{2}\left(\ln\left|\frac{1+(1-x)^{1/2}}{1-(1-x)^{1/2}}\right|\right)^2 \tag{8-240}$$

Hence a proper analytic continuation of $\ln$ yields for $s > 4\mu^2$.

$$\text{Re } F(s) = \frac{P}{\pi}\int_{4\mu^2}^{\infty} \frac{ds'}{s'-s}\,\text{Im } F(s')$$

$$= -\frac{1}{2\pi s}\left[\frac{1}{\mu^2} - \frac{\pi^2}{s} + \frac{1}{s}\left(\ln\left|\frac{1+[1-(4\mu^2/s)]^{1/2}}{1-[1-(4\mu^2/s)]^{1/2}}\right|\right)^2\right] \tag{8-241}$$

The decay rates for $K_1^0 \to \pi^+ + \pi^-$ and $K_1^0 \to 2\gamma$ are

$$\Gamma(K_1^0 \to \pi^+ + \pi^-) = \left(\frac{g^2}{4\pi}\right)\frac{1}{4M}\left(1 - 4\mu^2/M^2\right)^{1/2} \tag{8-242}$$

and

$$\Gamma(K_1^0 \to 2\gamma) = \left(\frac{g^2}{4\pi}\right)\frac{1}{4M}\,\alpha^2\left|\mu^2 M^2 F(M^2)\right|^2 \tag{8-243}$$

where M denotes the rest mass of the K_1^0 meson. Thus we get

$$\frac{\Gamma(K_1^0 \to 2\gamma)}{\Gamma(K_1^0 \to \pi^+ + \pi^-)} = \frac{\alpha^2}{(1 - 4\mu^2/M^2)^{1/2}} \left| \mu^2 M^2 F(M^2) \right|^2$$
$$\approx 2 \times 10^{-6} \qquad (8\text{-}244)$$

We have carried out the integration explicitly here since this is a typical dispersion integral.

8-11 THE MUSKHELISHVILI-OMNÈS EQUATION [12, 13]

As has been discussed in detail, all the troubles in the Feynman-Dyson theory occur exclusively in the dispersive parts, so that removal of divergences or of ambiguities in dispersion theory is rather simple. In the Feynman-Dyson theory, however, the amplitudes are usually not decomposed into the absorptive and dispersive parts so that removal of divergences or renormalization is carried out only in the power-series expansion; this prevents us from solving certain problems without recourse to perturbation theory. A typical example is the application of the ladder approximation to a vertex function, shown in Fig. 8-12. In each diagram we encounter a divergence; in order to renormalize the vertex function in a certain order we have to know the renormalized expression for the vertex function in the lower orders. Therefore, deterred by the divergence difficulty, we cannot solve the Bethe-Salpeter equation for the vertex function without recourse to the series expansion.

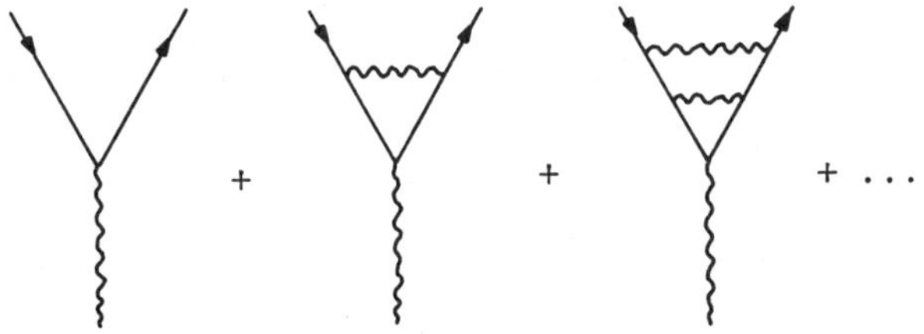

Figure 8-12. Ladder corrections to the vertex.

The situation is completely different in dispersion theory since we have no divergences in the absorptive part. The only additional step we have to take is to introduce a subtraction in the dispersion relation for the vertex function when it is necessary. The problem of summing contributions for a vertex function over an infinite set of diagrams within the framework of dispersion theory reduces to a new type of equation, called the Muskhelishvili-Omnès equation; its solution constitutes the object of this section.

We recall the unitarity condition for a vertex function F(s),

$$\frac{\operatorname{Im} F(s)}{(2p_0V)^{1/2}\,(2q_0V)^{1/2}} = -\frac{(2\pi)^4}{2}\frac{V}{(2\pi)^3}\int d^3p'\,\frac{V}{(2\pi)^3}\int d^3q'\langle p, q|T_{,}^{\dagger}|p', q'\rangle \times \delta^4(p+q-p'-q')\,\frac{F(s)}{(2p_0'V)^{1/2}\,(2q_0'V)^{1/2}} \tag{8-245}$$

The only approximation here is that only the meson-nucleon intermediate states are kept. This result is exact for energies below the threshold for inelastic scattering.

As we shall see in the next section T or $T^{\dagger}$ is related to the scattering amplitude $f(\theta)$ through

$$\begin{aligned}\langle p, q|\mathcal{T}|p', q'\rangle &= (2p_0V)^{1/2}\,(2q_0V)^{1/2}\,\langle p, q|T|p', q'\rangle(2p_0'V)^{1/2}\,(2q_0'V)^{1/2}\\ &= -8\pi\,W\,f(\theta)\end{aligned} \tag{8-246}$$

where W is the total barycentric energy. Then the above unitarity condition reduces to

$$\begin{aligned}\operatorname{Im} F(s) &= \frac{W}{4\pi}\int\frac{d^3p'}{p_0'}\int\frac{d^3q'}{q_0'}\,\delta^4(p+q-p'-q')\,f^*(\theta)\,F(s)\\ &= \frac{W}{4\pi}\cdot\frac{q}{W}\int d\Omega\,f^*(\theta)\,F(s)\\ &= \frac{q}{4\pi}\,F(s)\int d\Omega\,f^*(\theta)\end{aligned} \tag{8-247}$$

where the momentum q has been given in Section 8-8. Now using the partial wave decomposition of the scattering amplitude;

$$f(\theta) = \frac{1}{q}\sum_{\ell=0}^{\infty}(2\ell+1)e^{i\delta_\ell}\sin\delta_\ell\,P_\ell(\cos\theta) \tag{8-248}$$

the integral of (8-247) easily is evaluated to give

$$\text{Im } F(s) = \exp(-i\delta_0)\sin\,\delta_0\;F(s) \tag{8-249}$$

This shows that the phase of F(s) must be δ_0, so that

$$\text{Im } F(s) = \tan\,\delta_0(s)\cdot\text{Re } F(s)$$

Substituting this expression in the dispersion relation for F(s) we have an integral equation for F(s):

$$\text{Re } F(s) = \frac{P}{\pi}\int_{s_1}^{\infty} ds'\,\frac{\tan\,\delta_0(s')\,\text{Re } F(s')}{s' - s} \tag{8-250}$$

or, after a subtraction is made,

$$\text{Re } F(s) = F(s_0) + \frac{s - s_0}{\pi}\int_{s_1}^{\infty}\frac{ds'}{(s' - s_0)(s' - s)}\tan\,\delta_0(s')\cdot\text{Re } F(s')$$

This is a singular integral equation of Re F(s). We also can write it as

$$F(s) = \frac{1}{\pi}\int_{s_1}^{\infty}\frac{ds'}{s' - s - i\epsilon}\exp[-i\delta_0(s')]\sin\,\delta_0(s')\cdot F(s') \tag{8-251}$$

This type of equation is called the Muskhelishvili-Omnès equation. Although at first glance it looks complicated we can solve it easily. Physically, when we know the phase shift produced by the scattering we can determine the vertex function. In order to solve Eq. (8-251), let us set

$$\frac{F(s)}{F(s_1)} = G(s)$$

and assume the initial phase $\delta_0(s_1) = 0$; then the phase of G(s) is given by $\delta_0(s)$:

$$G(s) = \exp[i\delta_0(s)] \; |G(s)| \qquad s \geq s_1$$

$$= \exp[i\,\delta_0(s) + \Delta(s)] \tag{8-252}$$

We shall determine $\Delta(s)$ from the requirements (1) that G(s) satisfies a dispersion relation and (2) that $\Delta(s_1) = 0$.

The first requirement can be replaced by the equivalent one that G(s) is analytic in the upper half-plane. If the exponent is analytic in the upper half-plane, G(s) also will be analytic. The analyticity of the exponent is not a necessary condition but is sufficient to get a special solution.

The analyticity of $\Delta(s) + i\delta(s)$ implies that it satisfies a dispersion relation

$$\Delta(s) = \frac{P}{\pi} \int_{s_1}^{\infty} \frac{ds'}{s' - s} \; \delta_0(s') + \text{(polynomial of s)} \tag{8-253}$$

If we take a general polynomial of s, we encounter difficulty in writing a dispersion relation for G(s) because G(s) increases exponentially at $s \to \infty$; therefore, the only freedom is to take a constant:

$$\Delta(s) = \frac{P}{\pi} \int_{s_1}^{\infty} \frac{ds'}{s' - s} \; \delta_0(s') + C \tag{8-254}$$

The boundary condition $\Delta(s_1) = 0$ determines the constant. Therefore we get

$$\Delta(s) = \frac{s - s_1}{\pi} \int_{s_1}^{\infty} \frac{ds'}{(s' - s_1)(s' - s)} \; \delta_0(s') \tag{8-255}$$

or

$$\Delta(s) + i\delta_0(s) = \frac{s - s_1}{\pi} \int_{s_1}^{\infty} \frac{ds' \; \delta_0(s')}{(s' - s_1)(s' - s - i\epsilon)} \tag{8-256}$$

Using (8-252), the definition of G(s), we get

$$F(s) = F(s_1) \exp\left[\frac{s - s_1}{\pi} \int_{s_1}^{\infty} \frac{ds' \, \delta_0(s')}{(s' - s_1)(s' - s - i\epsilon)}\right] \tag{8-257}$$

This is a special solution; a more general one is obtained by multiplication of solution (8-257) by a real polynominal P(s):

$$F(s) = F(s_1) \frac{P(s)}{P(s_1)} \exp\left[\frac{s - s_1}{\pi} \int_{s_1}^{\infty} \frac{ds' \, \delta_0(s')}{(s' - s_1)(s' - s - i\epsilon)}\right] \tag{8-258}$$

In general the order of the polynomial is limited by the high energy behavior of F(s). By means of a direct calculation one can show that the high-energy behavior of the exponential function is governed by $\delta_0(\infty)$:

$$D_0^{-1}(s) = \exp\left[\frac{s - s_1}{\pi} \int_{s_1}^{\infty} \frac{ds' \, \delta_0(s')}{(s' - s_1)(s' - s - i\epsilon)}\right]$$

$$\propto s^{-\delta_0(\infty)/\pi} \qquad \text{for } s \to \infty \tag{8-259}$$

With this function D(s) we have

$$F(s) = P(s) \, D_0^{-1}(s) \tag{8-260}$$

where P(s) is normalized by $F(s_1) = P(s_1)$. The order of the polynomial P(s) is limited by the number of subtractions assumed for F(s) as well as by $\delta_0(\infty)$.

8-12 MØLLER'S FORMULA[14] AND THE OPTICAL THEOREM

In Section 4-8 a general formula for evaluating cross sections has been given and it is illustrated by several examples in Chapter 5. Use of the cross section rather than the transition probability is advantageous for two reasons: (1) the cross section is independent of the volume of quantization V provided that V is sufficiently large, and (2) the total cross section is a Lorentz invariant quantity. In the examples studied in Chapter 5 we have already encountered the first point, but it is worthwhile to study it in a systematic way.

In Section 4-8 we introduced the transition amplitude t by

$$\langle f | S | i \rangle = \delta_{fi} - i\delta^4(P_f - P_i) t_{fi}$$

then the transition probability is given by

$$dw_{fi} = \frac{1}{2\pi}\left(\frac{V}{(2\pi)^3}\, d^3 p_1\right)\cdots\left(\frac{V}{(2\pi)^3}\, d^3 p_n\right)\frac{V}{(2\pi)^3}\,\delta^4(P_f - P_i)\,\left|t_{fi}\right|^2$$

and the cross section is

$$d\sigma_{fi} = \frac{V}{v_{rel}}\, dw_{fi}$$

$$= \frac{4\pi^2}{v_{rel}}\left(\frac{V}{(2\pi)^3}\, d^3 p_1\right)\cdots\left(\frac{V}{(2\pi)^3}\, d^3 p_n\right)\left(\frac{V}{(2\pi)^3}\right)\delta^4(P_f - P_i)\left|t_{fi}\right|^2$$

There are many choices of the normalization of the transition amplitudes here we shall choose the one that is most convenient in discussing dispersion relations.
As we already have done, we can define T_{fi} as

$$\langle f|S|i\rangle = \delta_{fi} - i(2\pi)^4\,\delta^4(P_f - P_i)\,T_{fi}$$

then in the lowest-order perturbation theory, T is identical with the interaction Hamiltonian density and obviously $t_{fi} = (2\pi)^4 T_{fi}$.

The Cross-Section Formula

In order to see the cancellation of V in the cross section we write T_{fi} more explicitly for an n-particle final state,

$$T_{fi} = \langle p_1', \ldots, p_n'|T|p_1', p_2\rangle \tag{8-261}$$

and then consider the wave functions for the initial and final states. For spinless particles, incoming and outgoing particles have the wave functions

$$\langle 0|\varphi(x)|p\rangle = (2p_0 V)^{-1/2}\, e^{ipx} \qquad \langle p|\varphi^\dagger(x)|0\rangle = (2p_0 V)^{-1/2}\, e^{-ipx} \tag{8-262}$$

respectively. For spin-$\frac{1}{2}$ particles, the corresponding wave functions are

$$\langle 0|\psi(x)|p, r\rangle = V^{-1/2}\, u^{(r)}(p)\, e^{ipx}$$

$$\langle p, r|\bar\psi(x)|0\rangle = V^{-1/2}\, \bar u^{(r)}(p)\, e^{-ipx} \tag{8-263}$$

respectively. The Dirac spinor u is normalized by $u^* u = 1$. For antiparticles, the wave functions are given respectively, by

$$\langle 0|\bar{\psi}(x)|p, r\rangle = V^{-1/2}\bar{v}^{(r)}(p)e^{ipx}$$
$$\langle p, r|\psi(x)|0\rangle = V^{-1/2}v^{(r)}(p)e^{-ipx} \tag{8-264}$$

The spin summation is performed by use of Casimir operators

$$\sum_r u_\alpha^{(r)}(p)\bar{u}_\beta^{(r)}(p) = \frac{(-ip\gamma + m)_{\alpha\beta}}{2p_0} \tag{8-265}$$

$$\sum_r v_\alpha^{(r)}(p)\bar{v}_\beta^{(r)}(p) = \frac{(-ip\gamma - m)_{\alpha\beta}}{2p_0} \tag{8-266}$$

and the spinors u and v satisfy the Dirac equations

$$(ip\gamma + m)u(p) = \bar{u}(p)(ip\gamma + m) = 0$$
$$(-ip\gamma + m)v(p) = \bar{v}(p)(-ip\gamma + m) = 0 \tag{8-267}$$

We define an invariant matrix $\mathcal{T}_{fi}$ as

$$\langle p_1', \ldots, p_n'|\mathcal{T}|p_1, p_2\rangle$$
$$= (2(p_1')_0 V)^{1/2} \ldots (2(p_n')_0 V)^{1/2} \quad \langle p_1', \ldots, p_n'|T|p_1, p_2\rangle$$
$$\times (2(p_1)_0 V)^{1/2} \quad (2(p_2)_0 V)^{1/2} \tag{8-268}$$

This means that $\mathcal{T}$ is a Lorentz invariant matrix; in $\mathcal{T}_{fi}$ the boson wave function is effectively replaced by

$$\langle 0|\varphi(x)|p\rangle \rightarrow e^{ipx} \qquad \text{etc.} \tag{8-269}$$

and the fermion wave function becomes

$$\langle 0|\psi(x)|p, r\rangle \rightarrow (2p_0)^{1/2} \quad u^{(r)}(p)e^{ipx} \tag{8-270}$$

For the latter the corresponding Casimir operators become invariant:

$$\sum_r (2p_0)^{1/2} u_\alpha^{(r)}(p)(2p_0)^{1/2} \bar{u}_\beta^{(r)}(p) = (-ip\gamma + m)_{\alpha\beta} \tag{8-271}$$

$$\sum_r (2p_0)^{1/2} v_\alpha^{(r)}(p)(2p_0)^{1/2} \bar{v}_\beta^{(r)}(p) = (-ip\gamma - m)_{\alpha\beta} \tag{8-272}$$

It is worthwhile to mention that these Casimir operators are related to the operator $C(\partial)$ that we have encountered in Peierls' method of quantization. In Section 2-9 we defined $D(\partial)$ as

$$[\mathcal{L}]_{\psi_\alpha^\dagger} = D_{\alpha\beta}(\partial)\psi_\beta \tag{8-273}$$

for the Dirac field, but for a covariant treatment of the Dirac field it is more convenient to define $D_{\alpha\beta}(\partial)$ as

$$[\mathcal{L}]_{\bar{\psi}_\alpha} = D_{\alpha\beta}(\partial)\psi_\beta = -(\gamma\partial + m)_{\alpha\beta}\psi_\beta \tag{8-274}$$

The corresponding operator $C(\partial)$, satisfying the relation

$$D_{\alpha\beta}(\partial)\, C_{\beta\gamma}(\partial) = \delta_{\alpha\gamma}(\Box - m^2) \tag{8-275}$$

becomes

$$C_{\alpha\beta}(\partial) = (-\gamma\partial + m)_{\alpha\beta} \tag{8-276}$$

We recognize that the Casimir operators are expressible in terms of $C_{\alpha\beta}(\partial)$, that is,

$$\sum_r (2p_0)^{1/2} u_\alpha^{(r)}(p)\ (2p_0)^{1/2} \bar{u}_\beta^{(r)}(p) = C_{\alpha\beta}(ip) \tag{8-277}$$

$$\sum_r (2p_0)^{1/2} v_\alpha^{(r)}(p)\ (2p_0)^{1/2} \bar{v}_\beta^{(r)}(p) = -C_{\alpha\beta}(-ip) \tag{8-278}$$

(This is one reason why we have adopted the letter C for this operator.)

The total cross section is given, in terms of $\mathcal{T}$, by

$$\sigma = \frac{1}{4(p_1)_0 (p_2)_0 v_{rel}} (2\pi)^{4-3n} \frac{1}{(2S_1 + 1)(2S_2 + 1)}$$

$$\times \int \frac{d^3 p_1'}{2(p_1')_0} \cdots \int \frac{d^3 p_n'}{2(p_n')_0} \delta^4(P_f - P_i) \sum_{s_i} \sum_{s_f} |\mathcal{T}_{fi}|^2 \tag{8-279}$$

where S_1 and S_2 are the spins of the two colliding particles in the initial state, respectively. An average over initial spin states and sum over final spin states are taken. The summation over s_i and s_f denotes the sum over all possible spin orientations of the particles participating in this process. As has been discussed in Section 2-1 the phase space integrals are Lorentz invariant. The only remaining expression that has to be shown to be Lorentz invariant is the factor $(p_1)_0 (p_2)_0 v_{rel}$. In both the center-of-mass and the laboratory systems, v_{rel} is given by

$$v_{rel} = |\mathbf{v}_1 - \mathbf{v}_2| = \frac{|\mathbf{p}_1|}{(p_1)_0} + \frac{|\mathbf{p}_2|}{(p_2)_0} \tag{8-280}$$

so that

$$(p_1)_0 (p_2)_0 v_{rel} = (p_2)_0 |\mathbf{p}_1| + (p_1)_0 |\mathbf{p}_2| \tag{8-281}$$

There is an invariant generalization of this quantity, that is,

$$B = \left(|\mathbf{p}_1 (p_2)_0 - \mathbf{p}_2 (p_1)_0|^2 - |\mathbf{p}_1 \times \mathbf{p}_2|^2 \right)^{1/2}$$

$$= \left((p_1 p_2)^2 - m_1^2 m_2^2 \right)^{1/2} \tag{8-282}$$

Hence we find an invariant expression for the total cross section

$$\sigma = \frac{(2\pi)^{4-3n}}{4B} \int \frac{d^3 p_1'}{2(p_1')_0} \cdots \int \frac{d^3 p_n'}{2(p_n')_0} \delta^4(P_f - P_i) \frac{1}{(2S_1 + 1)(2S_2 + 1)}$$

$$\times \sum_{s_i} \sum_{s_f} |\mathcal{T}_{fi}|^2 \tag{8-283}$$

which is called Møller's formula.[14] When there are identical particles in the final state, this expression has to be divided by the factorial(s) of the number(s) of identical particles.

When there are only two particles in the final state we can simplify formula (8-283); using the manipulations already exhibited in Section 5-3 we can show that in the center-of-mass system

$$\int \frac{d^3 p_1}{(p_1)_0} \int \frac{d^3 p_2}{(p_2)_0} \delta^4(p_f - p_i) \cdots = \frac{p_f}{W} \int d\Omega \cdots$$

where W is the total barycentric energy, and p is the magnitude of the relative momentum in the final state $\frac{1}{2}|\mathbf{p}_1 - \mathbf{p}_2| = |\mathbf{p}_1| = |\mathbf{p}_2|$. Therefore, in this case the differential cross section in the center-of-mass system reduces to

$$\frac{d\sigma}{d\Omega} = \frac{1}{16B}\left(\frac{p_f}{W}\right) \frac{1}{(2\pi)^2} \frac{1}{(2S_1 + 1)(2S_2 + 1)} \sum_{s_i} \sum_{s_f} |\mathcal{T}_{fi}|^2$$

$$= \frac{1}{(8\pi W)^2}\left(\frac{p_f}{p_i}\right) \frac{1}{(2S_1 + 1)(2S_2 + 1)} \sum_{s_i} \sum_{s_f} |\mathcal{T}_{fi}|^2 \tag{8-284}$$

where the relation $B = p_i W$ in the center-of-mass system has been used. For elastic scattering there is the familiar formula

$$\frac{d\sigma}{d\Omega} = |f(\theta)|^2 \tag{8-285}$$

and $\mathcal{T}_{fi}$ is related to the scattering amplitude $f(\theta)$ through

$$\mathcal{T}_{fi} = -8\pi W f(\theta) \tag{8-286}$$

The sign of $f(\theta)$ is chosen to give a positive scattering amplitude for an attractive potential (negative $\mathcal{H}_{int}$) in the Born approximation.

The Lifetime Formula

We now shall consider the lifetime formula; in this case we have only one particle in the initial state and we evaluate the transition probability. The general lifetime formula is

$$\frac{1}{\tau} = w \quad \text{or} \quad \Gamma$$

$$= \frac{(2\pi)^{4-3n}}{2P_0} \frac{1}{2S_i + 1} \int \frac{d^3p_1}{2(p_1)_0} \cdots \int \frac{d^3p_n}{2(p_n)_0} \delta^4(P_f - P_i) \times \sum_{s_i} \sum_{s_f} |\mathcal{T}_{fi}|^2 \tag{8-287}$$

where S_i and P_0 denote the spin and energy, respectively, of the particle in the initial state. The lifetime τ transforms like energy P_0, so

$$\tau = \tau_0 (1 - \beta^2)^{-1/2} \tag{8-288}$$

where τ_0 is the lifetime of the particle at rest, and β is the velocity. For a two-body decay in the rest frame we get

$$\Gamma = \frac{p_f}{8\pi M^2} \frac{1}{2S_i + 1} \sum_{s_i} \sum_{s_f} \left| \mathcal{T}_{fi} \right|^2 \tag{8-289}$$

(Perhaps it is worthwhile to mention the connection between $\mathcal{T}_{fi}$ and another invariant amplitude $\mathfrak{M}$ introduced elsewhere.[10]

$$(2\pi)^4 \langle p_1, \ldots, p_n | \mathcal{T} | q_1, \ldots, q_m \rangle$$

$$= [2(2\pi)^3]^{(m+n)/2} \langle p_1, \ldots, p_n | \mathfrak{M} | q_1, \ldots, q_m \rangle$$

These normalizations are convenient for different purposes, but in this book we shall adhere to $\mathcal{T}$.)

The Optical Theorem

The scattering amplitude for two spinless particles has the partial wave decomposition

$$f(\theta) = \frac{1}{k} \sum_{\ell=0}^{\infty} (2\ell + 1) \exp(i\delta_\ell) \sin \delta_\ell \, P_\ell(\cos\theta)$$

where k denotes the magnitude of the relative momentum. Below the threshold energy for inelastic processes,

$$\sigma_{\text{total}} = \int d\Omega \left| f(\theta) \right|^2$$

$$= \frac{4\pi}{k^2} \sum_{\ell=0}^{\infty} (2\ell + 1) \sin^2 \delta_\ell$$

$$= \frac{4\pi}{k} \operatorname{Im} f(0) \qquad (8\text{-}290)$$

This relation is called the *optical theorem* and actually is valid under more relaxed conditions. We shall derive a general form of the optical theorem from the unitarity condition. The unitarity condition is expressed in terms of T by

$$\operatorname{Im} T_{fi} = -\frac{(2\pi)^4}{2} \sum_m T^\dagger_{fm} \, \delta^4(P_f - P_m) T_{mi}$$

where the left-hand side means the absorptive part of T_{fi}. We then set $|f\rangle = |i\rangle$ to get

$$\operatorname{Im} T_{ii} = -\frac{(2\pi)^4}{2} \sum_m \delta^4(P_i - P_m) \left| T_{mi} \right|^2 \qquad (8\text{-}291)$$

If we express this relation in terms of $\mathcal{T}$, we get

$$\operatorname{Im} \mathcal{T}_{ii} = -\sum_n \frac{(2\pi)^{4-3n}}{2} \int \frac{d^3 p_1}{2(p_1)_0} \cdots \int \frac{d^3 p_n}{2(p_n)_0} \, \delta^4(P_i - P_n) \left| \mathcal{T}_{ni} \right|^2 \qquad (8\text{-}292)$$

Summing over the final-state spin orientations and averaging over the initial spin orientations, we find

$$\frac{1}{(2S_1 + 1)(2S_2 + 1)} \sum_{s_i} \operatorname{Im} \mathcal{T}_{ii} = -2B \sum_n \sigma(n) \qquad (8\text{-}293)$$

where $\sigma(n)$ denotes the cross section for final states with n particles. The total cross section, including inelastic processes, is given by

$$\sigma_{\text{total}} = \sum_{n=2}^{\infty} \sigma(n) \tag{8-294}$$

The left-hand side will be denoted by $\langle \text{Im } \mathcal{T}_{ii} \rangle_{\text{spin}}$ meaning the average over the spin states. Then the general form of the optical theorem is given by

$$\langle \text{Im } \mathcal{T}_{ii} \rangle_{\text{spin}} = -2B\, \sigma_{\text{total}} \tag{8-295}$$

A similar formula holds for the total decay rate; by comparing the lifetime formula with the unitarity condition we immediately find

$$\Gamma_{\text{total}} = \frac{1}{\tau} = -\frac{1}{P_0} \langle \text{Im } \mathcal{T}_{ii} \rangle_{\text{spin}} \tag{8-296}$$

where $|i\rangle$ denotes an unstable particle state. In the rest frame we get

$$\frac{1}{\tau_0} = -\frac{1}{M} \langle \text{Im } \mathcal{T}_{ii} \rangle_{\text{spin}} \tag{8-297}$$

8-13 DISPERSION RELATIONS FOR SCATTERING AMPLITUDES

We have illustrated dispersion calculations for propagators and vertex functions, but so far the dispersion approach has been incomplete as a dynamical theory in the following sense.

In calculating propagators we had to borrow vertex functions from the Feynman-Dyson theory; in calculating vertex functions we had to borrow scattering amplitudes again from the Feynman-Dyson theory in perturbation expansion or from the phenomenological theory in the Muskhelishvili-Omnès approach.

In calculating scattering amplitudes we need scattering amplitudes; thus we can obtain integral equations to determine them. In other words, we can formulate a closed dynamical theory in an approximation neglecting inelastic scattering amplitudes. Then we can go back and determine vertex functions and propagators successively, starting with the scattering amplitudes determined from the approximate dynamical equations in the dispersion approach.

In order to carry out such a program, however, we need an elaborate type of machinery called the Mandelstam representation. Here, we shall confine ourselves to a more elementary project.

Crossing Symmetry

We shall consider meson-nucleon scattering for the scalar model; the scattering amplitude is given in terms of the τ function by

$$\langle p'q'|S|pq\rangle$$
$$= \int d^4x_1 d^4x_2 d^4y_1 d^4y_2 \langle p'|\Phi^\dagger(y_1)|0\rangle \langle q'|\varphi(x_1)|0\rangle$$
$$\times (-i) K_{x_1} (-i) K_{x_2} (-i) K_{y_1} (-i) K_{y_2}$$
$$\langle 0|T[\Phi(y_1)\Phi^\dagger(y_2)\varphi(x_1)\varphi(x_2)]|0\rangle$$
$$\times \langle 0|\varphi(x_2)|q\rangle \quad \langle 0|\Phi(y_2)|p\rangle \tag{8-298}$$

where

$$K_x = \Box_x - m^2 \qquad K_y = \Box_y - M^2 \tag{8-299}$$

In other words,

$$\langle p'q'|S|pq\rangle = -i(2\pi)^4 \frac{\langle p'q'|\mathcal{T}|pq\rangle\, \delta(P_f - P_i)}{(2p_0'V)^{1/2}(2p_0V)^{1/2}(2q_0'V)^{1/2}(2q_0V)^{1/2}}$$
$$= (2p_0'V)^{-1/2}(2p_0V)^{-1/2}(2q_0'V)^{-1/2}(2q_0V)^{-1/2}$$
$$\times \int d^4x_1 d^4x_2 d^4y_1 d^4y_2$$
$$\times \exp(-ip'y_1 + ipy_2 - iq'x_1 + iqx_2)$$
$$\times \bar{\tau}(y_1 y_2 x_1 x_2) \tag{8-300}$$

with $\bar{\tau}$ defined in Eq. (8-133). Since $\bar{\tau}$ is symmetric in x_1 and x_2 and since x_1 and x_2 are variables of integration, we can interchange x_1 and x_2; this practically amounts to the exchange

$$q \rightleftarrows -q' \tag{8-301}$$

Hence

$$\langle p'q'|\mathcal{T}|pq\rangle = \langle p', -q|\mathcal{T}|p, -q'\rangle \tag{8-302}$$

The origin of this symmetry, called crossing symmetry, can be traced back to the Bose statistics of the neutral scalar field. The relation (8-302) has a simple interpretation in terms of Feynman diagrams. In $\langle p', -q |\mathcal{T}| p, -q' \rangle$, the incoming meson of energy momentum $-q'$ is interpreted as an outgoing meson of energy momentum q'. Therefore, starting from a Feynman diagram representing the amplitude $\langle p', q' |\mathcal{T}| p, q \rangle$, we apply the crossing transformation $q \rightleftarrows -q'$, reinterpret the negative energies as prescribed in the preceding sentence, and draw a corresponding Feynman diagram; then we have a new diagram different from the original one. The sum of contributions to $\mathcal{T}$ from such a pair of diagrams makes up a crossing symmetric amplitude. The process is illustrated in Fig. 8-13.

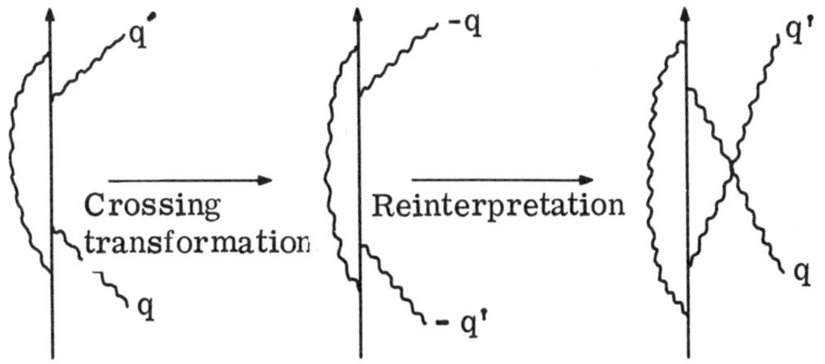

Figure 8-13. Application of the crossing transformation and reinterpretation to a new Feynman diagram.

Dispersion Relation in the Scalar Model

We have derived dispersion relations for vertex functions from the expressions given in the Feynman-Dyson theory. Here we shall calculate a scattering amplitude corresponding to a typical fourth-order Feynman diagram and show that it satisfies a dispersion relation.

The conservation of energy momentum is expressed by

$$p + q = p' + q' \tag{8-303}$$

whereas the mass-shell conditions are

$$p^2 + M^2 = p'^2 + M^2 = q^2 + m^2 = q'^2 + m^2 = 0 \tag{8-304}$$

Therefore from the four four-vectors p, p', q' and q' we can construct only two independent scalar products. Following Chew, Goldberger, Low, and Nambu,[15] we choose

$$\nu = -\frac{PQ}{M} \qquad \text{and} \qquad \varkappa^2 \tag{8-305}$$

as our independent scalar variables, where

$$P = \frac{1}{2}(p + p') \qquad Q = \frac{1}{2}(q + q') \qquad \varkappa = \frac{1}{2}(q - q') \tag{8-306}$$

we also introduce

$$\omega = \nu + \frac{\varkappa^2}{M} = -\frac{pq}{M} \tag{8-307}$$

which has the significance of being the incident meson energy in the laboratory system.

For the scalar model the invariant matrix $\mathcal{T}$ can be evaluated easily in perturbation theory. The Born term is given by

$$\mathcal{T}^{(2)} = -g^2 \left[\frac{1}{(p+q)^2 + M^2} + \frac{1}{(p-q')^2 + M^2}\right] \tag{8-308}$$

corresponding to the crossing conjugate pair of diagrams in Fig. 8-14.

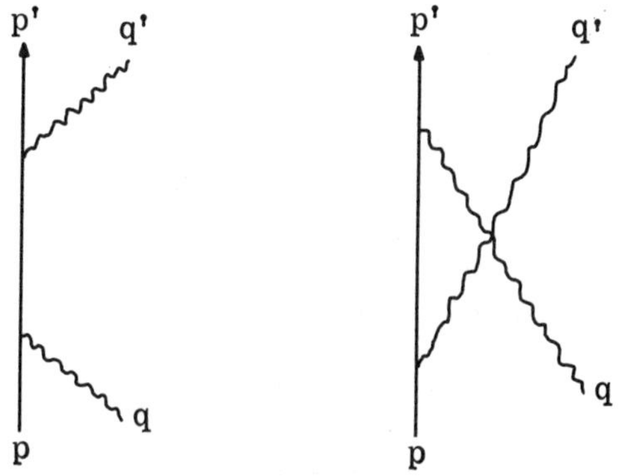

Figure 8-14. Crossing pair of diagrams for meson-nucleon scattering.

In terms of the invariant variables $\mathcal{T}^{(2)}$ is expressed as

$$\mathcal{T}^{(2)} = -\frac{g^2}{2M}\left[\frac{1}{\nu_B - \nu} + \frac{1}{\nu_B + \nu}\right] \tag{8-309}$$

where

$$\nu_B = -\frac{m^2}{2M} - \frac{\varkappa^2}{M} = \frac{qq'}{2M} \tag{8-310}$$

Under the crossing transformation $q \rightleftarrows -q'$, we find

$$P \to P \qquad Q \to -Q \qquad \varkappa \to \varkappa \tag{8-311}$$

and hence

$$\nu \to -\nu \qquad \varkappa^2 \to \varkappa^2 \tag{8-312}$$

The amplitude $\mathcal{T}^{(2)}$ is clearly invariant under $\nu \to -\nu$ so that it satisfies the requirement of crossing symmetry. When higher-order corrections are included, the scattering amplitude $\mathcal{T}$, which is an even function of ν, is known to obey the following dispersion relation:

$$\operatorname{Re} \mathcal{T}(\nu, \varkappa^2) = -\frac{g^2}{2M}\left(\frac{1}{\nu_B - \nu} + \frac{1}{\nu_B + \nu}\right) + \frac{P}{\pi}\int_{m - \frac{\varkappa^2}{M}}^{\infty} d\nu' \left(\frac{1}{\nu' - \nu} + \frac{1}{\nu' + \nu}\right) \operatorname{Im} \mathcal{T}(\nu', \varkappa^2) \tag{8-313}$$

The lowest-order terms are called Born, or pole, terms; the integral represents higher-order corrections. This dispersion relation can be verified easily for the following set of crossing conjugate pairs of diagrams in Fig. 8-15 since we already know the dispersion relations for the renormalized vertex function and propagator.

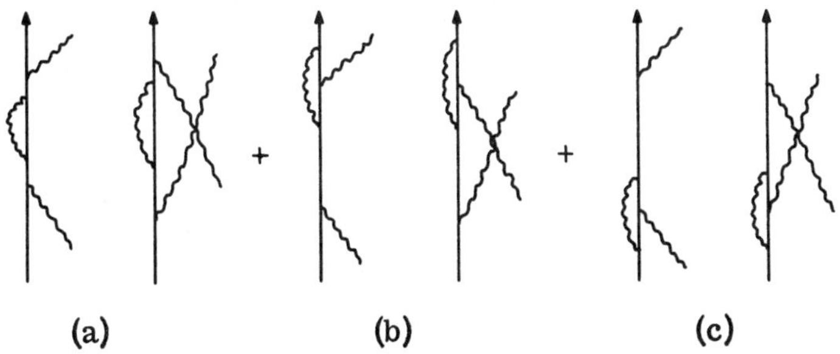

Figure 8-15. Crossing pairs of diagrams representing fourth-order meson nucleon scattering. (a) Propagator corrections. (b) and (c) Vertex corrections.

The only remaining pair of diagrams in the fourth order that has to be examined is shown in Fig. 8-16; the contribution of diagram (a) now will be discussed.

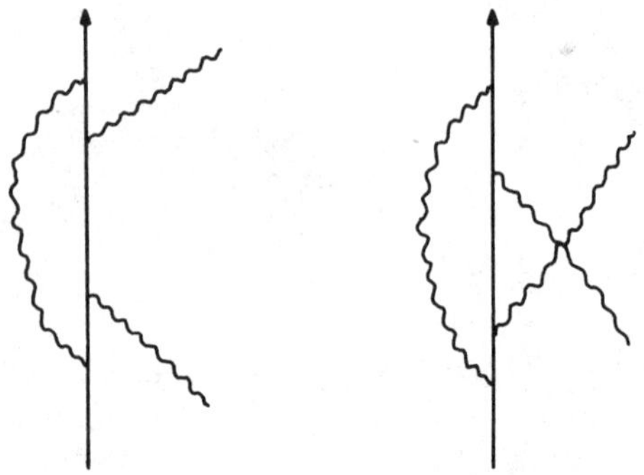

(a) (b)

Figure 8-16. Crossing pair of diagrams, representing fourth-order meson-nucleon scattering. See text for the respective contributions of (a) and (b).

Application of the Feynman-Dyson technique to Fig. 8-16a yields

$$\mathcal{T}_a = -\left(\frac{g^2}{4\pi}\right)^2 \int_0^1 dx_1 \int_0^{x_1} dx_2 \int_0^{x_2} dx_3 \, \frac{1}{(\Lambda - i\epsilon)^2} \tag{8-314}$$

with

$$\begin{aligned} \Lambda &= x_3 m^2 + (1 - x_1)(W^2 + M^2) \\ &\quad - [(x_2 - x_3)p + (x_1 - x_2)p' + (1 - x_1)W]^2 \\ W &= p + q = p' + q' \end{aligned} \tag{8-315}$$

In arriving at this result we have used Feynman's formula

$$\frac{1}{a_1 \ldots a_n} = (n-1)! \int_0^1 dx_1 \ldots \int_0^1 dx_n \, \frac{\delta(1 - \sum x_i)}{\left(a_1 x_1 + \ldots + a_n x_n\right)^n} \tag{8-316}$$

Furthermore, Λ can be expressed as a function of ω and $\varkappa^2$ using

$$\begin{aligned} W^2 &= (p + q)^2 = -(M^2 + m^2 + 2M\omega) \\ pW &= p'W = -(M^2 + M\omega) \\ pp' &= -(M^2 + 2\varkappa^2) \end{aligned} \tag{8-317}$$

that is,

$$\Lambda = [x_3 - x_1(1 - x_1)]\, m^2 + (1 - x_3)^2 M^2 + 4(x_2 - x_3)(x_1 - x_2)\varkappa^2 - 2x_3(1 - x_1) M\omega \tag{8-318}$$

Therefore, for a fixed value of $\varkappa^2$, $\mathcal{T}_a$ can be written in the form

$$\mathcal{T}_a = \int_a^\infty d\omega' \frac{\sigma(\omega' \varkappa^2)}{(\omega' - \omega - i\epsilon)^2} \tag{8-319}$$

where

$$\sigma(\omega' \varkappa^2) = -\left(\frac{g^2}{4\pi}\right)^2 \int_0^1 dx_1 \int_0^{x_1} dx_2 \int_0^{x_2} dx_3 \, [2x_3(1 - x_1)M]^{-2}$$

$$\delta\left(\omega' - \frac{[x_3 - x_1(1 - x_1)]\, m^2 + (1 - x_3)^2 M^2 + 4(x_2 - x_3)(x_1 - x_2)\varkappa^2}{2x_3(1 - x_1)M}\right) \tag{8-320}$$

and a is the minimum possible value of ω' for which σ is different from zero. Since σ is real and $\mathcal{T}_a$ is analytic in the complex ω plane cut along the real ω axis from a to ∞, it satisfies

$$\operatorname{Re} \mathcal{T}_a(\omega, \varkappa^2) = \frac{P}{\pi} \int_a^\infty \frac{d\omega'}{\omega' - \omega} \operatorname{Im} \mathcal{T}_a(\omega', \varkappa^2) \tag{8-321}$$

Thus the problem reduces to the determination of a. In order to minimize ω' we can set $\varkappa^2 = 0$ since ω' is an increasing function of $\varkappa^2$. Hence

$$a = \operatorname{Min} \frac{[x_3 - x_1(1 - x_1)]\, m^2 + (1 - x_3)^2 M^2}{2x_3(1 - x_1)M} \tag{8-322}$$

in the domain $1 \geq x_1 \geq x_3 \geq 0$. We find, by writing x_2 for x_3, that this is exactly the same problem as the one we studied in connection with the vertex function in Section 8-7. Thus we conclude for $M \geq m$ that $a = m$. Then switching our variables from ω to ν in the dispersion integral, we find

$$\operatorname{Re} \mathcal{T}_a(\nu, \varkappa^2) = \frac{P}{\pi} \int_{m-\frac{\varkappa^2}{M}}^{\infty} \frac{d\nu'}{\nu' - \nu} \operatorname{Im} \mathcal{T}_a(\nu', \varkappa^2) \tag{8-323}$$

We next consider $\mathcal{T}_b$, corresponding to diagram b of Fig. 8-16. We find that $\operatorname{Im} \mathcal{T}_b$ survives only for negative values of ν, but crossing symmetry yields

$$\operatorname{Re} \mathcal{T}_b(\nu, \varkappa^2) = \operatorname{Re} \mathcal{T}_a(-\nu, \varkappa^2) \tag{8-324}$$

Therefore,

$$\operatorname{Re} [\mathcal{T}_a(\nu, \varkappa^2) + \mathcal{T}_b(\nu, \varkappa^2)]$$

$$= \frac{P}{\pi} \int_{m-\frac{\varkappa^2}{M}}^{\infty} d\nu' \operatorname{Im} \mathcal{T}_a(\nu', \varkappa^2) \left(\frac{1}{\nu' - \nu} + \frac{1}{\nu' + \nu} \right)$$

$$= \frac{P}{\pi} \int_{m-\frac{\varkappa^2}{M}}^{\infty} d\nu' \operatorname{Im} [\mathcal{T}_a(\nu', \varkappa^2) + \mathcal{T}_b(\nu', \varkappa^2)]$$

$$\times \left(\frac{1}{\nu' - \nu} + \frac{1}{\nu' + \nu} \right) \tag{8-325}$$

on the basis of the relation $\operatorname{Im} \mathcal{T}_b(\nu, \varkappa^2) = 0$ for $\nu > 0$. Thus we essentially have verified the dispersion relation for $\mathcal{T}$ through the fourth order in g.

Dispersion Relations for Pion-Nucleon Scattering

The oldest and the most successful dispersion relations are those for pion-nucleon forward scattering. We must generalize the dispersion relation (8-325) to the realistic pion-nucleon scattering situation, because of kinematic complications arising from the presence of nucleon spin and isospin. The invariant scattering amplitude $\mathcal{T}$ can be written as:

$$\mathcal{T} = (2p_0')^{1/2} \, \bar{u}(p') \, T \, u(p) \, (2p_0)^{1/2} \tag{8-326}$$

where u is a Dirac spinor normalized by $u^*(p)u(p) = 1$ and T is a matrix defined in another paper.[15] The T is a matrix in both Dirac and isospin spaces and can be decomposed as

$$T = -A + i\gamma Q \cdot B \tag{8-327}$$

where A and B are Lorentz-invariant scalars but are still matrices in isospin space; they can be further decomposed as

$$A_{\beta\alpha} = \delta_{\beta\alpha} A^{(+)} + \frac{1}{2}[\tau_\beta, \tau_\alpha] A^{(-)}$$

$$B_{\beta\alpha} = \delta_{\beta\alpha} B^{(+)} + \frac{1}{2}[\tau_\beta, \tau_\alpha] B^{(-)} \tag{8-328}$$

where α and β are isospin indices of the incoming and outgoing pions, respectively. Under the crossing-symmetric operation

$$i\gamma Q \to -i\gamma Q, \quad [\tau_\beta, \tau_\alpha] \to -[\tau_\beta, \tau_\alpha] \tag{8-329}$$

since crossing requires the transformation $q \rightleftarrows -q'$ and, in this case, the transformation $\alpha \rightleftarrows \beta$.

The invariance of $\mathcal{T}$ under crossing implies

$$A^{(+)} \to A^{(+)}, \qquad A^{(-)} \to -A^{(-)}, \qquad B^{(+)} \to -B^{(+)}, \qquad B^{(-)} \to B^{(-)} \tag{8-330}$$

Therefore, $A^{(+)}$ and $B^{(-)}$ are even functions of ν, whereas $A^{(-)}$ and $B^{(+)}$ are odd functions of ν. The dispersion relations for these four amplitudes are given by

$$\mathrm{Re}\, A^{(\pm)}(\nu, \varkappa^2) = \frac{P}{\pi} \int_{\mu - \frac{\varkappa^2}{M}}^{\infty} d\nu' \,\mathrm{Im}\, A^{(\pm)}(\nu', \varkappa^2) \times \left(\frac{1}{\nu' - \nu} \pm \frac{1}{\nu' + \nu} \right) \tag{8-331}$$

$$\mathrm{Re}\, B^{(\pm)}(\nu, \varkappa^2) = \frac{G^2}{2M} \left(\frac{1}{\nu_B - \nu} \mp \frac{1}{\nu_B + \nu} \right) + \frac{P}{\pi} \int_{\mu - \frac{\varkappa^2}{M}}^{\infty} d\nu' \,\mathrm{Im}\, B^{(\pm)}(\nu', \varkappa^2) \times \left(\frac{1}{\nu' - \nu} \mp \frac{1}{\nu' + \nu} \right) \tag{8-332}$$

where μ denotes the pion rest mass and G the pion-nucleon coupling constant defined in Section 4-1. These dispersion relations are generalizations of (8-313), the one for the scalar model.

In order to discuss the forward-scattering dispersion relations it is convenient to introduce the following amplitudes:

$$C^{(\pm)}(\nu, \kappa^2) = A^{(\pm)}(\nu, \kappa^2) + \nu B^{(\pm)}(\nu, \kappa^2) \tag{8-333}$$

If we set $\kappa = 0$, then $\nu = \omega$, that is, ν becomes equal to the pion energy ω in the laboratory system. The dispersion relations for $C^{(\pm)}(\omega) = C^{(\pm)}(\nu, 0)$ then are given by

$$\begin{aligned} \operatorname{Re} C^{(\pm)}(\omega) &= \frac{G^2}{2M}\omega_0\left(\frac{1}{\omega_0 - \omega} \pm \frac{1}{\omega_0 + \omega}\right) \\ &\quad + \frac{P}{\pi}\int_{\mu}^{\infty} d\omega' \operatorname{Im} C^{(\pm)}(\omega')\left(\frac{1}{\omega' - \omega} \pm \frac{1}{\omega' + \omega}\right) \\ &\quad + \text{"subtraction"} \end{aligned} \tag{8-334}$$

where $\omega_0 = -\mu^2/2M$, and "subtraction" means that one has to write the dispersion relation in the once-subtracted form (8-183) when the original dispersion integral diverges. The high-energy behavior of the dispersion integrand for $C^{(+)}$ is worse than that for $C^{(-)}$, and we need a subtraction for $C^{(+)}$. We write, therefore, a once-subtracted dispersion relation for $C^{(+)}$.

$$\begin{aligned} &\operatorname{Re} C^{(+)}(\omega) - \operatorname{Re} C^{(+)}(\mu) \\ &= \frac{f^2}{M}\,\frac{\omega^2 - \mu^2}{\omega^2 - (\mu^2/2M)^2}\,\frac{1}{1 - (\mu/2M)^2} \\ &\quad + \frac{2(\omega^2 - \mu^2)}{\pi}\int_{\mu}^{\infty} d\omega' \frac{\omega' \operatorname{Im} C^{(+)}(\omega')}{(\omega'^2 - \omega^2)(\omega'^2 - \mu^2)} \\ &\approx \frac{f^2}{M}\,\frac{\omega^2 - \mu^2}{\omega^2 - (\mu^2/2M)^2} \\ &\quad + \frac{2(\omega^2 - \mu^2)}{\pi}\int_{\mu}^{\infty} d\omega' \frac{\omega' \operatorname{Im} C^{(+)}(\omega')}{(\omega'^2 - \omega^2)(\omega'^2 - \mu^2)} \end{aligned} \tag{8-335}$$

where $f = (\mu/2M)G$ and we have made the approximation $1 - (\mu/2M)^2 \approx 1$.

In order to test these dispersion relations experimentally we have to express the amplitudes $C^{(\pm)}$ in terms of more familiar quantities. First we shall study the absorptive part of the forward-scattering amplitude $\mathcal{T}(\omega, 0)$ with no change of nucleon isospin and hence $\beta = \alpha$.

$$\langle \mathcal{T}(\omega, 0)\rangle_{\text{spin}} = \langle (2p_0)^{1/2}\, \bar{u}(p)(-A + i\gamma Q \cdot B)\, u(p)(2p_0)^{1/2}\rangle_{\text{spin}}$$

$$= \frac{1}{2}\,\mathrm{Tr}\,[(-A + i\gamma Q \cdot B)(-i\gamma p + M)]$$

$$= -2MA + 2(pQ)B$$

$$= -2M(A + \nu B)$$

$$= -2MC \tag{8-336}$$

From the optical theorem (8-295) we know

$$\mathrm{Im}\,\langle \mathcal{T}(\omega, 0)\rangle_{\text{spin}} = -2B\,\sigma_{\text{total}}$$

$$= -2Mk\,\sigma_{\text{total}} \tag{8-337}$$

where k is the laboratory momentum of the incident pion, and σ_{total} is the total cross section of the pion with isospin index α on the proton. Hence

$$\mathrm{Im}\, C(\omega) = k\,\sigma_{\text{total}} = k\,\sigma(\omega) \tag{8-338}$$

From Eq. (8-286), the relation between the scattering amplitude $f(\theta)$ and the invariant amplitude $\mathcal{T}$, we can derive

$$\mathrm{Re}\, C(\omega) = 4\pi \frac{k}{q}\,\mathrm{Re}\, f(\omega, \theta = 0) \tag{8-339}$$

where q is the center-of-mass momentum of the incident pion.

Instead of dealing with mathematical pions with isospin indices $\alpha = 1, 2$, we can introduce charged pions with corresponding amplitudes:

$$C^{(+)}(\omega) = \frac{1}{2}\,[C_-(\omega) + C_+(\omega)]$$

$$C^{(-)}(\omega) = \frac{1}{2}\,[C_-(\omega) - C_+(\omega)] \tag{8-340}$$

where C_- and C_+ are the amplitudes, respectively, for

$$\pi^- + p \to p + \pi^-$$

$$\pi^+ + p \to p + \pi^+ \tag{8-341}$$

Hence,

$$\text{Re}\, C^{(\pm)}(\omega) = 2\pi \frac{k}{q}\, \text{Re}\, [f_-(\omega) \pm f_+(\omega)]$$

$$\text{Im}\, C^{(\pm)}(\omega) = \frac{k}{2}\, [\sigma_-(\omega) \pm \sigma_+(\omega)] \tag{8-342}$$

where the subscript (±) refers to the charge of the incident pion. Thus the dispersion integrals can be expressed in terms of experimentally measurable quantities. The dispersive parts at low energies can be expressed also in terms of experimentally measurable phase shifts. Following Goldberger, Miyazawa, and Oehme[16] we use the notation

$$D_\pm(\omega) = \frac{1}{4\pi}\, \text{Re}\, C_\pm(\omega) = \frac{k}{q}\, \text{Re}\, f_\pm(\omega) \tag{8-343}$$

We express $D_\pm(\omega)$ in terms of scattering amplitudes in pure isospin states, namely,

$$D_+(\omega) = D_3(\omega) \qquad D_-(\omega) = \frac{1}{3}\left[2D_1(\omega) + D_3(\omega)\right] \tag{8-344}$$

where $D_{2I}(\omega)$ denotes the dispersive part of the scattering amplitude in the pure isospin I state. From the partial wave decomposition of the scattering amplitude in the forward direction without the spin-flip terms we get

$$D_3(\omega) = \frac{k}{2q^2}(\sin 2\alpha_3 + 2 \sin 2\alpha_{33} + \sin 2\alpha_{31} + \cdots)$$

$$D_1(\omega) = \frac{k}{2q^2}(\sin 2\alpha_1 + 2 \sin 2\alpha_{13} + \sin 2\alpha_{11} + \cdots) \tag{8-345}$$

where α_{2I} denotes the S-wave phase shift in the isospin I state, and $\alpha_{2I,\,2J}$ denotes the P-wave phase shift in the isospin I and total angular momentum J state.

The dispersion relations (8-335) agree very well with experiment for the choice

$$\frac{f^2}{4\pi} \approx 0.08 \tag{8-346}$$

Use of Retarded Products

From the reduction formalism we get for pion-nucleon scattering the expression:

$$\langle p' ; q'\beta | S | p ; q\alpha \rangle$$

$$= \int d^4x_1 \, d^4x_2 \, \langle q' | \varphi_\beta (x_2) | 0 \rangle \, \langle 0 | \varphi_\alpha (x_1) | q \rangle$$

$$\times (-i) K_{x_1} (-i) K_{x_2} \, \langle p' | T [\varphi_\beta (x_2) \varphi_\alpha (x_1)] | p \rangle \qquad (8\text{-}347)$$

In particular, for the forward-scattering amplitude $p' = p$ the matrix element of the T product becomes a function of the difference $x = x_1 - x_2$ alone. Hence we find for the forward-scattering amplitude

$$\mathcal{T}(\omega) (2p_0)^{-2/2}$$

$$= i \int d^4x \, e^{iqx} (-i)K_x (-i)K_y \langle p | T[\varphi_\alpha(x), \varphi_\beta (y)] | p \rangle_{y=0, \, \beta=\alpha}$$

$$= i \int d^4x \, e^{-iqx} (-i)K_x (-i)K_y \langle p | T[\varphi_\alpha(y), \varphi_\beta (x)] | p \rangle_{y=0, \, \beta=\alpha} \qquad (8\text{-}348)$$

since this expression is an even function of q. When averaged over the proton spin the left-hand side is equal, according to the optical theorem (8-290), to -C in the laboratory system. Now we rewrite the T product as

$$T[\varphi_\alpha(y)\varphi_\beta (x)] = \theta(x_0 - y_0) [\varphi_\beta(x), \varphi_\alpha (y)] + \varphi_\alpha(y) \varphi_\beta(x) \qquad (8\text{-}349)$$

It can be shown that the second term does not give any contribution to the integral in (8-348) since we cannot find an intermediate state n that satisfies

$$\int d^4x \, e^{-iqx} \langle n | K_x \varphi_\beta(x) | p \rangle = (2\pi)^4 \, \delta^4 (p-p_n-q) \langle n | K_x \varphi_\beta(0) | p \rangle$$

$$\neq 0$$

Conservation of energy requires that $(p_n)_0 + q_0 = p_0 = M$ in the rest frame of the proton, but the energy of the intermediate state n, carrying unit baryon number, cannot be lower than that of the proton mass. Therefore, we can replace the T product by the retarded commutator.

$$\mathcal{T}(\omega)(2p_0)^{-1}$$
$$= -i \int d^4x\, e^{-iqx}\, K_x K_y\, \theta(x_0 - y_0)\langle p|[\varphi_\beta(x),\ \varphi_\alpha(y)]|p\rangle_{y=0,\ \alpha=\beta} \tag{8-350}$$

For a rigorous treatment of dispersion relations this commutator form is more convenient.

8-14 THE GOLDBERGER–TREIMAN RELATION

An approximate dynamical scheme in dispersion theory can be developed on the basis of the Mandelstam representation, but we can do a number of things before introducing this representation. As an example we shall discuss an application of dispersion theory to weak interactions; here we shall derive the so-called Goldberger-Treiman relation.

The *phenomenological* Hamiltonian density for beta decay or mu capture is given usually by the so-called V-A theory.

$$\mathcal{H} = (g_V \bar{\psi}_p \gamma_\lambda \psi_n + g_A \bar{\psi}_p \gamma_\lambda \gamma_5 \psi_n)\, \bar{\psi}_\ell\, \gamma_\lambda (1 + \gamma_5)\psi_\nu + \text{Hermitian conjugate} \tag{8-351}$$

where ψ_ℓ denotes the lepton field. We shall study only the axial-vector part of the Hamiltonian, which will be denoted by

$$\mathcal{H}_A = -A_\lambda \cdot \bar{\psi}_\nu i\gamma_\lambda (1 + \gamma_5)\, \psi_\ell - A_\lambda^\dagger\, \bar{\psi}_\ell i\gamma_\lambda (1 + \gamma_5)\psi_\nu \tag{8-352}$$

If A_λ is to be expressed in terms of field operators we should write an expression like the phenomenological one of (8-351), but we shall not do so since we do not know if it is possible. Instead of writing an explicit expression for A_λ, we shall determine its matrix elements by means of dispersion relations.

For example, let us consider the matrix element for charged-pion decay $\pi^+ \to \mu^+ + \nu$. In the first order of weak interactions we get

$$\langle \mu\nu|S|\pi\rangle = -i \int d^4x\, \langle \mu\nu|\mathcal{H}_A(x)|\pi\rangle \tag{8-353}$$

This problem reduces to the evaluation of the matrix element of A_λ between the vacuum and the single-pion state.

The decay rate is then determined from the matrix element

$$\langle \pi^-|A_\lambda(0)|0\rangle = (2q_0)^{-1/2}\, iq_\lambda f_\pi \tag{8-354}$$

where q_λ is the four-momentum of the π^- meson and we have set $V = 1$ for simplicity. This form of the matrix element follows from the Lorentz invariance of the theory. The decay rate for $\pi \to \mu + \nu$ then is given by

$$\frac{1}{\tau} = m_\pi \left(1 - \frac{m_\mu^2}{m_\pi^2}\right)^2 \frac{(m_\mu f_\pi)^2}{4\pi} \tag{8-355}$$

Let us now consider the divergence of A as a field operator $\Phi = \partial_\lambda A_\lambda$, then Φ is a pseudoscalar field and we get

$$\langle \pi^- | \Phi(0) | 0 \rangle = (2q_0)^{-1/2}\, q^2 f_\pi = -(2q_0)^{-1/2}\, \mu^2 f_\pi \tag{8-356}$$

where μ denotes m_π. A nucleon form factor associated with this field operator is introduced by

$$\langle n, \bar{p}\,(-) | \Phi(0) | 0 \rangle = \bar{u}\,(n)\, i\gamma_5\, v(\bar{p}) \cdot f_{n\bar{p}}\,(s) \tag{8-357}$$

where $s = -(n + \bar{p})^2$ is the total barycentric energy squared for the $n\bar{p}$ system. The u(n) and v(p) are the Dirac spinors introduced in Section 8-12. The phase convention of the state np is fixed by

$$|n\bar{p}(-)\rangle = a^\dagger(n)^{out}\, b^\dagger(\bar{p})^{out}\, |0\rangle \tag{8-358}$$

It is necessary also to introduce form factors associated with the original axial-vector A_λ:

$$\langle n\bar{p}(-) | A_\lambda(0) | 0 \rangle = \bar{u}(n)\, [a_{n\bar{p}}(s) i\gamma_\lambda \gamma_5 + b_{n\bar{p}}(s)\,(n + \bar{p})_\lambda\, \gamma_5]\, v(\bar{p}) \tag{8-359}$$

By using the Dirac equations for $\bar{u}(n)$ and $v(\bar{p})$ we easily can prove

$$f_{n\bar{p}}(s) = 2M\, a_{n\bar{p}}(s) + s\, b_{n\bar{p}}(s) \tag{8-360}$$

where M is the nucleon mass. The renormalized or observed axial-vector coupling constant g_A is defined as

$$g_A = a_{n\bar{p}}(0) \tag{8-361}$$

Combining Eqs. (8-360) and (8-361), we find

$$f_{n\bar{p}}(0) = 2M\, g_A \tag{8-362}$$

We next shall study the dispersion relation for $f_{n\bar{p}}(s)$. First we write the unitarity condition for the form factors associated with Φ,

$$\mathrm{Im}\,\langle\alpha(-)|\Phi(0)|0\rangle = -\frac{(2\pi)^4}{2}\sum_{\beta} T^{\dagger}_{\alpha\beta}\,\delta^4(P_\alpha - P_\beta)\langle\beta(-)|\Phi(0)|0\rangle \tag{8-363}$$

which has been derived in Sections 8-8 through 8-11.

For the nucleon-antinucleon channel $\alpha = n\bar{p}$, we keep only the least massive intermediate state $\beta = \pi^-$ in the unitarity condition, then

$$\mathrm{Im}\langle np(-)|\Phi(0)|0\rangle = -\frac{(2\pi)^4}{2}\int\frac{d^3q}{(2\pi)^3}\frac{1}{2q_0}\,\bar{u}(n)iG\sqrt{2}\,\gamma_5 v(\bar{p}) \times (-\mu^2)\,f_\pi\,\delta^4\,(n + \bar{p} - q) \tag{8-364}$$

where we have used

$$\langle n\bar{p}|T^{\dagger}|\pi^-\rangle = \bar{u}(n)iG\sqrt{2}\,\gamma_5 v(\bar{p})\cdot(2q_0)^{-1/2}$$

$$\langle\pi^-|\Phi(0)|0\rangle = (2q_0)^{-1/2}\,(-\mu^2)\,f_\pi \tag{8-365}$$

The evaluation of the integral of (8-364) is almost exactly the same as that of the Lehmann weight function for the pole term, so

$$\mathrm{Im}\,f_{n\bar{p}}(s) = \pi\sqrt{2}\,G\,f_\pi\mu^2\,\delta(s - \mu^2) \tag{8-366}$$

This results in a pole term in $f_{n\bar{p}}(s)$. The next-least-massive state is the three-pion state, where the dispersion relation for $f_{n\bar{p}}(s)$ must be of the form

$$f_{n\bar{p}}(s) = \frac{\sqrt{2}\,Gf_\pi\,\mu^2}{\mu^2 - s} + \frac{1}{\pi}\int_{(3\mu)^2}^{\infty} ds'\,\frac{\mathrm{Im}\,f_{n\bar{p}}(s')}{s' - s - i\epsilon} \tag{8-367}$$

We have assumed an unsubtracted dispersion relation for $f_{n\bar{p}}(s)$, which is the basis[17] for deriving the Goldberger-Treiman relation. Setting $s = 0$ in the dispersion relation (8-367) and recalling the relation (8-362), we get

$$2M\,g_A = \sqrt{2}\,G\,f_\pi + \frac{1}{\pi}\int_{(3\mu)^2}^{\infty} ds'\,\frac{\mathrm{Im}\,f_{n\bar{p}}(s')}{s'} \tag{8-368}$$

If we assume that the pole term is dominant for small values of s, namely $(3\mu)^2 \gg s \sim \mu^2$, and neglect the dispersion integral for s = 0, we get an approximate relationship between g_A and f_π,

$$2M\, g_A \approx \sqrt{2}\, G\, f_\pi \tag{8-369}$$

This relation was derived first by Goldberger and Treiman,[18] and has been named after them. This relation agrees with experiment within 10%. The foregoing derivation is based on the assumed unsubtracted dispersion relation for $f_{n\bar{p}}$, but the original derivation is based on a different assumption; we shall introduce other derivations in the discussion that follows.

Let us first write the dispersion relation for $f_{n\bar{p}}(s)$ in a once-subtracted form:

$$f_{n\bar{p}}(s) = 2M\, g_A + \frac{\sqrt{2}\, G f_\pi s}{\mu^2 - s} + \frac{s}{\pi} \int_{(3\mu)^2}^{\infty} ds' \frac{\text{Im}\, f_{n\bar{p}}(s')}{s'(s' - s - i\epsilon)} \tag{8-370}$$

We shall generalize this dispersion relation to an arbitrary channel α

$$f_\alpha(s) = f_\alpha(0) + \frac{G_\alpha f_\pi s}{\mu^2 - s} + \frac{s}{\pi} \int_{(3\mu)^2}^{\infty} ds' \frac{\text{Im}\, f_\alpha(s')}{s'(s' - s - i\epsilon)} \tag{8-371}$$

where

$$\langle \alpha(-)|\Phi(0)|0\rangle = c_\alpha f_\alpha(s) \qquad T^\dagger_{\alpha,\pi^-} = c_\alpha G_\alpha (2q_0)^{-1/2} \tag{8-372}$$

and c_α is an invariant in the channel α, and G_α is the coupling constant between the π^- meson and the channel α, for instance,

$$c_{n\bar{p}} = \bar{u}(n) i\gamma_5 v(\bar{p}) \qquad G_{n\bar{p}} = \sqrt{2}\, G \tag{8-373}$$

The dispersion relation (8-371) is reminiscent of those for $\Box\varphi$ discussed in Section 8-7; that is, if we define

$$\langle n\bar{p}(-)|\Box\varphi(0)|0\rangle = \bar{u}(n)\, i\gamma_5\, v(\bar{p}) \cdot h_{n\bar{p}}(s) \tag{8-374}$$

then the form factor $h_{n\bar{p}}(s)$ satisfies

$$h_{n\bar{p}}(s) = -\frac{\sqrt{2}\, Gs}{\mu^2 - s} + \frac{s}{\pi} \int_{(3\mu)^2}^{\infty} ds' \frac{\text{Im}\, h_{n\bar{p}}(s')}{s'(s' - s - i\epsilon)} \tag{8-375}$$

This dispersion relation is easily obtained from (8-165) for the form factor (8-163) describing the matrix element of $K_x\varphi(0)$; we soon shall come back to the latter form factors.

The generalization of the dispersion relation (8-375) to an arbitrary channel α is

$$h_\alpha(s) = -\frac{G_\alpha s}{\mu^2 - s} + \frac{s}{\pi}\int_{(3\mu)^2}^{\infty} ds' \frac{\operatorname{Im} h_\alpha(s')}{s'(s' - s - i\epsilon)} \tag{8-376}$$

Let us consider the combination

$$L_\alpha(s) = f_\alpha(s) + f_\pi h_\alpha(s) \tag{8-377}$$

then the new function $L_\alpha(s)$ satisfies a dispersion relation

$$L_\alpha(s) = f_\alpha(0) + \frac{s}{\pi}\int_{(3\mu)^2}^{\infty} ds' \frac{\operatorname{Im} L_\alpha(s')}{s'(s' - s - i\epsilon)} \tag{8-378}$$

There is no pole term in this dispersion relation; the poles of $f_\alpha(s)$ and $f_\pi h_\alpha(s)$ cancel each other. The quantity $h_\alpha(s)$ is rather unfamiliar, so we define the pion form factors $K_\alpha(s)$ as

$$\langle \alpha(-)|K_x\varphi(0)|0\rangle = c_\alpha K_\alpha(s) \tag{8-379}$$

It is clear that $h_\alpha(s)$ is related to $K_\alpha(s)$ by

$$h_\alpha(s) = \frac{s}{s - \mu^2} K_\alpha(s)$$

so that

$$L_\alpha(s) = f_\alpha(s) + \frac{s}{s - \mu^2} f_\pi K_\alpha(s) \tag{8-380}$$

Also,

$$\langle \alpha(-)|\Phi(0) + f_\pi \Box \varphi(0)|0\rangle$$

$$= \langle \alpha(-)|\partial_\lambda\left(A_\lambda(0) + f_\pi \partial_\lambda \varphi(0)\right)|0\rangle = c_\alpha L_\alpha(s) \tag{8-381}$$

Because of the simpler dispersion relations we can regard the set $L_\alpha(s)$ as being more fundamental than the set $f_\alpha(s)$. The function $K_\alpha(s)$ also satisfies a dispersion relation without the pion pole term.

$$K_\alpha(s) = G_\alpha + \frac{s - \mu^2}{\pi} \int_{(3\mu)^2}^{\infty} ds' \frac{\operatorname{Im} K_\alpha(s')}{(s' - \mu^2)(s' - s - i\epsilon)} \tag{8-382}$$

The decay constant f_π has been defined as

$$\langle \pi^- | \Phi(0) | 0 \rangle = \langle 0 | \Phi(0) | \pi^+ \rangle = -(2q_0)^{-1/2} \mu^2 f_\pi \tag{8-383}$$

but we can generalize this formula to the case of the off-shell pion "π^-" of the mass $s^{1/2}$ by

$$\langle "\pi^-"(-) | \Phi(0) | 0 \rangle = -(2q_0)^{-1/2} s f_\pi(s) \tag{8-384}$$

or more rigorously, with the help of the LSZ reduction formula, by

$$s f_\pi(s) = i \int d^4z\, e^{-iqz} (\Box_z - \mu^2) \langle 0 | T[\Phi(0)\varphi^\dagger(z)] | 0 \rangle \tag{8-385}$$

with $s = -q^2$. The function $f_\pi(s)$ is normalized by

$$f_\pi(\mu^2) = f_\pi \tag{8-386}$$

Taking the absorptive part of (8-385) we have

$$s \operatorname{Im} f_\pi(s) = \frac{(2\pi)^4}{2} \sum_\alpha K_\alpha^*(s) f_\alpha(s) c_\alpha^* c_\alpha \delta^4(q - P_\alpha) \tag{8-387}$$

Let us introduce a density function for the channel α by

$$\rho_\alpha(s) = (2\pi)^3 \sum_{n \in \alpha} c_\alpha^* c_\alpha \delta^4(q - P_n) \tag{8-388}$$

where the summation is taken over all the states in the channel α, then the absorptive part of $f_\pi(s)$ is given by

$$s \operatorname{Im} f_\pi(s) = \pi \sum_\alpha K_\alpha^*(s) f_\alpha(s) \rho_\alpha(s) \tag{8-389}$$

In their original derivation Goldberger and Treiman,[18] assumed *an unsubtracted* dispersion relation for $f_\pi(s)$, namely,

$$f_\pi(s) = \frac{1}{\pi} \int_{(3\mu)^2}^{\infty} ds' \frac{\operatorname{Im} f_\pi(s')}{s' - s - i\epsilon} \tag{8-390}$$

This assumption is important since otherwise we have to introduce an arbitrary subtraction constant and it becomes impossible to determine f_π. Now let us recall the definition of the Lehmann weight function $\sigma(s)$ for the pion field:

$$\begin{aligned}\sigma(s) &= (2\pi)^3 \sum_{\alpha,\ldots} \left| \frac{c_\alpha K_\alpha(s)}{s - \mu^2} \right|^2 \delta(P_\alpha - q) \\ &= \frac{1}{(s-\mu^2)^2} \sum_\alpha \left|K_\alpha(s)\right|^2 \sum_{n\epsilon\alpha} (2\pi)^3 c_\alpha^* c_\alpha \delta^4(P_n - q) \\ &= \frac{1}{(s-\mu^2)^2} \sum_\alpha \left|K_\alpha(s)\right|^2 \rho_\alpha(s)\end{aligned} \tag{8-391}$$

or

$$(s-\mu^2)^2 \sigma(s) = \sum_\alpha \left|K_\alpha(s)\right|^2 \rho_\alpha(s) \tag{8-392}$$

A similar function $\gamma(s)$ will be defined as

$$(s-\mu^2)^2 \gamma(s) = \sum_\alpha K_\alpha^*(s) L_\alpha(s) \rho_\alpha(s) \tag{8-393}$$

then we can express the right-hand side of the unitarity condition (8-389) for $f_\pi(s)$ in terms of $\sigma(s)$ and $\gamma(s)$.

$$s \operatorname{Im} f_\pi(s) = \pi(s-\mu^2)^2 \left[\gamma(s) - f_\pi \frac{s}{s-\mu^2} \sigma(s)\right] \tag{8-394}$$

or

$$\frac{1}{\pi} \frac{\operatorname{Im} f_\pi(s)}{s-\mu^2} = \frac{s-\mu^2}{s} \gamma(s) - f_\pi \sigma(s) \tag{8-395}$$

From the unsubtracted dispersion relation for $f_\pi(s)$ we find

$$f_\pi = f_\pi(\mu^2) = \int_{(3\mu)^2}^{\infty} ds \, \frac{s - \mu^2}{s} \, \gamma(s) - f_\pi \int_{(3\mu)^2}^{\infty} ds \, \sigma(s) \tag{8-396}$$

Solving this equation for f_π, we find

$$f_\pi = \int ds \, \frac{(s - \mu^2)}{s} \, \gamma(s) \Big/ [1 + \int \sigma(s) ds] \tag{8-397}$$

No approximation has been made in deriving this result, first obtained by Ida.[19] We notice at this point that the denominator is exactly equal to the wave-function renormalization constant as given in Section 8-1.

$$Z_\pi^{-1} = 1 + \int_{(3\mu)^2}^{\infty} \sigma(s) ds \tag{8-398}$$

In lowest-order perturbation theory for the diagram in Fig. 8-17

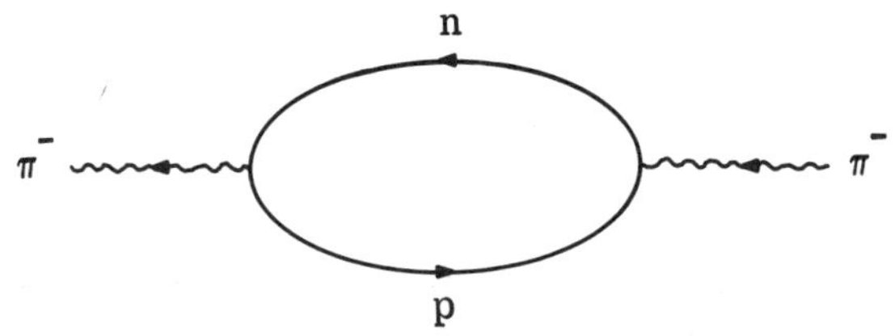

Figure 8-17. The lowest-order diagram giving contributions to σ (s).

we get

$$\sigma(s) = \frac{G^2}{4\pi^2} \, \frac{s^{1/2} (s - 4M^2)^{1/2}}{(s - \mu^2)^2} \tag{8-399}$$

This makes Z_π^{-1} divergent. Goldberger and Treiman assumed that the integrals in both the numerator and the denominator are very large, and approximated f_π by

$$f_\pi \approx \frac{\int ds \, \gamma(s)}{\int ds \, \sigma(s)} \tag{8-400}$$

In the approximation keeping only the nucleon-antinucleon channel, this ratio becomes very simple. In this approximation the unitary condition on the form factors $K_{n\bar{p}}(s)$ and $L_{n\bar{p}}(s)$ gives, as discussed in Section 8-11, the results

$$\begin{aligned} \text{Im } K_{n\bar{p}}(s) &= \tan \delta(s) \text{ Re } K_{n\bar{p}}(s) \\ \text{Im } L_{n\bar{p}}(s) &= \tan \delta(s) \text{ Re } L_{n\bar{p}}(s) \end{aligned} \tag{8-401}$$

where tan δ is the ratio of the imaginary to the real part of the amplitude for elastic $n\bar{p}$ scattering in the $^1S_0$ state.

The dispersion relations for the form factors (8-401) are

$$\begin{aligned} K_{n\bar{p}}(s) &= \sqrt{2}\, G + \frac{s - \mu^2}{\pi} \int_{4M^2}^{\infty} ds' \frac{\text{Im } K_{n\bar{p}}(s')}{(s' - \mu^2)(s' - s - i\epsilon)} \\ &= K_{n\bar{p}}(0) + \frac{s}{\pi} \int_{4M^2}^{\infty} ds' \frac{\text{Im } K_{n\bar{p}}(s')}{s'(s' - s - i\epsilon)} \end{aligned} \tag{8-402}$$

and

$$L_{n\bar{p}}(s) = 2Mg_A + \frac{s}{\pi} \int_{4M^2}^{\infty} ds' \frac{\text{Im } L_{n\bar{p}}(s')}{s'(s' - s - i\epsilon)} \tag{8-403}$$

Therefore, both K and L satisfy exactly the same Omnès equation except for normalization, so they should be proportional, that is,

$$\frac{L_{n\bar{p}}(s)}{K_{n\bar{p}}(s)} = \frac{L_{n\bar{p}}(0)}{K_{n\bar{p}}(0)} \approx \frac{L_{n\bar{p}}(0)}{K_{n\bar{p}}(\mu^2)} = \frac{2Mg_A}{\sqrt{2}\, G} \tag{8-404}$$

Hence

$$\frac{\gamma(s)}{\sigma(s)} = \frac{L_{n\bar{p}}(s)}{K_{n\bar{p}}(s)} \approx \frac{2Mg_A}{\sqrt{2}\, G} \tag{8-405}$$

Substituting this result in the ratio of the integrals gives an approximate expression for f_π,

$$f_\pi \approx \frac{\sqrt{2}\, Mg_A}{G} \tag{8-406}$$

The preceding derivation is closest to the original derivation by Goldberger and Treiman. The success of the Goldberger-Treiman relation stimulated many other investigations and we now shall discuss several of them.

The PCAC Hypothesis (Partially Conserved Axial-Vector Current)

Gell-Mann and Lévy[20] introduced an assumption that

$$\Phi \propto \varphi \tag{8-407}$$

The proportionality constant can be determined by taking the matrix element of this relation between the vacuum and the one-pion states.

$$\Phi = -\mu^2 f_\pi \varphi \tag{8-408}$$

This immediately leads to

$$f_\alpha(s) = -\frac{\mu^2 f_\pi K_\alpha(s)}{s - \mu^2} \tag{8-409}$$

and

$$L_\alpha(s) = f_\pi K_\alpha(s) \tag{8-410}$$

Then f_π is given by

$$f_\pi = \frac{L_\alpha(s)}{K_\alpha(s)} = \frac{L_\alpha(0)}{K_\alpha(0)} \approx \frac{L_\alpha(0)}{K_\alpha(\mu^2)} \tag{8-411}$$

This ratio, for the choice $\alpha = n\bar{p}$, reduces to $\sqrt{2}\, Mg_A/G$.

This hypothesis is used extensively in the current algebra. It should be emphasized that *the PCAC hypothesis necessarily implies a subtraction for $f_\pi(s)$ as shown below.*

The PCAC hypothesis leads to $\gamma(s) = f_\pi \sigma(s)$, so that

$$\frac{1}{\pi} \frac{\operatorname{Im} f_\pi(s)}{s - \mu^2} = -\frac{\mu^2}{s} f_\pi \sigma(s) \tag{8-412}$$

If an unsubtracted dispersion is assumed for $f_\pi(s)$, we get

$$f_\pi = -f_\pi \int ds \frac{\mu^2}{s} \sigma(s)$$

or

$$\left[1 + \int ds \frac{\mu^2}{s} \sigma(s)\right] f_\pi = 0 \qquad (8\text{-}413)$$

This equation implies $f_\pi = 0$, so that $f_\pi(s)$ needs a subtraction.

The Convergence Condition

Assume that Z_π^{-1} is divergent as is the case in perturbation theory; then the formula (8-397) for f_π in terms of $\sigma(s)$ and $\gamma(s)$ is meaningless. To overcome this difficulty, Ida gave an alternative form of f_π:

$$f_\pi = \frac{\int ds\ [(s - \mu^2)/s]\ [\gamma(s) - f_\pi \sigma(s)]}{1 + \int ds \frac{\mu^2}{s} \sigma(s)} \qquad (8\text{-}414)$$

In this case the denominator is convergent provided the Källén-Lehmann representation is valid, so the numerator must be convergent too. If we carry out the integration of the numerator the two terms are separately divergent. Therefore, in order to get a convergent result these two terms should cancel one another for large values of s, namely,

$$\gamma(s) - f_\pi \sigma(s) << \sigma(s) \qquad \text{for large s}$$

or

$$\lim_{s \to \infty} \frac{\gamma(s) - f_\pi \sigma(s)}{\sigma(s)} = 0 \qquad (8\text{-}415)$$

which implies

$$f_\pi = \lim_{s \to \infty} \frac{\gamma(s)}{\sigma(s)} \qquad (8\text{-}416)$$

In the nucleon-antinucleon approximation this reproduces the same result as that of Goldberger and Treiman. Therefore, if $\gamma(s)$ and $\sigma(s)$ are given, f_π is determined from (8-416). It is possible, however, to derive another formula for f_π by substituting the value of f_π determined in Ida's formula, (8-414). These two values of f_π must be equal! This is a nontrivial consistency condition and is supposed to impose a constraint on the choice of parameters in strong interactions.

Induced Pseudoscalar Coupling in μ-Capture

There is another application of the present formalism, that is, the evaluation of the induced pseudoscalar coupling in the capture of a negative lepton by a proton in the nucleus from the Bohr orbit. It is not hard to see that the dispersion relation for $b_{n\bar{p}}(s)$ is

$$b_{n\bar{p}}(s) = \frac{\sqrt{2}\, G f_\pi}{\mu^2 - s} + \frac{1}{\pi} \int_{(3\mu)^2}^{\infty} ds' \frac{\operatorname{Im} b_{n\bar{p}}(s')}{s' - s - i\epsilon} \qquad (8\text{-}417)$$

The effective pseudoscalar interaction results from the reduction of the effective Hamiltonian

$$-\langle n\bar{p}(-)|\, A_\lambda\, (0)|\, 0\rangle \bar{\psi}_\nu\, i\gamma_\lambda\, (1 + \gamma_5)\psi_\ell$$

$$= -\bar{u}(n)\, [a_{n\bar{p}}(s)\, i\gamma_\lambda \gamma_5 + b_{n\bar{p}}(s)\, (n + \bar{p})_\lambda\, \gamma_5\,] \times v(\bar{p})$$

$$\bar{\psi}_\nu\, i\gamma_\lambda (1 + \gamma_5)\, \psi_\ell \qquad (8\text{-}418)$$

Using the conservation of energy-momentum and the Dirac equation we get

$$(n + \bar{p})_\lambda\, \bar{\psi}_\nu i\gamma_\lambda (1 + \gamma_5)\psi_\ell = -m_\ell\, \bar{\psi}_\nu (1 - \gamma_5)\psi_\ell \qquad (8\text{-}419)$$

Therefore (8-148) is equal to

$$-\bar{u}(n) i\gamma_\lambda \gamma_5 v(\bar{p}) a_{n\bar{p}}(s) \bar{\psi}_\nu i\gamma_\lambda (1 + \gamma_5)\psi_\ell$$

$$+ \bar{u}(n)\, \gamma_5\, v(\bar{p})\, m_\ell\, b_{n\bar{p}}(s) \bar{\psi}_\nu (1 - \gamma_5)\psi_\ell \qquad (8\text{-}420)$$

For the lepton capture from a Bohr orbit $\ell^- + p \to n + \nu$, we can set $s = -m_\ell^2$ and the effective pseudoscalar coupling constant is given by

$$g_P = m_\ell\, b_{n\bar{p}}(-m_\ell^2) = \frac{\sqrt{2}\, G f_\pi m_\ell}{\mu^2 + m_\ell^2} + \frac{m_\ell}{\pi} \int_{(3\mu)^2}^{\infty} ds' \frac{\operatorname{Im} b_{n\bar{p}}(s')}{s' + m_\ell^2}$$

$$\approx \frac{\sqrt{2}\, G f_\pi m_\ell}{\mu^2 + m_\ell^2} \approx \frac{2 M m_\ell}{\mu^2 + m_\ell^2}\, g_A \qquad (8\text{-}421)$$

This expression is comparable with g_A only for the μ meson, and in this case

$$g_P \approx 7g_A \tag{8-422}$$

This result is consistent with experiment.

8-15 THE ADLER–WEISBERGER FORMULA[21]

The Goldberger-Treiman relation gives a relationship between two parameters g_A and f_π, both related to the axial-vector current in beta decay or mu capture. There is a similar relation, called the Adler-Weisberger formula, which gives the ratio g_A/g_V in terms of the total cross sections for pion-nucleon scattering.

The interaction Hamiltonian density for the strangeness-conserving semileptonic decay is given, *phenomenologically,* by

$$\mathcal{H} = (g_V \bar{\psi}_p \gamma_\lambda \psi_n + g_A \bar{\psi}_p \gamma_\lambda \gamma_5 \psi_n) \bar{\psi}_\ell \gamma_\lambda (1 + \gamma_5) \psi_\nu + \text{Hermitian conjugate} \tag{8-423}$$

Experimentally $g_A \approx 1.2\, g_V$; it is an interesting problem to study why the ratio is close to unity and how one can get the figure 1.2. In order to study this problem we need some preparation.

The CVC Hypothesis (Conserved Vector Current)

In Section 4-1 we formulated the pion-nucleon interaction in such a way that the theory becomes invariant under rotations in charge space. From the discussion in Section 1-3, it is clear that such an invariance requirement leads to a conservation law. In the case of invariance under rotations in ordinary space we are led to the conservation of angular momentum; in the present case we expect a similar conserved quantity. In analogy to the ordinary angular momentum, the conserved quantity in this case is called the isospin, denoted by I.

The three components of ordinary angular momentum form a set of infinitesimal generators of rotations in the ordinary three-dimensional space, and they satisfy the following commutation relations:

$$[J_x, J_y] = iJ_z \qquad [J_y, J_z] = iJ_x \qquad [J_z, J_x] = iJ_y \tag{8-424}$$

Similarly, the three components of isospin satisfy

$$[I_1, I_2] = iI_3 \qquad [I_2, I_3] = iI_1 \qquad [I_3, I_1] = iI_2 \tag{8-425}$$

It also was shown in Section 1-3 that the components of angular momentum can be expressed as space integrals of the angular-momentum densities in field theory. The same is true with isospin, and the three components of isospin can be expressed as

$$I_i = \int d^3x \; \mathcal{J}_{i0}(x) \qquad (i = 1,\ 2,\ 3) \tag{8-426}$$

where $\mathcal{J}_{i\lambda}$ denotes the λ-th space-time component of the isospin current density. The conservation of isospin is then expressed by

$$\partial_\lambda \mathcal{J}_{i\lambda} = 0 \tag{8-427}$$

In the case of the interacting pion-nucleon system, $\mathcal{J}_\lambda$ is given by

$$\mathcal{J}_\lambda = \frac{i}{2}\bar{\psi}\gamma_\lambda \tau \psi - \varphi \times \frac{\partial \varphi}{\partial x_\lambda} \tag{8-428}$$

where $\times$ in the second term denotes the vector product and τ is a vector in charge space whose components are the two-dimensional matrices given in Section 4-1. When fields other than the nucleon and pion fields are included, (8-428) is modified, but the commutation relations remain unchanged. Under the circumstances it seems wiser not to try to express $\mathcal{J}_\lambda$ in terms of field operators, but to specify their properties through commutation relations. Now, let us make the following replacement[22] in the strangeness-conserving semileptonic interaction given at the beginning of this section:

$$i\bar{\psi}_p \gamma_\lambda \psi_n = i(\bar{\psi}\gamma_\lambda \frac{\tau_1}{2}\psi + i\bar{\psi}\gamma_\lambda \frac{\tau_2}{2}\psi) \rightarrow \mathcal{J}_{1\lambda} + i\,\mathcal{J}_{2\lambda} \tag{8-429}$$

Then the vector part of the Fermi interaction becomes

$$\mathcal{H}_V = -g_V(\mathcal{J}_{1\lambda} + i\mathcal{J}_{2\lambda})\,\bar{\psi}_\ell i\gamma_\lambda(1+\gamma_5)\psi_\nu$$

$$+ \text{ Hermitian conjugate} \tag{8-430}$$

and the hadronic vector current satisfies

$$\partial_\lambda(\mathcal{J}_{1\lambda} + i\,\mathcal{J}_{2\lambda}) = 0 \tag{8-431}$$

For this reason the assumption that the vector current can be identified with an isospin current density is called the hypothesis of conserved vector current. The benefit of employing this assumption consists in the point that the energy-momentum-conserving matrix elements can be given solely from group-theoretical considerations so that they are determined without recourse to detailed knowledge of strong interactions. For instance, $|p\rangle$ and $|n\rangle$, having the same energy-momentum and the same spin orientation, form a basis of the two-dimensional irreducible representation of the group SU(2) generated by I_1, I_2, and I_3, and we have

$$\langle p | I_1 + iI_2 | n \rangle = 1 \tag{8-432}$$

so that

$$\langle p | \mathcal{J}_{10} + i\,\mathcal{J}_{20} | n \rangle = \frac{1}{V} \langle p | I_1 + iI_2 | n \rangle = \frac{1}{V} \tag{8-433}$$

and similarly

$$\langle \pi^+ | \mathcal{J}_{10} + i\,\mathcal{J}_{20} | \pi^0 \rangle = -\frac{\sqrt{2}}{V} \qquad \text{etc.} \tag{8-434}$$

Therefore, the matrix elements of $\mathcal{H}_V$ corresponding to

$$\begin{aligned} n &\to p + e^- + \nu \\ \pi^+ &\to \pi^0 + e^+ + \nu \end{aligned} \tag{8-435}$$

can be given without detailed knowledge about final-state interactions; this mechanism does not change the value of g_V by renormalization, since an identity analogous to Ward's exists also in the present case. This also explains why g_V is almost equal to the corresponding coupling constant in $\mu \to e + \nu + \nu$ despite the fact that renormalization effects would be likely to change the coupling constant in the former case even if the unrenormalized coupling constants were equal.

Current Algebra

In order to account for the relation $g_A \approx 1.2\, g_V$, Eq. (8-425) can be generalized to include the axial-vector current. Write the strangeness-conserving semileptonic Fermi interaction in the form

$$\mathcal{H} = -g_V [(\mathcal{J}_{1\lambda} + i\,\mathcal{J}_{2\lambda}) + (\mathcal{J}^5_{1\lambda} + i\,\mathcal{J}^5_{2\lambda})]\, \overline{\psi}_\ell i\gamma_\lambda (1 + \gamma_5)\, \psi_\nu$$

$$+ \text{ Hermitian conjugate} \tag{8-436}$$

Again, we do not specify the detailed structure of $\mathcal{J}_\lambda^5$; the three components of $\mathcal{J}_\lambda^5$ are characterized by commutation relations. The g_A is introduced through the relation

$$g_V (\mathcal{J}^5_{1\lambda} + i\,\mathcal{J}^5_{2\lambda}) = A_\lambda^\dagger \tag{8-437}$$

since the connection between g_A and $A_\lambda^\dagger$ is known. Define I_i^5 as

$$I_i^5 = \int d^3x \, \mathcal{J}_{i0}^5 (x) \qquad (i = 1, 2, 3) \tag{8-438}$$

and postulate

$$\begin{aligned} [I_i, I_j] &= i\epsilon_{ijk} I_k \\ [I_i, I_j^5] &= i\epsilon_{ijk} I_k^5 \\ [I_i^5, I_j^5] &= i\epsilon_{ijk} I_k \end{aligned} \tag{8-439}$$

where

$$\begin{aligned} \epsilon_{123} &= \epsilon_{312} = \epsilon_{231} = 1 \\ \epsilon_{132} &= \epsilon_{321} = \epsilon_{213} = -1 \\ &\text{all others} = 0 \end{aligned} \tag{8-440}$$

These 6 operators form a Lie algebra, which we will call a *current algebra*. Here we have started from the SU(2) algebra and extended it to SU(2) × SU(2) by including pseudoscalar generators; this is a subalgebra of Gell-Mann's current algebra SU(3) × SU(3).[23]

The Adler–Weisberger Formula

We start from the relation

$$[I_1^5 + iI_2^5, I_1^5 - iI_2^5] = 2I_3 \tag{8-441}$$

which follows from the set of commutation relations postulated in the preceding subsection. In the last section we introduced pseudoscalar fields Φ and $\Phi^\dagger$ by

$$\partial_\lambda A_\lambda = \Phi \qquad \partial_\lambda A_\lambda^\dagger = \Phi^\dagger \tag{8-442}$$

but in this section we redefine them as

$$\partial_\lambda A_\lambda = -\mu^2 f_\pi \Phi \qquad \partial_\lambda A_\lambda^\dagger = -\mu^2 f_\pi \Phi^\dagger \tag{8-443}$$

then both Φ and φ are properly normalized and qualified as pion-field operators in the sense of Section 8-6, namely,

$$\langle 0|\Phi(x)|q\rangle = \langle 0|\varphi(x)|q\rangle = (2q_0V)^{-1/2} e^{iqx} \qquad \text{etc.} \tag{8-444}$$

where $|q\rangle$ is a single-pion state. Let us introduce

$$\begin{aligned}\mathcal{a}_\lambda &= \mathcal{J}^5_{1\lambda} - i\,\mathcal{J}^5_{2\lambda} = A_\lambda/g_V \\ \mathcal{a}^\dagger_\lambda &= \mathcal{J}^5_{1\lambda} + i\,\mathcal{J}^5_{2\lambda} = A^\dagger_\lambda/g_V\end{aligned} \tag{8-445}$$

then the divergences of these axial-vector current densities are given by

$$\begin{aligned}\partial_\lambda \mathcal{a}_\lambda &= \partial_\lambda(\mathcal{J}^5_{1\lambda} - i\,\mathcal{J}^5_{2\lambda}) = \frac{\mu^2 f_\pi}{g_V}\Phi \\ \partial_\lambda \mathcal{a}^\dagger_\lambda &= \partial_\lambda(\mathcal{J}^5_{1\lambda} + i\,\mathcal{J}^5_{2\lambda}) = -\frac{\mu^2 f_\pi}{g_V}\Phi^\dagger\end{aligned} \tag{8-446}$$

The commutator relation under consideration now can be written as

$$\left[\int \mathcal{a}^\dagger_0(x, t)d^3x, \int \mathcal{a}_0(x', t)d^3x'\right]$$

$$= 2\int \mathcal{J}_{30}(x'', t)d^3x'' \tag{8-447}$$

By taking the expectation value of this equat ion in the proton state and setting V = 1, we get

$$\langle p\,|\left[\int \mathcal{a}^\dagger_0(x, t)d^3x,\ \mathcal{a}_0(0, t)\right]|\,p\rangle$$

$$= 2\langle p\,|\,\mathcal{J}_{30}(0, t)|\,p\rangle = 1 \tag{8-448}$$

where we have used $2I_3|p\rangle = |p\rangle$.

It is convenient to choose a special frame of reference; we take the rest frame of the proton. First, Eq. (8-448) can be cast in the form

$$\frac{1}{2\pi}\int dk_0 \int d^4x \exp(ik_0 t)\langle p|[\mathcal{a}^\dagger_0(x, t), \mathcal{a}_0(0, 0)]|p\rangle = 1 \tag{8-449}$$

Next, we replace $\mathcal{A}_0$ by Φ by using the relation between $\partial_\lambda \mathcal{A}_\lambda$ and Φ to get

$$\frac{1}{2\pi}\int \frac{dk_0}{k_0^2}\int d^4x \exp(ik_0t) \langle p|[\Phi^\dagger(\mathbf{x}, t), \Phi(0, 0)]|p\rangle$$

$$= \left(\frac{g_V}{\mu^2 f_\pi}\right)^2 \tag{8-450}$$

From the definition of Φ it is clear that the Fourier transform of its arbitrary matrix element vanishes for $k_0 = 0$ when $\mathbf{k} = 0$, so that we can replace k_0 by either $k_0 + i\epsilon$ or $k_0 - i\epsilon$, if necessary, without changing the result. We next write the commutator as

$$[\Phi^\dagger(\mathbf{x}, t), \Phi(0, 0)] = \theta(t)[\Phi^\dagger(\mathbf{x}, t), \Phi(0, 0)]$$
$$+ \theta(-t)[\Phi^\dagger(\mathbf{x}, t), \Phi(0, 0)] \tag{8-451}$$

When we substitute these terms in Eq. (8-450), we find that the Fourier transform of the first term is analytic in the upper half-complex k_0 plane, while that of the second term is analytic in the lower half-plane. Let us now replace k_0^{-2} in the integrand by $(k_0 - i\epsilon)^{-2}$. We choose the contour C_+ in evaluating the first integral, and C_- for the second integral as shown in Fig. 8-18.

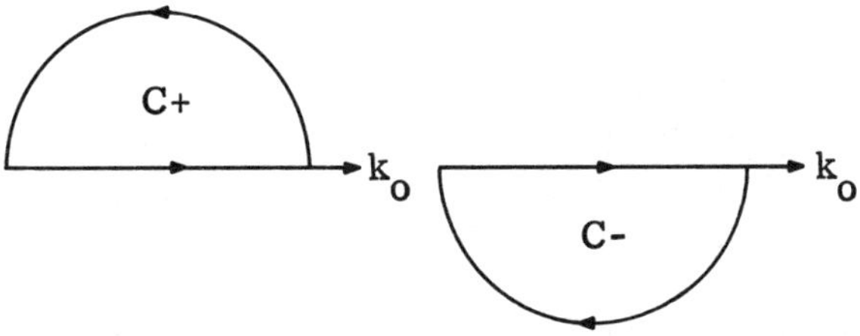

Figure 8-18. The contours C_+ and C_-.

Since $(k_0 - i\epsilon)^{-2}$ is analytic in the lower half-plane, the second term of modified Eq. (8-450) gives a vanishing contribution, provided the integral along the semicircle with an infinite radius vanishes. The retarded function is analytic in the upper half-plane, but $(k_0 - i\epsilon)^{-2}$ gives a double pole. Therefore, the integral on the left-hand side is given, according to Cauchy's theorem, by

$$i\left[\frac{d}{dk_0}\int d^4x \exp(ik_0t)\,\theta(t)\,\langle p|[\Phi^\dagger(\mathbf{x}, t), \Phi(0, 0)]|p\rangle\right]_{k_0=0} \tag{8-452}$$

We next introduce an amplitude $\mathfrak{M}(k)$, representing the pion-nucleon scattering amplitude, by

$$\mathfrak{M}(k) = i \int d^4x\, e^{-ikx}\, K_x K_y\, \theta(x_0 - y_0) \langle p | [\Phi^\dagger(x), \Phi(y)] | p \rangle_{y=0} \tag{8-453}$$

then (8-452) can be expressed in terms of $\mathfrak{M}(k)$ as

$$\frac{d}{dk_0}\left[\frac{\mathfrak{M}(k_0, \mathbf{k}=0)}{(k_0^2 - \mu^2)^2}\right]_{k_0=0} = \frac{1}{\mu^4}\left[\frac{d}{dk_0}\mathfrak{M}(k_0, \mathbf{k}=0)\right]_{k_0=0} \tag{8-454}$$

so that the sum rule assumes the form

$$\left[\frac{d}{dk_0}\mathfrak{M}(k_0, \mathbf{k}=0)\right]_{k_0=0} = \left(\frac{g_V}{f_\pi}\right)^2 \tag{8-455}$$

Since $\mathfrak{M}$ is an invariant function of pk and k^2, we write it as

$$\mathfrak{M}(\omega, m^2) = \mathfrak{M}\left(-\frac{pk}{M_p}, -k^2\right)$$

where ω is the laboratory energy of the pion and m is the mass of the incident and scattered pions. In the proton rest frame we have $\omega = k_0$, and $m = \omega$ since $\mathbf{k} = 0$. Therefore,

$$\left[\frac{d}{dk_0}\mathfrak{M}(k_0, \mathbf{k}=0)\right]_{k_0=0} = \left[\frac{\partial}{\partial\omega}\mathfrak{M}(\omega, m^2)\right]_{\substack{\omega=0\\ m=0}} + \left[2\omega\frac{\partial}{\partial m^2}\mathfrak{M}(\omega, m^2)\right]_{\substack{\omega=0\\ m=0}}$$

$$= \left[\frac{\partial}{\partial\omega}\mathfrak{M}(\omega, 0)\right]_{\omega=0} \tag{8-456}$$

Thus, we have obtained an invariant expression for the derivative of $\mathfrak{M}$ with respect to k_0. Let us evaluate the pole contribution corresponding to the diagram in Fig. 8-19.

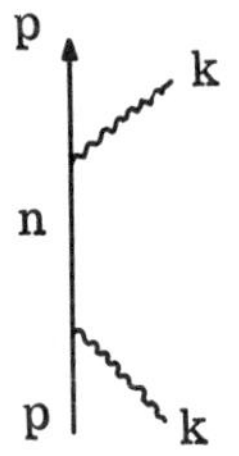

Figure 8-19. The diagram giving a pole term.

In evaluating the matrix element of Φ and $\Phi^\dagger$ between the proton and the intermediate neutron states we have to introduce a vertex correction; we introduce a vertex function $\Gamma(k_0^2)$ by

$$f_{n\bar{p}}(s) = \frac{\sqrt{2}\, G f_\pi \mu^2}{\mu^2 - s} + \frac{1}{\pi} \int_{(3\mu)^2}^{\infty} ds' \frac{\operatorname{Im} f_{n\bar{p}}(s')}{s' - s - i\epsilon}$$

$$= \frac{\sqrt{2}\, G f_\pi \mu^2}{\mu^2 - s} \Gamma(s) \tag{8-457}$$

then it is clear that $\Gamma(\mu^2) = 1$. The pole contribution to $\mathfrak{M}$ is given by

$$\mathfrak{M}_{\text{pole}} = -2G^2[\Gamma(k_0^2)]^2 \frac{\bar{u}(p)\, \gamma_5 [-i(p+k)\gamma + M_n]\, \gamma_5 u(p)}{(p+k)^2 + M_n^2} \tag{8-458}$$

By using the relations

$$p^2 + M_p^2 = 0 \qquad \mathbf{k} = 0 \qquad (ip\gamma + M_p)\, u(p) = 0$$

$$\text{and} \qquad \bar{u}(p) ik\gamma\, u(p) = -k_0$$

we find

$$\mathfrak{M}_{\text{pole}} = -2G^2 [\Gamma(k_0^2)]^2 \frac{1}{k_0 + M_n + M_p} \tag{8-459}$$

and hence

$$\left[\frac{d}{dk_0} \mathfrak{M}_{\text{pole}}\right]_{k_0 = 0} = \frac{2G^2[\Gamma(0)]^2}{(M_n + M_p)^2} \tag{8-460}$$

From here on let us set $M_n = M_p = M$. The sum rule is then

$$\left(\frac{g_V}{f_\pi}\right)^2 = \left[\frac{\partial}{\partial\omega}\mathfrak{M}(\omega, 0)\right]_{\omega=0} \tag{8-461}$$

With the understanding that the pole contribution is given by formula (8-460) this sum rule can be simplified. From the definition of the vertex function Γ we get

$$\left(\frac{g_V}{f_\pi}\right)^2 = \left(\frac{g_V}{g_A}\right)^2 \left(\frac{G\,\Gamma(0)}{\sqrt{2}\,M}\right)^2 \tag{8-462}$$

The pole contribution could be evaluated easily, but the continuum contribution is rather complicated. The sum rule reads

$$\left(\frac{g_V}{g_A}\right)^2 = \frac{2M^2}{G^2\,[\Gamma(0)]^2}\left[\frac{\partial}{\partial\omega}\mathfrak{M}(\omega, 0)\right]_{\omega=0}$$

$$= 1 + (\text{continuum contribution}) \tag{8-463}$$

We have seen that the pole contribution gives exactly 1, but the continuum (cont) contribution represents the scattering of a zero-mass pion from a proton, which is quite unknown. We introduce, therefore, an approximation by

$$\mathfrak{M}(\omega, 0)_{\text{cont}} \approx (\Gamma(0))^2\, \mathfrak{M}(\omega, \mu^2)_{\text{cont}} \tag{8-464}$$

This relation is exact in the case of the pole contribution but reduces to an approximate formula for the continuum. We first observe that

$$\mathfrak{M}(\omega, \mu^2) = i\int d^4x\, e^{-ikx}\, K_x K_y\, \theta(x_0 - y_0)\, \langle p|\,[\Phi^\dagger(x), \Phi(y)]\,|p\rangle_{y=0}$$

$$= i\int d^4x\, e^{-ikx}\, K_x K_y\, \theta(x_0 - y_0)\, \langle p|\,[\varphi^\dagger(x), \varphi(y)]\,|p\rangle_{y=0}$$

(8-465)

as a consequence of the argument in Section 8-6. This means that as far as *the mass shell values* are concerned it does not matter whether we use Φ or φ, and to this derivation the hypothesis of PCAC, $\Phi = \varphi$, is extraneous. Once we make the approximation (8-464), then $\mathfrak{M}(\omega, \mu^2)$ can be expressed in terms of the pion-nucleon scattering amplitude and can be evaluated easily. In fact, in terms of the amplitude $C^{(\pm)}(\omega)$ defined in Section 8-13,

$$\mathfrak{M}(\omega, \mu^2) = C^{(+)}(\omega) + C^{(-)}(\omega) \tag{8-466}$$

so that, from the dispersion relations for $C^{(\pm)}$ given before, we have

$$\begin{aligned}\operatorname{Re} \mathfrak{M}(\omega, \mu^2) &= \operatorname{Re} C^{(+)}(\omega) + \operatorname{Re} C^{(-)}(\omega) \\ &= \frac{G^2}{M}\left(\frac{\omega_0}{\omega_0 - \omega}\right) + \frac{P}{\pi}\int_\mu^\infty d\omega' \operatorname{Im} C^{(+)}(\omega')\left(\frac{1}{\omega' - \omega} + \frac{1}{\omega' + \omega}\right) \\ &\quad + \frac{P}{\pi}\int_\mu^\infty d\omega' \operatorname{Im} C^{(-)}(\omega')\left(\frac{1}{\omega' - \omega} - \frac{1}{\omega' + \omega}\right) + \text{subtraction}\end{aligned} \tag{8-467}$$

Hence

$$\left[\frac{\partial}{\partial\omega}\mathfrak{M}(\omega, \mu^2)_{\text{cont}}\right]_{\omega=0} = \left[\frac{\partial}{\partial\omega}(C^{(+)}(\omega) + C^{(-)}(\omega))_{\text{cont}}\right]_{\omega=0}$$

$$= \frac{P}{\pi}\int_\mu^\infty d\omega' \operatorname{Im} C^{(-)}(\omega')\frac{2}{\omega'^2} \tag{8-468}$$

The absorptive part of $C^{(-)}$ is given by the optical theorem as

$$\operatorname{Im} C^{(-)}(\omega) = \frac{k}{2}(\sigma_-(\omega) - \sigma_+(\omega)) \tag{8-469}$$

so that

$$(g_V/g_A)^2 \approx 1 + \frac{2}{\pi}(M/G)^2 \int_\mu^\infty \frac{k\,d\omega}{\omega^2}[\sigma_-(\omega) - \sigma_+(\omega)] \tag{8-470}$$

This relation is called the Adler-Weisberger formula and agrees with experiment. For the numerical value of g_A/g_V, Adler and Weisberger obtained 1.24 and 1.16, respectively, by using slightly different methods of evaluating the dispersion integral.

8-16 LANDAU-CUTKOSKY THEORY[24]

In writing dispersion relations for certain amplitudes, we have to know the positions of singularities such as poles and branch points. In the Feynman-Dyson approach we determine the lower limits of dispersion integrals by studying the Feynman denominators. In the examples given so far the lower limits are given by the least-massive-possible intermediate states in conformity with the unitarity condition.

We define T as before,

$$S_{fi} = \delta_{fi} - i(2\pi)^4 \delta^4(P_f - P_i)T_{fi} \tag{8-471}$$

then the unitarity condition for T is

$$\text{Im } T_{fi} = -\frac{(2\pi)^4}{2} \sum_n T_{fn}^\dagger \delta^4(P_f - P_n)T_{ni} \tag{8-472}$$

and the unitarity condition for form factors is

$$\text{Im} \langle\beta(-) \mid \Phi(0) \mid 0\rangle = -\frac{(2\pi)^4}{2} \sum_\alpha T_{\beta\alpha}^\dagger \delta^4(P_\beta - P_\alpha) \langle\alpha(-)|\Phi(0)|0\rangle \tag{8-473}$$

Suppose that, in both cases, we choose $s = -P^2$ as the variable in the dispersion integrals. Then the lower limit of integration is determined by the minimum possible value of $-P_n^2$ or $-P_\alpha^2$ consistent with various selection rules. This determination of the lower limit has been consistent with the other determination based on the Feynman-Dyson theory in the examples so far discussed; this is, however, not always the case. When both approaches give the same lower limit the dispersion relation is called a normal dispersion relation and the lower limit of the dispersion integral is called a normal threshold. Otherwise, we have anomalous dispersion relations and anomalous thresholds.

In order to discuss the problem of anomalous thresholds we start with a typical example, the deuteron form factor.

$$\langle d, \bar{d}(-)|j_\mu(0) \mid 0\rangle \tag{8-474}$$

We discuss the structure of this form factor corresponding to the diagram in Fig. 8-20.

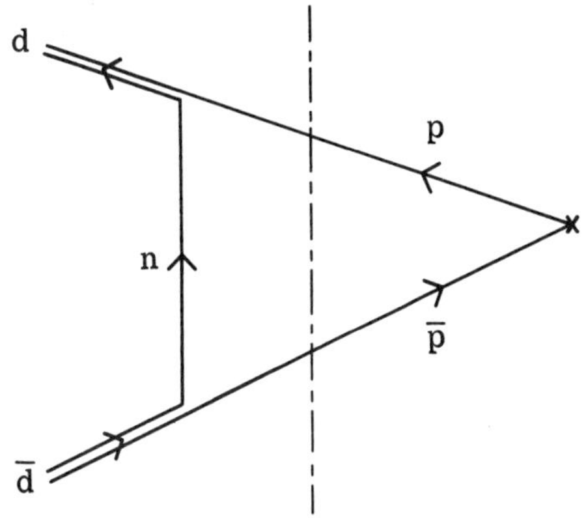

Figure 8-20. A diagram representing the deuteron form factor. The arrow indicates the direction of the baryon current, and the chain line (-··-) indicates where the intermediate states should be inserted in taking the absorptive part of the deuteron form factor.

To avoid inessential complications we again shall use a scalar model. In the unitarity condition let us set $\beta = d\bar{d}$, $\alpha = p\bar{p}$, and V = 1; then, to second order

$$\langle d\bar{d} | \mathcal{T}^{\dagger(2)} | p\bar{p}\rangle = (2\bar{d}_0)^{1/2}(2d_0)^{1/2}(2\bar{p}_0)^{1/2}(2p_0)^{1/2}\langle d\bar{d}| T^{\dagger(2)} | pp\rangle$$
$$= -\frac{g^2}{(d-p)^2+m^2} \qquad (8\text{-}475)$$

where m is the nucleon mass and M the deuteron mass. Writing

$$(2d_0)^{1/2}(2\bar{d}_0)^{1/2}\langle d\bar{d}(-)|j(0)|0\rangle = F_d(s) \qquad (8\text{-}476)$$

$$(2p_0)^{1/2}(2\bar{p}_0)^{1/2}\langle p\bar{p}(-)|j(0)|0\rangle = F_p(s) \qquad (8\text{-}477)$$

the unitarity condition reads

$$\text{Im } F_d(s) = \frac{(2\pi)^4}{2}\int\frac{d^3p}{(2\pi)^3}\int\frac{d^3\bar{p}}{(2\pi)^3}\,\delta^4(d+\bar{d}-p-p)$$
$$\times\frac{g^2}{(d-p)^2+m^2}\quad\frac{F_p(s)}{(2p_0)(2\bar{p}_0)} \qquad (8\text{-}478)$$

with

$$s = -(d + \bar{d})^2 = -(p + \bar{p})^2$$

The normal threshold s_n that we expect from the unitarity condition is given by

$$s_n = 4m^2 \tag{8-479}$$

Now by using a familiar technique we can write the unitarity condition as

$$\text{Im } F_d(s) = \frac{1}{2} \frac{g^2}{(2\pi)^2} \int d^4p \; \delta(p^2 + m^2)$$

$$\times \; \delta[(P - p)^2 + m^2] \; \frac{F_p(s)}{(d - p)^2 + m^2} \tag{8-480}$$

where $P = d + \bar{d}$. If we write $q = d - p$ we find

$$\text{Im } F_d(s) = \frac{1}{2} \frac{g^2}{(2\pi)^2} \int d^4q \; \delta[(d - q)^2 + m^2]$$

$$\times \; \delta[(\bar{d} + q)^2 + m^2] \; \frac{F_p(s)}{q^2 + m^2}$$

$$\equiv \frac{g^2}{16\pi} F_p(s) \cdot \mathcal{F}(s) \tag{8-481}$$

By carrying out the q integration we can find the explicit form of $\mathcal{F}(s)$.

Case I. $s > 4M^2$

$$\mathcal{F}_I(s) = \frac{1}{s^{1/2}(s - 4M^2)^{1/2}} \ln\left(\frac{s - 2M^2 + (s - 4M^2)^{1/2}(s - 4m^2)^{1/2}}{s - 2M^2 - (s - 4M^2)^{1/2}(s - 4m^2)^{1/2}}\right) \tag{8-482}$$

This result agrees with the Feynman-Dyson calculation.

Case II. $4M^2 > s > 4m^2$

Although the unitarity integral can be performed explicitly only for values above $4M^2$, we can get $\mathcal{F}$ (s) in this domain by simply continuing $\mathcal{F}_I$ (s) analytically.

$$\mathcal{F}_{II}(s) = \frac{2}{s^{1/2}(4M^2 - s)^{1/2}} \tan^{-1} \frac{(4M^2 - s)^{1/2}(s - 4m^2)^{1/2}}{s - 2M^2} \tag{8-483}$$

Let us study the behavior of this function in the neighborhood of $4m^2$, as s gradually decreases from $4M^2$ to $4m^2$. There are two possible cases:

(1) $4m^2 > 2M^2$ (normal threshold)

In this case we have $\mathcal{F}_{II}(4m^2) = 0$, showing that $4m^2$ is really the threshold.

(2) $2M^2 > 4m^2$ (anomalous threshold)

In this case, before the square-root factor vanishes at $s = 4m^2$ we get a pole at $s = 2M^2$, so that we write $\mathcal{F}_{II}$ (s) as

$$\frac{2}{s^{1/2}(4M^2 - s)^{1/2}} \left[\frac{\pi}{2} + \tan^{-1} \left(\frac{2M^2 - s}{(4M^2 - s)^{1/2}(s - 4m^2)^{1/2}}\right)\right] \tag{8-484}$$

At $s = 4M^2$, $\tan^{-1}(\ \ ) = \tan^{-1}(-\infty) = -(\pi/2)$, and at $s = 4m^2$, $\tan^{-1}(\ \ ) = \tan^{-1}(\infty) = \pi/2$. Therefore, $\mathcal{F}_{II}(4m^2) \neq 0$, showing that $4m^2$ is *not* the threshold. The expressions (8-483) and (8-484) for $\mathcal{F}_{II}$ (s), however, coincide with those of the Feynman-Dyson theory.

Case III. $4m^2 > s > s_a$

When $2M^2 > 4m^2$ the absorptive part $\mathcal{F}$(s) continues to the domain below $4m^2$, which is called the anomalous region. s_a is given by

$$s_a = 4M^2\left(1 - \frac{M^2}{4m^2}\right) = \frac{M^2}{m^2}(2m + M)\,B \approx 16mB \tag{8-485}$$

where B is the binding energy of the bound state $B = 2m - M$, so in the weak binding limit the last equality holds. The Feynman-Dyson theory gives

$$\mathcal{F}_{III}(s) = \frac{2\pi}{s^{1/2}(4M^2 - s)^{1/2}} \tag{8-486}$$

If we continue $\mathcal{F}_{II}(s)$ analytically to this region, $\mathcal{F}_{III}(s)$ becomes complex; furthermore we recognize that the expression (8-486) for $\mathcal{F}_{III}(s)$ is equal to the real part of this continued $\mathcal{F}_{III}(s)$.

What we have learned from this example may be summarized as follows: (1) The unitarity condition gives the correct absorptive part of the form factor above the normal threshold. (2) Depending on the relations among the rest masses of the particles participating in a process, the absorptive part sometimes fails to vanish at the normal threshold. Below the normal threshold the unitarity condition does not give the correct absorptive part.

Several questions are raised: (1) Under what conditions do anomalous thresholds show up? (2) How can one determine the positions of the anomalous thresholds? (3) In the anomalous region, how should the unitarity condition be modified? And so on. In order to answer these questions we shall introduce the Landau-Cutkosky theory. To answer the third question we shall give Cutkosky's prescription.

In the anomalous region, make the following replacement:

$$\frac{1}{(d-p)^2+m^2} \to 2\pi i\, \delta[(d-p)^2+m^2] \tag{8-487}$$

Then the unitarity condition gives, for $P_0 = d_0 + \bar{d}_0 > 0$, the result

$$\operatorname{Im} F_d(s) = \frac{1}{2}\frac{g^2}{(2\pi)^2}\int d^4q\, \delta[(q-d)^2+m^2]\,\delta[(q+\bar{d})^2+m^2]$$

$$\times\, \delta(q^2+m^2)(2\pi i)\, F_p(s)$$

$$= \frac{g^2 F_p(s)}{16\pi}\cdot\frac{2\pi}{s^{1/2}(4M^2-s)^{1/2}}\theta(s-s_a) \tag{8-488}$$

thereby reproducing the result of the Feynman-Dyson theory.

The formulation of the Landau-Cutkosky theory is based on the Feynman-Dyson theory. An arbitrary Feynman diagram represents a certain integral

$$\int \frac{B\, d^4k\, d^4\ell \ldots}{A_1 A_2 A_3 \ldots} \tag{8-489}$$

where

$$A_i = m_i^2 + q_i^2 \tag{8-490}$$

The q_i is a certain four-momentum, corresponding to an internal line in a diagram representing a particle of mass m_i, and B a certain polynominal of the vectors q_i. By using Feynman's technique of unifying denominators we can write

$$(A_1 A_2 \ldots A_n)^{-1} = (n-1)! \int_0^1 d\alpha_1 \ldots \int_0^1 d\alpha_n \, \delta(1 - \Sigma \alpha_i) \times (\alpha_1 A_1 + \ldots + \alpha_n A_n)^{-n} \tag{8-491}$$

Previously we have used a different form of parametrization, but for the present purpose this form is more convenient. The denominator $\alpha_1 A_1 + \cdots + \alpha_n A_n$ is a quadratic function of the variable internal four-momenta k, ℓ, ... and the terms linear in k, ℓ, ... always may be eliminated by means of linear transformations of the variables of integration, after which we obtain

$$\alpha_1 A_1 + \cdots + \alpha_n A_n = \varphi + K(k', \ell', \ldots) \tag{8-492}$$

where K is a homogeneous quadratic form of the new integration variables with coefficients depending only on the parameters α_i, and φ is a heterogeneous quadratic form of the vectors p_i describing the external momenta of the diagram under consideration. If we use Feynman's formula for k, ℓ, ... integrations we find

$$\int \frac{d^4k \, d^4\ell \cdots}{A_1 A_2 \cdots} \propto \int d\alpha_1 \cdots d\alpha_n \, \delta(1 - \Sigma \alpha_i) \frac{c(\alpha)}{\varphi^{n-2m}(\alpha)} \tag{8-493}$$

integrated over m independent internal momenta. More rigorously φ should be replaced by $\varphi - i\epsilon$. For a fixed set of external momenta, the φ is a function of the α. A singularity or branch point of φ occurs when the minimum value of φ becomes equal to zero:

$$\operatorname*{Min}_{\alpha} \varphi(\alpha, p_i p_j) = 0 \tag{8-494}$$

In determining the threshold for the vertex function in Section 8-7 we did exactly the same thing. There,

$$\varphi \propto s'(x_1, x_2) - s \tag{8-495}$$

and a singularity occurs where

$$s = \operatorname*{Min}_{x_1, x_2} s'(x_1, x_2) \tag{8-496}$$

The x_1 and x_2 are the Feynman parameters simply related to the α_i. To be more explicit, let us consider the case n = 3, then the domain of integration is given by a triangular region in Fig. 8-21. The sides of this triangle are represented by $\alpha_1 = 0$, $\alpha_2 = 0$, and $\alpha_3 = 0$. The minimum of φ should occur either on one of the boundaries $\alpha_i = 0$, or inside the triangle where

$$\frac{\partial \varphi}{\partial \alpha_1} = \frac{\partial \varphi}{\partial \alpha_2} = 0 \tag{8-497}$$

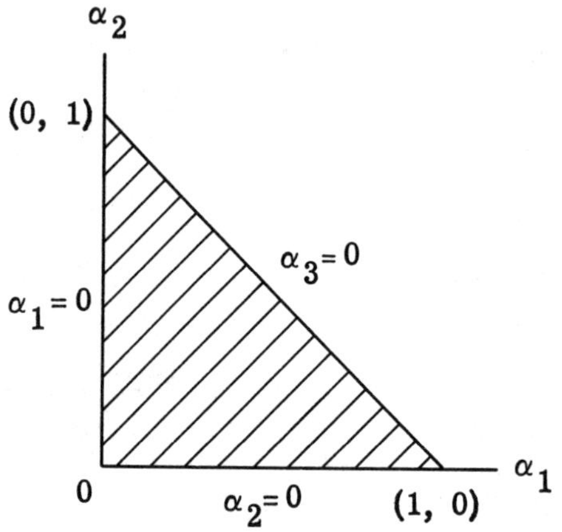

Figure 8-21. The triangular domain for the Feynman parameters α_1, α_2, and α_3.

It is clear that this argument is a generalization of the one given in Section 8-7.

It is desirable to treat all the α_i symmetrically, so we shall follow Landau's suggestion for doing so. We define

$$f \equiv \alpha_1 A_1 + \cdots + \alpha_n A_n = \varphi(\alpha) + K(\alpha; k'\ell' \cdots) \tag{8-498}$$

Since K is a homogeneous quadratic function of k', ℓ', ..., it is clear that for k', ℓ', $\cdots$ for which

$$\frac{\partial K}{\partial k'} = \frac{\partial K}{\partial \ell'} = \cdots = 0 \tag{8-499}$$

the K vanishes. Or, one can say

$$f = \varphi \tag{8-500}$$

for

$$\frac{\partial f}{\partial k'} = \frac{\partial f}{\partial \ell'} = \cdots = 0 \tag{8-501}$$

Since k' differs from k only by a constant vector, being a linear combination of external momenta, this condition may be replaced by

$$\frac{\partial f}{\partial k} = \frac{\partial f}{\partial \ell} = \cdots = 0 \tag{8-502}$$

Thus for the values of the vectors k, ℓ, $\cdots$ determined by Eqs. (8-502), f becomes equal to φ.

Since f is linear in α, we have

$$t\ f(\alpha_i\, ,\ p_i p_j) = f(t\alpha_i,\ p_i p_j) \qquad (t > 0) \tag{8-503}$$

and if Min $\varphi = 0$, then Min $t\varphi = 0$ and vice versa, since $t\varphi$ is the value of t f under consideration. Therefore, we shall consider $f(t\alpha_i,\ p_i p_j)$ instead of $f(\alpha_i,\ p_i p_j)$. For $t\alpha_i$, the restriction $\Sigma\, \alpha_i = 1$ is lifted, with the stipulation that t is positive. From here on let us call $t\alpha_i$ the new α_i, then the prescription is to solve the following simultaneous equations:

$$\begin{cases} \dfrac{\partial f}{\partial k} = \dfrac{\partial f}{\partial \ell} = \cdots = 0 \\ \\ \text{Min } f(\alpha) = 0 \end{cases} \tag{8-504}$$

Since the minimum of $f(\alpha)$ occurs either on the boundary ($\alpha_i = 0$) or where $\partial\varphi/\partial\alpha_i = \partial f/\partial\alpha_i = A_i = 0$, we have

$$\alpha_i (q_i^2 + m_i^2) = 0 \tag{8-505}$$

with the conditions

$$\frac{\partial f}{\partial k} = \sum_i \alpha_i \frac{\partial A_i}{\partial k} = 2 \sum_i \alpha_i q_i \frac{\partial q_i}{\partial k} = 0 \tag{8-506}$$

where

$$\frac{\partial q_i}{\partial k} = \begin{cases} 1 & \text{if } q_i \text{ is in the same loop as k and is running in the same direction} \\ \\ -1 & \text{as above, but is running in the opposite direction.} \\ \\ 0 & \text{otherwise} \end{cases}$$

Landau's rules may be summarized by

$$\alpha_i (q_i^2 + m_i^2) = 0$$

$$\sum \alpha_i q_i = 0 \qquad \text{with} \quad \alpha_i \geq 0 \tag{8-507}$$

where in the second equation summation should be carried out along each closed loop with all q_i chosen to run in the same direction.

We shall illustrate Landau's rules by simple examples.

Example (8-1) **Propagation function**

The Landau equations for the propagation function in lowest order are

$$q_1^2 + M^2 = 0 \qquad q_2^2 + m^2 = 0 \qquad \alpha_1 q_1 + \alpha_2 q_2 = 0 \tag{8-508}$$

with the conservation of energy-momentum $q_1 - q_2 = P$ (see Fig. 8-22).

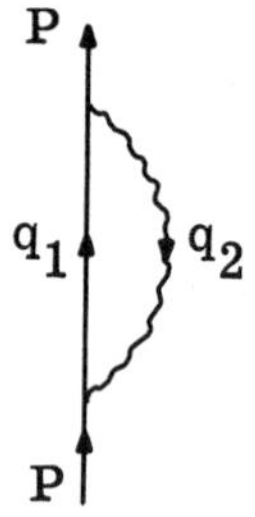

Figure 8-22. A propagator diagram labeled with momenta.

We find the solution

$$q_1 = \frac{\alpha_2}{\alpha_1 + \alpha_2} P \qquad q_2 = -\frac{\alpha_1}{\alpha_1 + \alpha_2} P \tag{8-509}$$

and applying the mass-shell condition we get, with $s = -P^2$,

$$M^2 = \left(\frac{\alpha_2}{\alpha_1 + \alpha_2}\right)^2 s \qquad m^2 = \left(\frac{\alpha_1}{\alpha_1 + \alpha_2}\right)^2 s$$

or

$$M = \frac{\alpha_2}{\alpha_1 + \alpha_2} s^{1/2} \qquad m = \frac{\alpha_1}{\alpha_1 + \alpha_2} s^{1/2}$$

Hence

$$s^{1/2} = \frac{\alpha_2}{\alpha_1 + \alpha_2} s^{1/2} + \frac{\alpha_1}{\alpha_1 + \alpha_2} s^{1/2} = M + m \qquad (8\text{-}510)$$

This is exactly the threshold we have found for the Lehmann representation.

Example (8-2) **Vertex function**

The lowest order nontrivial diagram for the vertex function is illustrated in Fig. 8-23. If the Q line is omitted, by setting the corresponding α equal to zero, and two vertices are joined, we get the same situation as we have discussed in the preceding example, namely

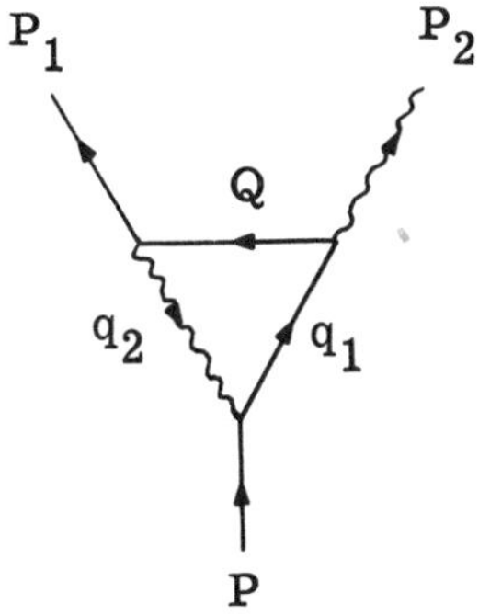

Figure 8-23. A vertex diagram labeled with momenta.

$$s = -P^2 = (M + m)^2 \qquad (8\text{-}511)$$

Both examples give normal thresholds consistent without intuitive picture of the unitarity condition.

In order to proceed further, the concept of Landau diagrams, representing the Landau equations diagrammatically, is useful. These are obtained from the four-dimensional timelike vectors p_i, by construction of spacelike vectors $\mathbf{p}_i$ in such a way that

$$(p_i, p_j) = -\mathbf{p}_i \cdot \mathbf{p}_j \qquad (8\text{-}512)$$

In the case of a vertex function we can do this; the Landau equations guarantee that all these vectors must be coplanar. The Landau diagram corresponding to the Feynman diagram in Fig. 8-23 is given in Fig. 8-24.

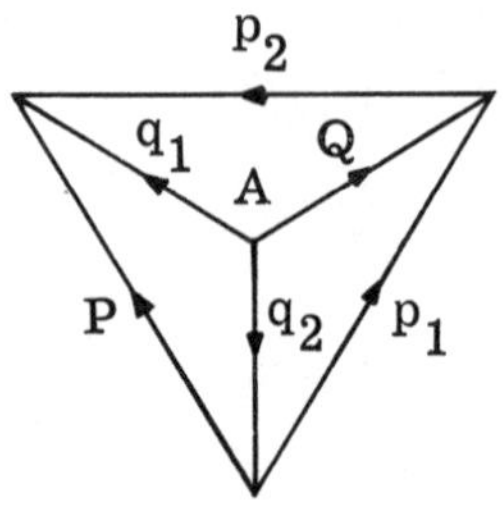

Figure 8-24. The Landau diagram corresponding to Fig. 8-23.

The condition $\alpha_i > 0$ (i = 1, 2, 3) means that the point A must be inside the triangle, but this is impossible since

$$|q_1| = |Q| = |p_1| = M$$

$$|q_2| = |p_2| = m$$

Therefore, there is no anomalous threshold in this vertex function; in other words, at least one of the α's must be zero.

Example (8-3) **Deuteron form factor**

For the deuteron form factor, discussed earlier in this section, the Landau diagram is shown in Fig. 8-25. The geometry easily gives the threshold value of s

$$s_a = 4M^2 \left(1 - M^2/4m^2\right) \tag{8-513}$$

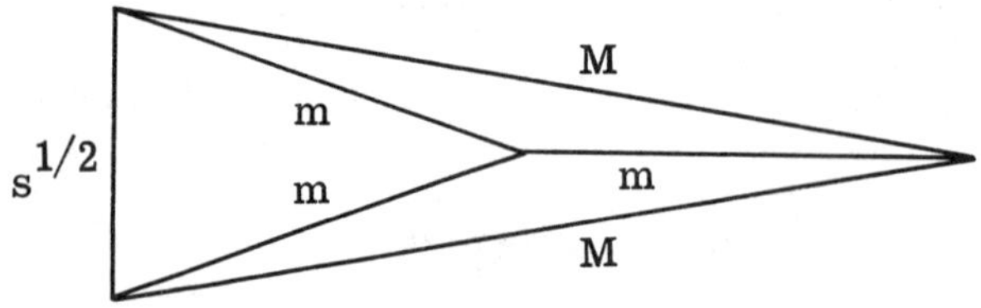

Figure 8-25. The Landau diagram for the deuteron form factor.

Let us consider a slightly more general vertex function in Fig. 8-26 and study the condition for the absence of the anomalous threshold.

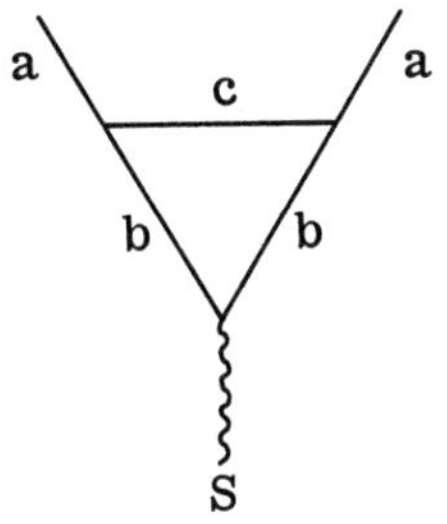

Figure 8-26. A typical vertex diagram labeled with particle species a, b, and c.

Then, as can be seen from the Landau diagram in Fig. 8-27 for a critical case, the condition for the absence of the anomalous threshold is

$$M_a^2 < M_b^2 + M_c^2 \qquad (8\text{-}514)$$

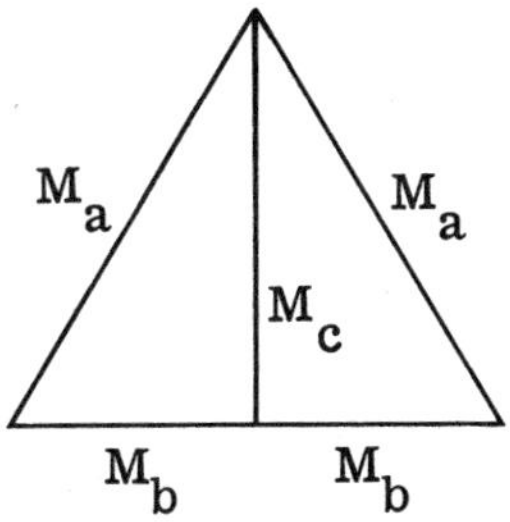

Figure 8-27. The Landau diagram corresponding to Fig. 8-26 in a critical configuration.

If $M_a^2 > M_b^2 + M_c^2$, the particle a is regarded as unstable or composite in dispersion theory.

The next problem is how we can evaluate the absorptive part of an amplitude in the presence of an anomalous threshold. To answer this question we cannot use the unitarity condition, but we must go back to the original Feynman-Dyson theory. This problem has been investigated by Cutkosky, but his arguments are rather involved; therefore, we shall not try to reproduce them, but give only an interpretation.

When we have a denominator of the form

$$\frac{1}{A - i\epsilon} \qquad (8\text{-}515)$$

its change or discontinuity across a branch cut is given by

$$2\pi i \delta(A) \tag{8-516}$$

The change of a product of the form

$$\frac{1}{(A_1 - i\epsilon)\cdots(A_N - i\epsilon)} \tag{8-517}$$

across a branch cut, when all N factors vanish simultaneously, is given by

$$(2\pi i)^N \delta(A_1)\cdots\delta(A_N) \tag{8-518}$$

Such is really the case when the energy variable is above a Landau singularity and the corresponding Landau diagram can be drawn. Therefore, the change of the Feynman integral

$$F = \int \prod_i d^4k_i \frac{B}{A_1\cdots A_N} \tag{8-519}$$

is given, for crossing a branch cut starting on a Landau singularity with m lines (Fig. 8-28) on the mass-shell, by

$$F_m = (2\pi i)^m \int B \prod_i d^4k_i\, \delta_p(q_1{}^2 + M_1{}^2)\cdots\delta_p(q_m{}^2 + M_m{}^2) \times \frac{1}{A_{m+1}\cdots A_N} \tag{8-520}$$

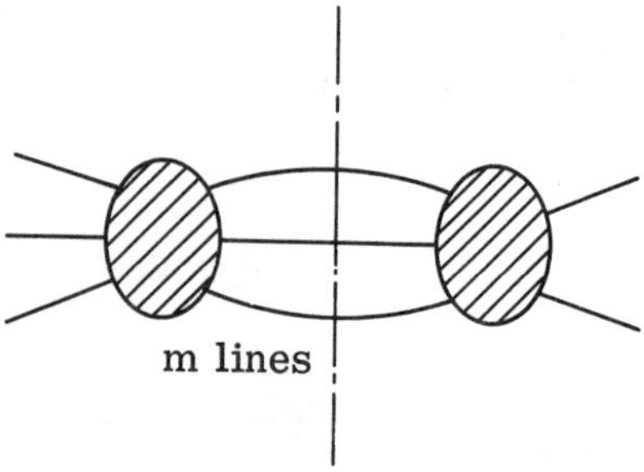

Figure 8-28. The intermediate states in the normal threshold case.

where δ_p means that the sign of q_0 is fixed by the Landau equations. When the diagram is decomposed into two disconnected parts by removing these m lines, the Landau singularity corresponds to a normal threshold, and formula (8-520) is identical with the ordinary unitarity condition giving the absorptive part of F.

When the diagram is decomposed into more than two disconnected parts as illustrated in Fig. 8-29, the Landau singularity corresponds to an anomalous threshold, and the discontinuity formula (8-520) is a generalization of the unitarity condition.

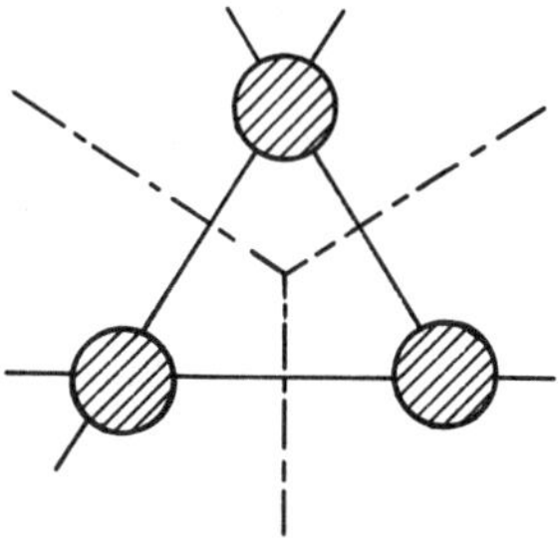

Figure 8-29. The intermediate states in the anomalous threshold cas(

8-17 THE MANDELSTAM REPRESENTATION

We have discussed so far the determination of vertex functions from dispersion theory, but the most fundamental subject in dispersion theory is the determination of the scattering amplitude. In order to discuss this subject the Mandelstam representation[25] is useful.

Let us consider a Feynman diagram with four external lines, as in Fig. 8-30. The reaction can be written formally as

$$a + b + c + d \rightarrow \text{vacuum} \tag{8-521}$$

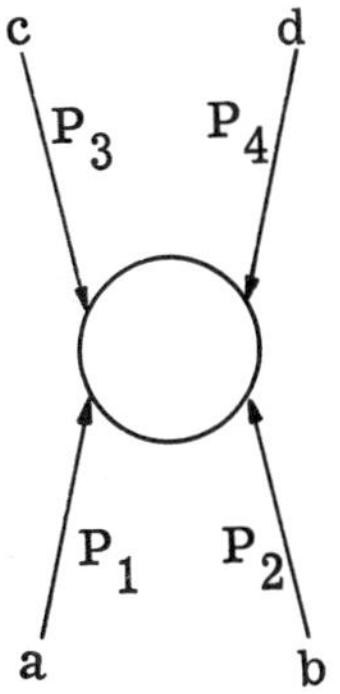

Figure 8-30. A diagram with four external lines.

or more realistically in terms of possibly observable reactions

$$a + b \rightarrow \bar{c} + \bar{d} \qquad a + c \rightarrow \bar{b} + \bar{d} \qquad a + d \rightarrow \bar{b} + \bar{c} \tag{8-522}$$

The conservation of energy and momentum implies

$$p_1 + p_2 + p_3 + p_4 = 0 \tag{8-523}$$

and all four vectors are subject to the mass-shell conditions, that is,

$$p_i^2 + m_i^2 = 0 \qquad (i = 1, 2, 3, 4) \tag{8-524}$$

We can form two independent scalar products out of four momenta.

$$\begin{aligned} s_1 \quad &\text{or} \quad s = -(p_1 + p_2)^2 = -(p_3 + p_4)^2 \\ s_2 \quad &\text{or} \quad t = -(p_1 + p_3)^2 = -(p_2 + p_4)^2 \\ s_3 \quad &\text{or} \quad u = -(p_1 + p_4)^2 = -(p_2 + p_3)^2 \end{aligned} \tag{8-525}$$

each of which represents the square of the total barycentric energy of a corresponding process given in (8-522). These three scalar products are not independent but satisfy a relation

$$\sum_i s_i = s + t + u = \sum_i m_i^2 = m_1^2 + m_2^2 + m_3^2 + m_4^2 \tag{8-526}$$

In order to represent a set of variables s, t, and u we use the so-called Dalitz plot. For simplicity we shall use $\sum m_i^2 = M^2$. We draw an equilateral triangle (Fig. 8-31) whose height is M^2. The sum of lengths of the perpendiculars to the sides from a point P is equal to M^2, that is, $s + t + u = M^2$. When the point P is outside the triangle we assign negative values to several of the variables so that (8-526) is algebraically satisfied. When the variables s, t, and u are so chosen that one of the processes in (8-522) is physically realizable, we say that we are in the s, t, or u channel, respectively.

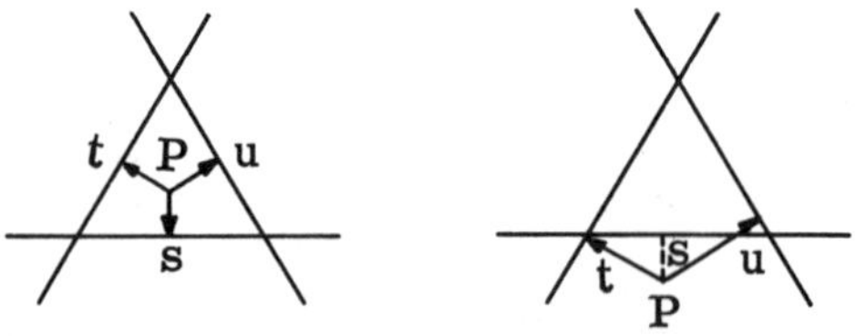

Figure 8-31. The Dalitz plot, or the Mandelstam plot, of a set of variables s, t, and u.

The physical domains of these channels can be plotted on a two-dimensional graph (Fig. 8-32) as in Fig. 8-31 (The Mandelstam plot).

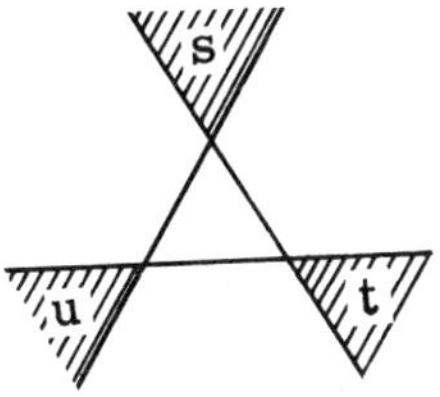

Figure 8-32. The physical domains for the s, t, and u channels.

Different processes in (8-522) correspond to different domains on this plot. For instance, for all $m_i = m$, the shaded domains of Fig. 8-32 correspond to the three different processes mentioned in (8-525). In order to find the precise form of the physical domains one has to study the kinematics.

The invariant scattering amplitude $\mathcal{T}$ becomes an invariant function of s, t, and u; we define a function F(s, t, u) which represents $\mathcal{T}$ in the physical domains. Here we shall consider meson-nucleon scattering in the scalar model and shall identify

$$p_1 = p \qquad p_2 = q \qquad p_3 = -p' \qquad p_4 = -q' \tag{8-527}$$

then

$$\begin{aligned} s &= -(p+q)^2 = -(p'+q')^2 \\ t &= -(p-p')^2 = -(q'-q)^2 \\ u &= -(p-q')^2 = -(p'-q)^2 \end{aligned} \tag{8-528}$$

Let us study the structure of the contribution of a typical fourth-order diagram, in Fig. 8-33.

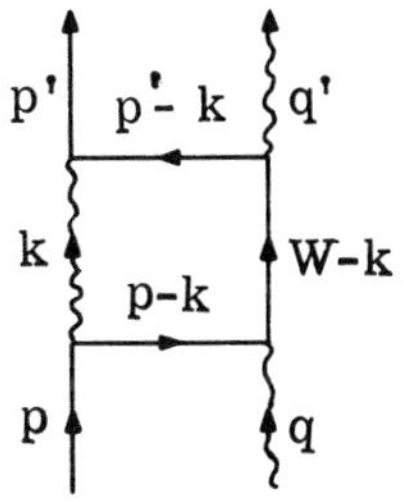

Figure 8-33. A typical fourth-order diagram for meson-nucleon scattering.

$$W = p + q = p' + q'$$

$$s = -W^2$$

The contribution of this diagram has already been studied in Section 8-13, so we can immediately write the expression for F

$$F = \frac{ig^4}{(2\pi)^4} \int d^4k \frac{1}{[k^2 + \mu^2][(W-k)^2 + M^2][(p'-k)^2 + M^2][(p-k)^2 + M^2]} \tag{8-529}$$

Let us regard F as a function of s by fixing t and calculate the discontinuity of F across the branch cut starting from the Landau singularity caused by the condition that both intermediate particles are on the mass shell. By use of Cutkosky's rule, we get

$$\Delta_s F = F(s + i\epsilon) - F(s - i\epsilon)$$

$$= \frac{ig^4}{(2\pi)^4} \int d^4k (2\pi i)^2 \, \delta_p(k^2 + \mu^2) \, \delta_p[(W - k)^2 + M]$$

$$\times \frac{1}{[(p' - k)^2 + M^2][(p - k)^2 + M^2]}$$

$$= \frac{ig^4}{(2\pi)^4} \int d^4k d^4k' \, (2\pi i)^2 \, \delta_p(k^2 + \mu^2) \, \delta_p(k'^2 + M^2) \, \delta^4(k + k' - W)$$

$$\times \frac{1}{[(p' - k)^2 + M^2][(p - k)^2 + M^2]}$$

$$= \frac{-i}{(2\pi)^2} \int \frac{d^3k}{2k_0} \int \frac{d^3k'}{2k'_0} \, \delta^4(k + k' - W) \, \frac{-g^2}{(p' - k)^2 + M^2}$$

$$\times \frac{-g^2}{(p - k)^2 + M^2}$$

$$= \frac{-i}{16\pi^2} \int \frac{d^3k}{k_0} \int \frac{d^3k'}{k'_0} \, \delta^4(k + k' - W) \, F^*_{fn} \, F_{ni} \tag{8-530}$$

where F_{ba} denotes the second-order invariant scattering amplitude for $a \to b$. In the s channel $\Delta_s F = 2i\,\mathrm{Im}\,F$, so we get

$$\mathrm{Im}\,F_{fi} = -\frac{1}{32\pi^2}\int \frac{d^3p_n}{(p_n)_0}\int \frac{d^3q_n}{(q_n)_0}\,\delta^4(P_n - P_i)\,F^*_{fn}\,F_{ni} \tag{8-531}$$

which is exactly the unitarity condition we obtained in Section 8-12. From this example we see that Cutkosky's prescription is a generalization of the unitarity condition.

In order to show that the left-hand side of (8-531) is the absorptive part in the s channel we should write

$$\Delta_s F = 2i\,\mathrm{Im}_s F = F(s + i\epsilon, t, u) - F(s - i\epsilon, t, u) \tag{8-532}$$

Then we can write a dispersion relation for F in s as well as in t

$$F(s, t) = \frac{1}{\pi}\int_{(M+\mu)^2}^{\infty} \frac{ds'}{s' - s - i\epsilon}\,\mathrm{Im}_s\,F(s', t)$$

$$= \frac{1}{\pi}\int_{(2M)^2}^{\infty} \frac{dt'}{t' - t - i\epsilon}\,\mathrm{Im}_t\,F(s, t') \tag{8-533}$$

The absorptive part can be computed again by use of Cutkosky's rule; now we write the dispersion relations for F and $\Delta_s F$:

$$F(s, t) = \frac{1}{2\pi i}\int_{(M+\mu)^2}^{\infty} \frac{ds'}{s' - s - i\epsilon}\,\Delta_s F(s', t) \tag{8-534}$$

$$\Delta_s F(s', t) = \frac{1}{2\pi i}\int_{(2M)^2}^{\infty} \frac{dt'}{t' - t - i\epsilon}\,\Delta_t \Delta_s\,F(s', t') \tag{8-535}$$

and by combining them get

$$F(s, t) = \frac{1}{(2\pi i)^2}\int_{(M+\mu)^2}^{\infty} \frac{ds'}{s' - s - i\epsilon}\int_{(2M)^2}^{\infty} \frac{dt'}{t' - t - i\epsilon}\,\Delta_t \Delta_s\,F(s', t') \tag{8-536}$$

where

$$\Delta_t \Delta_s F(s, t) = \frac{ig^4}{(2\pi)^4} (2\pi i)^4 \int d^4k \delta_p (k^2 + \mu^2) \delta_p [(W - k)^2 + M^2]$$

$$\times \delta_p [(p' - k)^2 + M^2] \delta_p [(p - k)^2 + M^2]$$

$$= ig^4 \int d^4k \delta_p (k^2 + \mu^2) \delta_p [(W - k)^2 + M^2]$$

$$\times \delta_p [(p' - k)^2 + M^2] \delta_p [(p - k)^2 + M^2] \tag{8-537}$$

It is clear, however, that simultaneous discontinuity does not occur in the physical region. Therefore, the function (2-537) survives only in the unphysical region. Let us denote the value of the integral by $\frac{1}{16}(-D)^{-1/2}$ for later convenience; then

$$\Delta_t \Delta_s F = \frac{ig^4}{16(-D)^{1/2}} \equiv \frac{g^4}{16(D)^{1/2}} \tag{8-538}$$

Hence we find

$$F(s, t) = \frac{1}{\pi^2} \int \frac{ds'}{s' - s} \int \frac{dt'}{t' - t} \rho(s', t') \tag{8-539}$$

where

$$\rho(s, t) = \frac{-g^4}{64D^{1/2}} \tag{8-540}$$

In defining the physical amplitude in the s channel we must take

$$\lim_{\epsilon \to 0} F(s + i\epsilon, t) \tag{8-541}$$

and a corresponding expression in the t or u channel.

The discontinuity integral can be performed as follows:

$$\frac{1}{16(-D)^{1/2}} = \int d^4k\, \delta_p (k^2 + \mu^2) \delta_p [(W - k)^2 + M^2]$$

$$\times \delta_p [(p' - k)^2 + M^2] \delta_p [(p - k)^2 + M^2]$$

$$= \int d^4k\, \delta_p (k^2 + \mu^2) \delta_p [W^2 + M^2 - 2kW - \mu^2]$$

$$\times \delta_p [2p'k + \mu^2] \delta_p [2pk + \mu^2] \tag{8-542}$$

We make a transformation of the variables of integration

$$k_1,\ k_2,\ k_3,\ k_0 \rightarrow k^2,\ kW,\ p'k,\ pk \tag{8-543}$$

so we get

$$\begin{aligned}\frac{1}{16}(-D)^{-1/2} &= \int \left| \frac{\partial(k_1, k_2, k_3, k_0)}{\partial(k^2, kW, kp', kp)} \right| \frac{1}{8}\, dk^2 d(2kW)\, d(2kp')\, d(2kp) \\ &\quad \times \delta_p(k^2 + \mu^2)\, \delta_p[W^2 + M^2 - 2kW - \mu^2]\, \delta_p(2p'k + \mu^2) \\ &\quad \times \delta_p(2pk + \mu^2) \\ &= \frac{1}{8} \left| \frac{\partial(k^2, kW, kp', kp)}{\partial(k_1, k_2, k_3, k_0)} \right|^{-1}\end{aligned} \tag{8-544}$$

Hence

$$D^{-1} = \begin{vmatrix} k^2 & kW & kp' & kp \\ kW & W^2 & Wp' & Wp \\ kp' & Wp' & p'^2 & p'p \\ kp & Wp & p'p & p^2 \end{vmatrix}^{-1} \tag{8-545}$$

The scalar products involving k should be replaced by those not depending on k by setting the arguments of the four δ functions equal to zero. Then D is given explicitly in terms of external variables:

$$D = \begin{vmatrix} -\mu^2 & \frac{1}{2}(M^2 - s - \mu^2) & -\frac{1}{2}\mu^2 & -\frac{1}{2}\mu^2 \\ \frac{1}{2}(M^2 - s - \mu^2) & -s & \frac{1}{2}(\mu^2 - s - M^2) & \frac{1}{2}(\mu^2 - s - M^2) \\ -\frac{1}{2}\mu^2 & \frac{1}{2}(\mu^2 - s - M^2) & -M^2 & -M^2 + \frac{1}{2}t \\ -\frac{1}{2}\mu^2 & \frac{1}{2}(\mu^2 - s - M^2) & -M^2 + \frac{1}{2}t & -M^2 \end{vmatrix}$$

$$
= \frac{t}{2} \begin{vmatrix} \mu^2 & \frac{1}{2}(s + \mu^2 - M^2) & \frac{1}{2}\mu^2 \\ \frac{1}{2}(s + \mu^2 - M^2) & s & \frac{1}{2}(s + M^2 - \mu^2) \\ \mu^2 & s + M^2 - \mu^2 & 2M^2 - \frac{1}{2}t \end{vmatrix}
$$

(8-546)

The double discontinuity function is different from zero in a domain where

$$D > 0 \qquad s > (M + \mu)^2 \qquad t > 4M^2 \tag{8-547}$$

as is clear from its derivation. If we set $M = \mu$ for simplicity, the boundary curve is described by

$$(s - 4\mu^2)(t - 4\mu^2) = 4\mu^4 \tag{8-548}$$

and the domain for the discontinuity (support) shown in Fig. 8-34, is given by

$$(s - 4\mu^2)(t - 4\mu^2) > 4\mu^4 \tag{8-549}$$

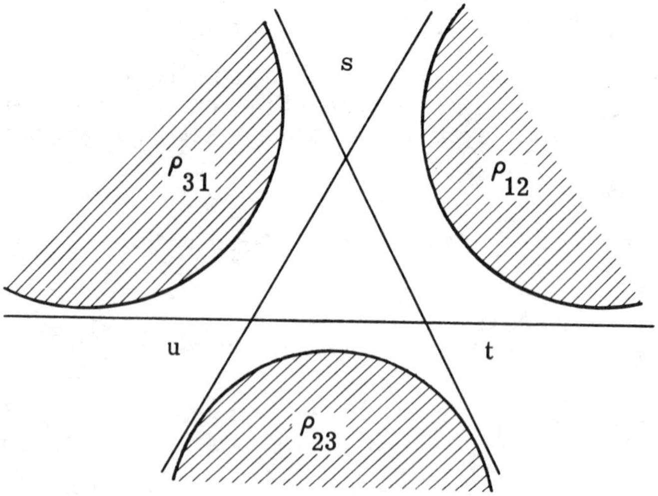

Figure 8-34. Supports of the double spectral functions.

We can show through the fourth order by an explicit calculation that the most general form of F is given by

$$F(s, t, u) = \frac{1}{\pi}\int ds' \frac{\rho_1(s')}{s' - s} + \frac{1}{\pi}\int dt' \frac{\rho_2(t')}{t' - t} + \frac{1}{\pi}\int du' \frac{\rho_3(u')}{u' - u}$$
$$+ \frac{1}{\pi^2}\iint ds'\, dt' \frac{\rho_{12}(s', t')}{(s' - s)(t' - t)}$$
$$+ \frac{1}{\pi^2}\iint dt'\, du' \frac{\rho_{23}(t', u')}{(t' - t)(u' - u)}$$
$$+ \frac{1}{\pi^2}\iint du'\, ds' \frac{\rho_{31}(u', s')}{(u' - u)(s' - s)} \tag{8-550}$$

This integral representation is called the Mandelstam representation; it gives explicitly the analyticity properties of the amplitude F as a function of two invariant variables.

As we have already mentioned, the physical amplitude in the s channel is given as the boundary value of the function F by

$$\lim_{\epsilon \to 0} F(s + i\epsilon, t, u)$$

Let us next study the consequences of the crossing symmetry. The crossing transformation is

$$q \rightleftarrows -q' \qquad \text{or} \qquad p_2 \rightleftarrows p_4$$

and in terms of the s, t, u variables we get

$$s \rightleftarrows u \qquad t \to t \tag{8-551}$$

This shows that F is symmetric in s and u, namely,

$$F(s, t, u) = F(u, t, s) \tag{8-552}$$

Finally, we shall reproduce the dispersion relation for meson-nucleon scattering in the scalar model starting from the Mandelstam representation.

Assume that t is negative and fixed, then this domain includes both the s and u channels. We shall further split this domain in two according to whether $s > u$ or $u > s$. The absorptive part of the amplitude in the s channel is given by

$$\text{Im}_s\, F(s, t, u) = \rho_1(s) + \frac{1}{\pi}\int dt' \frac{\rho_{12}(s, t')}{t' - t} + \frac{1}{\pi}\int du' \frac{\rho_{31}(u', s)}{u' - u}$$

$$\text{for} \quad s > u \tag{8-553}$$

and in the u channel by

$$\mathrm{Im}_u\ F(s, t, u) = \rho_3(u) + \frac{1}{\pi}\int dt' \frac{\rho_{23}(t', u)}{t' - t} + \frac{1}{\pi}\int ds' \frac{\rho_{31}(u, s')}{s' - s}$$

$$\text{for} \quad u > s \tag{8-554}$$

From these relations we get for $t < 0$ the expression

$$F(s, t, u) = \frac{1}{\pi}\int_{s'>u'} ds' \frac{\mathrm{Im}_s\ F(s', t, u')}{s' - s - i\epsilon} + \frac{1}{\pi}\int_{u'>s'} du' \frac{\mathrm{Im}_u\ F(s', t, u')}{u' - u - i\epsilon} \tag{8-555}$$

provided that $\rho_2(t) = 0$. In carrying out the s' and u' integrations it should be noticed that s' and u' are not independent since

$$s' + u' = \sum m_i^2 - t \tag{8-556}$$

If we use crossing symmetry we find that the two dispersion integrals are related to one another through the transformation $s \leftrightarrows u$, so that

$$F(s, t, u) = \frac{1}{\pi}\int_{s'>u'} ds' \frac{1}{s' - s}\ \mathrm{Im}_s F(s', t, u') + (s \leftrightarrows u)$$

$$= \frac{1}{\pi}\int_{s'>u'} ds' \left(\frac{1}{s' - s} + \frac{1}{s' - u}\right) \mathrm{Im}_s F(s', t, u') \tag{8-557}$$

In the notation of Section 8-13 we have

$$\mathrm{Re}\ \mathcal{T}(\nu, \varkappa^2) = -\frac{g^2}{2M}\left(\frac{1}{\nu_B - \nu} + \frac{1}{\nu_B + \nu}\right)$$

$$+ \frac{P}{\pi}\int_{\mu - \frac{\varkappa^2}{M}}^{\infty} d\nu' \left(\frac{1}{\nu' - \nu} + \frac{1}{\nu' + \nu}\right) \mathrm{Im}\ \mathcal{T}(\nu', \varkappa^2) \tag{8-558}$$

8-18 THE CINI–FUBINI APPROXIMATION

The analyticity properties of the scattering amplitudes as functions of two variables manifest themselves through the Mandelstam representation. When we combine the Mandelstam representation with unitarity in various channels we find a coupled set of nonlinear integral equations in two variables. This is an extremely complicated mathematical problem and we have to find some means to reduce the number of variables. The introduction of partial wave dispersion relation fits this purpose and the Mandelstam representation provides the appropriate basis for their derivation. In this section we shall discuss the problem à la Cini and Fubini.

Let us first consider meson-meson scattering and denote the meson mass by μ. The Mandelstam variables in this case satisfy

$$s + t + u = 4\mu^2 \tag{8-559}$$

The s channel is characterized by

$$4\mu^2 < s < \infty \qquad 4\mu^2 - s < t < 0 \tag{8-560}$$

If we write the four-momenta as

$$p_1 = (q, \omega_q) \qquad p_2 = (-q, \omega_q)$$

$$p_3 = (-q', -\omega_q) \qquad p_4 = (q', -\omega_q) \tag{8-561}$$

with

$$qq' = q^2 \cos\theta \equiv \nu \cos\theta$$

$$\omega_q = \left(q^2 + \mu^2\right)^{1/2} \equiv \left(\nu + \mu^2\right)^{1/2} \tag{8-562}$$

then

$$s = 4(\nu + \mu^2) \qquad t = -2\nu(1 - \cos\theta) \qquad u = -2\nu(1 + \cos\theta) \tag{8-563}$$

Similarly the domains

$$4\mu^2 < t < \infty \qquad 4\mu^2 - t < u < 0 \tag{8-564}$$

and

$$4\mu^2 < u < \infty \qquad 4\mu^2 - u < s < 0 \tag{8-565}$$

characterize the physical regions of the t and u channels, respectively. The Mandelstam representation can be written as

$$F(s, t, u) = \int_{4\mu^2}^{\infty} dx \int_{4\mu^2}^{\infty} dy\, A(x, y) \times \left[\frac{1}{(x - s)(y - t)} + \frac{1}{(x - t)(y - u)} + \frac{1}{(x - u)(y - s)}\right] \tag{8-566}$$

where A (x, y) is a real symmetric function corresponding to $\pi^{-2}\rho(x, y)$. The lower limit $4\mu^2$ is determined by the lowest possible mass in the intermediate state which can be reached by the two-meson system.

Now let us assume that the neutral meson under consideration is pseudoscalar so that reactions of the type

odd number of mesons $\leftrightarrows$ even number of mesons

are forbidden. This also excludes the one-meson pole terms in the Mandelstam representation. An important consequence of this assumption is that no two of the variables of integration reach the lower limit $4\mu^2$ at the same time. In order to see this let us insert a cut into a scattering diagram; then the various possible intermediate states involve 2, 4, 6, . . . particles as shown in Fig. 8-35.

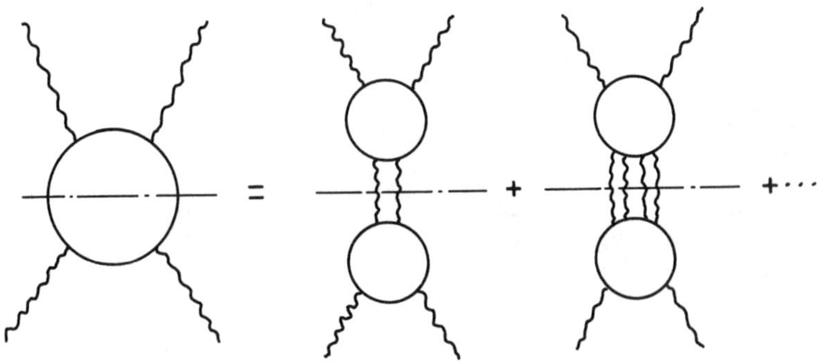

Figure 8-35. Possible intermediate states in the s channel.

Of these diagrams only the first one can reach the lower limit $4\mu^2$ in the s channel; but if we cut this diagram again in the t or u channel, we find that the intermediate states now must have 4, 8, 12, ... particles because of the conservation of parity as seen from Fig. 8-36. This shows that if the lower limit $4\mu^2$ is reached in one of the variables of integration, the lower limit for the other is $16\mu^2$.

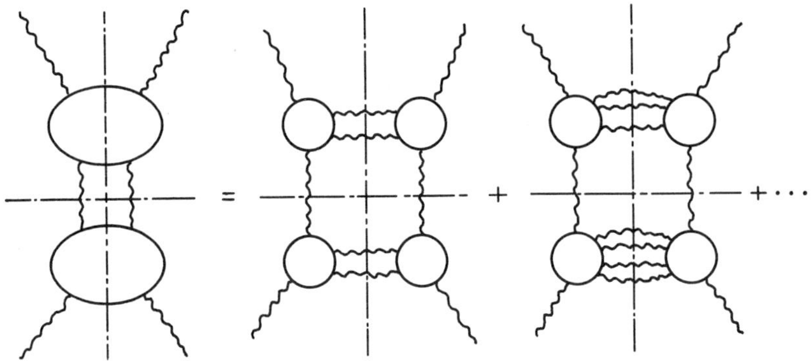

Figure 8-36. Possible intermediate states in both s and t channels.

If we take the two alternative diagrams to compute the boundary curves for the support of the double spectral function in Fig. 8-37 we get two intersecting curves in Fig. 8-38. We can compute the boundary curves by the method studied in the preceding section:

$$A_1(x, y) = 0 \qquad \text{if} \qquad y < \frac{16\mu^2 x}{x - 4\mu^2}$$

$$A_2(x, y) = 0 \qquad \text{if} \qquad x < \frac{16\mu^2 y}{y - 4\mu^2} \tag{8-567}$$

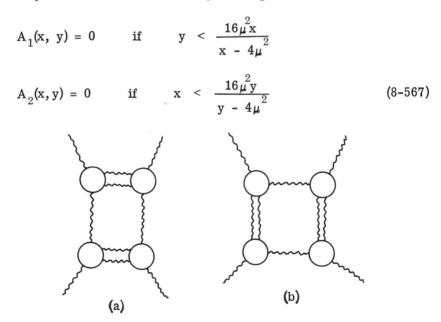

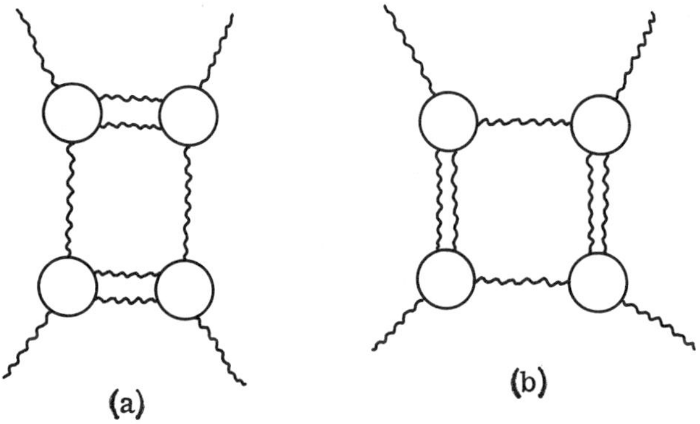

Figure 8-37. Structure of two diagrams with intermediate states of lowest masses in the s and t channels.

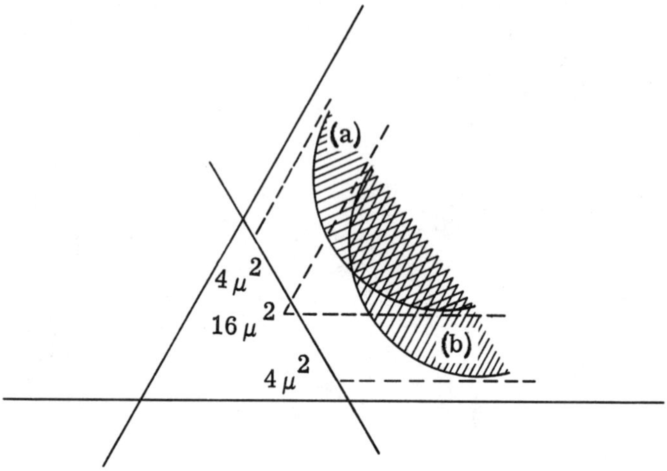

Figure 8-38. Boundary curves for the supports of the double spectral function for the diagrams in Fig. 8-37.

Therefore, we shall write each of the three integrals in the Mandelstam representation in the form

$$\int dx \int dy \frac{A(x, y)}{(x - s_i)(y - s_j)} = \frac{1}{2}\int_{4\mu^2}^{\infty} dx \int_{16\mu^2}^{\infty} dy \frac{A_1(x, y)}{(x - s_i)(y - s_j)}$$

$$+ \frac{1}{2}\int_{4\mu^2}^{\infty} dy \int_{16\mu^2}^{\infty} dx \frac{A_2(x, y)}{(x - s_i)(y - s_j)}$$

(8-568)

where s_i, s_j = s, t, u and $A_1(x, y) = A_2(y, x)$.

For the present purpose this representation is useful; the Mandelstam representation consists of three pairs of terms, each term having a cut in one variable starting at $4\mu^2$ and another cut in the other variable starting at $16\mu^2$. Now it is convenient to introduce the new variables.

$$z_1 = t - u \qquad z_2 = u - s \qquad z_3 = s - t \tag{8-569}$$

In the s channel we have

$$z_1 = 4\nu \cos\theta \tag{8-570}$$

We recombine six terms in the Mandelstam representation as follows:

$$F(s, t, u) = \alpha(s, z_1) + \alpha(t, z_2) + \alpha(u, z_3) \tag{8-571}$$

with

$$\alpha(s, z) = \int_{4\mu^2}^{\infty} \frac{dx}{x - s} \int_{16\mu^2}^{\infty} dy\, A_1(x, y) \times \left[(2y + s - 4\mu^2 + z)^{-1} + (2y + s - 4\mu^2 - z)^{-1} \right] \tag{8-572}$$

As long as we deal with elastic scattering below the threshold energy for inelastic processes, the variables s, t, and u are all smaller than $16\mu^2$, and the denominators in the integrals starting at $16\mu^2$ never vanish. Therefore, we introduce an expansion of the denominators to obtain an approximation valid in the elastic region.

$$\alpha(s, z) \simeq \int_{4\mu^2}^{\infty} \frac{dx}{x - s} \int_{16\mu^2}^{\infty} \frac{dy}{y} A_1(x, y) \times \left[1 + \frac{4\mu^2 - s}{2y} + \frac{(4\mu^2 - s)^2}{4y^2} + \frac{z^2}{4y^2} + \cdots \right] \tag{8-573}$$

First, let us keep only the first term in the expansion, then

$$\alpha(s, z) \simeq \int_{4\mu^2}^{\infty} \frac{dx}{x - s} \rho_0(x) \tag{8-574}$$

therefore,

$$F(s, t, u) \simeq \int_{4\mu^2}^{\infty} \frac{dx}{x - s} \rho_0(x) + \int_{4\mu^2}^{\infty} \frac{dx}{x - t} \rho_0(x) + \int_{4\mu^2}^{\infty} \frac{dx}{x - u} \rho_0(x) \tag{8-575}$$

In order to determine the unknown function $\rho_0(x)$ we have to use the unitarity condition in terms of partial waves, recalling the relation

$$F = -8\pi W f(\theta) \tag{8-576}$$

or for identical particles the modified relation

$$F = -8\pi\ W[f(\theta) + f(\pi - \theta)] \tag{8-577}$$

Then we see that

$$\begin{aligned} h_\ell(\nu) &= \frac{1}{2}\int_{-1}^{1} d(\cos\theta)\ P_\ell(\cos\theta)\ F(\nu, \cos\theta) \\ &= -16\pi \left(\frac{\nu + \mu^2}{\nu}\right)^{1/2} \exp^{(i\delta_\ell)} \sin\delta_\ell\,[1 + (-1)^\ell] \end{aligned} \tag{8-578}$$

If we use the approximate one-dimensional representation only the first term has a nonvanishing absorptive part in the s channel, so we keep only the first term at low energies.

$$h_0(\nu) \approx \int_{4\mu^2}^{\infty} dx\ \frac{\rho_0(x)}{x - 4\mu^2 - 4\nu - i\epsilon} \tag{8-579}$$

Thus

$$\text{Im}\ h_0(\nu) \approx \pi\,\rho_0(s) \tag{8-580}$$

$$\text{Im}\ h_\ell(\nu) \approx 0 \qquad \text{for}\ \ \ell > 0 \tag{8-581}$$

By introducing this approximation into the one-dimensional representation we get

$$\begin{aligned} F(\nu, \cos\theta) &= \frac{1}{\pi}\int_0^\infty d\nu'\,\frac{\text{Im}\ h_0(\nu')}{\nu' - \nu - i\epsilon} \\ &+ \frac{4}{\pi}\int_0^\infty d\nu'\ \text{Im}\ h_0(\nu') \\ &\times \left(\frac{1}{4\nu' + 4\mu^2 + 2\nu(1 - \cos\theta)} + \frac{1}{4\nu' + 4\mu^2 + 2\nu(1 + \cos\theta)}\right) \end{aligned} \tag{8-582}$$

This equation shows that only the S-wave term has a nonvanishing absorptive part so that this approximation is valid only when $\sin \delta_\ell$ is small as compared with $\cos \delta_\ell$ for $\ell > 0$.

By taking the S-wave projection of F we obtain an equation for $h_0(\nu)$:

$$h_0(\nu) = \frac{1}{\pi} \int_0^\infty d\nu'' \frac{\text{Im } h_0(\nu')}{\nu' - \nu - i\epsilon}$$

$$+ \frac{2}{\pi} \int_{-1}^{1} d(\cos \theta) \int_0^\infty d\nu' \text{ Im } h_0(\nu')$$

$$\times \left[\frac{1}{4\nu' + 4\mu^2 + 2\nu(1 - \cos \theta)} + \frac{1}{4\nu' + 4\mu^2 + 2\nu(1 + \cos \theta)} \right] \qquad (8\text{-}583)$$

It is also possible to write this equation in the form

$$h_0(\nu) = \frac{1}{\pi} \int_0^\infty d\nu' \frac{\text{Im } h_0(\nu')}{\nu' - \nu - i\epsilon} + \frac{1}{\pi} \int_{-\infty}^{-\mu^2} d\nu' \frac{f(\nu')}{\nu' - \nu} \qquad (8\text{-}584)$$

with

$$f(\nu') = \frac{2}{\nu'} \int_0^{-\nu' - \mu^2} d\nu'' \text{ Im } h_0(\nu'') \qquad (8\text{-}585)$$

The latter form shows that $h_0(\nu)$ has two cuts, one starting from 0 and continuing along the positive real axis and the other along the negative real axis, as shown in Fig. 8-39.

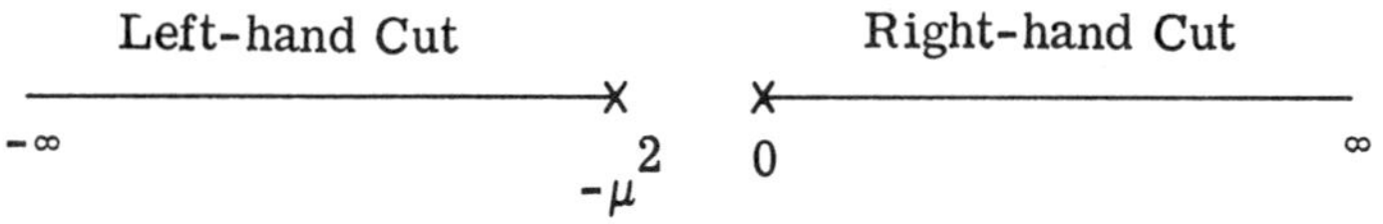

Figure 8-39. The cut structure of the amplitude $h_0(\nu)$ in the complex ν plane.

Equation (8-584) is satisfactory in that the unitarity condition for the S wave can be satisfied in all three channels. It is necessary, however, to introduce a subtraction in order to exclude the trivial solution $h_0 = 0$. Therefore we fix F(s, t, u) at the symmetrical point $s = t = u = \frac{4}{3} \mu^2 = s_0$:

$$F(s_0, s_0, s_0) = \lambda \qquad (8\text{-}586)$$

This defines a coupling constant for the effective interaction of the φ^4 type. Making the subtraction, we find that the one-dimensional representation is modified

$$F(s, t, u) = \lambda + \sum_{i=1}^{3} (s_i - s_0) \int_{4\mu^2}^{\infty} dx \frac{\rho_0(x)}{(x - s_0)(x - s_i)} \tag{8-587}$$

with $s_1 = s$, $s_2 = t$, and $s_3 = u$. Or, if we keep the second term $(4\mu^2 - s)/2y$ in the expansion (8-573), we automatically get a subtracted form,

$$\alpha(s, z) \simeq \int_{4\mu^2}^{\infty} \frac{dx}{x - s} \rho_0(x) + (s - 4\mu^2) \int_{4\mu^2}^{\infty} \frac{dx}{x - s} \rho_1(x)$$

$$= \alpha_0 + (s - s_0) \int_{4\mu^2}^{\infty} \frac{dx}{(x - s_0)(x - s)} \alpha(x) \tag{8-588}$$

with

$$\alpha(x) = \rho_0(x) + (x - 4\mu^2) \rho_1(x) \qquad \alpha_0 = \alpha(s_0, 0) \tag{8-589}$$

From the subtracted form of F we get

$$h_0(\nu) = \lambda + \frac{1}{\pi}(\nu + \frac{2}{3}\mu^2) \int_0^{\infty} \frac{d\nu' \operatorname{Im} h_0(\nu')}{(\nu' + \frac{2}{3}\mu^2)(\nu' - \nu - i\epsilon)}$$

$$- \frac{1}{\pi} \int_{-1}^{1} d(\cos\theta)[2\nu(1 - \cos\theta) + \frac{4}{3}\mu^2]$$

$$\times \int_0^{\infty} \frac{d\nu' \operatorname{Im} h_0(\nu')}{(\nu' + \frac{2}{3}\mu^2)(4\nu' + 4\mu^2 + 2\nu(1 - \cos\theta))} \tag{8-590}$$

so Eq. (8-584) becomes

$$h_0(\nu) = a_0 + \frac{1}{\pi}\left(\nu + \frac{2}{3}\mu^2\right)\int_0^{\infty} \frac{d\nu' \ \mathrm{Im}\ h_0(\nu')}{(\nu' + \frac{2}{3}\mu^2)(\nu' - \nu - i\epsilon)}$$

$$+ \frac{1}{\pi}\left(\nu + \frac{2}{3}\mu^2\right)\int_{-\infty}^{-\mu^2} \frac{d\nu' \ f(\nu')}{(\nu' + \frac{2}{3}\mu^2)(\nu' - \nu)} \qquad (8\text{-}591)$$

with

$$a_0 = \lambda + \frac{2}{\pi}\int_0^{\infty} d\nu' \ \mathrm{Im}\ h_0(\nu')\left[\frac{3}{\mu^2}\ln\left(\frac{\nu' + \mu^2}{\nu' + \frac{1}{3}\mu^2}\right) - \frac{1}{\nu' + \frac{2}{3}\mu^2}\right]$$

$$(8\text{-}592)$$

and $f(\nu')$ defined in Eq. (8-585).

These equations were first derived by Chew and Mandelstam[27]; by solving them we can determine the scattering amplitude without recourse to the Feynman-Dyson theory. The advantage of the partial-wave dispersion relations lies in the fact that the number of variables we have to deal with has been reduced to only one as compared with two in the original Mandelstam representation.

8-19 THE PARTIAL-WAVE DISPERSION RELATIONS

In the preceding section we have discussed a dynamical formulation of the scattering problem based on the Cini-Fubini approximation. In this section we shall show that the partial-wave dispersion relations are valid in general without the necessity of making a particular approximation. We choose the problem of nucleon-nucleon scattering for the scalar model to illustrate this technique.

The choice of the Mandelstam variables is made as follows:

$$p_1 = p \qquad p_2 = n \qquad p_3 = -p' \qquad p_4 = -n' \qquad (8\text{-}593)$$

Then we get

$$s = -(p + n)^2 \qquad t = -(p - p')^2 \qquad u = -(p - n')^2 \qquad (8\text{-}594)$$

which correspond, respectively, to the channels

$$p + n \to p' + n' \qquad p + \bar{p}' \to n' + \bar{n} \qquad p + \bar{n}' \to p' + \bar{n} \qquad (8\text{-}595)$$

In the present model only the t channel has a pole arising from the one (neutral)-meson intermediate state. Therefore, in analogy with the analysis of the preceding section we can write the amplitude as

$$F(s, t, u) = \frac{g^2}{t - \mu^2} + \frac{1}{\pi^2}\int_{4M^2}^{\infty} ds' \int_{4\mu^2}^{\infty} dt' \frac{\rho_{12}(s', t')}{(s' - s)(t' - t)}$$

$$+ \frac{1}{\pi^2}\int_{4\mu^2}^{\infty} dt' \int_{4M^2}^{\infty} du' \frac{\rho_{23}(t', u')}{(t' - t)(u' - u)}$$

$$+ \frac{1}{\pi^2}\int_{4M^2}^{\infty} du' \int_{4M^2}^{\infty} ds' \frac{\rho_{31}(u', s')}{(u' - u)(s' - s)} \tag{8-596}$$

In addition we should add single-integral terms but we shall not write them explicitly. Then in the center-of-mass system of the s channel we introduce the relative momentum q and the scattering angle θ as in the preceding section, and recall the partial wave expansion of the amplitude

$$F = -8\pi\, s^{1/2} f(\theta) \tag{8-597}$$

$$f(\theta) = \frac{1}{q}\sum_{\ell} (2\ell + 1)\exp(i\,\delta_\ell)\sin\delta_\ell\, P_\ell(\cos\theta) \tag{8-598}$$

The partial-wave amplitude h_ℓ is then defined, as in the preceding section, by

$$h_\ell = \frac{1}{2}\int_{-1}^{1} d(\cos\theta)\, P_\ell(\cos\theta)\, F(q^2, \cos\theta)$$

$$= -16\pi \left(\frac{q^2 + M^2}{q^2}\right)^{1/2} \exp(i\,\delta_\ell)\sin\delta_\ell \tag{8-599}$$

and the Mandelstam variables are

$$s = 4(M^2 + q^2) = 4(M^2 + \nu)$$

$$t = -2q^2(1 - \cos\theta) = -2\nu(1 - \cos\theta)$$

$$u = -2q^2(1 + \cos\theta) = -2\nu(1 + \cos\theta) \tag{8-600}$$

Now we shall study the analytic structure of $F(q^2, \cos\theta)$ or $h_\ell(q^2)$. There are four kinds of denominators in the Mandelstam representation.

(1) $s' - s = s' - 4M^2 - 4\nu$

The s' runs from $4M^2$ to ∞, so that this denominator vanishes for

$$0 \leq \nu \leq \infty$$

giving rise to the *right-hand cut.*

(2) $t' - t = t' + 2\nu(1 - \cos\theta)$ with $t' \geq 4\mu^2$.

This denominator vanishes for

$$\nu = -\frac{t'}{2(1 - \cos\theta)} \qquad \text{or} \qquad -\nu \geq \frac{t'}{4} \geq \frac{4\mu^2}{4} = \mu^2$$

or

$$-\infty \leq \nu \leq -\mu^2$$

which produces the *left-hand cut.*

(3) $u' - u = u' + 2\nu(1 + \cos\theta)$ with $u' \geq 4M^2$.

In this case we get a *left-hand cut* beginning at $-M^2$.

(4) The pole term has the denominator

$$t - \mu^2 = -2\nu(1 - \cos\theta) - \mu^2$$

which generates a *left-hand cut* beginning at

$$-\frac{\mu^2}{4}$$

The complete cut situation is illustrated in Fig. 8-40.

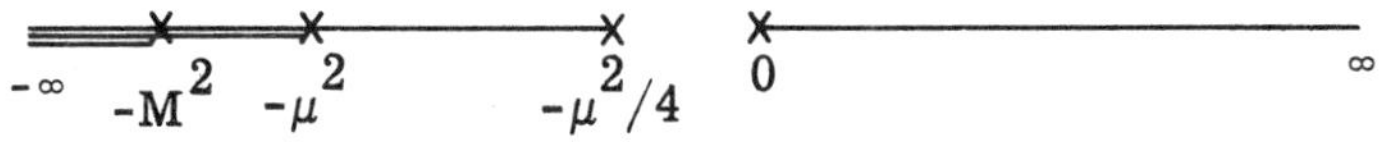

Figure 8-40. The cut structure of the partial-wave amplitude $h_\ell(\nu)$ in the complex ν plane.

Hence Im h_ℓ vanishes for $-\mu^2/4 < \nu < 0$ and we can write

$$h_\ell(\nu) = \frac{1}{\pi} \int_{-\infty}^{-\mu^2/4} d\nu' \frac{\text{Im } h_\ell(\nu')}{\nu' - \nu - i\epsilon} + \frac{1}{\pi} \int_0^\infty d\nu' \frac{\text{Im } h_\ell(\nu')}{\nu' - \nu - i\epsilon} \tag{8-601}$$

We also can give an explicit form of the pole contribution to the partial wave amplitude:

$$h_\ell(\nu)_{\text{pole}} = \frac{1}{2} \int_{-1}^{1} dx\, P_\ell(x) \frac{g^2}{-2\nu(1-x) - \mu^2}$$

$$= -\frac{g^2}{2} \int_{-1}^{1} dx \frac{P_\ell(x)}{2\nu(1-x) + \mu^2} \tag{8-602}$$

The right-hand cut corresponds to the contributions from intermediate states in the s channel (N-N scattering) and the left-hand cut results from those in the t and u channels (N-$\overline{\text{N}}$ scattering). The explicit form of the left-hand cut contributions can be given only after the N-$\overline{\text{N}}$ scattering problem is solved. We shall simply assume here, however, that the result is known and shall write it as $f_\ell(\nu)$. Then

$$h_\ell(\nu) = \frac{1}{\pi} \int_0^\infty d\nu' \frac{\text{Im } h_\ell(\nu')}{\nu' - \nu - i\epsilon}$$

$$+ \frac{1}{\pi} \int_{-\infty}^{-\mu^2/4} d\nu' \frac{f_\ell(\nu')}{\nu' - \nu} \tag{8-603}$$

If $f_\ell(\nu)$ is known, this integral equation determines $h_\ell(\nu)$.

As a starting point we often replace the left-hand cut by the pole contribution; then an approximate equation is

$$h_\ell(\nu) = \frac{1}{\pi} \int_0^\infty d\nu' \frac{\text{Im } h_\ell(\nu')}{\nu' - \nu - i\epsilon} + h_\ell(\nu)_{\text{pole}} \tag{8-604}$$

In the problem discussed in the preceding section this approximation cannot be used since there is no pole term in that process, but the unknown left-hand cut can be expressed by the same function occurring on the right-hand cut. In general

$$h_\ell(\nu)_{\text{pole}} = -\frac{g^2}{4}\int_{-\infty}^{-\mu^2/4} \frac{d\nu'}{\nu'-\nu}\,\frac{1}{\nu'}\,P_\ell\left(1+\frac{\mu^2}{2\nu'}\right)$$

and in particular,

$$h_0(\nu)_{\text{pole}} = -\frac{g^2}{4\nu}\,\ell n\left(1+\frac{4\nu}{\mu^2}\right)$$

In order to determine $h_\ell(\nu)$ we have to taken account of unitarity, the result of which can be seen from the expression for $h_\ell(\nu)$ in terms of the ℓ-th partial-wave-phase shift.

$$\operatorname{Im} h_\ell(\nu) = -\frac{1}{16\pi}\left(\frac{\nu}{\nu+M^2}\right)^{1/2}\left|h_\ell(\nu)\right|^2 \qquad \text{for} \quad \nu > 0 \tag{8-605}$$

This form is obtained by neglect of the contributions from inelastic channels, and for that reason this relation is called the elastic unitarity. Upon inserting this into the dispersion integral of (8-604) we obtain a nonlinear integral equation for the partial-wave amplitude $h_\ell(\nu)$.

8-20 THE N/D METHOD

In order to simplify the unitarity condition we introduce the partial-wave amplitude $F_\ell(\nu)$ by

$$F_\ell(\nu) = -\frac{1}{16\pi}\,h_\ell(\nu) = \left(\frac{\nu+M^2}{\nu}\right)^{1/2}\exp(i\delta_\ell)\sin\delta_\ell \tag{8-606}$$

then the elastic unitarity assumes the form

$$\operatorname{Im} F_\ell(\nu) = \left(\frac{\nu}{\nu+M^2}\right)^{1/2}\left|F_\ell(\nu)\right|^2 \qquad \text{for} \quad \nu > 0 \tag{8-607}$$

The scattering equation then reads

$$F_\ell(\nu) = \frac{1}{\pi}\int_0^\infty d\nu'\,\frac{\operatorname{Im} F_\ell(\nu')}{\nu'-\nu-i\epsilon} + F_\ell(\nu)_{\text{pole}} \tag{8-608}$$

In order to linearize the equations we introduce the N/D method devised by Chew and Mandelstam.[27] We write the amplitude as the quotient of two functions:

$$F_\ell(\nu) = \frac{N_\ell(\nu)}{D_\ell(\nu)} \tag{8-609}$$

where $N_\ell(\nu)$ has only a left-hand cut and is real for $\nu > 0$, and $D_\ell(\nu)$ has only a right-hand cut and is real for $\nu < 0$. The elastic unitarity can be written in the form

$$\text{Im}\left(F_\ell(\nu)\right)^{-1} = -\left(\frac{\nu}{\nu + M^2}\right)^{1/2} \equiv -\rho(\nu) \qquad \text{for} \quad \nu > 0 \tag{8-610}$$

For $\nu > 0$, this gives

$$\text{Im}\left(F_\ell(\nu)\right)^{-1} = \text{Im}\,\frac{D_\ell(\nu)}{N_\ell(\nu)} = \frac{\text{Im}\,D_\ell(\nu)}{N_\ell(\nu)} = -\rho(\nu)$$

or

$$\text{Im}\,D_\ell(\nu) = -\rho(\nu)\,N_\ell(\nu) \qquad \text{for} \quad \nu > 0 \tag{8-611}$$

For $\nu < 0$,

$$\text{Im}\,F_\ell(\nu) = \frac{\text{Im}\,N_\ell(\nu)}{D_\ell(\nu)}$$

or

$$\text{Im}\,N_\ell(\nu) = D_\ell(\nu)\,\text{Im}\,F_\ell(\nu) \approx D_\ell(\nu)\,\text{Im}\,F_\ell(\nu)_{\text{pole}} \qquad \text{for} \quad \nu < 0 \tag{8-612}$$

We define yet another function, which will in general be known as

$$\text{Im}\,F_\ell(\nu)_{\text{pole}} = -\frac{1}{16\pi} f_\ell(\nu) \equiv v_\ell(\nu) \tag{8-613}$$

For the example (8-602) considered in the preceding section

$$v_\ell(\nu) = -\frac{1}{16\pi}\left(-\frac{\pi g^2}{4}\right)\frac{1}{\nu}\,P_\ell\left(1 + \frac{\mu^2}{2\nu}\right)$$

$$= \frac{g^2}{64}\,\frac{1}{\nu}\,P_\ell\left(1 + \frac{\mu^2}{2\nu}\right) \qquad \text{for} \quad \nu < -\frac{\mu^2}{4}$$

Then

$$\text{Im } D_\ell(\nu) = -\rho(\nu)\, N_\ell(\nu) \qquad \text{for} \quad \nu > 0$$

$$\text{Im } N_\ell(\nu) = v_\ell(\nu)\, D_\ell(\nu) \qquad \text{for} \quad \nu < 0 \tag{8-614}$$

Let us normalize N_ℓ and D_ℓ by $D_\ell(0) = 1$, and write the once-subtracted dispersion relation for $D_\ell(\nu)$:

$$D_\ell(\nu) = 1 + \frac{\nu}{\pi} \int_0^\infty \frac{d\nu' \;\text{Im } D_\ell(\nu')}{\nu'(\nu' - \nu - i\epsilon)}$$

$$= 1 - \frac{\nu}{\pi} \int_0^\infty d\nu' \frac{\rho(\nu')\, N_\ell(\nu')}{\nu'(\nu' - \nu - i\epsilon)} \tag{8-615}$$

Then $N_\ell(0) = F_\ell(0)$, and the subtracted dispersion relation for $N_\ell(\nu)$ is

$$N_\ell(\nu) = F_\ell(0) + \frac{\nu}{\pi} \int_{-\infty}^0 d\nu' \frac{\text{Im } N_\ell(\nu')}{\nu'(\nu' - \nu - i\epsilon)}$$

$$= F_\ell(0) + \frac{\nu}{\pi} \int_{-\infty}^0 d\nu' \frac{v_\ell(\nu')\, D_\ell(\nu')}{\nu'(\nu' - \nu - i\epsilon)} \tag{8-616}$$

Together the dispersion relations for N_ℓ and D_ℓ form a coupled set of linear integral equations. We also may assume that $N_\ell(\nu)$ satisfies an unsubtracted dispersion relation.

$$N_\ell(\nu) = \frac{1}{\pi} \int_{-\infty}^0 d\nu' \frac{v_\ell(\nu')\, D_\ell(\nu')}{\nu' - \nu - i\epsilon} \tag{8-617}$$

With the help of this method we have succeeded in linearizing the original nonlinear integral equation. The next step consists in transforming the singular equation into a nonsingular equation.

Combining the dispersion relations we can eliminate $N_\ell(\nu)$.

$$D_\ell(\nu) = 1 + \frac{\nu}{\pi^2}\left[\int_{-\infty}^{-\mu^2/4} d\nu'' \frac{1}{\nu - \nu''}\right.$$

$$\left. \times \int_0^\infty d\nu' \frac{\rho(\nu')}{\nu'}\left(\frac{1}{\nu' - \nu} - \frac{1}{\nu' - \nu''}\right)\right] v_\ell(\nu'')D_\ell(\nu'') \tag{8-618}$$

If we solve this equation for negative values of ν, then $D_\ell(\nu)$ is real and there is no singularity since

$$D_\ell(\nu) = 1 + \frac{\nu}{\pi^2}\int_{-\infty}^{-\mu^2/4} d\nu'' \frac{K(\nu) - K(\nu'')}{\nu - \nu''} v_\ell(\nu'')D_\ell(\nu'') \tag{8-619}$$

with

$$K(\nu) = \int_0^\infty d\nu' \frac{\rho(\nu')}{\nu'(\nu' - \nu)} = \int_0^\infty d\nu' \frac{1}{\left[\nu'(\nu' + M^2)\right]^{1/2}} \cdot \frac{1}{\nu' - \nu} \tag{8-620}$$

Once $D_\ell(\nu)$ is known for negative values of ν, we can compute $N_\ell(\nu)$ for all values of ν and then $D_\ell(\nu)$, using dispersion relations. Let us set $\nu = -x$, $D_\ell(-x) = D(x)$, and $v_\ell(-x) = v(x)$, in order to discuss the integral equation for negative values of ν. The integral equation for D is

$$D(x) = 1 - \frac{x}{\pi^2}\int_{\mu^2/4}^{\infty} dx'' \frac{K(-x) - K(-x'')}{x'' - x} v(x'')D(x'') \tag{8-621}$$

Defining the symmetric kernel

$$K(x, x'') = \frac{K(-x) - K(-x'')}{x'' - x}$$

$$= \int_0^\infty d\nu' \frac{1}{\left[\nu'(\nu' + M^2)\right]^{1/2}} \frac{1}{(\nu' + x)(\nu' + x'')} \tag{8-622}$$

we have

$$D(x) = 1 - \frac{1}{\pi^2} \int_{\mu^2/4}^{\infty} dx'' \; xK(x, x'') \, v(x'') \, D(x''). \tag{8-623}$$

In case v has a definite sign we can immediately transform this equation into the standard form. Take, for instance, the S-wave amplitude for n - p scattering in the scalar model, then

$$v_0(-x) = -\frac{g^2}{64} \cdot \frac{1}{x}$$

Assume that $v(x'')$ is negative definite, and write

$$v(x) = - |v(x)| \tag{8-624}$$

then the integral equation can be transformed into

$$\left(\frac{|v(x)|}{x}\right)^{1/2} D(x) = \left(\frac{|v(x)|}{x}\right)^{1/2} + \frac{1}{\pi^2} \int_{\mu^2/4}^{\infty} dx' \left(x \, |v(x)|\right)^{1/2}$$

$$\times K(x, x'') \left(x'' \, |v(x'')|\right)^{1/2} \left(\frac{|v(x'')|}{x''}\right)^{1/2} D(x'') \tag{8-625}$$

which is an integral equation of the Fredholm type.

To conclude, we have just overcome three major difficulties step by step. First, we have reduced the number of variables from two to one by introducing the partial-wave dispersion relations. Second, we have transformed the original nonlinear integral equations into linear ones on the basis of the N/D method. Third, we have reduced the linear but singular integral equations into the nonsingular Fredholm type.

The Fredholm equation is nonsingular and is subject to various methods of solution. Thus the scattering problem in dispersion theory can be formulated in principle without recourse to the Feynman-Dyson theory.

8-21 FURTHER DISCUSSION ON THE SCATTERING EQUATION

In the preceding section we have studied a general method of solving the scattering equation of the form

$$F_\ell(\nu) = \frac{1}{\pi}\int_0^{\infty} d\nu' \frac{\text{Im } F_\ell(\nu')}{\nu' - \nu - i\epsilon} + \frac{1}{\pi} \int_{\infty}^{-\mu^2/4} d\nu' \frac{v_\ell(\nu')}{\nu' - \nu - i\epsilon} \tag{8-626}$$

with

$$\text{Im } F_\ell(\nu) = \rho(\nu) \, |F_\ell(\nu)|^2 \qquad \text{for} \quad \nu > 0 \tag{8-627}$$

We have exploited the N/D method to linearize the equation and eliminate the singular kernel from the equation. Because of the nonlinearity, however, it happens that the solution discussed in the preceding section is not unique, and occasionally it is not even the solution of the original equation.

Before discussing these points we shall study the relation between the D function and the phase shift. The function $D_\ell(\nu)$ satisfies a dispersion relation of the form

$$D_\ell(\nu) = 1 + \frac{\nu}{\pi} \int_0^\infty d\nu' \, \frac{\text{Im } D_\ell(\nu')}{\nu'(\nu' - \nu - i\epsilon)} \tag{8-628}$$

In order to evaluate Im D_ℓ let us recall the relation

$$D_\ell = \frac{N_\ell}{F_\ell}$$

and also the fact that N_ℓ is real for $\nu > 0$. Thus we have

$$\frac{\text{Im } D_\ell}{\text{Re } D_\ell} = -\frac{\text{Im } F_\ell}{\text{Re } F_\ell} = -\tan \delta_\ell \tag{8-629}$$

or

$$\frac{D_\ell^*}{D_\ell} = \frac{F_\ell}{F_\ell^*} = \exp(2i\delta_\ell) = S_\ell \tag{8-630}$$

Combining the dispersion relation with

$$\text{Im } D_\ell(\nu) = -\tan \delta_\ell(\nu) \cdot \text{Re } D_\ell(\nu) \tag{8-631}$$

we get the standard Muskhelishvili-Omnès equation for $D_\ell(\nu)$. The solution is

$$D_\ell(\nu) = \exp \left[-\frac{\nu}{\pi} \int_0^\infty \frac{d\nu' \, \delta_\ell(\nu')}{\nu'(\nu' - \nu - i\epsilon)} \right] \tag{8-632}$$

In the discussion of the vertex function in Section 8-11 we found that the vertex function $F(\nu)$ has a ν dependence of the form

$$F(\nu) = D_0(\nu)^{-1} \, F(0) \tag{8-633}$$

In the present scattering problem we are using the same D function, but the formal expression for D_ℓ does not help since we are trying to determine the phase shifts themselves. The most interesting subjects in this connection are the zeros and poles of the D function.

In order to discuss these subjects let us consider a simple example:

$$v_0(\nu) = -\pi \Gamma \delta(\nu + \nu_i) \qquad (\nu_i > 0) \tag{8-634}$$

The integral equation (8-617) reduces to an alegebraic equation as follows:

$$N_0(\nu) = \frac{1}{\pi} \int_{-\infty}^{0} d\nu' \frac{v_0(\nu')\, D_0(\nu')}{\nu' - \nu - i\epsilon} = \frac{\Gamma}{\nu_i + \nu} D_0(-\nu_i) \tag{8-635}$$

Instead of normalizing D_0 by $D_0(0) = 1$ we may choose an alternative normalization $D_0(-\nu_i) = 1$ then

$$N_0(\nu) = \frac{\Gamma}{\nu_i + \nu} \tag{8-636}$$

and

$$D_0(\nu) = 1 - \frac{\nu + \nu_i}{\pi} \int_0^\infty d\nu' \frac{\rho(\nu')}{\nu' + \nu_i} \cdot \frac{N_0(\nu')}{\nu' - \nu - i\epsilon}$$

$$= 1 - \frac{\Gamma}{\pi}(\nu + \nu_i) \int_0^\infty d\nu' \left(\frac{\nu'}{\nu' + M^2}\right)^{1/2} \frac{1}{(\nu' + \nu_i)^2\, (\nu' - \nu - i\epsilon)}$$

$$\cong 1 - \frac{\Gamma}{2M} \frac{\nu + \nu_i}{\nu_i^{1/2}\left(\nu_i^{1/2} + (-\nu - i\epsilon)^{1/2}\right)^2} \tag{8-637}$$

where we have evaluated the dispersion integral in the nonrelativistic approximation, that is, $\nu << M^2$. This expression is certainly real for $\nu < 0$, but it develops an imaginary part for $\nu > 0$.

In the physical region $\nu > 0$, we find

$$\frac{\operatorname{Re} D_0(\nu)}{N_0(\nu)} = \rho(\nu) \cot \delta_0 \approx \frac{\nu^{1/2}}{M} \cot \delta_0$$

$$= \left(\frac{\nu_i}{\Gamma} - \frac{\nu_i^{1/2}}{2M}\right) + \nu \left(\frac{1}{\Gamma} + \frac{1}{2M\, \nu_i^{1/2}}\right) \tag{8-638}$$

Comparing this formula with the standard nonrelativistic effective range formula

$$q \cot \delta_0 = \frac{1}{a} + \frac{1}{2} r q^2 \qquad (q^2 = \nu) \tag{8-639}$$

we see that

$$\begin{aligned} \frac{1}{a} &= \frac{M}{\Gamma}\nu_i - \frac{1}{2}(\nu_i)^{1/2} \\ \frac{1}{2} r &= \frac{M}{\Gamma} + \frac{1}{2(\nu_i)^{1/2}} = \frac{1}{(\nu_i)^{1/2}} + \frac{1}{a\nu_i} \end{aligned} \tag{8-640}$$

If $\Gamma \geq 2M(\nu_i)^{1/2}$, we can find a solution of the equation

$$D_0(\nu) = 0 \tag{8-641}$$

that is,

$$-\nu = \alpha^2 = \nu_i \left(\frac{\Gamma - 2M(\nu_i)^{1/2}}{\Gamma + 2M(\nu_i)^{1/2}} \right)^2 \tag{8-642}$$

This determines the position of the bound state, since the zeros of $D_\ell(\nu)$ are the poles of $F_\ell(\nu)$ or $h_\ell(\nu)$, and we are forced to accept such states. When such is the case then $\nu = -\alpha^2$ represents a pole, which is not present in the original dispersion relation.

There is another subject concerning the poles of $D_\ell(\nu)$. Assume that $D_\ell(\nu)$ has poles at ν_i (i = 1, 2, ..., n); then ν_i appears as zeros of the amplitude $F_\ell(\nu)$. The zeros are not singularities so that $D_\ell(\nu)$ can have poles without modifying the dispersion relation for $F_\ell(\nu)$. Therefore, the equation

$$\operatorname{Im} D_\ell(\nu) = -\rho(\nu) N_\ell(\nu) \qquad (\nu > 0) \tag{8-643}$$

does not determine the dispersion relation for $D_\ell(\nu)$ uniquely, for example, we may write it as

$$D_\ell(\nu) = 1 - \nu \left[\frac{1}{\pi} \int_0^\infty d\nu' \frac{\rho(\nu')}{\nu'} \frac{N_\ell(\nu')}{\nu' - \nu - i\epsilon} + \sum_{i=1}^{n} \frac{c_i}{\nu_i - \nu} + A \right] \tag{8-644}$$

The reality condition for $D_\ell(\nu)$, when $\nu < 0$, implies that all c_i, ν_i, and A be real. This kind of nonuniqueness was first discussed by Castillejo, Dalitz and Dyson,[29] and these points are called CDD zeros. Whether or not the A term is present depends on the convergence of the unsubtracted dispersion relation for $N_\ell(\nu)$. The term A is associated with a CDD zero at $\nu = \infty$.[30]

One of the important conditions that have to be fulfilled is that $D_\ell(\nu)$ should not vanish between the branch cuts, otherwise this zero would show up as a pole in the amplitude $F_\ell(\nu)$ which originally does not have a pole in this domain, for instance, for the simple scalar model of Section 8-19.

$$D_\ell(\nu) \neq 0 \qquad \text{for} \qquad 0 > \nu > -\mu^2/4 \tag{8-645}$$

Herglotz Functions

In order to illustrate that the solution of the scattering equation is not unique we shall consider the scattering equation in the absence of the left-hand cut. We have chosen this example since it is clear that the only solution in perturbation theory is $F_\ell = 0$.

The dispersion relation, combined with unitarity, for each partial-wave amplitude gives, in this case, the equation

$$F(\nu) = \frac{1}{\pi} \int_0^\infty d\nu' \, \frac{\rho(\nu')\,|F(\nu')|^2}{\nu' - \nu - i\epsilon} \tag{8-646}$$

For complex values of ν we get

$$\frac{\text{Im}\, F(\nu)}{\text{Im}\, \nu} = \frac{1}{\pi} \int_0^\infty d\nu' \, \rho(\nu') \, \frac{|F(\nu')|^2}{|\nu' - \nu|^2} > 0 \tag{8-647}$$

Now we shall define the inverse of F and study its properties.

$$H(\nu) = \frac{1}{F(\nu)} \qquad \text{then} \qquad \frac{\text{Im}\, H(\nu)}{\text{Im}\, \nu} < 0 \tag{8-648}$$

Functions, for which $\text{Im}\, F(\nu)/\text{Im}\, \nu$ has a definite sign, are called Herglotz functions. Let us determine the form of $H(\nu)$ on unitarity and analyticity. We already have seen that

$$\text{Im}\, H(\nu) = -\rho(\nu) \qquad \text{for} \qquad \nu > 0 \tag{8-649}$$

The $H(\nu)$ has a cut along the positive real axis, and in addition it can have poles corresponding to zeros of F, so in general

$$H(\nu) = -\frac{1}{\pi}\int_0^\infty d\nu' \frac{\rho(\nu')}{\nu' - \nu - i\epsilon} - \sum_i \frac{c_i}{\nu_i - \nu} + A \tag{8-650}$$

or

$$H(\nu) = -\frac{\nu}{\pi}\int_0^\infty \frac{d\nu' \rho(\nu')}{\nu'(\nu' - \nu - i\epsilon)} - \sum_i \frac{c_i}{\nu_i - \nu} + A - B\nu \tag{8-651}$$

with A, B real. Then

$$\frac{\operatorname{Im} H(\nu)}{\operatorname{Im} \nu} = -\left[\frac{1}{\pi}\int_0^\infty d\nu' \frac{\rho(\nu')}{|\nu' - \nu|^2} + \sum_i \frac{c_i}{|\nu_i - \nu|^2} + B\right] \tag{8-652}$$

Therefore, $H(\nu)$ is a Herglotz function provided ν_i is real, $c_i \geq 0$, and $B \geq 0$. For this form of $H(\nu)$ to be correct it should not have zeros which introduce poles in F. The function $H(\nu)$ now has the following features:

(1) For complex ν, $H(\nu)$ has a nonvanishing imaginary part because of the Herglotz property, so $H(\nu) \neq 0$.
(2) For positive ν, $\operatorname{Im} H(\nu) = -\rho(\nu) \neq 0$, so $H(\nu) \neq 0$.
(3) For negative ν, we get

$$H(\nu) = \frac{|\nu|}{\pi}\int_0^\infty d\nu' \frac{\rho(\nu')}{\nu'(\nu' + |\nu|)} - \sum_i \frac{c_i}{\nu_i + |\nu|} + A + B|\nu| \tag{8-653}$$

We have to choose parameters in such a way that $H(\nu)$ does not vanish along the negative real axis. We first notice for $\nu < 0$ that

$$\frac{dH(\nu)}{d\nu} = -\frac{1}{\pi}\int_0^\infty d\nu' \frac{\rho(\nu')}{(\nu' + |\nu|)^2} - \sum_i \frac{c_i}{(\nu_i + |\nu|)^2} - B < 0 \tag{8-654}$$

provided $\nu_i > 0$. Hence $H(\nu)$ is a monotonic, decreasing function of ν along the negative real axis. Therefore, if $H(0)$ and $H(-\infty)$ are of the same sign, $H(\nu)$ has no zero for $\nu < 0$.

$$H(0) = -\sum_i \frac{c_i}{\nu_i} + A$$

and

$$H(-\infty) = \lim_{|\nu| \to \infty} |\nu| \left(\frac{1}{\pi} \int_0^\infty d\nu' \frac{\rho(\nu')}{\nu'(\nu' + |\nu|)} + B \right) + A$$

$$= +\infty$$

Therefore, we require

$$A > \sum_i \frac{c_i}{\nu_i} \tag{8-655}$$

The divergence of H at infinity is a necessary condition for the convergence of F at infinity. The above form of H is really a solution of the original equation provided

$$c_i > 0 \qquad \nu_i > 0 \qquad B \geq 0 \qquad \text{and} \qquad A > \sum_i \frac{c_i}{\nu_i} \tag{8-656}$$

The Lehmann-Symanzik-Zimmermann Theorem on the Vertex Functions [31]

Let us consider the propagation function for the pion field,

$$\langle 0 | T[\varphi_i(x), \varphi_j(y)] | 0 \rangle = \delta_{ij} \Delta'_F(x - y) \tag{8-657}$$

where i and j are isospin indices. Let us call the Fourier transform of Δ'_F simply $\Delta(s)$. The Lehmann representation reads

$$\Delta(s) = \frac{1}{\mu^2 - s} + \int_{(3\mu)^2} \frac{\sigma(s')ds'}{s' - s - i\epsilon} \tag{8-658}$$

where

$$\delta_{ij}\sigma(s) = (2\pi)^3 \sum_\alpha \langle 0 | \varphi_i(0) | \alpha \rangle \langle \alpha | \varphi_j(0) | 0 \rangle \delta^4(P_\alpha - k) \tag{8-659}$$

with $s = -k^2$. We shall use the same notation as in the derivation of the Goldberger-Treiman relation, with the addition of isospin indices. Thus

$$\langle \alpha, (-) | K_x \varphi_i(0) | 0 \rangle = c_{\alpha i} K_\alpha(s) \tag{8-660}$$

with $c_{N\bar{N}, i} = \bar{u}(N) i\gamma_5 \tau_i v(\bar{N})$, and so on.

$$\begin{aligned} \langle \alpha, (-) | \varphi_i(0) | 0 \rangle &= c_{\alpha i} \frac{K_\alpha(s)}{s - \mu^2} \\ &= c_{\alpha i} \left(\frac{1}{s - \mu^2} + \int \frac{\sigma(s')}{s - s'} ds' \right) \Gamma_\alpha(s) \\ &= -c_{\alpha i} \Delta(s) \Gamma_\alpha(s) \end{aligned} \tag{8-661}$$

where Γ_α is the so-called amputated vertex function and $\gamma_5 \Gamma_{N\bar{N}}(s)$ corresponds to the vertex $e\Gamma_\mu$ in electrodynamics introduced in Section 6-5. Substituting this form of the matrix element in the expression $\sigma(s)$, we have

$$\delta_{ij} \sigma(s) = \delta_{ij} |\Delta(s)|^2 \sum_\alpha \rho_\alpha(s) |\Gamma_\alpha(s)|^2 \tag{8-662}$$

with ρ_α defined in Section 8-14. In particular, we have

$$\rho_{N\bar{N}}(s) = \frac{1}{4\pi^2} s^{1/2} (s - 4M^2)^{1/2} \theta(s - 4M^2) \tag{8-663}$$

We can write

$$\pi \sigma(s) = k(s) |\Delta(s)|^2 \tag{8-664}$$

where

$$k(s) = \pi \sum_\alpha \rho_\alpha(s) |\Gamma_\alpha(s)|^2 > 0 \tag{8-665}$$

Inserting this into the Lehmann representation we get

$$\Delta(s) = \frac{1}{\mu^2 - s} + \frac{1}{\pi} \int_{(3\mu)^2}^{\infty} ds' \frac{k(s') |\Delta(s')|^2}{s' - s - i\epsilon} \tag{8-666}$$

This can be regarded as an integral equation for $\Delta(s)$ with a structure similar to that of a scattering equation. The $\Delta(s)$ is a Herglotz function as one can easily verify.

In order to solve this function we introduce

$$H(s) \equiv \frac{1}{\Delta(s)} \tag{8-667}$$

From the pole term in the Lehmann representation we find

$$H(\mu^2) = 0 \qquad H'(\mu^2) = -1 \tag{8-668}$$

The unitarity condition is

$$\operatorname{Im} \Delta(s) = \pi\sigma(s) = k(s)\,|\Delta(s)|^2 \tag{8-669}$$

or

$$\operatorname{Im} H(s) = -k(s) \tag{8-670}$$

Combining this with the boundary conditions we get, as the simplest solution,

$$\begin{aligned} H(s) &= -\frac{s-\mu^2}{\pi}\int \frac{ds'\,k(s')}{(s'-\mu^2)(s'-s-i\epsilon)} + (\mu^2-s)Z \\ &= (\mu^2-s)\left[Z + \frac{1}{\pi}\int \frac{ds'\,k(s')}{(s'-\mu^2)(s'-s-i\epsilon)}\right] \end{aligned} \tag{8-671}$$

Since $\Delta(s)$ is a Herglotz function H(s) must be one also, but it is true only when $Z \geq 0$. The H(s) of (8-671) certainly satisfies $H(\mu^2) = 0$, but $H'(\mu^2) = -1$ implies

$$Z + \frac{1}{\pi}\int \frac{ds'\,k(s')}{(s'-\mu^2)^2} = 1$$

or

$$Z + \sum_\alpha \int ds\,\frac{\rho_\alpha(s)\,|\Gamma_\alpha(s)|^2}{(s-\mu^2)^2} = 1 \tag{8-672}$$

Since every term on the left-hand side is positive, the contribution of one channel must be less than unity. Take, for instance, the channel $\alpha = N\overline{N}$, then

$$\frac{1}{4\pi^2}\int_{4M^2}^{\infty} ds\, \frac{s^{1/2}(s-4M^2)^{1/2}\left|\Gamma_{N\overline{N}}(s)\right|^2}{(s-\mu^2)^2} < 1 \tag{8-673}$$

This implies that the vertex function $\Gamma_{N\overline{N}}(s)$ vanishes in the high-energy limit

$$\lim_{s\to\infty} \Gamma_{N\overline{N}}(s) = 0 \tag{8-674}$$

This result does not imply that the form factor $K_{N\overline{N}}(s)$ defined by

$$K_{N\overline{N}}(s) = (\mu^2 - s)\,\Delta(s)\,\Gamma_{N\overline{N}}(s) \tag{8-675}$$

vanishes in the limit $s \to \infty$, since

$$\lim_{s\to\infty} K_{N\overline{N}}(s) = Z_\pi^{-1}\,\Gamma_{N\overline{N}}(\infty) \tag{8-676}$$

and Z_π^{-1} is likely to be divergent.

Inclusion of the CDD zeros would modify H(s) to

$$H(s) = (\mu^2 - s)\left[Z + \sum_i \frac{c_i}{(s_i - \mu^2)(s_i - s)} + \frac{1}{\pi}\int \frac{ds'\, k(s')}{(s'-\mu^2)(s'-s-i\epsilon)}\right] \tag{8-677}$$

with the requirement

$$Z + \sum_i \frac{c_i}{(s_i - \mu^2)^2} + \frac{1}{\pi}\int \frac{ds'\, k(s')}{(s'-\mu^2)^2} = 1 \tag{8-678}$$

This does not modify our conclusion concerning the vertex function $\Gamma_{N\overline{N}}$, which develops poles at the CDD zeros.

8-22 MULTI-CHANNEL INTEGRAL EQUATIONS

We have studied two types of integral equations arising from dispersion relations and unitarity, namely, the Muskhelishvili-Omnès equation and the nonlinear scattering equation or the N/D equation.

In the preceding treatments we always have considered only one unknown function, but in general many channels are coupled through unitarity so that we have to solve coupled integral equations. In this section we shall study a formal method of solving these equations.

The Multichannel N/D Method

The scattering amplitudes for a given partial wave connecting channels α and β are to be denoted by $F_{\ell\alpha\beta}$. In the following discussion we shall express the resulting scattering matrix by F, deleting the partial-wave subscript. What time reversal invariance amounts to is that the matrix F can be chosen to be symmetric.

$$F^T = F \tag{8-679}$$

This is easily seen when all the particles participating in this matrix are spinless, otherwise we have to use the helicity amplitudes.[32] If we express the kinematic factor $\rho_\alpha(\nu)$ by a diagonal matrix ρ with $\rho_{\alpha\beta} = \rho_\alpha \delta_{\alpha\beta}$, so that

$$\rho^T = \rho \tag{8-680}$$

the unitarity condition can be expressed in matrix form as

$$\text{Im } F = F^\dagger \rho F = F \rho F^\dagger \qquad \text{for} \quad \nu > 0 \tag{8-681}$$

Since F is symmetric $F^\dagger = F^*$. Then let us express the unitarity condition in terms of F^{-1}; from the identity

$$\begin{aligned} 0 &= F \cdot \text{Im } F^{-1} + \text{Im } F \cdot (F^{-1})^* \\ &= F \cdot \text{Im } F^{-1} + F \rho F^* (F^{-1})^* \\ &= F(\text{Im } F^{-1} + \rho) \end{aligned}$$

we find that

$$\text{Im } F^{-1} = -\rho \qquad \text{for} \quad \nu > 0 \tag{8-682}$$

In order to express the analyticity let us set

$$F = ND^{-1} \tag{8-683}$$

where both N and D are matrices. We write the dispersion relation for F:

$$F(\nu) = \frac{1}{\pi}\int_0^\infty d\nu' \frac{\text{Im } F(\nu')}{\nu' - \nu - i\epsilon} + \frac{1}{\pi}\int_{-\infty}^{-\nu_L} d\nu' \frac{v(\nu')}{\nu' - \nu - i\epsilon} \tag{8-684}$$

The matrix $v(\nu)$ is also symmetric, namely,

$$v^T = v \tag{8-685}$$

Then the analyticity property is expressed in terms of N and D by

$$\begin{aligned} \text{Im } N(\nu) = \text{Im } F(\nu) \cdot D(\nu) &= v(\nu)D(\nu) \qquad \text{for} \quad \nu < -\nu_L \\ &= 0 \qquad \text{otherwise} \end{aligned} \tag{8-686}$$

$$\begin{aligned} \text{Im } D(\nu) = (\text{Im } F(\nu)^{-1})\, N(\nu) &= -\rho(\nu)\, N(\nu) \qquad \text{for} \quad \nu > 0 \\ &= 0 \qquad \text{otherwise} \end{aligned} \tag{8-687}$$

Equation (8-687) incorporates the unitarity condition. Therefore, the integral equations we have to solve are

$$\begin{aligned} N(\nu) &= \frac{1}{\pi}\int_{-\infty}^{-\nu_L} d\nu' \frac{\text{Im } N(\nu')}{\nu' - \nu - i\epsilon} \\ &= \frac{1}{\pi}\int_{-\infty}^{-\nu_L} d\nu' \frac{v(\nu')D(\nu')}{\nu' - \nu - i\epsilon} \end{aligned} \tag{8-688}$$

and

$$D(\nu) = 1 - \frac{\nu}{\pi}\int_0^\infty d\nu' \frac{\rho(\nu')\, N(\nu')}{\nu'(\nu' - \nu - i\epsilon)} \tag{8-689}$$

The final form of the equations is exactly the same as for the one-channel case, but here we have matrix equations.[33] In deriving these equations we have assumed the symmetry of F, so that for the consistency of this method the symmetry of the solution must be guaranteed. The symmetry of F based on that of v has been proved by Bjorken and Nauenberg,[34] and we reproduce it in the following passage.

Since $F - F^T = ND^{-1} - (D^T)^{-1} N^T$, we have

$$\begin{aligned}
\mathrm{Im}[D^T(F - F^T)D] &= D^T(\mathrm{Im}\ F)D - D^T(\mathrm{Im}\ F^T)D \\
&= D^T v D - D^T v^T D \\
&= 0 \\
&\qquad \text{for} \quad \nu < -\nu_L \quad \text{since} \quad v^T = v \\
&= -N^T \rho^T N + N^T \rho N = 0, \\
&\qquad \text{for} \quad \nu > 0 \quad \text{since} \quad \rho^T = \rho
\end{aligned}$$

From the analyticity we conclude that its dispersive part is also equal to zero, provided that the matrix describing subtraction constants for F, if any, is also symmetric. Hence

$$D^T(F - F^T)D = 0 \qquad \text{or} \qquad F = F^T \tag{8-690}$$

Coupled Integral Equations for Vertex Functions

Let us consider a set of vertex functions f with components f_α for different channels α. We treat f as a column vector and write the unitarity condition in the matrix form

$$\mathrm{Im}\ f = F^\dagger \rho f \tag{8-691}$$

The analyticity of f is quite different from the case of the scattering amplitude F, since f has only a right-hand cut; the subtracted dispersion relation is

$$f(\nu) = f_0 + \frac{\nu}{\pi} \int_0^\infty \frac{d\nu'\ \mathrm{Im}\ f(\nu')}{\nu'(\nu' - \nu - i\epsilon)} \tag{8-692}$$

The solution of this problem was first given by Bjorken[33] and then by MacDowell.[35] Here we shall present a simplified version.

We start with the observation

$$F = F^T = (ND^{-1})^T = (D^T)^{-1} N^T$$
$$F^* = (D^{*T})^{-1} N^{*T} \tag{8-693}$$

In particular, for $\nu > 0$, the N is real therefore,

$$F^* = (D^{*T})^{-1} N^T \qquad \text{for} \quad \nu > 0 \tag{8-694}$$

Then we recall that

$$\text{Im}(AB) = (\text{Im } A)B + A^* \text{ Im } B$$

With these preliminaries we shall show that

$$\text{Im}(D^T f) = 0$$

It is clear from the definitions of f and D that this is true for $\nu < 0$, so we shall prove it only for $\nu > 0$, in which case we have

$$\begin{aligned} \text{Im}(D^T f) &= (\text{Im } D^T) f + D^{*T} \text{ Im } f \\ &= -N^T \rho f + D^{*T} F^* \rho f \\ &= -N^T \rho f + (D^{*T})(D^{*T})^{-1} N^T \rho f \\ &= 0 \end{aligned} \tag{8-695}$$

Therefore, if we assume all the necessary properties of D and N, this result is equivalent to the unitarity condition. Since both D and f have only a right-hand cut we get

$$D^T f = P(\nu)$$

a real constant or a real polynomial of ν. Hence

$$f(\nu) = D^T(\nu)^{-1} P(\nu) \tag{8-696}$$

Inhomogeneous Single Channel Muskhelishvili–Omnès Equation

Suppose we have a once-subtracted dispersion relation for $F(\nu)$:

$$F(\nu) = F(0) + \frac{\nu}{\pi} \int_0^\infty d\nu' \, \frac{\text{Im } F(\nu')}{\nu'(\nu' - \nu - i\epsilon)} \tag{8-697}$$

and a unitarity condition of the form

$$\text{Im } F(\nu) = \tan \delta(\nu) \cdot \text{Re } F(\nu) + f(\nu) \qquad \nu > 0 \tag{8-698}$$

The solution without the inhomogeneous term f (ν) is given by

$$F(\nu) = D(\nu)^{-1} F(0)$$

therefore, we consider the combination $D(\nu)\ F(\nu)$, and evaluate its absorptive part using $D(\nu) = |D(\nu)| \exp(-i\delta)$:

$$\begin{aligned} \text{Im}(DF) &= |D| \ \text{Im}(e^{-i\delta}F) \\ &= |D| \ (-\sin\delta \cdot \text{Re } F + \cos\delta \cdot \text{Im } F) \\ &= |D| \ (\cos\delta) f \end{aligned}$$

or

$$\text{Im}(DF) = f \text{ Re } D \tag{8-699}$$

thus

$$D(\nu) F(\nu) = F(0) + \frac{\nu}{\pi} \int_0^\infty d\nu' \ \frac{f(\nu') \text{ Re } D(\nu')}{\nu'(\nu' - \nu - i\epsilon)}$$

or

$$F(\nu) = D(\nu)^{-1} \left[F(0) + \frac{\nu}{\pi} \int_0^\infty d\nu' \ \frac{f(\nu') \text{ Re } D(\nu')}{\nu'(\nu' - \nu - i\epsilon} \right] \tag{8-700}$$

This solution reduces to the homogeneous one in the case $f = 0$ as it should, and it can be verified that the boundary condition at $\nu = 0$ is satisfied. This formula has many applications in dispersion theory.

PROBLEMS

8-1. For an interaction characterized by

$$\mathcal{H}_{\text{int}} = G \, \bar{\psi}\psi \, \varphi$$

calculate the Lehmann spectral function for the scalar field φ to second order in G. Use M and μ for the rest masses of the spinor field ψ and scalar field φ.

8-2. In the calculation of Δ_F' in Section 8-2 we have identified B with the wave-function renormalization constant Z_2 through

$$Z_2 = 1 + \frac{g^2}{(2\pi)^4} B + \cdots$$

or

$$Z_2^{-1} = 1 - \frac{g^2}{(2\pi)^4} B + \cdots$$

Show that the latter expression is equal to

$$1 + \int \sigma_r(\kappa^2)\, d\kappa^2$$

by a direct comparison. Show also that the second-order unrenormalized propagator satisfies the dispersion relation

$$\Delta_F'(s)_{unr} = \frac{1}{M^2 - s - i\epsilon}$$

$$+ \frac{1}{(M^2 - s - i\epsilon)^2} \int_{(M+m)^2}^{\infty} d\kappa^2 \frac{(\kappa^2 - M^2)^2}{\kappa^2 - s - i\epsilon} \sigma_r(\kappa^2)$$

8-3 This problem concerns cluster expansion of the S matrix. In Section 7-6 we have introduced two generating functionals $\mathcal{T}[J]$ and $\mathcal{R}[J]$ related to each other through

$$\mathcal{T}[J] = \exp \mathcal{R}[J]$$

This equation gives the relationship between the τ and ρ functions. Using the above relation prove the formula

$$S = :\exp \Theta:$$

where

$$\Theta = \sum_{l=1}^{\infty} \frac{(-i)^{\ell}}{\ell!} \int d^4y_1 \ldots d^4y_{\ell}\, K_{y_1} \ldots K_{y_{\ell}}$$

$$\times\ \rho(y_1 \ldots y_{\ell})\ :\varphi^{in}(y_1) \ldots \varphi^{in}(y_{\ell}):$$

REFERENCES

1. H. Umezawa and S. Kamefuchi, Progr. Theor. Phys. (Kyoto) **6**, 543 (1951).
2. G. Källén, Helv. Phys. Acta, 25, 417 (1952).
3. H. Lehmann, Nuovo Cimento, 11, 342 (1954).
4. M. Gell-Mann and F. E. Low, Phys. Rev. 95, 1300 (1954).
5. H. Lehmann, K. Symanzik, and W. Zimmermann, Nuovo Cimento, 1, 205 (1955).
6. K. Nishijima, Phys. Rev. **122**, 298 (1961).
7. R. Haag, Phys. Rev. **112**, 669 (1958).
8. K. Nishijima, Phys. Rev. **111**, 995 (1958).
9. W. Zimmermann, Nuovo Cimento, 10, 597 (1958).
10. K. Nishijima, *Fundamental Particles* (W. A. Benjamin, Inc., New York, 1963).
11. F. J. Yndurain, Progr. Theor. Phys. (Kyoto) 46,990 (1971).
12. N. I. Muskhelishvili, *Singular Integral Equations*, (P. Noordhoff, Groningen, 1953).
13. R. Omnès, Nuovo Cimento 8, 316 (1958).
14. C. Møller, Det. K. Danske Vidensk. Selsk. Mat-Fys. Medd. **23**, 1 (1945).
15. G. F. Chew, M. L. Goldberger, F. E. Low, and Y. Nambu, Phys. Rev. 106, 1337 and 1345 (1957).
16. M. L. Goldberger, H. Miyazawa, and R. Oehme, Phys. Rev. 96, 986 (1955).
17. Y. Nambu, Phys. Rev. Letters 4, 380 (1960).
18. M. L. Goldberger and S. B. Treiman, Phys. Rev. 110, 1178 (1958); 111, 354 (1958).
19. M. Ida, Phys. Rev. 132, 401 (1963). Here we follow a modified version by K. Nishijima, Phys. Rev. 133, B1092 (1964).
20. M. Gell-Mann and M. Lévy, Nuovo Cimento 16, 705 (1960).
21. W. I. Weisberger, Phys. Rev. Letters, 14, 1047.(1965).
    S. L. Adler, Phys. Rev. Letters, 14, 1051 (1965).
    S. Fubini and G. Furlan, Physics 1, 229 (1965).
22. R. P. Feynman and M. Gell-Mann, Phys. Rev. 109, 193 (1958).
    M. Gell-Mann, Phys. Rev. 111, 362 (1958).
23. M. Gell-Mann, Physics 1, 63 (1964).
24. L. D. Landau, Nucl. Phys. 13, 181 (1959).
    R. E. Cutkosky, J. Math. Phys. 1, 429 (1960).
25. S. Mandelstam, Phys. Rev. 112, 1344 (1958).
26. M. Cini and S. Fubini, Ann. Phys. (N. Y.) 3, 352 (1960).
27. G. F. Chew and S. Mandelstam, Phys. Rev. 119, 467 (1960).
28. G. F. Chew, *S-Matrix Theory of Strong Interactions* (W. A. Benjamin, Inc., New York, 1961), p. 52.
29. L. Castillejo, R. H. Dalitz, and F. J. Dyson, Phys. Rev. 101, 453 (1956).
30. G. Frye and R. L. Warnock, Phys. Rev. 130, 473 (1963).
31. H. Lehmann, K. Symanzik, and W. Zimmermann, Nuovo Cimento 2, 425 (1955).
32. M. Jacob and G. C. Wick, Ann. Phys. 7, 404 (1959).
33. J. D. Bjorken, Phys. Rev. Letters 4, 473 (1960).
34. J. D. Bjorken and M. Nauenberg, Phys. Rev. 121, 1250 (1961).
35. S. W. MacDowell, Phys. Rev. Letters 6, 385 (1961).

INDEX